Autodesk Revit 2021 Architectural Command Reference

Daniel John Stine

Jeffrey L. Hanson

SDC
PUBLICATIONS

SDC Publications

P.O. Box 1334

Mission, KS 66222

913-262-2664

www.SDCpublications.com

Publisher: Stephen Schroff

Examination Copies

Books received as examination copies are for review purposes only and may not be made available for student use. Resale of examination copies is prohibited.

Electronic Files

Any electronic files associated with this book are licensed to the original user only. These files may not be transferred to any other party.

Trademarks

Autodesk Revit is a registered trademark of Autodesk, Inc. All other trademarks are trademarks of their respective holders.

The author and publisher of this book have used their best efforts in preparing this book. These efforts include the development, research and testing of the material presented. The author and publisher shall not be liable in any event for incidental or consequential damages with, or arising out of, the furnishing, performance, or use of the material.

ISBN-13: 978-1-63057-355-3

ISBN-10: 1-63057-355-8

Printed and bound in the United States of America.

Foreword

The intent of this book is to provide the reader with an easy to use command reference. This command reference can be used as you are working in the software to help you understand what each command does and how it may be used in your overall workflow.

The book is organized in the same way the Revit user interface is presented. Each tab of the Ribbon is represented as a chapter in the book. Within the chapter each button is represented in the book as it appears on the Ribbon from left to right. By organizing the book in this way, we hope it makes it easy to locate each command in the book and understand its use.

On the entries for each command you will see a brief description of what the tool will do, how it is used, and the options you will be given as you use the tool. In some cases, the author's suggestions or tips about the use of the tool will also be presented. As you learn the tools in Revit you may not need to read the full entry on the tool. To help facilitate this, many of the tools include a "Quick Steps" section to explain the tools and options in outline form.

We hope this book facilitates your learning of the Revit interface and the commands. For more experienced users, the command reference may introduce you to commands you have not used before or help you with commands you use less frequently. Whatever level of user you are, we hope this command reference becomes a valuable resource to you as you work with Revit.

Videos:

Nearly 100 videos are provided online and can be accessed using the code and instructions found on the inside front cover.

Trial and Student Software:

This book is based on Autodesk *Revit 2021*. A **30-day** trial may be downloaded from Autodesk's website. Additionally, qualifying students may download the free **3-year** student version of the software from students.autodesk.com. Both are fully functional versions of the software. Much of this book applies to Autodesk Revit LT as well, but no attempt has been made to qualify the limitations of LT.

About the Authors:

Jeff Hanson works for Autodesk and has been part of the Revit development team since 2006. He works directly as a Technical Writer/Subject Matter Expert developing learning content for Revit and other applications. As part of the greater UX (User Experience) team, Jeff assists the product design teams as they develop features in Revit. Before joining Autodesk, Jeff worked as an Architect for Blumentals Architecture in Minneapolis, MN, for 8 years. While there he worked on projects in scale from small residential additions up to 100 unit multi-family housing projects. Jeff currently teaches a beginner level Revit class and directs the BIM curriculum at New Hampshire Technical Institute (NHTI) and has taught in the past at The Boston Architectural College (BAC). Jeff has a Master's Degree in Architecture from the University of Minnesota and a Bachelor's Degree in Architecture from the University of New Mexico.

Daniel John Stine AIA, CSI, CDT, is a registered architect with over twenty years of experience in the field of architecture. He is the BIM Administrator at LHB, a 250-person full service design firm. In addition to providing training and support for four offices, Dan implemented BIM-based lighting analysis using ElumTools, early energy modeling using Autodesk Insight, virtual reality (VR) using the HTC Vive/Oculus Rift along with Fuzor & Enscape, Augmented Reality (AR) using the Microsoft Hololens, and the Electrical Productivity Pack for Revit (sold by CTC Express Tools). Dell, the world-renowned computer company, created a video highlighting his implementation of VR at LHB.

Dan has presented internationally on BIM in the USA, Canada, Ireland, Denmark, Slovenia, Australia and Singapore. He was ranked multiple times as a top-ten speaker by attendees at Autodesk University, RTC/BILT, Midwest University, AUGI CAD Camp, NVIDIA GPU Technology Conference, Lightfair, and AIA-MN Convention. By invitation, he spent a week at Autodesk's largest R&D facility in Shanghai, China, to beta test and brainstorm new Revit features in 2016.

Committed to furthering the design profession, Dan teaches graduate architecture students at North Dakota State University (NDSU) and has lectured for interior design programs at NDSU, Northern Iowa State, and University of Minnesota, as well as Dunwoody's new School of Architecture in Minneapolis. As an adjunct instructor, Dan previously taught AutoCAD and Revit for twelve years at Lake Superior College. Dan is a member of the American Institute of Architects (AIA), Construction Specifications Institute (CSI), and Autodesk Developer Network (ADN), and is a Construction Document Technician (issued by CSI). He has presented live webinars for ElumTools, ArchVision, Revizto and NVIDIA. Dan writes about design on his blog, BIM Chapters, and in his textbooks published by SDC Publications:

- *Autodesk Revit for Architecture Certified User Exam Preparation (Revit 2021 Edition)*

- *Residential Design Using Autodesk Revit 2021*
- *Commercial Design Using Autodesk Revit 2021*
- *Design Integration Using Autodesk Revit 2021 (Architecture, Structure and MEP)*
- *Interior Design Using Autodesk Revit 2021 (with co-author Aaron Hansen)*
- *Residential Design Using AutoCAD 2021*
- *Commercial Design Using AutoCAD 2013*
- *Chapters in Architectural Drawing (with co-author Steven H. McNeill, AIA, LEED AP)*
- *Interior Design using Hand Sketching, SketchUp and Photoshop (also with Steven H. McNeill)*
- *SketchUp 8 for Interior Designers; Just the Basics*

You may contact the publisher with comments or suggestions at service@SDCpublications.com.

Social Media:

Students can use social media, such as Twitter and LinkedIn, to start developing professional contacts and knowledge. Follow the author on social media for new articles, tips and errata updates. Also, consider following the design firms and associations (AIA, CSI, etc.) in your area; this could give you an edge in an interview!

 Twitter
@ADSKJeffHanson

 LinkedIn
https://www.linkedin.com/in/autodesk-jeff-hanson

 Twitter
@DanStine_MN

 LinkedIn
https://www.linkedin.com/in/danstinemn

BIM Chapters:
Daniel Stine's blog: http://bimchapters.blogspot.com/

Many thanks go out to Stephen Schroff and SDC Publications for making this book possible!

Notes:

Table of Contents

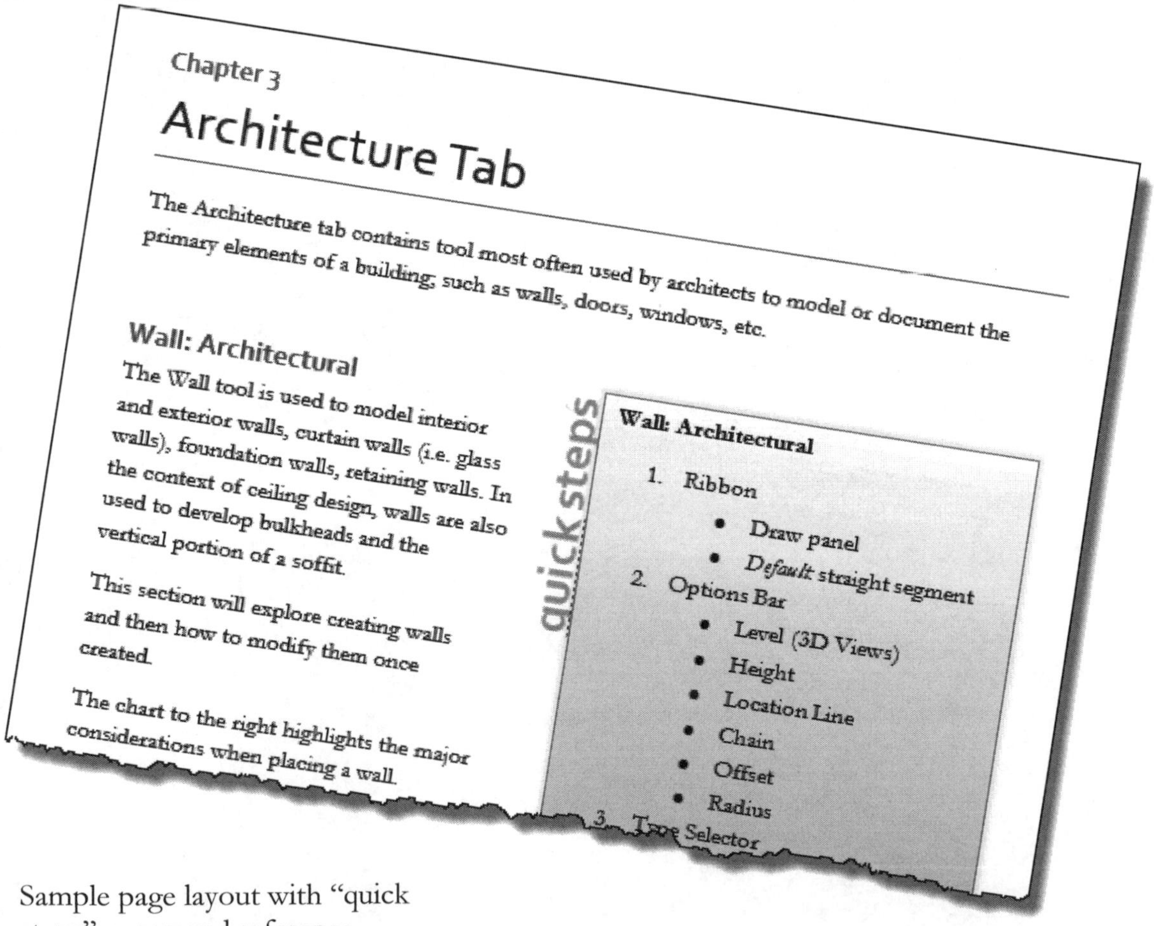

Chapter 3

Architecture Tab

The Architecture tab contains tool most often used by architects to model or document the primary elements of a building, such as walls, doors, windows, etc.

Wall: Architectural

The Wall tool is used to model interior and exterior walls, curtain walls (i.e. glass walls), foundation walls, retaining walls. In the context of ceiling design, walls are also used to develop bulkheads and the vertical portion of a soffit.

This section will explore creating walls and then how to modify them once created.

The chart to the right highlights the major considerations when placing a wall.

quick steps

Wall: Architectural

1. Ribbon
   - Draw panel
   - *Default* straight segment
2. Options Bar
   - Level (3D Views)
   - Height
   - Location Line
   - Chain
   - Offset
   - Radius
3. Type Selector

Sample page layout with "quick steps" command reference

Exclusive Online Content: Videos

Several videos are provided with this book and can be accessed with the code and instructions found on the inside front cover of this book.

Chapter 1

1.1 Introduction

What is Autodesk Revit?

Autodesk Revit (Architecture, Structure and MEP) is the world's first fully parametric building design software. This revolutionary software, for the first time, truly takes architectural computer aided design beyond simply being a high-tech pencil. Revit is a product of Autodesk, makers of AutoCAD, Civil 3D, Inventor, 3DS Max, Maya and many other popular design programs.

Revit can be thought of as the foundation of a larger process called **Building Information Modeling** (BIM). The BIM process revolves around a virtual, information rich 3D model. In this model all the major building elements are represented and contain information such as manufacturer, model, cost, phase and much more. Once a model has been developed in Revit, third-party add-ins and applications can be used to further leverage the data. Some examples are Facilities Management, Analysis (Energy, Structural, Lighting), Construction Sequencing, Cost Estimating, Code Compliance and much more!

Revit can be an invaluable tool to designers when leveraged to its full potential. The iterative design process can be accomplished using special Revit features such as *Phasing* and *Design Options*. Material selections can be developed and attached to various elements in the model, where one simple change adjusts the wood from oak to maple throughout the project. The power of schedules may be used to determine quantities and document various parameters contained within content (this is the "I" in BIM, which stands for Information). Finally, the three-dimensional nature of a Revit-based model allows the designer to present compelling still images and animations. These graphics help to more clearly communicate the design intent to clients and other interested parties. This book will cover many of these tools and techniques to **assist** in the creative process.

What is a parametric building modeler?

Revit is a program designed from the ground up using state-of-the-art technology. The term parametric describes a process by which an element is modified, and an adjacent element(s) is automatically modified to maintain a previously established relationship. For example, if a wall is moved, perpendicular walls will grow, or shrink, in length to remain attached to the related wall. Additionally, elements attached to the wall will move, such as wall cabinets, doors, windows, air grilles, etc.

Revit stands for **Rev**ise **In**stantly; a change made in one view is automatically updated in all other views and schedules. For example, if you move a door in an interior elevation view, the floor plan will automatically update. Or, if you delete a door, it will be deleted from all other views and schedules.

A major goal of Revit is to eliminate much of the repetitive and mundane tasks traditionally associated with CAD programs allowing more time for design, coordination and visualization. For example: all sheet numbers, elevation tags and reference bubbles are updated automatically when changed anywhere in the project. Therefore, it is difficult to find a mis-referenced detail tag.

The best way to understand how a parametric model works is to describe the Revit project file. A single Revit file contains your entire building project. Even though you mostly draw in 2D views, you are actually modeling in 3D. In fact, the entire building project is a 3D model. From this 3D model you can generate 2D elevations, 2D sections and perspective views. Therefore, when you delete a door in an elevation view you are actually deleting the door from the 3D model from which all 2D views are generated and automatically updated.

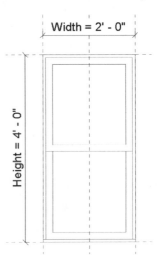

Another way in which Revit is a parametric building modeler is that parameters can be used to control the size and shape of geometry. For example, a window model can have two parameters set up which control the size of the window. Thus, from a window's properties it is possible to control the size of the window without using any of the drawing modify tools such as Scale or Move. Furthermore, the parameter settings (i.e., width and height in this example) can be saved within the window model (called a Family).

Window model size controlled by parameters *Width* and

You could have the 2′ x 4′ settings saved as "Type A" and the 2′ x 6′ as "Type B." Each saved list of values is called a Type within the Family. Thus, this one double-hung window Family could represent an unlimited number of window sizes! You will learn more about this later in the book.

Student Software

Students can download a free 3-year version of Revit at www.students.autodesk.com. Be sure to use your school email address. When using the educational version and logged in to Autodesk A360, using your school email address, you have access to unlimited cloud credits for rendering.

> The student version of the software is intended for learning purposes only; it is against the licensing agreement to use a student version for any professional work. If found using student licenses to do professional work, you may be subject to legal action from Autodesk.

3D model of lunchroom created in Interior Design using Autodesk Revit 2021

64bit Architecture

Revit is designed for 64bit operating systems (OS) and computers. Be sure your installed version of Windows is 64bit.

Why use Revit?

Many people ask the question, why use Revit versus other programs? The answer can certainly vary depending on the situation and particular needs of an individual or organization.

Generally speaking, this is why companies use Revit:

- Many designers and drafters are using Revit to streamline repetitive drafting tasks and focus more on designing and detailing a project.

- Revit is a very progressive program and offers many features for designing buildings. Revit is constantly being developed and Autodesk provides incremental upgrades and patches on a regular basis.
- Revit was designed specifically for architectural design and includes features like:
  - Photo Realistic Rendering
  - Phasing (different design over time)
  - Design Options (different designs within the same time period)
  - Live Schedules
    - Quantity Schedules
    - Material Takeoff Schedules
    - Sheet Index Schedule
  - Cloud Rendering via *Autodesk 360*
  - Analysis via *Autodesk 360*
    - Conceptual Energy Analysis
    - Daylighting Analysis
    - Structural Analysis

File Types and their extensions:

Revit has four primary types of files that you will work with as a Revit user. Each file type, as with any Microsoft Windows based program, has a specific three letter file name extension; that is, after the name of the file on your hard drive you will see a period and three letters:

.RVT   Revit project files; the file most used

    Backup files .RVT**.0001**, .RVT**.0002**, etc.

.RFA   Revit family file; loadable content for your project

.RTE   Revit template; a project starter file with office standards preset

.RFT   Revit family template; a family starter file with parameters

---

A few basic Revit concepts:

The following is meant to be a brief overview of the basic organization of Revit as a software application. You should not get too hung up on these concepts and terms as they will make more sense as you work through this book. This discussion is simply laying the groundwork so you have a general frame of reference on how Revit works.

The Revit platform has three fundamental types of elements:

- Model Elements
- Datum Elements
- View-Specific Elements

Model Elements

Think of *Model Elements* as things you can put your hands on once the building has been constructed. They are typically 3D but can sometimes be 2D. There are two types of *Model Elements*:

- **Host Elements (aka System Family)** Walls, floors, slabs, roofs, ceilings. These are items that can only exist within a project—they cannot be loaded from a file.

- **Model Components** (Stairs, Doors, Furniture, Beams, Columns, Pipes, Ducts, Light Fixtures, Model Lines) – Options vary depending on the "flavor" of Revit.

  o Some *Model Components* require a host before they can be placed within a project. For example, a window can only be placed in a host, which could be a wall, roof or floor depending on how the element was created. If the host is deleted, all hosted or dependent elements are automatically deleted.

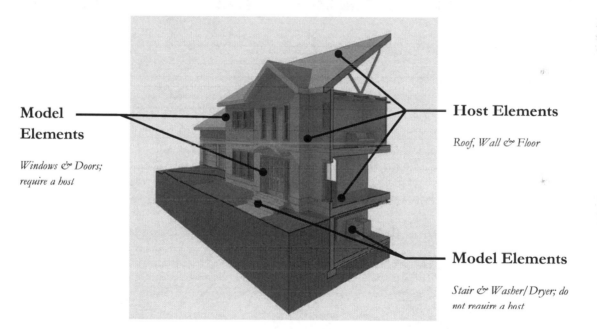

Model Elements

Windows & Doors; require a host

Host Elements

Roof, Wall & Floor

Model Elements

Stair & Washer/Dryer; do not require a host

Datum Elements

Datum Elements are reference planes within the building that graphically and parametrically define the location of various elements within the model. These features are available in all "flavors" of Revit. These are the three types of *Datum Elements*:

- **Grids**
  - ❖ Typically laid out in a plan view to locate structural elements such as columns and beams, as well as walls. Grids show up in plan, elevation and section views. Moving a grid in one view moves it in all other views as it is the same element. (See the next page for an example of a grid in plan view.)

- **Levels**
  - ❖ Used to define vertical relationships, mainly surfaces that you walk on. Many elements are placed relative to a *Level*; when the *Level* is moved those elements move with it (e.g., doors, windows, casework, ceilings). *WARNING: If a Level is deleted, those same "dependent" elements will also be deleted from the project! A warning will be displayed if you do this.*

- **Reference Planes**
  - ❖ These are similar to grids in that they show up in plan and elevation or sections. They do not have reference bubbles at the end like grids. Revit breaks many tasks down into simple 2D tasks which result in 3D geometry. *Reference Planes* are used to define 2D surfaces on which to work within the 3D model. They can be placed in any view, either horizontally or vertically.

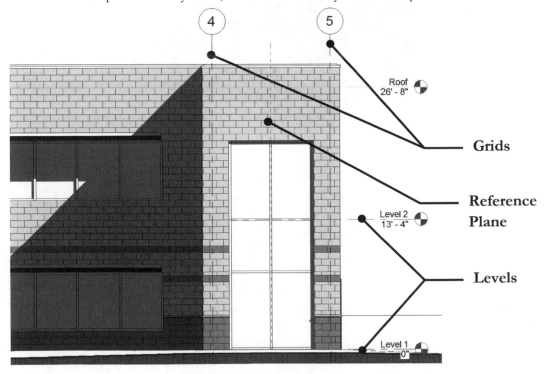

View-Specific Elements

As the name implies, the items about to be discussed only show up in the specific view in which they are created. For example, notes and dimensions added in the architectural floor plans will not show up in the structural floor plans. These elements are all 2D and are mainly communication tools used to accurately document the building for construction or presentations.

- **Annotation elements** (text, tags, symbols, dimensions)
  - Size automatically set and changed based on selected drawing scale

- **Details** (detail lines, filled regions, 2D detail components)

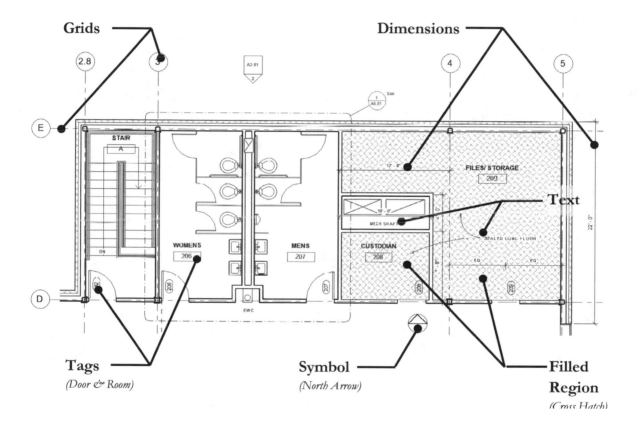

Grids   **Dimensions**

Text

Tags
(Door & Room)

Symbol
(North Arrow)

Filled Region
(Cross Hatch)

The Revit platform has three fundamental ways in which to work with the elements (for display and manipulation):

- Views
- Schedules
- Sheets

The following is a cursory overview of the main ideas you need to know. This is not an exhaustive study on views, schedules and sheets.

Views

Views, accessible from the *Project Browser* (see Page 5-4), is where most of the work is done while using Revit. Think of views as slices through the building, both horizontal (plans) and vertical (elevations and sections).

- **Plans**

  A *Plan View* is a horizontal slice through the building. You can specify the location of the **cut plane** which determines if certain windows show up or how much of the stair is seen. A few examples are architectural floor plan, reflected ceiling plan, site plan, structural framing plan, HVAC floor plan, electrical floor plan, lighting [ceiling] plan, etc. The images below show this concept; the image on the left is the 3D BIM. The middle image shows the portion of building above the cut plane removed. Finally, the last image on the right shows the plan view you work in and place on a sheet.

  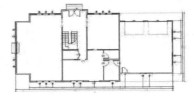

- **Elevations**

  Elevations are vertical slices, but where the slice lies outside the floor plan as in the middle image below. Each elevation created is listed in the *Project Browser*. The image on the right is an example of a South exterior elevation view, which is a "live" view of the 3D model. If you select a window here and delete it, the floor plans will update instantly.

- **Sections**

  Similar to elevations, sections are also vertical slices. However, these slices cut through the building. A section view can be cropped down to become a wall section or even look just like an elevation. The images below show the slice, the portion of building in the foreground removed, and then the actual view created by the slice. A setting exists, for each section view, to control how far into that view you can see. The example on the right is "seeing" deep enough to show the doors on the interior walls.

- **3D and Camera**

  In addition to the traditional "flattened" 2D views that you will typically work in, you are able to see your designs more naturally via 3D and Camera views.

  A **3D view** is simply an axonometric view; i.e., three-dimensional but without perspective.

  A **Camera view** is a true perspective view; cameras can be created both in and outside of the building. Like the 2D views, these 3D/Camera views can be placed on a sheet to be printed. Revit provides a number of tools to help explore the 3D view, such as Section Box, Steering Wheel, Temporary Hide and Isolate, and Render.

  You can use the View Cube in 3D views to toggle the view between axonometric and perspective projections

  The image on the left is a 3D view set to "shaded mode" and has shadows turned on. The image on the right is a camera view set up inside the building; the view is set to "hidden line" rather than shaded, and the camera is at eye level.

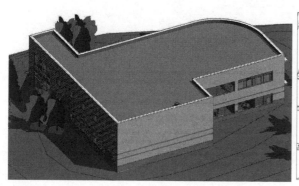

Schedules

Schedules are lists of information generated based on content that has been placed, or modeled, within the project. A schedule can be created, such as the door schedule example shown below, that lists any of the data associated with each door that exists in the project. Revit allows you to work directly in the schedule views. Any change within a schedule view is a change directly to the element being scheduled. Again, if a door were to be deleted from this schedule, that door would be instantly deleted from the project.

Think of schedules as a "non-graphical" view of model elements. You are viewing the model; you are just not seeing the graphic part of the model. You are only viewing the information about an element, not the graphic representation.

DOOR AND FRAME SCHEDULE												
DOOR NUMBER	DOOR				FRAME		DETAIL				FIRE RATING	HDWR GROUP	
	WIDTH	HEIGHT	MATL	TYPE	MATL	TYPE	HEAD	JAMB	SILL	GLAZING			
1000A	3' - 8"	7' - 2"	WD		HM		11/A8.01	11/A8.01					
1046	3' - 0"	7' - 2"	WD	D10	HM	F10	11/A8.01	11/A8.01 SIM				34	
1047A	6' - 0"	7' - 10"	ALUM	D15	ALUM	SF4	6/A8.01	6/A8.01	1/A8.01 SIM	1" INSUL		2	CARD READER N. LEAF
1047B	8' - 0"	7' - 2"	WD	D10	HM	F13	12/A8.01	11/A8.01 SIM			60 MIN	85	MAG HOLD OPENS
1050	3' - 0"	7' - 2"	WD	D10	HM	F21	8/A8.01	11/A8.01		1/4" TEMP		33	
1051	3' - 0"	7' - 2"	WD	D10	HM	F21	8/A8.01	11/A8.01		1/4" TEMP		33	
1052	3' - 0"	7' - 2"	WD	D10	HM	F21	8/A8.01	11/A8.01		1/4" TEMP		33	
1053	3' - 0"	7' - 2"	WD	D10	HM	F21	8/A8.01	11/A8.01		1/4" TEMP		33	
1054A	3' - 0"	7' - 2"	WD	D10	HM	F10	8/A8.01	11/A8.01		1/4" TEMP	-	34	
1054B	3' - 0"	7' - 2"	WD	D10	HM	F21	8/A8.01	11/A8.01		1/4" TEMP	-	33	
1055	3' - 0"	7' - 2"	WD	D10	HM	F21	8/A8.01	11/A8.01		1/4" TEMP	-	33	
1056A	3' - 0"	7' - 2"	WD	D10	HM	F10	9/A8.01	9/A8.01			20 MIN	33	
1056B	3' - 0"	7' - 2"	WD	D10	HM	F10	11/A8.01	11/A8.01			20 MIN	34	
1056C	3' - 0"	7' - 2"	WD	D10	HM	F10	20/A8.01	20/A8.01			20 MIN	33	
1057A	3' - 0"	7' - 2"	WD	D10	HM	F10	8/A8.01	11/A8.01			20 MIN	34	
1057B	3' - 0"	7' - 2"	WD	D10	HM	F30	9/A8.01	9/A8.01		1/4" TEMP	20 MIN	33	
1058A	3' - 0"	7' - 2"	WD	D10	HM	F10	9/A8.01	9/A8.01			-	33	

Sheets

You can think of sheets as the pieces of paper on which your views and schedules will be printed. Views and schedules are placed on sheets and then arranged. Once a view has been placed on a sheet, its reference bubble is automatically filled out and that view cannot be placed on any other sheet. The setting for each view, called "view scale," controls the size of the drawing on each sheet; view scale also controls the size of the text, tags and dimensions.

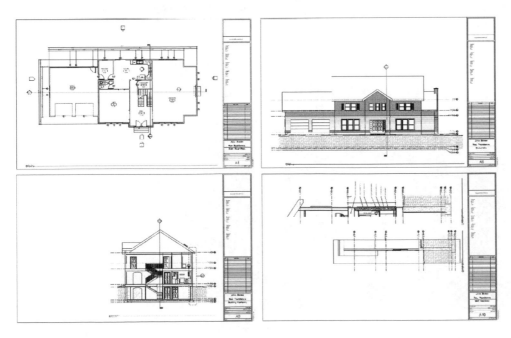

Chapter 2
File Tab
and User Interface (UI)

This chapter will take a detailed look at the User Interface, which is often referred to as "the UI." As with all computer programs, all access to the Revit model, by the designer, is done through the UI. The ability to create, edit and print a model is done using various clicks and selections with the mouse on the computer screen. Clicking one graphic starts a tool and clicking another closes the view. Understanding the UI is a fundamental first step in learning to use Revit effectively and efficiently.

The goal of this chapter is to provide a systematic overview of the UI. A person new to Revit would benefit from reading this chapter in its entirety.

The image to the right is from the author's *Interior Design Using Autodesk Revit* textbook. The image was rendered using Enscape, *www.Enscape3d.com*, which is free for students.

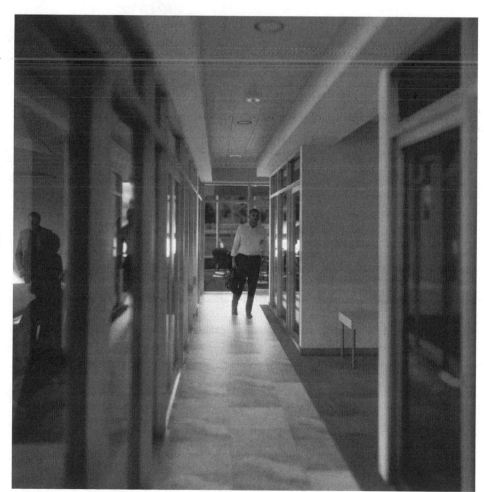

2.1 User Interface

Revit is a powerful and sophisticated program. Because of its powerful feature set, it has a measurable learning curve, though its intuitive design makes it easier to learn than other CAD or BIM based programs. However, like anything, when broken down into smaller pieces, we can easily learn to harness the power of Revit. That is the goal of this book.

This section will walk through the different aspects of the User Interface (UI). As with any program, understanding the user interface is the key to using the program's features.

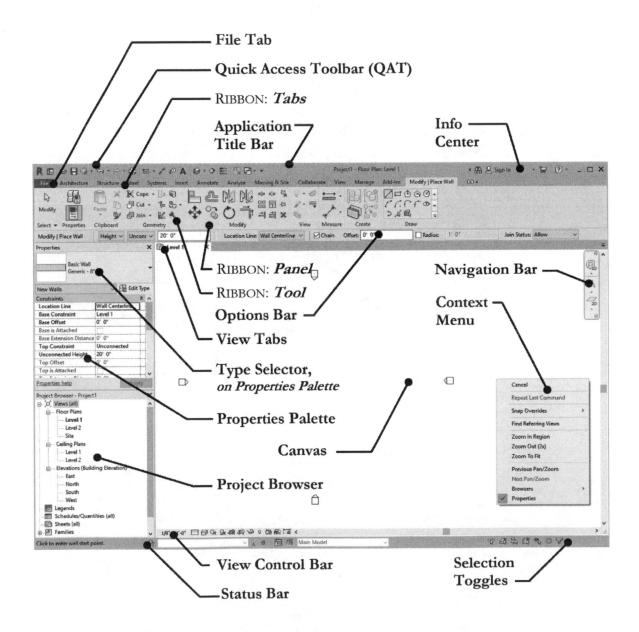

Figure 2.1-1
Revit User Interface

Application Title Bar:

In addition to the *File Tab*, *Quick Access Toolbar* and *Info Center*, which are all covered in the next few sections, you are also presented with the product name, version and the current file-view in the center. As previously noted already, the version is important as you do not want to upgrade unless you have coordinated with other staff and/or consultants; everyone must be using the same version of Revit.

File Tab:

Access to *File* tools such as *Save*, *Plot*, *Export* and *Print* (both hardcopy and electronic printing). You also have access to tools which control the Revit application as a whole, not just the current project, such as *Options* (see the end of this section for more on *Options*).

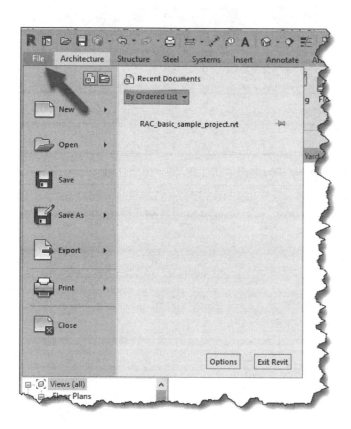

Recent and Open Documents:

These two icons (from the *File Tab*) toggle the entire area on the right to show either the recent documents you have been in (icon on the left) or a list of the documents you currently have open.

In the *Recent Documents* list you click a listed document to open it. This saves time as you do not have to click *Open* → *Project* and browse for the document (*Document* and *Project* mean the same thing here). Finally, clicking the "Pin" keeps that project from getting bumped off the list as additional projects are opened.

In the *Open Documents* list, the "active" project you are working in is listed first; clicking another project switches you to that open project.

File Tab & UI

The list on the left, in the *File Tab* shown above, represents three different types of buttons: *button*, *drop-down button* and *split button*. Save and Close are simply **buttons**. Save-As and Export are **drop-down buttons**, which means to reveal a group of related tools. If you click or hover your cursor over one of these buttons, you will get a list of tools on the right. Finally, **split buttons** have two actions depending on what part of the button you click on; hovering over the button reveals the two parts (see bottom image to the right). The main area is the most used tool; the arrow reveals additional related options.

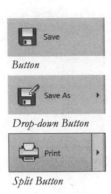

Button

Drop-down Button

Split Button

Quick Access Toolbar:

Referred to as *QAT* in this book, this single toolbar provides access to often used tools (*Open, Save, Undo, Redo,* Print, *Measure, Tag,* etc.). It is always visible regardless of what part of the *Ribbon* is active.

The *QAT* can be positioned above or below the *Ribbon* and any command from the *Ribbon* can be placed on it; simply right-click on any tool on the *Ribbon* and select *Add to Quick Access Toolbar.* Moving the *QAT* below the *Ribbon* gives you a lot more room for your favorite commands to be added from the *Ribbon.* Clicking the larger down-arrow to the far right reveals a list of common tools which can be toggled on and off.

Some of the icons on the *QAT* have a down-arrow on the right. Clicking this arrow reveals a list of related tools. In the case of *Undo* and *Redo,* you have the ability to undo (or redo) several actions at once.

Ribbon – Architecture Tab:

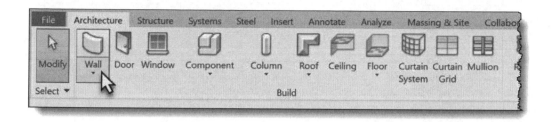

The *Architecture* tab on the *Ribbon* contains most of the tools the architect needs to model a building, essentially the things you can put your hands on when the building is done. The specific discipline versions of Revit omit some of the other discipline tabs.

Each tab starts with the *Modify* tool, i.e., the first button on the left. This tool puts you into "selection mode" so you can select elements to modify. Clicking this tool cancels the current tool and unselects elements. With the *Modify* tool selected you may select elements to view their properties or edit them. Note that the *Modify* tool, which is a button, is different than the *Modify* tab on the *Ribbon*.

The *Ribbon* has three types of buttons: *button*, *drop-down button* and *split*, as covered on the previous page. In the image above you can see the *Wall* tool is a **split button**. Most of the time you would simply click the top part of the button to draw a wall. Clicking the down-arrow part of the button, for the *Wall* tool example, gives you the option to draw a *Wall*, *Structural Wall*, *Wall by Face*, *Wall Sweep*, and a *Reveal*.

TIP: The Model Text tool is only for placing 3D text in your model, not for adding notes!

Ribbon – Annotate Tab:

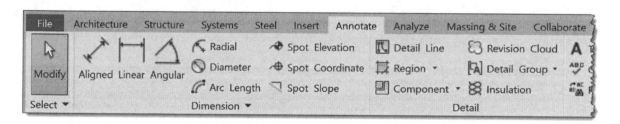

To view this tab, simply click the label "Annotate" near the top of the *Ribbon*. This tab presents a series of tools which allow you to add notes, dimensions and 2D "embellishments" to your model in a specific view, such as a floor plan, elevation, or section. All of these tools are **view specific**, meaning a note added in the first-floor plan will not show up anywhere else, not even another first floor plan: for instance, a first floor electrical plan.

File Tab & UI

Notice, in the image above, that the *Dimension* panel label has a down-arrow next to it. Clicking the down-arrow will reveal an **extended panel** with additional related tools.

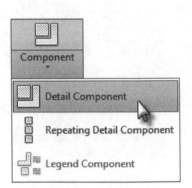

Finally, notice the *Component* tool in the image above; it is a **split button** rather than a *drop-down button*. Clicking the top part of this button will initiate the *Detail Component* tool. Clicking the bottom part of the button opens the fly-out menu revealing related tools.

Ribbon – Modify Tab:

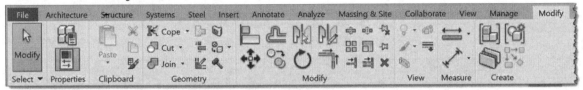

Several tools which manipulate and derive information from the current model are available on the *Modify* tab. Additional *Modify* tools are automatically appended to this tab when elements are selected in the model (see *Modify Contextual Tab* on the next page).

> **TIP:** *Do not confuse the Modify tab with the Modify tool when following instructions in this book.*

Ribbon – View Tab:

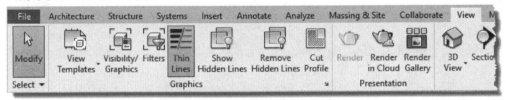

The tools on the *View* tab allow you to create new views of your 3D model; this includes views that look 2D (e.g., floor plans, elevations and sections) as well as 3D views (e.g., isometric and perspective views).

The *View* tab also gives you tools to control how views look, everything from what types of elements are seen (e.g., Plumbing Fixture, Furniture or Section Marks) to line weights.

> *NOTE: Line weights are controlled at a project wide level but may be overridden on a view by view basis.*

Finally, notice the little arrow in the lower-right corner of the *Graphics* panel. When you see an arrow like this you can click on it to open a dialog box with settings that relate to the panel's tool set (*Graphics* in this example). Hovering over the arrow reveals a tooltip which will tell you what dialog box will be opened.

Ribbon – Modify Contextual Tab:

The *Modify* tab is appended when certain tools are active or elements are selected in the model; this is referred to as a *contextual tab*. The first image below shows the *Place Wall* tab which presents various options while adding walls. The next example shows the *Modify Walls* contextual tab which is accessible when one or more walls are selected.

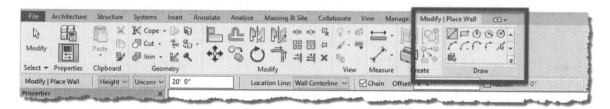

Place Wall contextual tab – visible when the Wall tool is active.

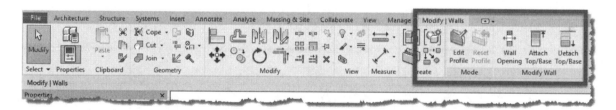

Modify Walls contextual tab – visible when a wall is selected.

File Tab & UI

Ribbon – Customization:

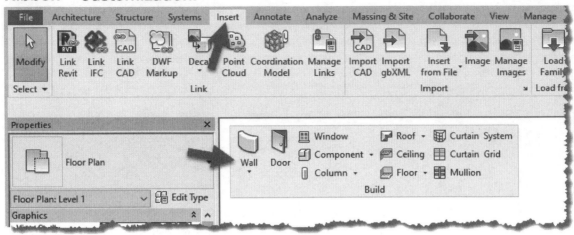

There is not too much customization that can be done to the *Ribbon*. One of the only things you can do is pull a panel off the *Ribbon* by clicking and holding down the left mouse button on the titles listed across the bottom. This panel can be placed within the *drawing window* or on another screen if you have a dual monitor setup.

The image above shows the *Build* panel, from the *Architecture* tab, detached from the *Ribbon* and floating within the drawing window. Notice that the *Insert* tab is active. Thus, you have constant access to the *Build* tools while accessing other tools. Note that the *Build* panel is not available on the *Architecture* tab as it is literally moved, not just copied.

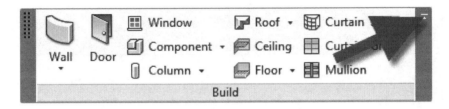

When you need to move a detached panel back to the *Ribbon* you do the following: hover over the detached panel until the sidebars show up and then click the "Return panels to ribbon" icon in the upper right (identified in the image above).

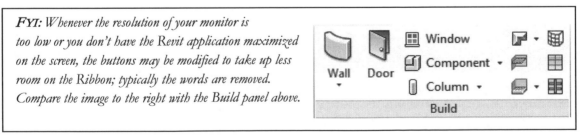

FYI: Whenever the resolution of your monitor is too low or you don't have the Revit application maximized on the screen, the buttons may be modified to take up less room on the Ribbon; typically the words are removed. Compare the image to the right with the Build panel above.

If you install an **add-in** for Revit on your computer, you will likely see a new tab appear on the Ribbon. Some add-ins are free while others require a fee. If an add-in only has one tool, it will likely be added to the catch-all tab called Add-Ins (shown in the image below).

This is not really customizing the User Interface, but in the Options dialog there are several adjustments one can make – such as turning off tabs and tools not used. However, turning off, for example, the electrical tools, in turn limits the number of electrical parameter types that can be created.

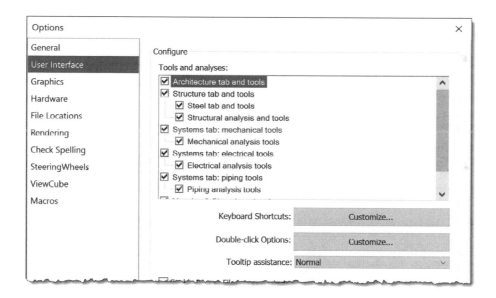

File Tab & UI

Ribbon – States:

The *Ribbon* can be displayed in one of four states:

- Full Ribbon (default)
- Minimize to Tabs
- Minimize to Panel Tiles
- Minimize to Panel Buttons

The intent of this feature is to increase the size of the available drawing window. It is recommended, however, that you leave the *Ribbon* fully expanded while learning to use the program. The images in this book show the fully expanded state. The images below show the other three options. When using one of the minimized options you simply hover (or click) your cursor over the Tab or Panel to temporarily reveal the tools.

> *FYI:* Double-clicking on a Ribbon tab will also toggle the states.

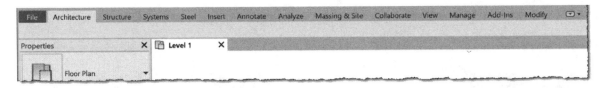

Minimize to Tabs

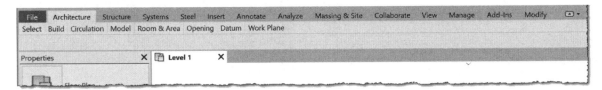

Minimize to Panel Tiles

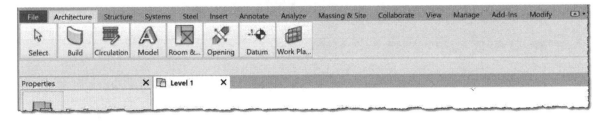

Minimize to Panel Buttons

Options Bar:

This area dynamically changes to show options that complement the current operation. The *Options Bar* is located directly below the *Ribbon*. When you are learning to use Revit you should keep your eye on this area and watch for features and options appearing at specific times. The image below shows the *Options Bar* example with the *Wall* tool active.

Properties Palette – Element Type Selector:

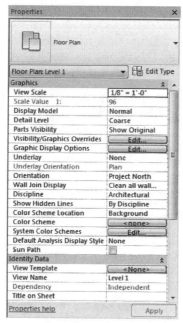

Properties Palette; nothing selected

The *Properties Palette* provides instant access to settings related to the element selected or about to be created. When nothing is selected, it shows information about the current view. When multiple elements are selected, the common parameters are displayed.

The *Element Type Selector* is an important part of the *Properties Palette*. Whenever you are adding elements or have them selected, you can select from this list to determine how a wall to be drawn will look, or how a wall previously drawn should look (see image to left). If a wall type needs to change, you never delete it and redraw it; you simply select it and pick a new type from the *Type Selector*.

The **Selection Filter** drop-down list below the *Type Selector* lets you know the type and quantity of the elements currently selected. When multiple elements are selected you can narrow down the properties for just one element type, such as *wall*. Notice the image to the left shows four walls are in the current selection set.

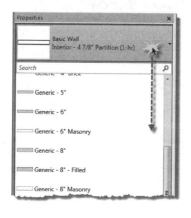

Type Selector; Wall tool active or a Wall is selected

Selection Filter; multiple elements selected

Selecting **Walls (4)** will cause the *Palette* to only show *Wall* properties even though many other elements are selected (and remain selected).

The width of the *Properties Palette* and the center column position can be adjusted by dragging the cursor over that area. You may need to do this at times to see all the information. However, making the *Palette* too wide will reduce the usable drawing area.

The *Properties Palette* should be left open; if you accidentally close it you can reopen it by **View** → **Window** → **User Interface** → **Properties** or by typing **PP** on the keyboard.

Project Browser:

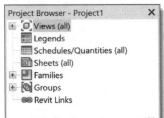

The *Project Browser* is the "Grand Central Station" of the Revit project database. All the views, schedules, sheets and content are accessible through this hierarchical list. The first image to the left shows the seven major categories; any item with a "plus" next to it contains sub-categories or items.

Double-clicking on a View, Legend, Schedule or Sheet will open it for editing; the current item open for editing is bold (**Level 1** in the example to the left). Right-clicking will display a pop-up menu with a few options such as *Delete* and *Copy*.

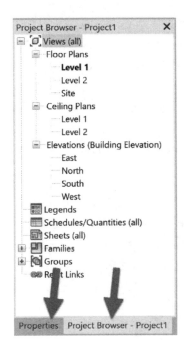

Right-click on *Views (all)*, at the top of the *Project Browser*, and you will find a **Search** option in the pop-up menu. This can be used to search for a *View*, *Family*, etc., a very useful tool when working on a large project with 100s of items to sift through.

Like the *Properties Palette*, the width of the *Project Browser* can be adjusted. When the two are stacked above each other, they both move together. You can also stack the two directly on top of each other; in this case you will see a tab for each at the bottom as shown in the second image to the left.

The *Project Browser* should be left open; if you accidentally close it by clicking the "X" in the upper right, you can reopen it by
View → **Window** → **User Interface** → **Project Browser**.

The *Project Browser* and *Properties Palette* can be repositioned on a second monitor, if you have one, when you want more room to work in the drawing window.

> If the **Project Browser** or **Properties Palette** are accidentally closed, open it via **View** → **User Interface**—checked options here are open/visible.

View Tabs

Revit displays a tab for each open view. Clicking a tab is a quick way to switch between open views. Click the "X" in each view tab to close that view. Drag a tab to change its position. As you drag a view to a different position, a preview of the view position will be shown before you dock it in place. Tabs can also be pulled outside of the main Revit application window – even to a second monitor. In the second image below, the schedule could be filling a second computer screen while reviewing the same information in the floor plan.

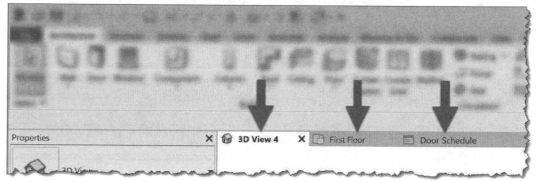

View Tabs; one for each open view

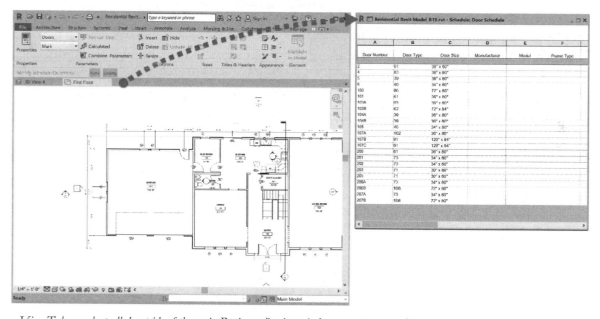

View Tabs can be pulled outside of the main Revit application window, even to a second screen

Status Bar:

This area will display information, on the far left, about the current command or list information about a selected element. The right-hand side of the *Status Bar* shows the number of elements selected. The small funnel icon to the left of the selection number can be clicked to open the *Filter* dialog box, which allows you to reduce your current selection to a specific category; for example, you could select the entire floor plan, and then filter it down to just the doors. This is different than the *Selection Filter* in the *Properties Palette* which keeps everything selected.

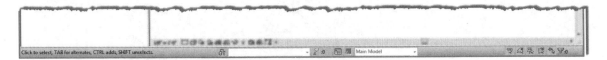

On the *Status Bar*, the five icons on the right, shown in the image below, control how elements are selected. These are, from left to right:

- Select Links
- Select Underlay Elements
- Select Pinned Elements
- Select Elements by Face
- Drag Elements on Selection

Hover your cursor over an icon for the name and for a brief description of what it does. These are toggles that are on or off; **the red 'X' in the upper right of each icon means you cannot select that type of element within the model**. These controls help prevent accidentally moving or deleting things. Keep these toggles in mind if you are having trouble selecting something; you may have accidentally toggled one of these on.

Finally, the two drop-down lists towards the center of the *Status Bar* control **Design Options** and **Worksets** (see image on previous page). The latter is not covered in this book, but *Design Options* are. *Worksets* relate to the ability for more than one designer to be in the model at a time.

View Control Bar:

$$1/8" = 1'-0"$$

This is a feature which gives you convenient access to tools which control each view's display settings (i.e., scale, shadows, detail level, graphics style, etc.). The options vary slightly between view types: 2D View, 3D view, Sheet and Schedule. The important thing to know is that these settings only affect the current view, the one listed on the *Application Title Bar*. Most of these settings are available in the *Properties Palette*, but this toolbar cannot be turned off like the *Properties Palette* can. Some of the tools act as a "mode" for the view. When a "mode" is turned on, the button will be highlighted on the view control bar, and the border of the view will be highlighted in a similar color.

Context Menu:

The *context menu* appears near the cursor whenever you right-click on the mouse (see image at right). The options on that menu will vary depending on what tool is active or what element is selected.

Cancel
Repeat [Wall: Architectural]
Recent Commands ▶
Change wall's orientation
Select Joined Elements
Hide in View ▶
Override Graphics in View ▶
Create Similar
Edit Family
Select Previous
Select All Instances ▶
Delete
Find Referring Views
Zoom In Region
Zoom Out (2x)
Zoom To Fit
Previous Pan/Zoom
Next Pan/Zoom
Browsers ▶
√ Properties

Context menu example with a wall selected

Drawing Window:

This is where you manipulate the Building Information Model (BIM). Here you will see the current view (plan, elevation or section), schedule or sheet. Any changes made are instantly propagated to the entire database.

Elevation Marker:

This item is not really part of the Revit UI but is visible in the drawing window by default via the various templates you can start with, so it is worth mentioning at this point. The four elevation markers point at each side of your project and ultimately indicate the drawing sheet on which you would find an elevation drawing of each side of the building. All you need to know right now is that you should draw your floor plan generally in the middle of the four elevation markers that you will see in each plan view; DO NOT delete them as this will remove the related view from the *Project Browser*.

Info Center:

The upper left portion of the application title bar has two items to be aware of.

About

The About command (see image below) provides access to information about the current Revit build. It is important that everyone on a project team uses the same build. Autodesk provides free service packs and updates for subscription customers via their website.

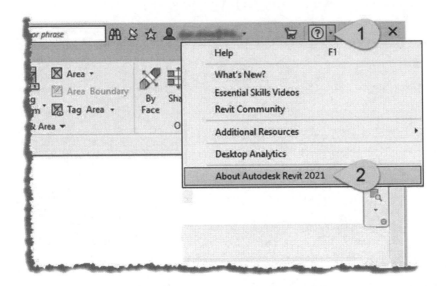

The image below points out the current Revit build.

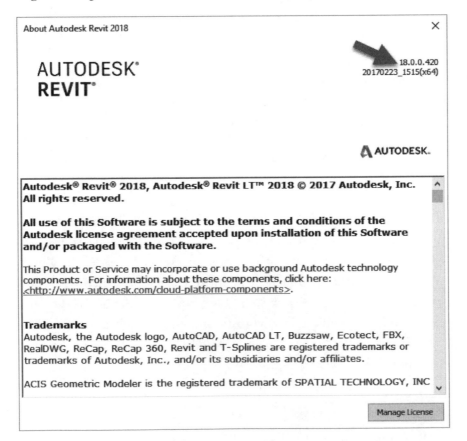

Licensing

Clicking the Manage License button shown in the previous image opens the licensing dialog.

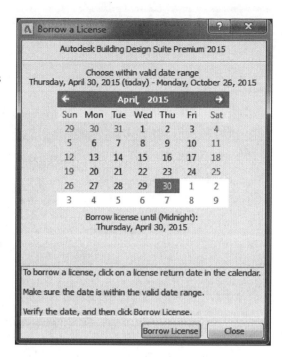

Networked licenses can be borrowed; this option is only available when using a networked version of Revit—not for standalone installations. A network license is more expensive than a standalone, but it can be used by multiple staff. If a firm has 10 network licenses, and 15 staff who use Revit, only ten of those 15 may use Revit at any given time. When the eleventh person tries to access Revit, they will receive a warning message indicating there are no available licenses. That person must wait or ask someone to close Revit, which would instantly free up a license.

Borrow License

Used to borrow a license in order to work away from the office, often on a laptop to work on a project site or at home. In the *Borrow License* dialog, simply select a date and click the **Borrow License** button shown in the image to the right.

If an office has 10 network licenses, and one is checked out, the office now has 9 licenses available for use in the office until the license is returned.

Do not borrow a license for more time than is needed or for months at a time. If the computer is stolen or crashes and needs to be reformatted, the license cannot be returned. Time would have to run its course for the checked-out license to be returned. Autodesk will not be able to fix this as it would be a way for some to cheat and get extra licenses for free.

> When a firm is on subscription, they may request special **Home Use** serial numbers. The full version of the software may be installed on an employee's home computer. This is separate and in addition to the software installed in the office. Autodesk allows one Home Use license for each subscription license.

Return License Early

When a license does not need to be checked out, it can/should be returned early, rather than letting the time run out. This will make the asset available to others in the firm.

On occasion, Autodesk products will not allow the license to be returned early. This is a glitch in the system. In this case, the time will have to run its course to return the license.

2.2 File Tab

Introduction

Although part of the User Interface, the **File Tab** (formally the Application Menu) will be covered separately here as it is an important starting point. Without knowing how to open, save, close or print a drawing, the rest of the UI is not of much use.

The File Tab is the primary connection between Revit and the Windows operating system. From here your Revit model may be opened, saved or copied (Save As). The model can also be exported to various formats for use in other applications. When a drawing needs to be printed, this is where that task is initiated.

The File Tab also provides access to the **Options** dialog. This dialog contains many high-level settings which control things like graphics quality and performance, default file locations and more.

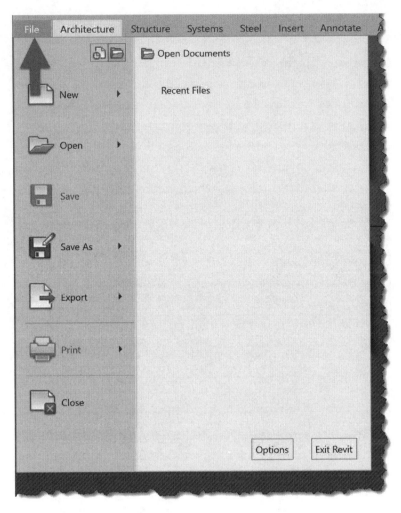

Figure 2.2-1 File Tab

In this section, each of these features will be explored in detail.

To access the File Tab, simply click on the **colored tab** (not the blue "R" like in previous releases) in the upper left corner of the Revit window. When a tool is selected, the menu will automatically close. To close the menu without clicking a tool, **press ESC**.

New → Project (Ctrl + N)

Use this tool to start a new project.

Clicking **New → Project** opens the New Project dialog (Figure 2.2-3). Here you choose which template to start your new project from. A template is essentially an empty project file which gets copied and becomes your new project.

This action may also be started by simply clicking on the word "New," which is part of a split button or the New… link under Projects on the Recent Files startup screen (see the UI section in this chapter for more on Split Buttons).

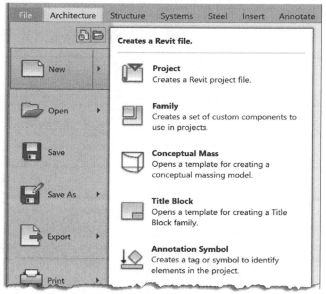

Figure 2.2-2 File Tab – New fly-out menu

> TIP: Selecting the correct template is extremely important. A template can have many things pre-loaded, such as sheets with custom title blocks, sheet list, standard drawings and much more.

Template file

The **drop-down list** presents a convenient set of templates from which to choose. The templates listed are based on the File Location settings in the Options dialog covered later in this section.

The **Browse button** will open the Choose Template dialog (Figure 2.2-4) focused on the folder containing the selected template in the drop-down list.

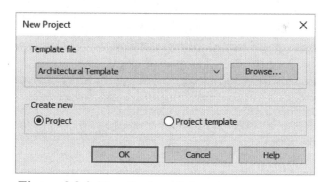

Figure 2.2-3 New Project dialog

For a default Revit installation, the same folder will open for all options. This folder contains additional templates not listed in the drop-down.

> FYI: A template file has an **.RTE** extension—for example: <u>Architectural Template.RTE</u>. A Revit project file has an **.RVT** extension—for example: <u>Lincoln Elementary School.RVT</u>.

File Tab & UI

Create new

In the New Project dialog, under the Create New label, the **Project** option is selected by default, as this is what would be used 99% of the time (Figure 2.2-3). However, if a new template needs to be developed, the **Project Template** option is selected, creating a file with an .RTE extension. Notice the "Files of type" setting, in the dialog shown below, is only looking for .RTE files.

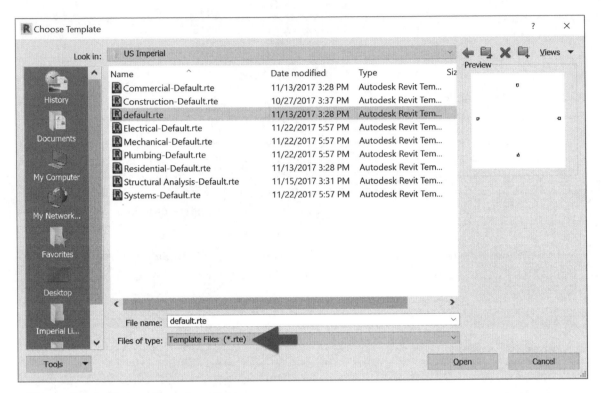

Figure 2.2-4 Choose Template dialog

Additional Things to Consider

Be careful not to start a project in a template file (RTE rather than RVT). Template files cannot have Worksharing enabled—thus multiple people cannot work in the same model. Also, it is not possible to do a Save As from a template file to a project file, nor can you simply rename the file as can be done with other programs. If a project is accidentally started as an RTE file, do the following:

- **Close** the RTE file (which was accidentally used to start a new project in)
- Use the **New → Project** tool
- **Browse** to the RTE file containing the project
- Start a new project (RVT file) based on the template file
- **Save** the file
- **Delete** the RTE file to avoid confusion

New → Family

Use this command to start a new Family in Revit—this is the content to be placed into a Revit project; e.g. furniture, doors, toilets, etc. This command may also be initiated from the Recent Files startup screen when Revit is first opened.

Immediately after clicking New → Family, the **New Family – Select Template File** dialog opens (Figure 2.2-5).

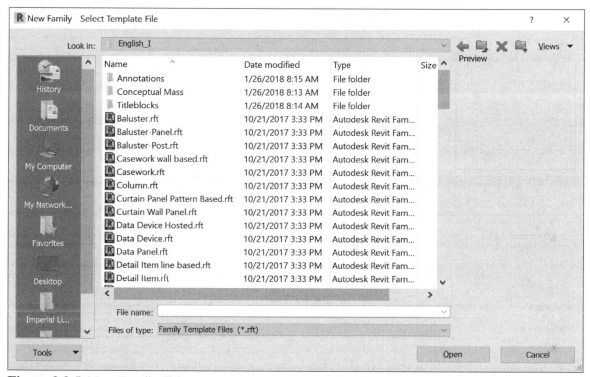

Figure 2.2-5 New Family dialog

This dialog opens in the folder containing several Family template files from which to start a new Family.

> *Default path:* C:\ProgramData\Autodesk\RVT 2021\Family Templates\English_I
> The folder location is defined in **Options → File Locations → Default path for family template files**. This path may be changed to point to a folder with a customized version of these files. FYI: A firm may create custom family templates to ensure consistent parameters and sub-categories are employed.

Notice the **Files of type** setting; this dialog is hardwired to only look for files with an **.RFT** extension—which stands for Revit Family Template. When a Family template file is selected and opened, the template file is copied and used as the starting point for the new Family. When the new Family file is saved, it will have the file extension **.RFA**.

New → Conceptual Mass

Use this command to start a new conceptual mass in Revit.

This command is nearly identical to the previous—new family. It opens to the same Family templates folder, just in the Conceptual Mass folder seen in Figure 2.2-5. In this folder is a single Family template file named **Mass.rft**. The main reason Conceptual Mass exists as a separate option is to highlight this relatively new and powerful feature.

A **Conceptual Mass**, along with **Adaptive Components**, allows the designer to create complex parametric geometry which may be loaded into the Revit Project environment for documentation.

Additional Things to Consider

The image to the right shows an elevation view within the Project environment. This high-rise building started as a Conceptual Mass in the Family Editor, where an Adaptive Component-based surface pattern (Rhomboid pattern in this example) was applied to the mass. When the mass is adjusted the surface pattern updates as well. Next, the Conceptual Mass is loaded into the Project environment. Here the Mass is selected and the Mass Floors tool on the Ribbon is used to define floors where the Revit Levels intersect the Mass. Finally, a schedule is created to total the Mass Floor area for the entire building.

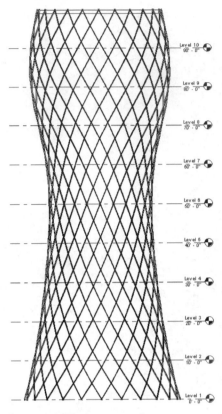

Conceptual mass example:
High-rise building with adaptive component curtain wall system. Adjusting the mass causes the curtain wall to update along with the floor area schedule for the entire facility.

New → Title Block

Create a new Title Block, in the Family Editor, to be used on Sheets in the Revit Project environment.

The **New Title Block** dialog opens in the Titleblocks folder shown in Figure 2.2-5. This folder provides a few sizes to use as a starting point for a new title block. The outermost line in the title block family defines the paper size for the sheet the title block is placed on.

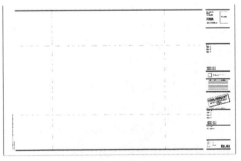

Custom title block family

New → Annotation Symbol

Used to create a tag, in the Family Editor, to graphically display information in the model, or a symbol which changes size when the view scale is adjusted.

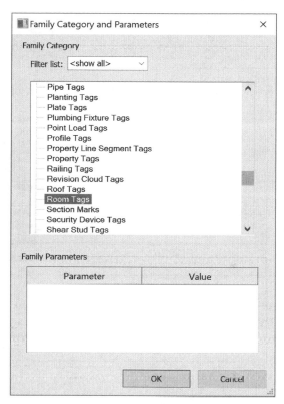

The **New Annotation Symbol** dialog opens in the Annotations folder shown in Figure 2.2-5. This folder provides several templates from which to start a tag or symbol.

Additional Things to Consider

Most of the templates are nearly identical. The main difference is the **Family Category** is set appropriately (Figure 2.2-6). This determines which elements the tag will work on in the project environment.

Labels are used within Tag families to define what information will be displayed in the Project environment. Based on what Family Category is selected, the **Edit Label** dialog (Figure 2.2-7) lists the standard parameters available on the left. The parameter added/listed on the right defines what information will be displayed for a given element when loaded and used in the Project. Multiple parameters may be listed on the right; e.g. the Width and Height of a door could be shown in a tag with an "X" automatically added between.

Figure 2.2-6 Select Family Category

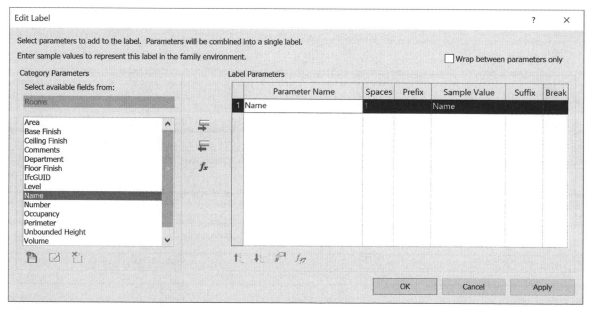

Figure 2.2-7 Edit label dialog – used within a tag family

Open → Project

Used to open a Revit Project file.

This command may also be started by simply clicking the word Open, which is part of the split-button. Additionally, the Open link, under Projects on the Recent Files startup screen works, too.

When using the **Open** split-button or Recent Files link (rather than Open → Project) the file filter looks for all Revit files—including families and templates. Thus, this open tool can be used to open any Revit file.

> TIP: Never open a Revit project file from the Recent Files startup or by double-clicking on a file via Windows Explorer if using Worksharing.

Figure 2.2-8 File Tab – Open fly-out menu

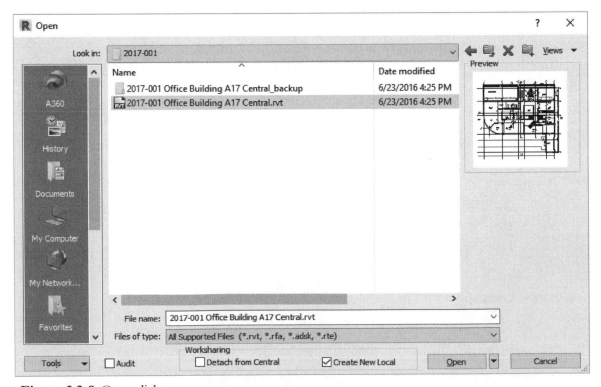

Figure 2.2-9 Open dialog

Open dialog: Tools

The **Tools** drop-down allows shortcut links to be added to the Places pane and/or to Windows favorites.

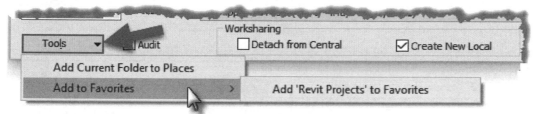

Figure 2.2-10 Open dialog – tools drop-down options

Selecting **Add Current Folder to Places** adds a shortcut to the left side of the Open dialog (Figure 2.2-11). Clicking one of these shortcuts opens the specified folder. A designer can add the current projects they are working on to efficiently access Revit files for each project. If the folder is deleted, or not available due to network connection, the icon will become a plain white folder. The link may be repositioned by dragging it or removed via the right-click menu.

Selecting **Add to Favorites** will add a link in Windows Internet Explorer to the specified folder (Figure 2.2-10).

Figure 2.2-11
Open dialog – Places pane with new custom link added

Open dialog: Audit

Used to find and correct problems within the Revit project (Figure 2.2-8).

If a file will not open due to **corruption**, try opening the file by first selecting this option. Revit will not indicate if it found anything. However, if the file may be opened then you know it fixed something.

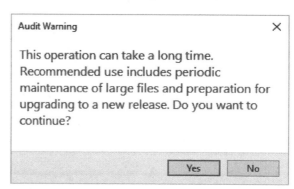

Depending on the project size, this process can take a long time, as indicated in the **warning** received when the option is checked (Figure 2.2-12).

Figure 2.2-12 Audit option warning

As indicated in the warning, an Audit should be run regularly on large projects (e.g. weekly on a file over 80MB) and prior to upgrading a project to a new version of Revit.

Open dialog: Detach from Central

The **Detach from Central** option (Figure 2.2-13) will open a Central file (that is one which has Worksharing enabled to allow multiple people to work in the same model) as **a new unnamed/unsaved central file with no ties to the original file**.

> Warning: This is an extremely important option to understand for those not familiar with Worksharing projects. Significant problems can be created for the design team by someone thinking they are just "looking around."

Additional Things to Consider

A Central file saves the file path to itself within the Revit database, i.e., the project file. If a central file is copied using Windows Explorer, usually from a network location to one's desktop, the newly copied file becomes a Local file which is attached to the original Central file. Thus, keep in mind that you cannot just copy a file out of a project folder.

When opening a Revit project, always use the Open dialog so you can either detach or create a local file. The Detach from Central option will work on both Local and Central files.

Figure 2.2-13 Open dialog options

Open dialog: Create New Local

This option is used to create a Local file based on the selected Central file.

When a file is selected within the open dialog box, Revit scans the file to see if it is a Central file. If it is, the Detach and Create Local options become active. Create New Local is checked by default when a Central (i.e., Worksharing) file is selected. With this option checked, once Open is clicked, Revit copies the Central file to the user's hard drive and adds the user's name to the end of the file name—this is the Local file the user works in.

For example:

- *Central File:* Lincoln Elementary School A21.rvt
- *Local File:* Lincoln Elementary School A21_**dstine**.rvt

The **A16** in the file name above is a common way to alert users as to what the discipline and Revit version are. In this case, **A** represents Architecture and **21** means Revit 2021. Projects should never be upgraded until the entire project team, including all internal and external disciplines, is ready to do so.

If the **Create New Local** is greyed out and not active after selecting a Revit project file, one of two things is true:

- The selected file is not a central file
    - ○ Worksharing has never been enabled
    - ○ A Local file was selected rather than a Central file
- The selected file is not the same version of Revit being used currently

Additional Things to Consider

Creating a new local file every day will keep your Revit file on your hard drive leaner and more efficient. This is because the database maintenance (e.g. **Compact Central**) done to the Central file does not always extend back to the Local file. This also prevents starting in an outdated file. Keep in mind local files need to be saved on a local, *i.e. C: Drive of the computer you are working on,* and not on a network resource or a folder that is syncing to another location.

Recommendation: Create a new Local file every day.

The username added to the end of the Local file name is set in **Options → General → Username**. The default username is based on the user login name for Windows. However, the first time the user logs into **Autodesk A360 Cloud Services**, the username will change to match the A360 username. This is done to support cloud collaboration within Revit.

Once a username is changed in Revit, new local files must be created. It is not possible to work in any old Local files—Revit will immediately show a warning if the file is opened. If a new local file is created each day, as recommended, this will not be a problem.

Open dialog: Open button drop-down options

Additional Workset options are available when a Worksharing (i.e., Central) file is selected.

Large projects, such as a new 400,000SF hospital or correction facility, often employ multiple worksets to segregate model elements into groups. Revit allows users to control which worksets are open. Not opening all worksets can take a significant burden off the computer's RAM, CPU and graphics card.

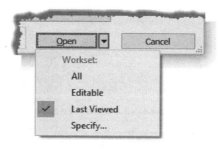

Figure 2.2-14 Open dialog – open

The default option is **Last Viewed**, which works for most users. For large projects, selecting **Specify** (and then Open) will bring up an intermediate dialog listing the worksets in the selected project.

Additional Things to Consider

Before moving on to the next topic, here are a few more points related to the Open dialog.

Searching for files

In Revit, and most Windows programs, it is possible to search the listed files on the screen—resulting in a shortened file list. To do this, simply type the desired word between two asterisk symbols (*) and then press Enter as shown in Figure 2.2-15 below.

> This example is using the default sample files folder. To try this, select "R" → Open → Sample Files.

In this example, ***advanced***, an asterisk in a search means any number of characters can occur before or after the search word—thus, the word "advanced" can occur anywhere in the file name. Changing the search to **rac*** would return only file names which start with "rac."

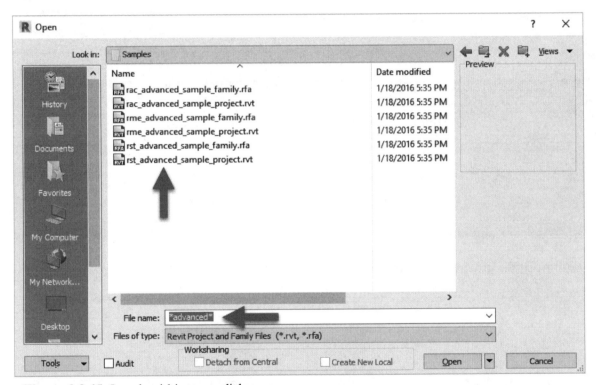

Figure 2.2-15 Search within open dialog

Copy/Paste File Path from Windows Explorer

Another navigation option within the Open dialog is to Copy/Paste a folder path into the **File name** field and press Enter to make that location current; for example, if the location has already been found using Windows Explorer, but the Revit Open dialog is pointing somewhere else. Simply click within the address field at the top of Windows Explorer, and with the folder path highlighted press **Ctrl+C** (Figure 2.2-16). Now, in Revit's Open dialog, click within the File name field and press **Ctrl+V** and then Enter (Figure 2.2-17). The Open dialog will switch to that specified location, allowing a file to be selected.

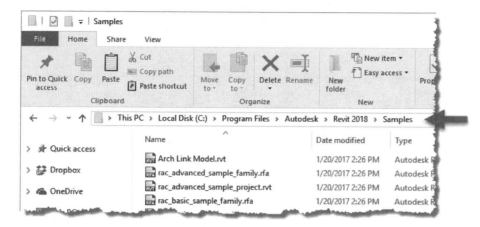

Figure 2.2-16 Windows Explorer example

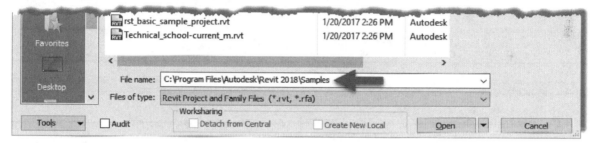

Figure 2.2-17 Paste path from Windows Explorer and press enter

Thumbnail view of files

If the Revit Family is saved with a decent preview image, the Open dialog can be toggled to Thumbnails mode (Figure 2.2-18). Simply click the **Views** drop-down in the upper right and select **Thumbnails** from the menu.

> TIP: To increase the size of the thumbnail image, click within the file preview area and while holding down the **Alt** key, spin the **wheel** on the mouse. This tip works in most Windows programs.

Also note that the size of the Open dialog may be enlarged by dragging its edges. With larger preview images and a larger open dialog, the designer is able to more quickly find the content needed.

File Tab & UI

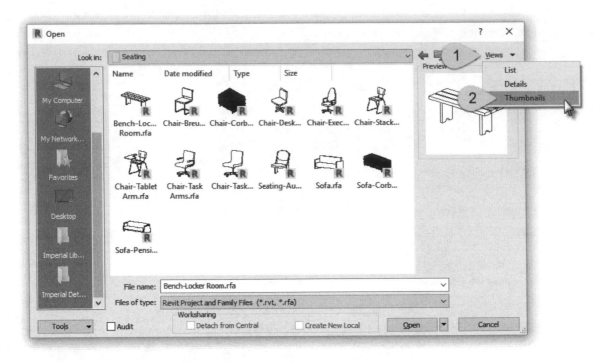

Figure 2.2-18 Open dialog – switch to thumbnail preview mode

Open → Family

Used to open a Revit Family file for editing.

The file filter is set to only look for RFA or ADSK files. The file extension for a native Revit Family is **RFA**. Less often used is the **ADSK** file type—a special "exchange" format used to share data with Autodesk Inventor.

By default Revit opens to the family library folder; notice the **Imperial Library** (USA installation) shortcut on the left is highlighted because it is the current "place" (Figure 2.2-19). This default location is specified in **Options → File Locations → Places…** (Button). The first item in the list determines the location the Open Family dialog starts.

Once a design firm has a decent Revit library, they will modify Revit for all staff to point to their content rather than the Out Of The Box (aka OOTB) content provided with the Autodesk software. The OOTB content is not installed with Revit. It can be downloaded and installed from the following URL:

> TIP: It is recommended that the OOTB content be kept separate from the firm's custom content. They can be next to each other (on a common network) and organized the same way. This helps keep the content separate during Revit upgrades. Sometimes the Revit content is modified or renamed. It would be difficult to replace the OOTB content if the firm's custom content was mixed within it.

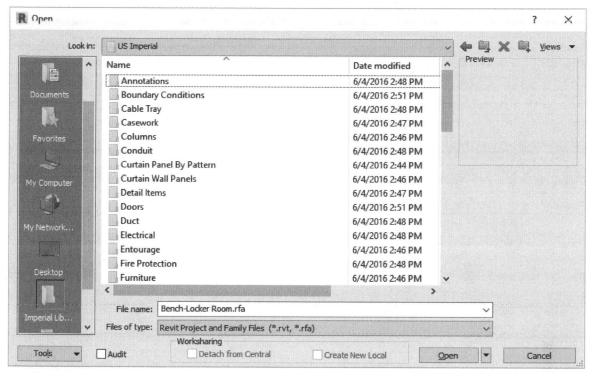

Figure 2.2-19 Open Family dialog

For additional information about the Open dialog, see the previous section on Open → Project.

Open → Revit File

Used to open any Revit file type; formats: RVT, RFA, ADSK, RTE, RFT.

This can also be achieved by simply clicking directly on the Open button in the File Tab or the Open link in the Recent Files startup screen.

Open → Building Component

Used to open a Building Component file; Format: ADSK.

This is a special "exchange" format to share manufacturer data from Autodesk Inventor with Revit. This can include geometry and MEP connectors. The geometry cannot be modified in Revit.

Open → IFC

Used to open an Industry Foundation Class (IFC) file; formats: ifc, ifcXML, ifcZIP.

The IFC format is an open standard BIM exchange format, which is often used in clash detection on large complex projects where various design teams are using different BIM applications. This format is needed due to the significant differences in program design between competing software companies such as Autodesk and Bentley.

File Tab & UI

| Some clients require IFC files be provided at the end of a project.

When an IFC file is opened Revit actually converts it to an RVT file—thus, once opened the only save option is to Save As a Revit Project file. This is also what is done in the background when the **Link IFC** tool is used in a Revit project.

To read more about IFC, see Revit help and visit this URL:
http://www.iso.org/iso/catalogue_detail.htm?csnumber=51622

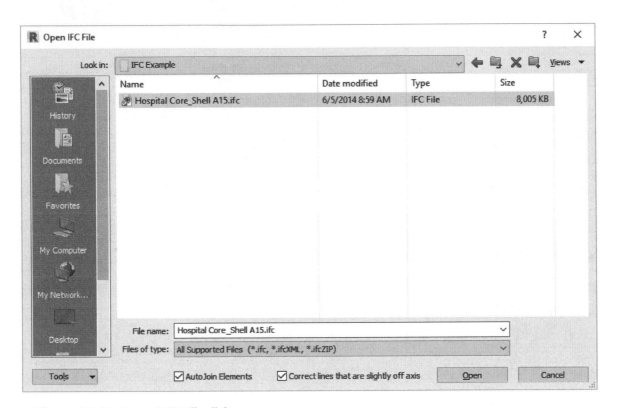

Figure 2.2-20 Open IFC File dialog

Auto Join Elements

This option is checked by default (Figure 2.2-20). When checked, walls are automatically joined to other walls and to columns. Unchecking this option may improve performance but not look as good graphically.

Correct lines that are slightly off axis

This option is checked by default (Figure 2.2-20). When checked, lines that are less than 0.1 degrees off of horizontal or vertical will be changed to 0 degrees. Uncheck this option for site plans as they naturally have various angles.

Open → IFC Options

Used to map Revit categories to the standard IFC naming convention (Figure 2.2-21).

An essential component in a common exchange format between dissimilar BIM applications is a standard naming convention for real-world building components. This dialog provides a default mapping which will generally work for most but may be customized if required.

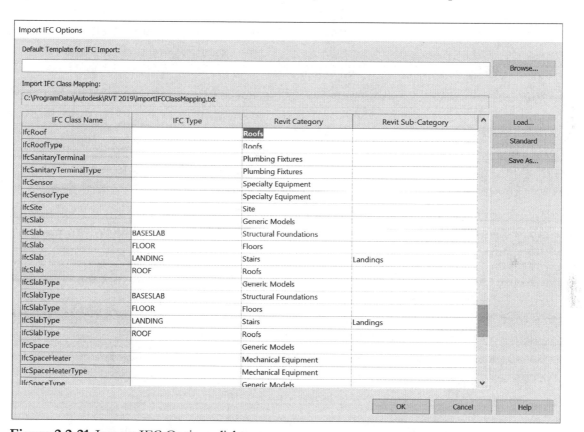

Figure 2.2-21 Import IFC Options dialog

Open → Sample Files

Used to quickly point the Open dialog at the folder containing sample files provided with the software.

These sample files are great for testing new families or schedules. These are the same sample files that are shown on the Recent Files startup screen when Revit is opened for the first time. These links end up being replaced as project files are opened. This command allows access to the sample files at any time in the future.

File Tab & UI

Save (Ctrl + S)

Use to save the current Family or Project file to disk.

Save regularly to avoid losing work if Revit crashes or in case of power failure. If working in a Worksharing project (i.e., a Central file) this Save common is the "Save Local" option. A **"Sync with Central"** would also be required occasionally; see the Quick Access Toolbar section for more on Sync with Central.

Be sure to back up your files regularly. Consider using Cloud Storage such as Autodesk A360 or Dropbox.

Backup files

Each time a non-Worksharing file is saved a backup copy is made in the same folder as the Revit Project or Family being edited. The file name has a ".0001" added as a suffix and has the same file extension (RVT or RFA). Because the file has the same extension, it can be difficult to quickly determine which file is current and which is the backup. Revit makes multiple backup files based on the **Maximum Backups** setting (See Save As → Options). Each file represents the last handful of successive saves, which provides an easy way to restore a file that has become corrupt.

> For another backup option, consider turning on Windows **Previous Versions** feature. This can be enabled on local hard drives and network drives. If enabled on a firm network, staff are able to right-click on a folder or file, select Properties → Previous Versions and then select from time stamped backups from 10am, noon and 3pm for as many days back as the designated disk space will allow. This can take a significant burden off of the IT department when files need to be restored. Of course, this does not replace the need for a firm to do traditional nightly backups.

A backup file for Families and non-Worksharing files are automatically saved here:
C:\Users*your user name*\AppData\Local\Autodesk\Revit\Autodesk Revit 2021\Journals
If your file becomes corrupt or it gets deleted, try looking here for a backup copy. For computers with low disk space, consider deleting the older files in this folder from time to time.

See the Options section for information on setting **Save Notifications**. By default, Revit will remind the user to save every 30 minutes. Note that Revit does not have an auto save option.

Save As → Project

Used to save the current Project as a new file with a different name and/or location.

Notice Save As, in the File Tab, is a **fly out button** rather than a split-button like New and Open. This just means you must click an option after running the cursor over Save As.

When this command is selected, the Save As dialog opens. The user is prompted to provide a name and location for the new file. Once complete, the NEW file is open and the original file now closed.

It is important to understand that the original file is not automatically saved as part of the Save As command.

Figure 2.2-22 File Tab – Save As fly out

Therefore, if the intention is to duplicate the current state of the file, as an archive perhaps, a regular Save should be done first.

For a project file, the Save As result varies depending on the Worksharing status of the file. Note that by default a new project is non-Worksharing—meaning only one person can work in the file at a time.

- Save As for a non-Worksharing project file: simply creates a new file based on the current project open. This new file has no connection to the original file and is still non-Worksharing.
- Save As for a Worksharing project file: a new local file is created and always has a connection back to the central file. This is true for both **Central** and **Local** files – that is, a Save As will create a local file. Caution should be used by those who are not familiar with Worksharing as significant problems can be created in the central file. See the information on the **Options** button in the Save As dialog for information on how to create a new Central file which is NOT connected to the original Central file.

A Save As **for a family** will always create a separate new file with no ties to the original file.

> **If the Central file becomes corrupt**, any user's Local file can become the new Central file. First find the user with the most current working file (i.e., this user has the most to lose). Open their Local file "Detached from Central," then do a Save As and overwrite the Central file. Finally, everyone on the project team needs to create a new Local file (all original Local files are now invalid and WILL NOT work with the new Central file).

File Tab & UI

File Save Options

Clicking the **Options** button in the Save As dialog opens the **File Save Options** dialog (Figure 2.2-23). This provides access to three Worksharing options (if applicable) and a few other options not found anywhere else.

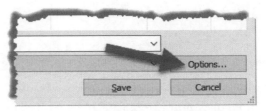

Figure 2.2-23 Save As – Options button

Maximum backups tells Revit how many backup files are required. Each time the model is saved, the current RVT is copied and renamed with a **.0001** suffix. Then, the next time the file is saved, the **.0001** file is copied and renamed to **.0002** and the current RVT file becomes **.0001**. This process repeats up to the number selected for Maximum backups. Once the maximum number is reached, the oldest file is deleted. This option only applies to non-Worksharing files as backups are handled differently for Worksharing models.

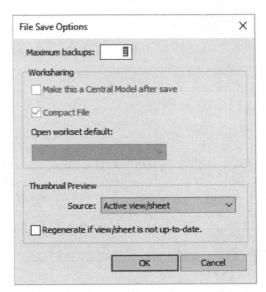

Figure 2.2-24 File Save Options dialog

Worksharing

As previously mentioned, doing a Save As on a Central **or** Local file will create a Local file. If the original Central file has become corrupt and needs to be replaced with a good Local file, or if a new Central file is required for a separate/new project or for testing, be sure to check **Make this a Central Model after save**. Doing so will create a new Central file with no connection to the original.

If a Central file was opened 'Detached from central' (and with preserve Worksets picked when prompted) then the **Make this a Central file after save** is grayed out and already checked.

> **BIM Manager Tip:** When providing support for various design teams, open the Revit project file 'Detached From Central' or use the 'Make this a Central Model after save' when investigating a reported problem. This allows for more aggressive experimentation when searching for the problem without affecting the actual project (which may have staff working in it). Things like deleting a wall or roof or applying a View Template to reset a view can be done without worry. When the problem is found, then the actual model can be opened and corrected.

The **Compact Central** is used to do routine maintenance on the database (i.e, worksharing file). The worksharing file retains a lot of history and backup information which can lead to reduced performance and potential stability issues. This feature is also available within the 'Synchronize with Central' command. This should be done approximately once a week on large complex projects or in the final few weeks of any project. This command can be run any time; it just takes longer to save which can affect several people, as only one person may save at a time.

The **Open Workset Default** option provides access to the same options covered previously in the Open dialog; specifying which Workset(s) should open by default for this model. For large projects, this can significantly reduce the amount of computer resources needed/used as large chunks (i.e., Worksets) are not being loaded.

Thumbnail Preview

This option determines what the thumbnail preview will look like—i.e., the images seen in Figure 2.2-18. The default Active view/sheet means the preview will be based on the last current view when the file is Saved/Closed. Selecting from the list provides the opportunity for a more meaningful image; for example, an elevation view of door families or a 3D view of a furniture family. In a project, it might be a 3D view rather than the last floor edge detail someone on the project recently worked on.

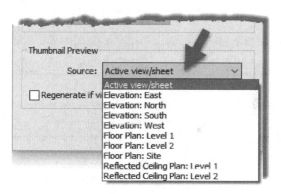

Figure 2.2-25 Select thumbnail preview

When a specific view is selected as the thumbnail preview for the file (Figure 2.2-25), it has no effect on which view is actually opened. The view the project opens to is still based on the setting in **Manage → Starting View**.

The **Regenerate if view/sheet is not up to date** will force the preview to update, which consumes a lot of computer resources. It is unchecked by default.

Save As → Family

Used to save the current Family as a new file with a different name and/or location.

When this command is selected, the Save As dialog opens. The user is prompted to provide a name and location for the new file. Once complete, the NEW file is open and the original file now closed.

It is important to understand that the original file is not automatically saved as part of the Save As command. Therefore, if the intention is to duplicate the current state of the file, as an archive perhaps, a regular Save should be done first.

Options button: see the previous section on Save As → Project for a detailed discussion on this topic.

Save As → Template

Save the current Revit project file as a template file.

This command allows a Project (RVT file extension) to become a template (RTE file extension) on which future projects may be based—using the New command. An example of this might be a firm or a specialized group within a firm, e.g. a Healthcare Group, finishes their first Revit project and wants to turn it into a template for future projects.

This command cannot be used on Worksharing files—it must be opened with 'Detach from Central' selected and then 'Discard Worksets' first.

Note that Revit can save a Project file as a Template, but **Revit will not allow a Template file to be saved as a Project file**. To save a Template file as a Project file, simply use the New command. The New command, in essence, copies a Template file and turns it into a Project file.

> **BIM Manager Tip:** Modify Revit to only show custom firm templates via **Options →** **File Locations** and adjust the 'Project template files listing'. Delete the OOTB template files provided with the software. This will help ensure staff always starts with the correct template when creating a new project.

Save As → Library → Family

Used as a quick way to save all families or a single selected family **in a project** to the library.

After selecting the command (Figure 2.2-26), the **Save Family** dialog opens (Figure 2.2-27).

The main option here is the **Family to Save** setting—the default being <All Families>. When the All Families option is selected, the File Name is set to **Same as family name**. Thus, the file name will match the family name.

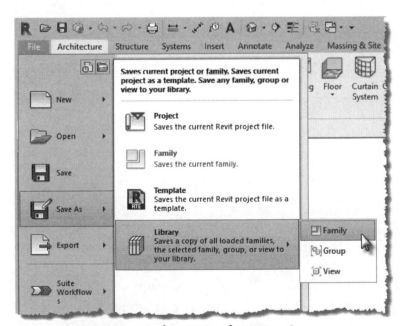

Figure 2.2-26 File Tab → Save As → Library fly out

Clicking the drop-down list next to **Family to save,** all the Families in the current model are listed. Selecting a Family from the list will tell Revit to only export that Family to a file—and will also allow a different name to be entered if desired.

> Saving *All Families* in the current project file may take a long time.

The **Save Family** dialog opens to the default Family library location specified in Options. However, what may be confusing to some is that the families will not automatically go into their appropriate sub-folders; e.g. Furniture, Doors, etc. They all get dumped into the current folder. So it may be advantageous to save all the families to a temporary folder so they can be manually sorted out and not confused with other files.

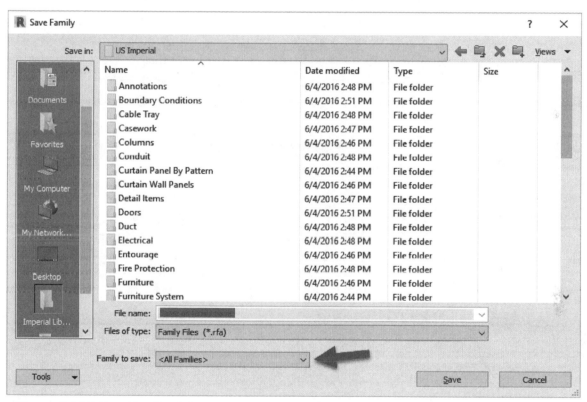

Figure 2.2-27 Save Family dialog

The Save Family command can also be selected by right-clicking a Family in the Project Browser. With the main heading Families selected in the Project Browser, the default is to save 'All Families'. When an individual file is selected, only it can be saved to file from the project via the Project Browser.

Save As → Library → Group

Used to save a specific Model or Detail Group to a file, outside of the current project.

This command provides a convenient way to quickly save Groups to an external file for use in other Revit projects. The result is a stripped-down Revit project file with just the contents of the Group in the model (but not actually Grouped).

In the Save Group dialog (Figure 2.2-28) simply select the Group to be saved from the **Group to save** list. If the selected Group is a Model Group, the Include attached Detail Group is active and may be selected.

> Model groups contain 3D elements, and Detail groups contain 2D and view specific elements such as Detail Lines and Annotation.

On large multi-building projects, Groups are a good way of managing similar content between files. This is partially a manual process but is a good option given the options. When a Group is saved to file it may be inserted into other projects using **Insert → Load as Group** command.

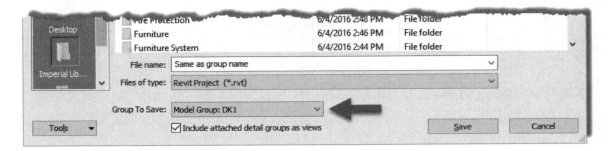

Figure 2.2-28 Save Group dialog

If the Group already exists, a prompt will be provided (Figure 2.2-29) with an option to replace the current, i.e. old, group or just load with a different name—thus, retaining the original group within the current project.

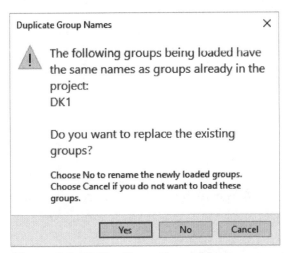

Figure 2.2-29 Duplicate Group Names prompt

Save As → Library → View

Used to save 2D and view specific information to file in the default library location.

This tool is a quick way to save a Drafting View to the default library location. For example, at the end of a project, there may be several details which can be used on future projects. Using this tool, all the details can be exported at once.

Once this command is selected, the **Save Views** dialog appears (Figure 2.2-30). Select the view(s) desired and click OK. A file name and location must be provided. All selected Views are saved in the same stripped-down Revit project file. Thus, it may be best to save each view separately.

> These new details can be inserted into future projects using the **Insert → Insert from File → Insert Views from File**.

Although the tool is intended for exporting 2D information, if a view or sheet is selected that has 3D content, the 3D information will be exported to the new file.

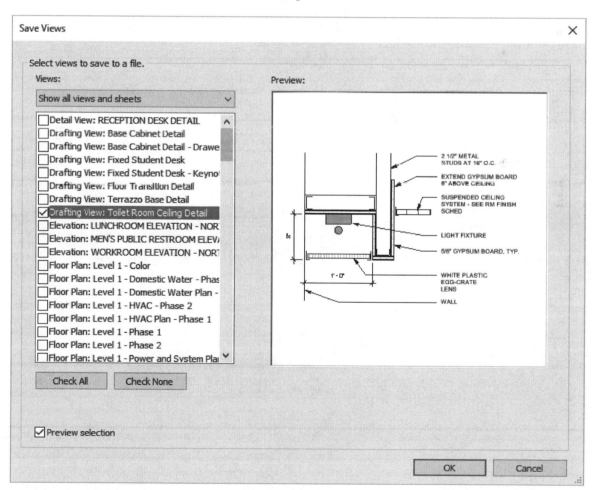

Figure 2.2-30 Save Views dialog – 2D drafting view selected

Export fly out (Intro)

Autodesk Revit provides several export options, which are mainly geared towards sharing data with other software programs.

To see all the export options, the small arrow-bar near the bottom (or top) needs to be clicked multiple times – pointed out in Figure 2.2-31.

Some export options are grayed out based on the current view. For example, the CAD export options are not available if the current view is a schedule—and vice versa. If the focus is not on the main drawing/canvas area, an export tool may be grayed out. Simply click within the canvas area and then the export tool will be available.

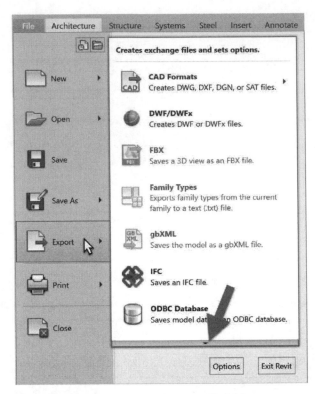

Figure 2.2-31 Export options – File Tab

Export → CAD Formats (Intro)

The CAD formats export option provides the opportunity to save the Revit model in a number of different formats used by CAD (Computer Aided Drafting) programs such as Autodesk's

AutoCAD, Bentley's Microstation and several others that are able to read their industry standard format (Figure 2.2-32).

> A distinction is often made between **CAD vs. BIM** in the AEC (Architecture, Engineering & Construction) industry. CAD is used to refer to traditional drafting software which is largely used for 2D drafting. BIM (Building Information Modeling) is used, in

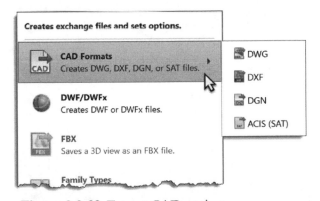

Figure 2.2-32 Export CAD options

contrast, to describe newer software/workflows which mainly involve creating 3D models with element specific information provided.

Export → CAD Formats → DWG

Used to create a 2D or 3D file in the industry standard AutoCAD DWG format.

This export option, and export Image, are often the most used of all the export options. Given the large market share of AutoCAD in the AEC industry and that all major CAD programs are able to read and/or write to the DWG format, it is the "go to" format for most designers when a CAD file has been requested of them.

The DWG export feature has a lot of options. Often, the default settings will be sufficient. The default options will map the Revit categories to the equivalent **AIA Layer Standard**, and also make the DWG file look as close as possible to the Revit view being exported.

When the DWG export option is selected, the **DWG Export** dialog appears (Figure 2.2-33). By default, only the current view will be exported. The view name is listed on the right and a preview shown on the left. The **Export Setup** options in the upper left will be covered later in this chapter under Export → Cad Formats → Export Options → Export Setup DWG/DXF.

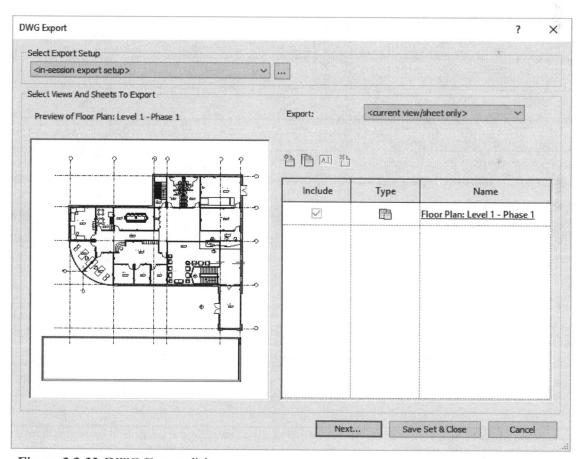

Figure 2.2-33 DWG Export dialog

Exporting Multiple Views/Sheets

Revit can export multiple Views and/or Sheets at once. Using this feature will save a lot of time if an entire set of construction document sheets need to be exported to CAD format.

Changing the **Export** drop-down list to **<in-session view/sheet set>** will reveal another drop-down list, named **Show in list**, to filter the list (Figure 2.2-34). The list can show just sheets, just views or both. This list can also show saved sheets listed previously created in the Print dialog.

> FYI: If a View is selected, and then a "sheets only" option is selected, the View will still be selected even though it is not visible. Use the Check None button before switching the filter.

Clicking **Save Set and Close** will save the current settings and close the export dialog without actually exporting anything. If a View/Sheet Set is selected, this will update and save it—such that the update will affect the future printing (not just exporting).

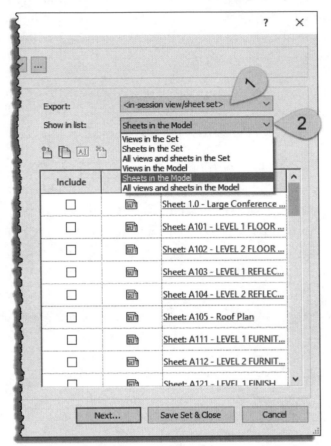

Figure 2.2-34 Export multiple CAD views/sheets

When all the settings are good, click Next, which opens the **Export CAD Formats – Save to Target Folder** dialog shown in Figure 2.2-35—this is the last step before the export begins. Here you have several important, but easy to miss, settings.

The obvious thing here is to specify the location for the exported files. However, the file version, naming and if the exported files should have any links may also be specified.

File Tab & UI

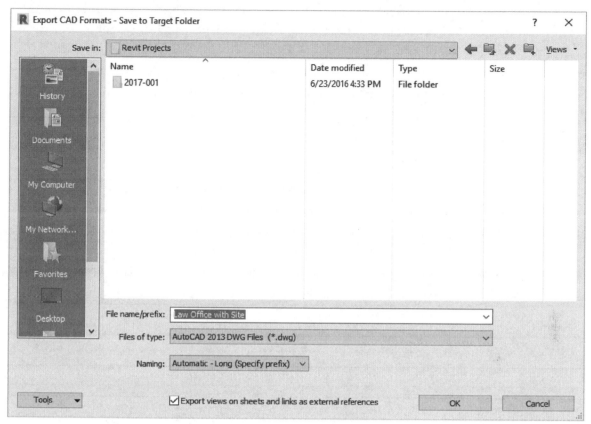

Figure 2.2-35 Export CAD Formats – Save to Target Folder dialog

Specify AutoCAD Version

The **Files of type** should be changed to correspond to the required AutoCAD version. Traditionally, Autodesk changes the file format for the DWG file every few years such that previous versions are not able to read the new format. For example, AutoCAD 2011 cannot read a file saved in AutoCAD 2021. This is done primarily to support new features in AutoCAD. If a client, contractor, etc. do not have the most current version of their CAD (or SketchUp Pro) software, they will need the DWG files provided in an older format.

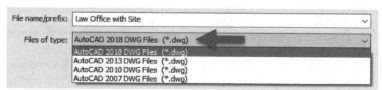

Figure 2.2-36 Selecting DWG format to be exported

Specify File Naming

The default file name includes the filename, the view type and the view name (as seen in Figure 2.2-35). If the project file has been opened **Detached from Central**, which is sometimes done prior to printing, rendering or exporting to conserve system resources, the file name is not included (because the current file is unnamed). Changing the Naming setting to **Automatic – Short** will omit the filename—this is useful when exporting a sheet and only the sheet number is desired for the filename. When a single file is selected for export, a third option is available in the list: **Manual (specify name)**.

External References

The default option for sheets is to export views as **External References** (aka Xrefs). This creates one DWG file for the sheet and one for each view. The sheet file has each view xref'ed in (similar to a link in Revit). If the view files are moved or deleted, they will not appear in the sheet file. This default setting is similar to how a firm would organize their DWG files when using AutoCAD to draw the project. For example, all sheets for a large floor plan spilt up, over multiple sheets, would reference the same floor plan DWG file.

> Unlike Revit, AutoCAD does not store the latest version of the DWG file internal to the host DWG file. Thus, if the xref'ed files are moved, renamed or deleted, they will be completely lost.

If the drawings are being exported as final deliverables to a client or for a contractor or manufacturer, it may be desirable to make each sheet self-contained—i.e., without any external references.

Additional things to consider

For large projects and when several views/sheets are selected, the export process can take a long time. **For worksharing project files**, if the file name is not required in the exported file name, consider opening the file **Detached from Central**; this will conserve computer resources—especially useful if the computer will be used to do something else during the export process.

All views are exported at 2D except 3D and Camera views. If a 2D view is required for a 3D view, export a sheet with the 3D view and it will be 2D.

There are some slight graphical differences between how a View vs. a Sheet is exported. Sometimes it is worth trying both to see if one looks (or is formatted) better.

Export → CAD Formats → DXF

Used to export project Views/Sheets to a generic CAD format supported by many programs—even non-CAD programs.

A DXF file is a text-based file which describes the vector based drawing—as such, these files can become very large compared to a DWG file which is a proprietary compressed format. This feature works exactly like the DWG export option just covered. Please see the Export → CAD Formats → DWG section for details on using this export option.

Export → CAD Formats → DGN

Used to create a Microstation compatible DGN drawing file.

Bentley's Microstation is similar to Autodesk's AutoCAD; both are a traditional CAD program, and neither is considered a BIM application. The DGN file format is a proprietary file format controlled by Bentley. The DGN export workflow is identical to the DWG/DXF workflow previously covered in this chapter.

It should be noted that a **Level** in Microstation is equivalent to a **Layer** in AutoCAD; both are used to control visibility, color, linetype, etc. Also, just for reference, a **Cell** in Microstation is equivalent to a **Block** in AutoCAD; both are symbols somewhat similar to clipart in a word processing program.

> This author has had better luck exporting to DWG and opening the DWG file in Microstation, and then saving as a DGN. Of course, this requires access to the Microstation software.

See **Export → CAD Formats → DWG** for detailed coverage on the export DGN workflow and options (covered prior to this section). Also see **Export → CAD Formats → Options → Export Setups DGN** for information on saved related export settings.

Export → CAD Formats → ACIS (SAT)

Used to export a generic text-based file of 3D geometry (ACIS) which can be imported by many CAD programs.

While the DXF format was primarily developed for 2D information, the SAT is geared towards 3D geometry. The SAT file contains ACIS solids, which are smooth looking surfaces (**not** made up of several triangles which produce lots of unwanted extra lines – aka Polymesh).

> The DWG exporter can create Polymesh or ACIS solids based on the setup options selected.

File Tab & UI

Upon selecting this command, the SAT Export Settings dialog appears (Figure 2.2-37). This format does not contain the same information as the other export formats, thus the options are fewer.

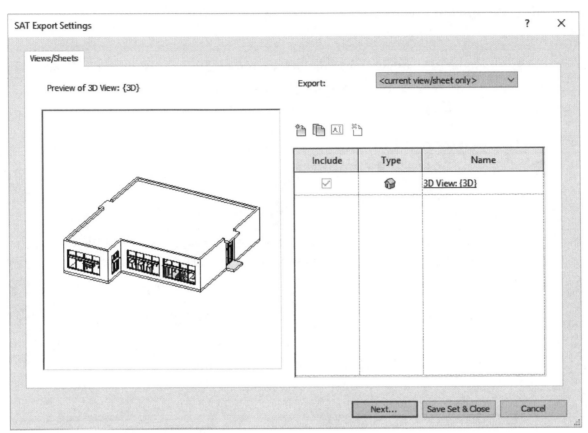

Figure 2.2-37 SAT Export Settings dialog

Export → DWF/DWFx

Used to export a Revit View/Sheet to the Autodesk Drawing Web Format (DWF/DWFx).

The DWF format is meant to be a lightweight, non-editable file that may be easily shared, similar to a PDF file. This format is used to post drawings on websites and for secondary applications—such as construction administration, facilities management and/or cost estimating.

A DWF/DWFx format may contain 2D, 3D drawings information (or both). The DWFx format is newer than the DWF format and is based on the Microsoft **XML Paper Specification** (XPS). The DWFx format can be viewed using the free Microsoft XPS viewer (Figure 2.2-38). Autodesk provides a free viewer/redline tool called **Autodesk Design Review 2013**. DWFs can also be viewed in Autodesk's cloud-based **A360 Viewer**: https://a360.autodesk.com/viewer/.

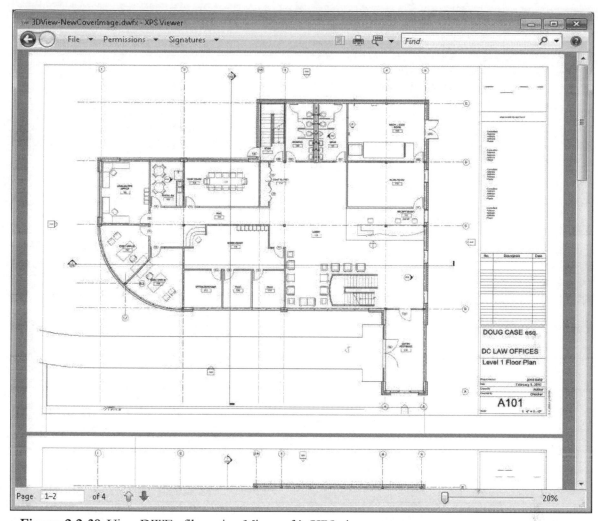

Figure 2.2-38 View DWFx files using Microsoft's XPS viewer

Once the **Export → DWF/DWFx** command is selected, the DWF Export Settings dialog appears (Figure 2.2-39). The options for selecting one or more views/sheets are the same as the **Export → CAD Formats → DWG** workflow—please see that section for more information.

The second and third tabs (DWF Properties and Project information) are self-explanatory; be sure to review them before continuing.

Once the settings are correct, click **Next**.

File Tab & UI

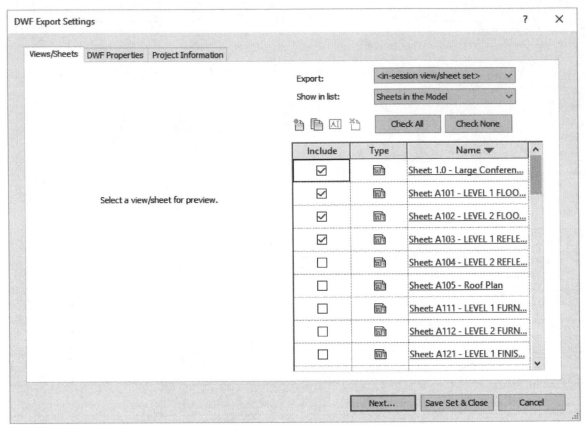

Figure 2.2-39 DWF Export Settings dialog

After clicking **Next**, the **Export DWF – Save to Target Folder** dialog appears (Figure 2.2-40). In addition to specifying the location, it is also recommended to review the Naming and checkbox options. The naming options are similar to the previous export tools covered. The checkbox determines if the views/sheets are exported into a single file (usually preferred) or individual files.

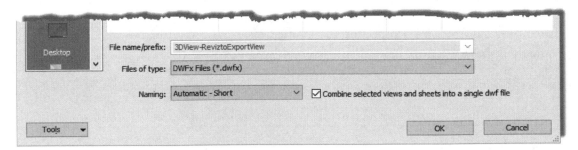

Figure 2.2-40 Export DWF – Save to Target Folder

Export → FBX

Used to export the Revit project model with materials, lights and a single camera view for use in **Autodesk 3ds Max** or any application compatible with the FBX format.

The FBX file format is a proprietary format owned by Autodesk. This format is used to streamline the process of setting up a Revit model in 3ds Max (or **3ds Max Design** which is tweaked for architects) to produce a **high quality rendered image or animation**. Although Revit can produce acceptable rendered images for most, 3ds Max can create even better rendered images and animations with moving objects—Revit's animations cannot have moving objects.

> The word **Rendered** is used to describe the computer process to create a photo-realistic image.

Once the **Export → FBX** command is selected, the **Export 3ds Max (FBX)** dialog appears (Figure 2.2-41). This is a simple dialog with three options:

- **File naming** options drop-down
- **Use LOD**
  - LOD means "Level of Detail"
  - When this option is checked the model geometry is more faceted, resulting in a smaller file size which will be more efficient in 3ds Max.
  - Checking LOD may work well for an exterior rendering a distance away from the building.
- **Without boundary edge**
  - This option will remove lines/edges between surfaces which are aligned in the same plane.

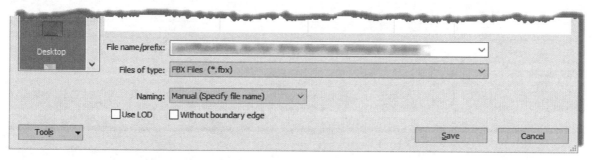

Figure 2.2-41 Export 3DS Max (FBX) dialog

File Tab & UI

Export → Family Types

Used to create a text file listing all the differences between multiple types with a Family.

To use this command, a Family view must be current on screen in the Family Editor. The command is simple; once activated the only thing to do is provide a file name and location. The created text file will list all of the parameters in the Family (Type and Instance) and then list the value for each Type in the current family. The example below is an export for a standard door family as seen in Notepad (Figure 2.2-42).

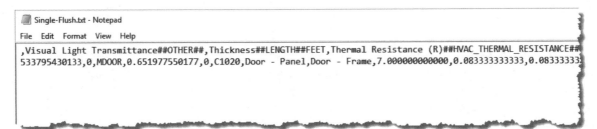

Figure 2.2-42 Exported family types viewed in Notepad

The created text file can then be used as a **Type Catalog** (Figure 2.2-43). A Type Catalog is used when a Family has a lot of Types, e.g. more than 6. This feature allows the user to select which Types to load into the Family. For example, it is not practical to load all of the wide flange beam types as the Type Selector becomes too hard to navigate for day-to-day work.

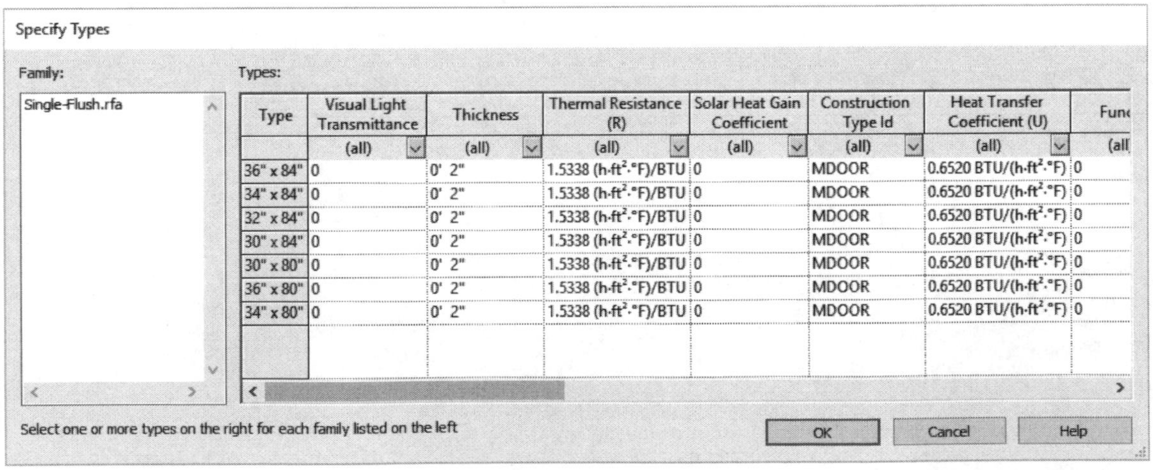

Specify Types

Family: Single-Flush.rfa

Types:

Type	Visual Light Transmittance	Thickness	Thermal Resistance (R)	Solar Heat Gain Coefficient	Construction Type Id	Heat Transfer Coefficient (U)	Fun
(all)	(all)	(all)	(all)	(all)	(all)	(all)	(all
36" x 84"	0	0' 2"	1.5338 (h·ft²·°F)/BTU	0	MDOOR	0.6520 BTU/(h·ft²·°F)	0
34" x 84"	0	0' 2"	1.5338 (h·ft²·°F)/BTU	0	MDOOR	0.6520 BTU/(h·ft²·°F)	0
32" x 84"	0	0' 2"	1.5338 (h·ft²·°F)/BTU	0	MDOOR	0.6520 BTU/(h·ft²·°F)	0
30" x 84"	0	0' 2"	1.5338 (h·ft²·°F)/BTU	0	MDOOR	0.6520 BTU/(h·ft²·°F)	0
30" x 80"	0	0' 2"	1.5338 (h·ft²·°F)/BTU	0	MDOOR	0.6520 BTU/(h·ft²·°F)	0
36" x 80"	0	0' 2"	1.5338 (h·ft²·°F)/BTU	0	MDOOR	0.6520 BTU/(h·ft²·°F)	0
34" x 80"	0	0' 2"	1.5338 (h·ft²·°F)/BTU	0	MDOOR	0.6520 BTU/(h·ft²·°F)	0

Select one or more types on the right for each family listed on the left

[OK]   [Cancel]   [Help]

Figure 2.2-43 Type Catalog as seen during the Load Family process

For a Type Catalog to work, the TXT file just needs to be in the same folder as the Family. When a Family is loaded, if Revit sees a TXT file with the same name, e.g. **Single-Flush.rfa** and **Single-Flush.txt**, the Type Catalog will be shown.

> **Tip:** The exported text file can be opened in Microsoft Excel, making it more readable and editable.

If the exported TXT file will be used as a Type Catalog, it is recommended that all the Types be deleted out of the original Family to avoid confusion (as this information will be ignored). Once a Family Type is created, it is not possible to delete them

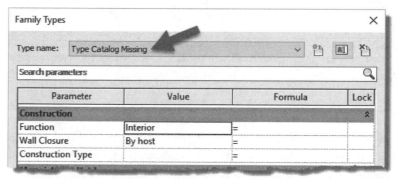

Figure 2.2-44 Family type renamed

completely—at least one must remain. This is not a problem, as it is recommended that one Type exist and be named something like **Type Catalog Missing** (Figure 2.2-44). This will alert anyone loading the Family from a location that does not contain the Type Catalog. This can happen with manufacturer content downloaded from the internet.

Export → gbXML

Used to create a "green building" XML file containing building information for use in external energy analysis tools such as Autodesk Green Building Studio or Trane Trace 700.

When the **Export → gbXML** command is selected, the **Export gbXML** dialog appears; select from the following options:

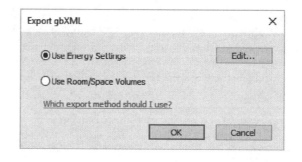

- Use Energy Settings
- Use Room/Space Volumes

Both options require certain conditions apply before a proper gbXML file can be created.

Energy Use Settings

This method creates the gbXML file based on the results of Revit's **Create Energy Model** tool, on the Analysis tab, which creates a simplified energy analysis model (EAM) based on more detailed Revit elements.

Before the EAM can be properly created, the Energy Settings dialog must be edited. At the very least, the project Location must be set. Clicking the **Edit...** button (shown in the image above) is a convenient way to open the **Energy Settings** dialog. However, all dialogs must be closed so the Create Energy Model tool can be run; at the conclusion of this command, click the "Continue working" options rather than "Run energy simulation"-- the latter is not required for a gbXML export.

Once the EAM is created, the gbXML file can be created via the Use Energy Settings option. This method does not require Rooms or Spaces exist, but if they do the Name and Number data will be used. For Spaces, additional information is included, as well as some information from MEP Zones.

This option creates a more accurate gbXML file than the method described next. See Autodesk Help for more information on this common option.

Use Room/Space Volumes

Selecting the Use Room/Space Volumes option causes the **Export gbXML – Settings** dialog to appear (Figure 2.2-45). The preview on the right shows the Rooms or Spaces to help the user understand what will be exported—namely, room information rather than geometry. If the model has both Rooms and Spaces, the Export Category should be set to the appropriate options: Rooms or Spaces.

> **Rooms** are used by the architects to hold information about finishes and track contents. **Spaces** are used by engineers to track electrical, lighting, energy and HVAC information. Both are able to define area and volume.

To learn more about the gbXML format visit: http://www.gbxml.org.

> **It is** difficult, actually impossible in most cases, to accurately define all enclosed spaces within a building using the Room or Space tool. There are many conditions, such as plenums, enclosed space below stairs, etc., which cannot be defined by a Room/Space element. Thus, this method is less accurate than the "Use Energy Settings" option previously defined.

The two tabs in the Export gbXML - Settings dialog will be expounded upon now.

General Tab

The **General Tab** lists the Energy Settings for the building. This may be adjusted here or set ahead of time via **Analysis** (tab) → **Energy Settings**.

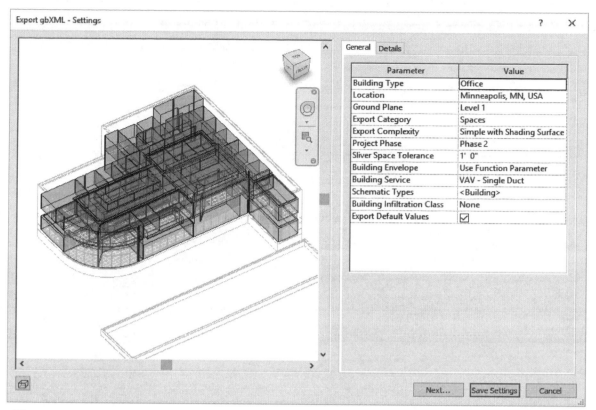

Figure 2.2-45 Export gbXML – Settings dialog

Details Tab

The **Details Tab** lists each Room/Space and allows the user to override the settings for an individual area (Figure 2.2-46). For example, use the General tab to specify the settings for the majority of the building, and then use the Details tab to specify unique rooms such as a loading dock area with suspended unit heaters.

It is possible for a Room/Space to exist in the project but not in the model—these unplaced elements are marked with a warning symbol.

The "details" information can also be edited in the Properties Pallet when a Space is selected.

When the dialog box settings are correct, click the **Next** button. The next dialog simply prompts for a file name and location; the file extension will be XML. The XML file may be viewed in the default web browser by double-clicking on it. However, the file is not really meant to be viewed directly.

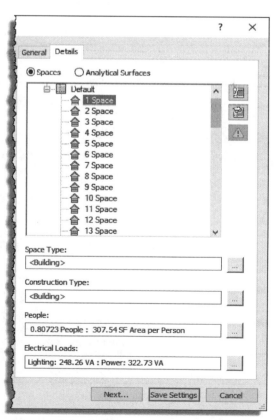

Figure 2.2-46 Export gbXML – Settings dialog

Although Revit can perform **Conceptual Energy Analysis** itself, there are many other options—some are by Autodesk and some are not. The image below shows an example of an exported gbXML from a Revit project, which has been imported into **Autodesk Green Building Studio** which is more sophisticated and customizable (Figure 2.2-47).

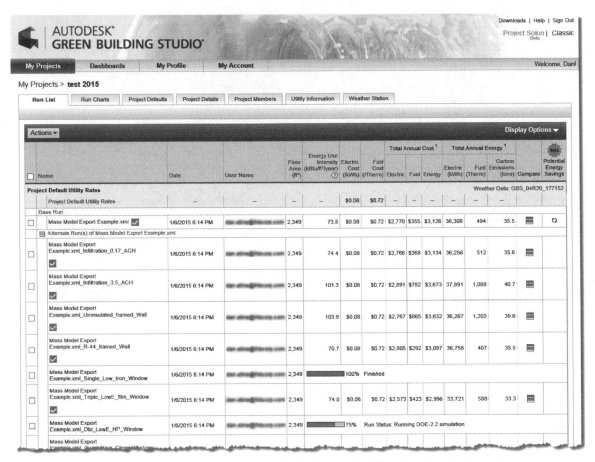

Figure 2.2-47 gbXML file uploaded to Autodesk Green Building Studio cloud service

Some analysis programs that can import the gbXML format can also export them. Thus, once the analysis is completed, the data may be saved to a gbXML file which can then be imported back into Revit using the **Insert → Import gbXML** command.

Export → IFC

Used to create an Industry Foundation Class (IFC) file to use in another application or share with a client/consultant outside the design firm.

This export feature allows the design team to share a [mostly] non-editable version of their model for coordination, clash detection, etc. The image below shows the two dialogs associated with this export feature (Figure 2.2-48).

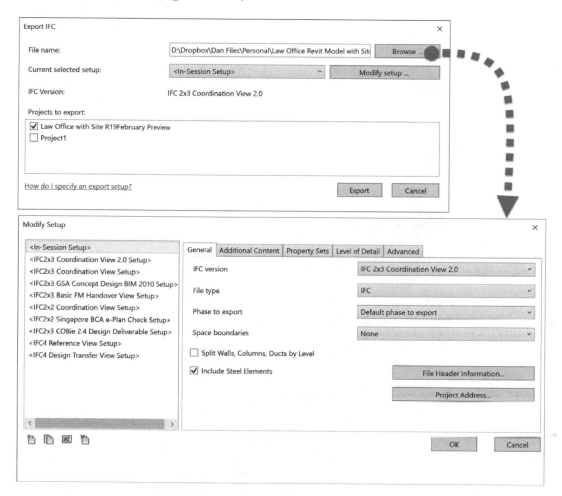

Figure 2.2-48 Export IFC dialog

These dialogs offer a number of settings to control how the IFC is created and what is included. In the Modify Setup dialog, several predefined setups are listed—notice this list can be modified with the icons below.

Here are just a few of the settings one can control:

- Current view only
  - Limits the elements exported to those visible in the current view – with some exceptions
  - Rooms are not visible in a 3D view, thus they will not be exported from 3D views
- Split walls and columns by story
  - Split occurs at levels designated as a building story
- Export base quantities

        o   Raw data about the exported model
- Include space boundaries

The exported results can be seen via **File Tab → Open → IFC**. The elements will no longer be native Revit elements. The Revit properties are visible in the properties palette when something is selected—but in the IFC format.

The export, and open, process can take several minutes on large complex projects. The final file can be very large in terms of data size on disk.

The image below shows the IFC model opened in Revit. Notice, the selected wall is recognized as a Wall element in the Properties Pallet (Figure 2.2-49). This wall can even be swapped out with the wall types already defined in the current Revit project.

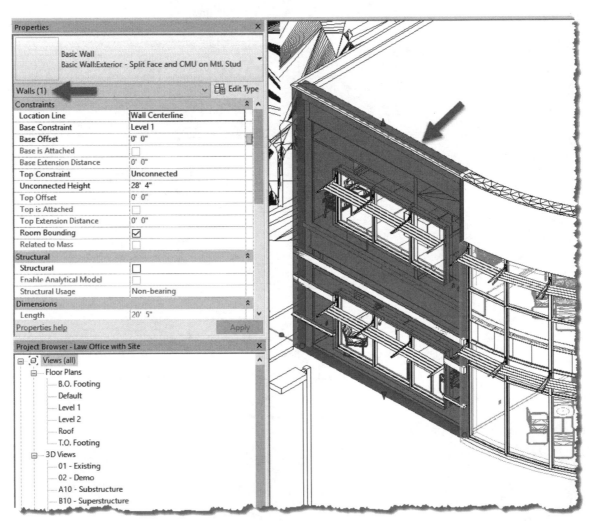

Figure 2.2-49 Exported IFC file opened in Revit via Open → IFC file

Export → ODBC Database

Used to export the Revit project to a file which can be viewed in a database program such as Microsoft Access.

Search Revit help for more information on this topic.

Export → Images and Animations → Walkthrough

Used to save a Revit project Walkthrough as an AVI file which can be played back in Microsoft's Media Player or most other video playback applications. Unfortunately, Revit cannot create the more common MP4 video format.

A Walkthrough view must be current or this command is grayed out in the File Tab.

Once the export command is selected, the **Length/Format** dialog appears (Figure 2.2-50). The overall length, defined by "total time," is directly related to how smooth the final animation plays—time and file size. The Visual Style can be anything from wireframe to fully rendered.

The Dimensions determine how large the image is on the screen. Although the final animation can be maximized on the screen, if the dimensions are too small (i.e., resolution) the animation will be pixilated and not look very good.

When the final AVI file is very large, e.g. 2-1GB, the video cannot be played due to memory limitations. Using a conversion tool, e.g. Format Factory at http://www.pcfreetime.com/, the file can be converted from an uncompressed AVI to an MP4 file which is smaller and can be played on many more programs and devices.

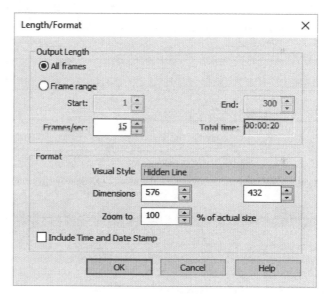

Figure 2.2-50 Export walkthrough settings dialog

> Each frame of the animation can be saved to individual files if the **Files of type** is changed from AVI to JPG, PNG, etc.

Export → Images and Animations → Solar Study

Used to create an animation of a single-day or multi-day solar study.

For the Solar Study command to be available, i.e., not grayed out, the current view must have the Sun Settings set to Single-day or Multi-day Solar Study **AND** Shadows must be turned on.

When the export command is selected, the same **Length/Format** dialog for Animation appears (Figure 2.2-50). The result of a Solar Study is an animation where the building is stationary and the sun (not visible) and shadows are moving.

> Checking the **Include Time and Date Stamp** is helpful for Solar Study animations.

The look of the shadows can be adjusted in the view's **Graphic Display dialog, Options** under Lighting. Also note that the Solar Study export can be done in plan and elevation views, not just 3D views.

Export → Images and Animations → Image

Use to create a raster image file of one or more Revit view/sheet(s).

When images are needed for presentations, marketing or proposals, the export image command can be used.

Starting the command opens the **Export Image** dialog (Figure 2.2-51). The default settings will make a rather low-quality image.

For a higher quality image, set the **Image Size** to "Zoom to" and 100%. Also, set the **Format** to PNG and the **Raster Image Quality** to 300 – 600 [dpi].

> In a 3D or Camera view, selecting the View Port and then the **Size Crop** on the Ribbon allows the user to adjust the size of the crop. This is directly related to the image size here and for rendering. The larger the image, the larger it can be printed without looking pixilated.

The **Options** section is self-explanatory.

The **Export Range** offers three options:

- **Current window** (default)
    - Everything in current window
    - Type ZF (zoom to fit) to see everything that will be printed
    - "Floating" elements or linked CAD files with large extents can be a problem – use Crop to better control the extents of what will export

- **Visible portion of current window**
    - This option is a little misleading; depending on the proportions of the view on screen and the selected paper size, additional information, other than what is visible on screen, often prints.
    - TIP: Tile the document view (so the view is not maximized within the Revit application, and then adjust the Window to the desired area to be printed). This often works better at limiting what prints.

- **Selected views/sheets**
  - ○ This option allows multiple views or even all the sheets to be quickly exported to image files.
  - ○ When this option is selected, **the Create browsable web site with a linked HTML page or each view** option becomes checkable. The results can be viewed in your internet browser by double-clicking on one of the files.
  - ○ Saved sets can be created for specific views/sheets which will need to be exported/printed often. For more on this see the Printing section.

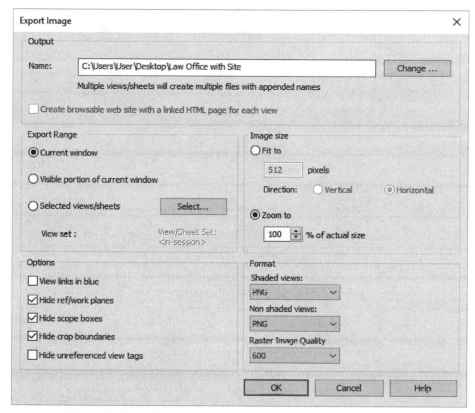

Figure 2.2-51 Export Image dialog

Export → Reports → Schedule

Used to export the values in a schedule to a text file—which can be opened in a spreadsheet application such as Microsoft Excel or Google Sheets.

A schedule view must be current, and active, for this command to be available. If the command is grayed out, try clicking within the schedule and then the command should be available.

Once the command is selected, the user is prompted to provide a name and location for

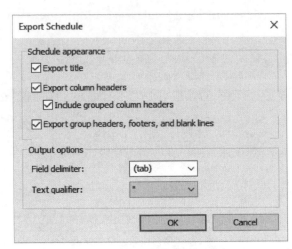

Figure 2.2-52 Export Schedule dialog

the text file (*.txt format). Next, the **Export Schedule** dialog appears (Figure 2.2-52). These options control how much information is included and what constitutes a column.

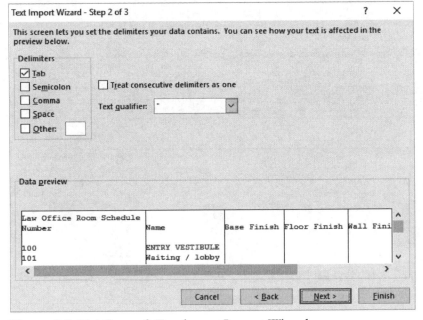

Figure 2.2-53 Microsoft Excel - text Import Wizard

To open the exported file in Microsoft Excel, do this:

- File → Open
- Change the filter from **All Excel files** to **All files**
- Browser/select the file and click Open
- With "Delimited" selected, click **Next**
- In Step 2 of 3, be sure the **Delimiters/Text Qualifier** match the settings used in Revit (Figure 2.2-53)
  - o Notice the preview is showing the columns correctly
- Click **Next**.

- Adjust the **data format** for each column if desired – they are defaulted to General
- Click **Finish**

The data exported from Revit is now in Excel in the proper columns/rows (Figure 2.2-54). Some of the formatting is lost, but overall the schedule is intact. Excel is still working with the text file, doing a Save As to an Excel format provides the most options going forward.

> No formulas are exported, just raw data.

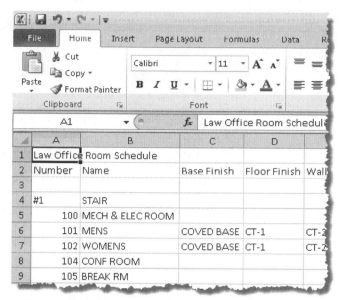

Figure 2.2-54 Revit schedule exported to Excel

Export → Reports → Room / Area Report

Used to create a triangulation or integration report for each room in the selected view(s), and its associated level. The created area report is viewed in a web browser. According to the Revit Help page on this command, this feature is mainly geared towards **European users**.

When the command is selected, the **Views** dialog appears (Figure 2.2-55). Select the views which contain rooms to be reported—holding CTRL or Shift to select multiple Views.

Next, provide a name and location for the report (Figure 2.2-56). It is recommended that an empty folder is selected, as multiple files and folders will be created. Clicking the Options button opens the **Area Report Settings** dialog (Figure 2.2-57). These options control the formatting of the final HTML documents.

The **Files of type** drop-down has the option of completely changing the output: triangulation vs. integration reports. The default is a triangulation report.

Figure 2.2-55 Views to export room/area

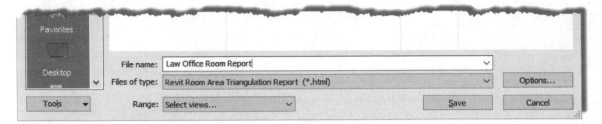

Figure 2.2-56 New Project dialog

The **Range** drop-down allows the user to go back to the Views dialog and make changes to the selected views.

One HTML file and one folder are created for each level (Figure 2.2-58). The folder contains images used in the HTML file.

> HTML is a text-based web browser file format.

An example of a final triangulation report can be seen in Figure 2.2-59.

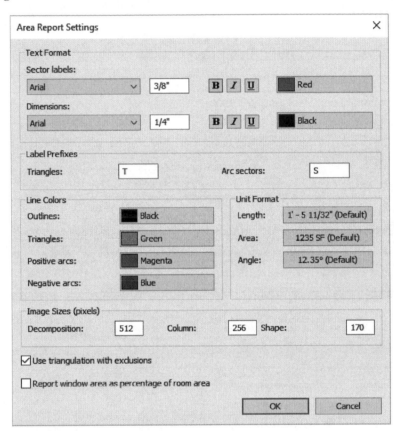

Figure 2.2-57 Area Report Settings dialog

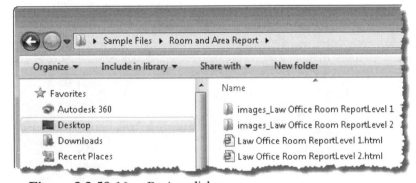

Figure 2.2-58 New Project dialog

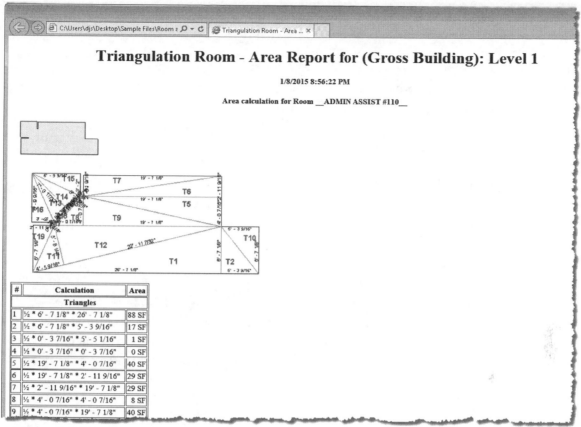

Figure 2.2-59 Room Area Report Example in web browser

Room Data Sheets

As originally mentioned, this feature finds its application in the European market – likely to comply with regulatory requirements. In the US, and perhaps elsewhere, there is a need to create **Room Data Sheets**, which show an isolated view of each room with selected information printed below the image.

Revit cannot do this directly but there are a few add-ins and workflows developed that can achieve this goal.

Add-ins used to create Room Data Sheets:

- dRofus – www.drofus.no
- CodeBook - http://codebookinternational.org.uk/
- eRDS - http://roomdata.co.uk/

Workflow: do a web search on **revit "room data sheets" using mail merge**.

Export → Options → Export Setups DWG/DXF

Used to predetermine the settings for DWG and DXF exports.

Revit offers many adjustments to how a DWG file is created using the Export → DWG command. The defaults settings will often work most of the time. Exceptions are for matching client CAD standards, or specific 3D modeling needs.

Layers

The default settings create a DWG file that most closely matches the view in Revit. This includes using **overrides by entity** which is not a common AutoCAD drafting practice but done here for maximum graphical fidelity – notice the Export layer options drop-down in the image below (Figure 2.2-60). This option can also be changed to "do not override" and "create new layers for overrides."

The American Institute of Architects (AIA) layer standard is the default mapping between Revit categories and Layers. Most AEC firms using AutoCAD have a Layer standard pretty close to this. If the recipient of the CAD files uses color to define line weight, which is common, the colors listed for each Layer may need to be revised. The color used to define a specific line weight usually varies between AEC firms – this is defined in their AutoCAD **Plot Style Table** (*.CTB).

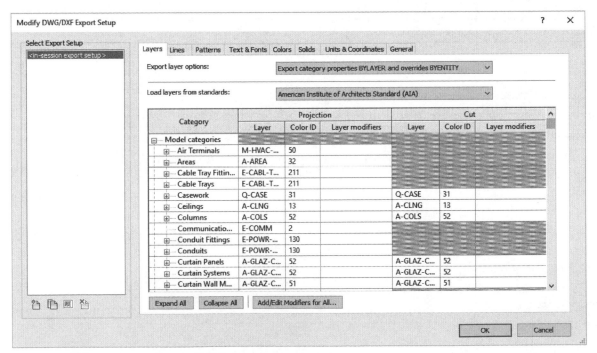

Figure 2.2-60 Modify DWG/DXF Export Setup

Text & Fonts

The drop-down for **Text treatment** is **Exact** (default) or **Approximate**. Exact means the visual fidelity will be maintained at the expense of ease of future formatting and editing in AutoCAD.

AutoCAD has a special font format called SHX. Revit cannot use or export those. Rather it only uses Window True Type fonts. There are a few True Type fonts created to match the most standard SHX fonts (e.g. RomanS and Simplex) but these do not print as well.

Colors

Most AEC firms use **AutoCAD Index color (255 colors)** to define color in their drawings—this is the default setting in Revit. The **True color (RGB values)** are used to specify color when everything else is to be printed as black lines. The True Colors cannot be controlled by the AutoCAD Plot Style Table – which is used to define the line weight and color (usually black) each color will be printed.

Solids

This defines the type of solid created for 3D views. The default is **Polymesh**, which has a lot of extra lines on the surface. If the CAD application supports **ACIS solids**, which AutoCAD does, then that option is preferred as the on-screen geometry looks better.

Units & Coordinates

The default setting is **Project internal**, which will create a drawing with the same orientation as most views in the model—with the AutoCAD drawing origin (i.e. 0,0,0) at the Project Base Point.

The **Shared** option can be used to export a CAD file where the origin and rotation match the civil drawings—meaning the civil engineers can xref the DWG file at 0,0 without needing to rotate or move the DWG once in their drawing. For this to work, the Revit **Survey Point** must be properly adjusted in the model.

> The **Project Base Point** and **Survey Point** can be turned on in any view via **VV →**
> **Model Categories** (tab) **→ Site** (expanded to view sub-categories).

General

Most of the options here are self-explanatory. Note that AutoCAD has an option to make a certain Layer visible but will not plot—Revit does not have such a feature. Also, the **AutoCAD/DWG format**, or file version, can be specified here. This can be important if the recipient of the DWG files has an older version of AutoCAD or their software has not been updated to read the latest DWG file format. For example, when Autodesk changes the DWG format, another software company, such as **SketchUp**, will not be able to read that format until they come out with a service pack or a new version.

Export → Options → Export Setups DGN

Used to predetermine the settings for DGN exports.

The settings are similar to the DWG export options – AutoCAD and Microstation are fairly similar CAD programs (Figure 2.2-61). A Seed file is a template the exported DGN file will be based on.

> What AutoCAD calls a Layer, Microstation calls a Level; **Layer = Level**.

Multiple export settings can be saved using the icons in the lower left.

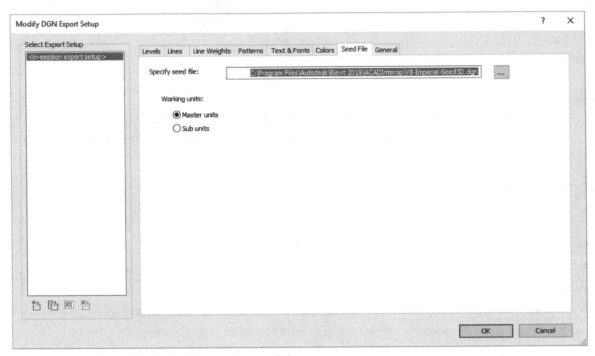

Figure 2.2-61 Modify DGN Export Setup dialog

Export → Options → IFC Options

Used to predetermine the mapping of Revit categories and sub-categories to IFC Class Name and IFC Type.

IFC Export classes dialog opens when the IFC Options command is selected (Figure 2 -62). Notice the file location listed in the dialog title bar at the top—the saved mapping file can be saved on a server or shared externally with clients and consultants. The Load button is used to replace the current mapping with one previously saved to a file.

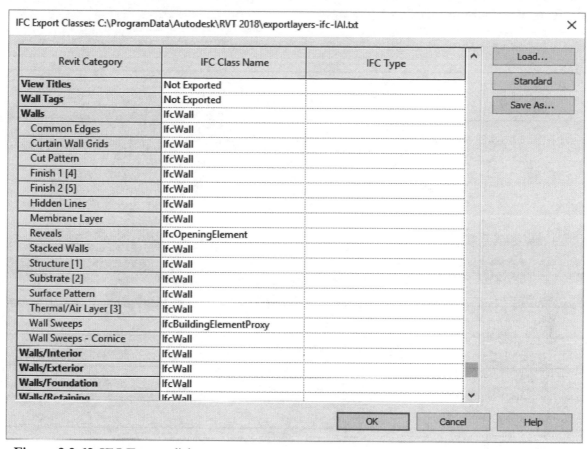

Figure 2.2-62 IFC Export dialog

File Tab & UI

Print → Print (Ctrl + P)

Used to create electronic and hardcopy drawings of Revit views and sheets.

This command can be started by clicking on the first **Print** button in the File Tab. Once selected, the Print dialog appears (Figure 2.2-63).

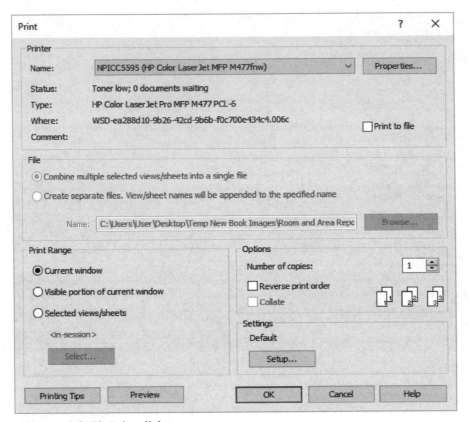

Figure 2.2-63 Print dialog

Basic steps to print a view/sheet:

- Make view or sheet current
    - o May need to click within the view to make sure it is current (has focus)
- Select the printer (name) first
    - o This affects the paper size options
- Select what to print (Print Range)
    - o Current window
    - o Visible portion of current window
    - o Selected views/sheets
- Setup button
    - o Select sheet size
    - o Orientation
    - o Scale

Printer Name

The printers installed on the computer are listed here. The list may vary from computer to computer. Clicking the **Properties** button opens the Windows printer settings for the selected printer/plotter.

> If the **Print Range** and **Setup** button are grayed out and inactive, try changing the default printer in Windows to another installed printer. This author has seen this problem multiple times – mainly with Xerox WorkCenter office printers.

File

The File options are available when printing to a file, e.g. a PDF driver. The main feature here is to combine all the drawings into a single PDF file. PDF editing software, such as Adobe Acrobat, is still often needed to further manipulate the PDF file. For example, the sheets may not be sorted properly; the Civil sheets (C101, C102, etc.) are placed before the Architectural sheets (A101, A102, etc.) in a set of construction drawings. Revit cannot sort the drawings this way; the drawings are printed in alpha-numeric order. Also, sheets provided by an external consultant need to be manually combined with the main PDF file.

There is also a **Print to file** checkbox which will make this section active. This method creates a PRN file type which is not common and should be avoided in most cases.

Print Range

Current Window – the extents of the entire view will be printed, including elements not visible on the screen.

Visible portion of current window – only what is currently visible on the screen will print. This is not entirely accurate, as the proportions between the screen and printed page do not match, thus extra information often appears on the final print. Other CAD/BIM programs have a Print Window option when printing, which allows the user to select a specific area. One trick to getting closer to just what is desired to print is to tile the view so its edges can be adjusted. Adjusting the view to match what is wanted in the print gets a lot closer than when the view (aka document window) is maximized within the Revit application.

Selected views/sheets – print the entire set of drawings in the current project at once. When this option is selected, the **Select** button becomes available.

> Clicking the Select button opens the **View/Sheet Set** dialog (Figure 2.2-64). Here, one selects the Views and/or Sheets to print. If just printing Sheets, uncheck the Views toggle at the bottom to just show Sheets. Before unchecking the Sheets or Views filter, click the **Check None** options to ensure nothing is selected. If a view is selected, and then the Views toggle is unchecked so Views are not listed, the View will still be selected even though it is not visible.

The View/Sheet Set dialog allows specific groupings of Sheets and Views to be saved. Notice the saved name **Full Drawing Set** in the example shown. Once saved, this list can be quickly selected in the future, from the drop-down list at the top, which saves multiple steps. Any new views/sheets added to the project will require the list to be updated.

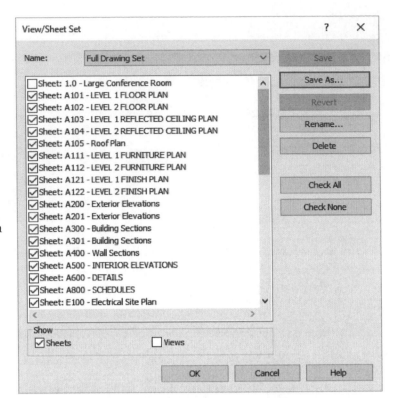

Figure 2.2-64 View/Sheet dialog

A saved view/sheet set will remain the default selection until changed. The main Print dialog will indicate the saved set (Figure 2.2-65). Seeing the saved set indicated here, the user does not need to click on the Select button to verify settings (again, unless new sheets have been added). Seeing **Default** indicates no saved list is selected and requires manual selection.

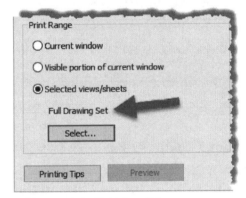

Figure 2.2-65 Print dialog

Options

Select **number of copies**; default is 1. When more than one is selected, the **Collate** option becomes available. Check this to print complete sets. Not checking this often results in faster printing, as the loaded print data can print the same sheet multiple times more quickly. However, the time required to shorten the sheets may not be worth it.

Some plotters finish with the printed sheet face up. In this case, select the **Reverse print order** so the last sheet prints first, and ends up on the bottom.

Settings

Lists a saved settings name, **Full Size** in Figure 2.2-67, or **Default** in Figure 2.2-63 which means manual adjustments are required.

Clicking the **Setup** button opens the **Print Setup** dialog (Figure 2.2-66).

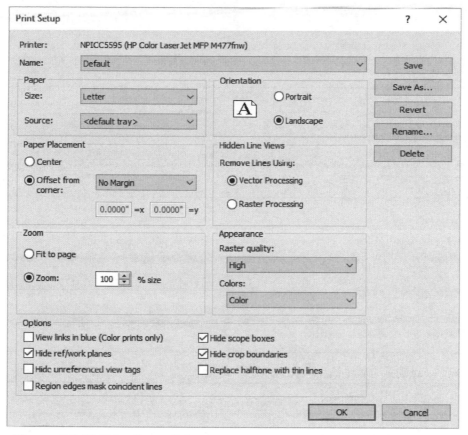

Figure 2.2-66 Print Setup dialog

File Tab & UI

Printer

The selected printer is listed here for reference. However, the printer can only be changed in the Print dialog.

Name (Print Setup):

All the settings in the Print Setup dialog can be saved. This is saved in the project and all users will have access. Two common saved settings are Full Size and Half Size plots (Figure 2.2-67). To save a group of settings, make the desired changes to the dialog, click the **Save As** button, and then provide a name. In the future, the saved settings will be listed in the Name drop-down.

When a saved setup is selected, it will be listed in the main Print dialog (Figure 2.2-68). Seeing the correct saved setup listed here means the Setup dialog does not need to be opened and verified. **TIP:** Add named setups to the firm template so they are always available and consistent in both settings and naming convention.

Figure 2.2-67 Print Setup dialog – saved settings

Paper

Select the paper size to be printed to.

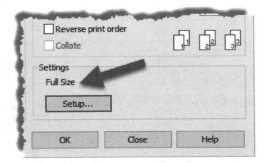

Figure 2.2-68 Print dialog – saved settings

> **Custom Paper Size:** Many printers and plotters have options, via Windows printer settings, to turn on (or off) blocks of papers sizes; e.g. ANSI or ARCH paper sizes. They also will typically have the ability to apply custom paper sizes to a "custom" size, which can be edited as needed.

The **Source** option allows the user to select a specific tray/roll in the printer/plotter. The options vary, so the desired printer must be selected first. For example, a printer might have expensive high-quality glossy paper in Tray 5. To ensure this paper is used, the user would select Tray 5 from the list. When this option is set to <default tray> the default setting in the Windows printer settings dialog will be used.

Paper Placement

This allows users to control how the view/sheet is positioned on the paper. For example, a view printed for reference might be centered on the paper.

Zoom

The drawing scale is set only one time, in the view. When printing, the user can print **Fit to page** or enter a **percentage** based on the view scale; i.e., 100% means the drawing will be printed at the view scale. For a half size drawing, the view scale would be set to 50% - thus, a ¼"=1'-0" drawing would be printed at 1/8" = 1'-0".

Orientation

Select Portrait or Landscape.

Hidden Line Views

Use Vector Processing whenever possible as it is usually faster and produces a smaller print file to send to the printer. Raster Processing will be used automatically when the view/sheet contains an image/decal/shadows/point cloud.

Appearance

Raster Quality: When Raster Processing is used, this option determines the resolution.

Colors: There are three options:

- <u>Black Lines:</u> Everything prints black, except:
  - o Raster images and solid fill patters print grayscale
  - o Anything set to **Halftone** also prints grayscale. This includes "underlay" elements set by the view's discipline toggle or overriding an element in a view.
- <u>Grayscale</u>: Everything prints black or shades of gray. A green line would be a darker shade of gray and a yellow line based on the hue/tone of the two colors.
- <u>Color</u>: Everything seen as color on the screen prints in color. When printing Color to a non-color printer, the result will be similar to the Grayscale setting above.

Options

Most of these options are self-explanatory.

Hiding unreferenced view tags means an elevation, section or callout not placed on a sheet will be hidden because the reference bubble is empty. This may not be a good idea as seeing the empty bubble will remind the design team that the view still needs to be placed on a sheet. Also, the person reviewing the documents might think views are missing and add redlines to include something that is already there.

Printing Tips

Click the Printing Tips button to access helpful printing information in the Help system.

When printing large sets in a worksharing model, open a copy of the model Detached from Central prior to printing. This will conserve local and network resources.

Preview

When printing a single view/sheet, use the Preview option to verify everything looks correct before sending to the printer; this saves time and paper.

It is not possible to print schedule views. To print a schedule it must be placed on a sheet, and the sheet is printed, or the schedule is exported to a comma delimited file and opened in a spreadsheet program as covered previously in this chapter.

Print → Print Preview

Used to verify the current print settings prior to sending to printer.

While in preview mode, a series of navigation buttons are provided in the upper left. Clicking **Close** will end the Print Preview and clicking **Print** will open the Print dialog.

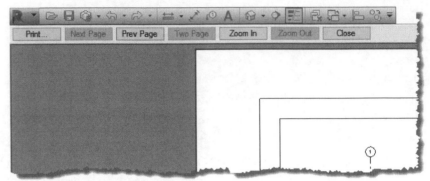

Figure 2.2-69 Print Preview

Print → Print Setup

Provides direct access to the Print Setup options for the current Revit session or the current Save/Named setup. This also allows the user to create multiple saved setups which may then be accessed from any view.

Close

Used to close the current Revit Family or Project. It does not matter how many views are open, the project will be closed. If the project has not been saved, a prompt to save will be provided.

The Revit project is also closed by clicking the "X" in the upper right, for the document window (not the application). When the last view, for a given project, is closed the project also closes.

> Closing Families and/or Projects when they are no longer needed conserves computer resources.

Exit Revit

Closes the entire Revit application; a prompt is provided for each unsaved file.

2.3 Options Dialog

The Options dialog contains several settings related to the Revit Application, rather than a project file. These settings are persistent among all Revit project files **per user**. Some firms will adjust these settings when the software is deployed or manually at each desk for consistency.

Options → General

Selecting General on the left will display the related settings on the right (Figure 2.3-1).

Notifications

Revit does not automatically save; however, a reminder appears based on the **Notifications** settings. The top option is for a non-worksharing file or a local file, and the bottom option is a reminder to synchronize the local file with the central file in a worksharing environment.

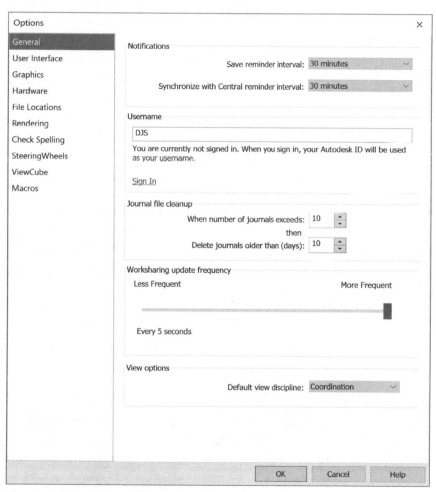

Figure 2.3-1 Options dialog – General tab

Username

The username is used to identify each user in a worksharing project environment, i.e. a way in which multiple designers can work in the same Revit project file. When creating a Local file from the Open dialog, Revit copies the Central file to the C drive and adds the Username to the file name as a suffix. There cannot be two users with the same username. If the Username changes, previously created Local files will no longer be valid – new local files will need to be created.

> Recommendation: Create a new local file each day when working in a Worksharing model.

When a user logs into Autodesk A360 Cloud Services through Revit, or any locally installed Autodesk cloud services, the Username will automatically change to match the A360 username (only if the user is not currently in a worksharing Local or Central model). When this happens, previously created Local files will be invalid. It may be best to log in to A360 via the provided link so the Username matches the A360 username to avoid future problems. The default username matches the user's login username for Windows.

Journal File Cleanup

A Journal file is a text-based file that records computer system statistics and every command the user performs. In theory, the journal file can be dragged onto the Revit icon, on the desktop, and Revit will repeat all the steps to create a corrupt or lost file. In this author's experience, this process has a low success rate. On the other hand, the Journal file is very useful in troubleshooting problems with the computer or Revit model. Autodesk support often requests these files.

This option tells Revit how many journal files to save. However, it is good practice to manually purge this folder occasionally as Revit will also automatically save non-worksharing RVT (project) and RFA (family) files in this folder – making the folder very large over time; these are backups of the original.

The files are located here:
C:\Users**username**\AppData\Local\Autodesk\Revit\Autodesk Revit 2021\Journals
username = Windows login username

Worksharing Update Frequency

This setting tells Revit how often to update the status of worksets and checked out elements in a Central file, which in turn updates the user's Worksharing Display Mode if active.

View Options

Each model view in Revit has a Discipline setting, which controls several automatic visibility settings and Project Browser organization. This setting predetermines which discipline new views are associated with—unless a default View Template exists and has a discipline specified.

Options → User Interface

This section customized the User Interface (UI) and input-related options (i.e., mouse and keyboard). These adjustments only apply to the user logged into the computer (Fig. 2.3-2). Note that the **Persona-based wizard** turns some of these options on/off when Revit is first started for each user, based on answers provided to the prompts.

Configure

The first part simply hides tools on the Ribbon. The analysis tools are on the Analyze tab, while the others are on their respective tabs.

Turning off the structural tools will actually remove the Structure tab from the Ribbon. Turning off one of the three MEP options will just remove those tools, but the Systems tab will still be visible— while turning off all three will remove the tab.

> **Caution:** turning off tools here will also limit the types of Project Parameters one can create. For example, after turning off the electrical tools, it is not possible to create a parameter based on Electrical Potential—but Wattage is still available.

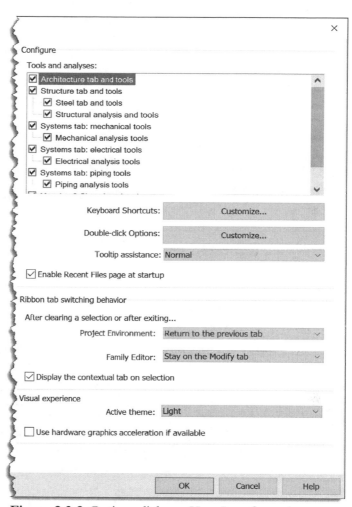

Figure 2.3-2 Options dialog – User Interface tab

It is recommended that all tabs and tools are left on while learning the software. Even after that, many of these tools are occasionally needed. For example, a consulting MEP firm (i.e., they only provide MEP services, not architecture or Structural) may have a project doing work in a mechanical room that does not involve/require other disciplines. Someone must draw the walls and columns in the model.

The **Active theme** allows for a Light (default) or Dark color palette for the UI.

Power users often use **keyboard shortcuts** rather than icons and right-click commands. Many tools can be started by typing two letters (no Enter required). If a tool has a keyboard shortcut, the two-letter shortcut will appear in the tooltip. Those same options can be seen and edited here, by clicking the **Customize...** button. Figure 2.3-3 shows the Keyboard Shortcuts dialog.

> The customized settings can be exported and then imported on other computers.

The **Double-click Settings** does exactly what the name implies. Some users are confused, or annoyed, that the Family Editor is opened when a family is double-clicked on. This dialog (Figure 2.3-4) can fix that problem.

Tooltips can be completely turned off here if they are distracting to more experienced users. Turning this feature off may also help with performance on slower computers.

The last option in this section, **Enable Recent Files page at startup**, provides a way to skip the normal startup screen, when Revit is opened, and go directly to the default template. Most do not change this as the new/empty project often just needs to be closed without saving as an existing project needs to be opened.

Figure 2.3-3 Keyboard Shortcuts dialog

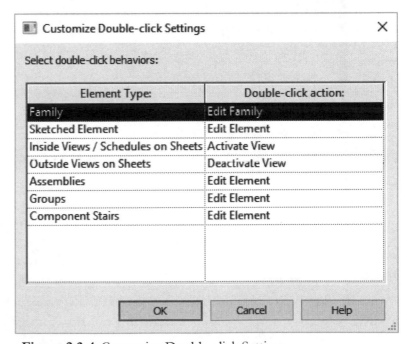

Figure 2.3-4 Customize Double-click Settings

Tab Switching Behavior

This section controls what tab is selected when a selection is cleared or a command is finished. The options here are self-explanatory.

Options → Graphics

This section of the Options dialog is important to understand as it can have a significant impact on overall Revit performance.

View Navigation Performance

Allow Navigation during redraw lets you start zooming or panning before a view is done refreshing the screen.

Simplify display during view navigation temporarily simplifies and hides model elements to improve performance.

Graphics Mode

Smooth lines with anti-aliasing makes angled lines smoothed. However, this requires Revit to "think" harder so it should be reserved for presentation views. This is more applicable to views with lots of angled lines such as 3D and Camera views, rather than plan, elevation or section views as their

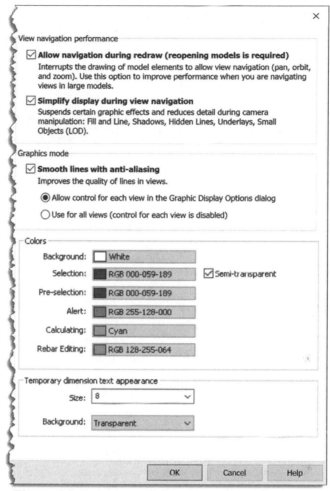

Figure 2.3-5 Options dialog – Graphics tab

lines are mostly orthogonal with the screen (i.e. the lines align with the pixels on your computer's monitor). However, if the building has a large wing at an angle, then this feature might be used (for presentation drawings). When set to "Allow control for each view…" an individual view's lines can be smoothed via that view's Graphic Display Options dialog.

Colors

The color settings are self-explanatory. Adjustments here are just for the current user.

Temporary Dimension Text Appearance

These settings control how the temporary dimension text looks when an element is selected in the model.

Options → Hardware

This section of the Options dialog is important to understand as it can also have a significant impact on overall Revit performance.

Hardware setup

The computer's **Video Card** is listed here for reference. This is helpful when updating the video driver. Notice, **Driver Version** is also listed.

If the driver **Status** is "Untested" and you are experiencing regular Revit crashes or slowness, steps should be taken to align the installed video card driver software with the version tested and approved by Autodesk.

For optimal performance, it is best to leave **Use hardware acceleration** and **Draw visible elements** checked. However, turning these off is a good way to rule out the video card/driver when Revit is crashing or running slowly.

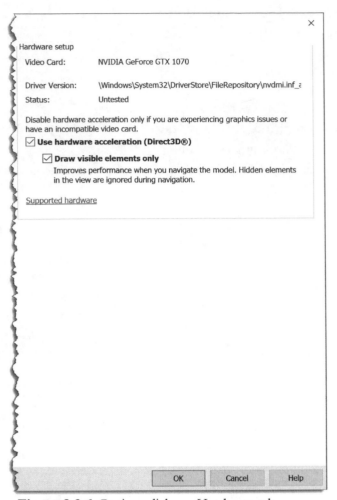

Hardware setup

Video Card: NVIDIA GeForce GTX 1070

Driver Version: \Windows\System32\DriverStore\FileRepository\nvdmi.inf_a
Status: Untested

Disable hardware acceleration only if you are experiencing graphics issues or have an incompatible video card.

☑ **Use hardware acceleration (Direct3D®)**

☑ **Draw visible elements only**
Improves performance when you navigate the model. Hidden elements in the view are ignored during navigation.

Supported hardware

OK   Cancel   Help

Figure 2.3-6 Options dialog – Hardware tab

The 'Supported Hardware' link will access Autodesk's website, showing the most current approved and certified computer hardware and video drivers.

Options → File Locations

The **File Locations** tab provides a way to customize where Revit looks for certain files—such as templates, families, etc. (Figure 2.3-7).

At the top of the File Locations tab, new project template files are listed in the order they will appear on the Recent File page and the New Project dialog (Figure 2.3-8). Clicking on a row will activate the **Up** and **Down** buttons, allowing them to be repositioned. The **Plus** and **Minus** buttons allow the list to be modified.

> In an office, this list should be replaced with the firm's template file(s). This will prevent staff from accidentally starting a project with the wrong template.

The next three items allow the default path to be set for folders (not files) for Project files, Family templates and point cloud data.

> Consider changing the User Files to a sub-folder called Revit files to separate Revit files from other documents on your local drive.

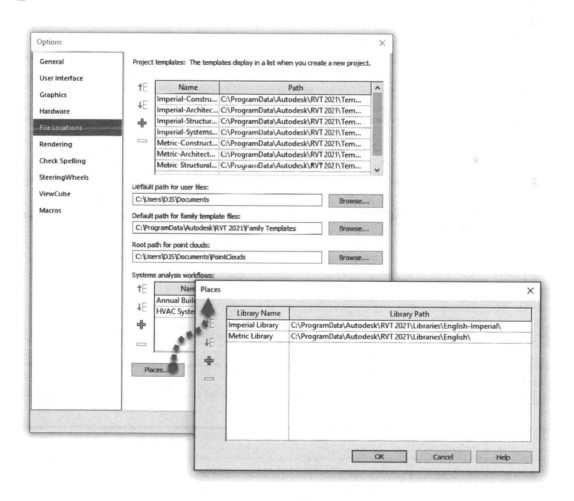

Figure 2.3-7 Options dialog – File Locations tab and Places dialog

The Places button opens the Places dialog shown in Figure 2.3-7. This feature allows the shortcut links in the file management dialogs, i.e., Open, Save, Save-As, to be modified. Figure 2.3-9 highlights the two items listed in Figure 2.3-7. Notice there are several shortcuts which cannot be modified. Some users will add shortcuts to major projects they are working on, making it easier to browse to those folders.

> In addition to adding shortcuts to Places here, it is also possible to drag folders from the Open/New dialog into the Places pane. These shortcuts do not appear in the Places dialog. They can only be deleted via a right-click on the shortcut.

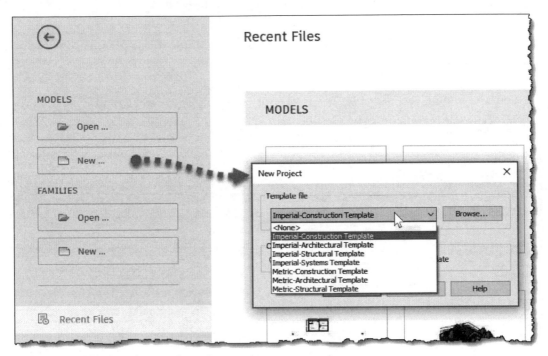

Figure 2.3-8 Template options for starting a new project

Missing Content

During installation of the Revit software, the content and families are sometimes not installed. The content is downloaded from the internet during installation, so connection issues cause this problem.

If this happens, go to:

- Windows **Programs and Features** (aka Add/Remove Programs)
- Select **Autodesk Revit Content Libraries 2021**
- Click **Add or Remove Features**

In this final dialogue, content can be selected and installed.

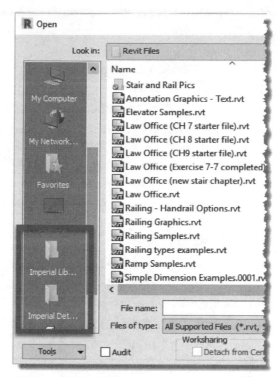

Figure 2.3-9 Places section of Open dialog

Options → Rendering

When a texture image, i.e., a raster image file (jpg, png), is attached to an image, the path and name of the file cannot change or Revit will not be able to find it and include it for Realistic or rendered views. This is a problem when a model is shared between offices with different file structures.

When a rendering is started the **Rendering Progress** dialog will list any missing textures under the Warnings heading.

To deal with this issue, Revit provides a way to manually point to one or more folders for textures when not found via the material settings.

Simply click the green plus to add a path, and the minus to remove one. The Up and Down buttons are

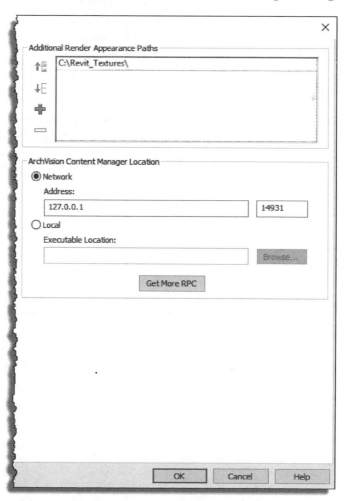

Figure 2.3-10 Options dialog – Rendering tab

provided to determine the order the folders are searched. Once a match is found, Revit will stop looking (Figure 2.3-10).

ArchVision Content Manager Location

This section is obsolete and should not be modified. The third-party company, ArchVision, now uses a different method (their standalone **Dashboard** or **Avail** program) to validate an active license when using their subscription-based entourage in a rendering.

For more information, go to www.archvision.com.

Options → Check Spelling

The options on this dialog are mostly self-explanatory. The Additional Dictionaries could be located on a server so all users have the same custom words added – for example, a client name or unique building product (Figure 2.3-11).

Remaining Tabs

There are three more tabs:

- Steering Wheel
- ViewCube
- Macros

The **Steering Wheel** appears on the Navigation Bar. This tab, in Options, allows related options to be modified.

The **ViewCube** allows click and drag orbiting of a 3D view. The Options dialog has several settings related to this tool, including whether it even shows up.

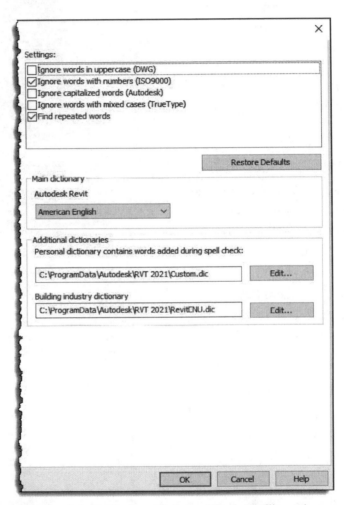

Figure 2.3-11 Options dialog – Check Spelling tab

The Macros tab allows the user to block custom programs.

Chapter 3
Architecture Tab

The Architecture tab, the first tab on the Ribbon, contains tools most often used by architects to model a building in Autodesk® Revit®. This chapter will provide detailed information on each tool on this tab.

3.1 Architecture tab: Build panel

The Build panel, on the Architecture tab, contains the primary elements of any building: Walls, Floors, Roofs, etc.

Wall: Architectural

The Wall tool is used to model interior and exterior walls, curtain walls (i.e., glass walls), foundation walls and retaining walls. In the context of ceiling design, walls are also used to develop bulkheads and the vertical portion of a soffit.

This section will explore creating walls.

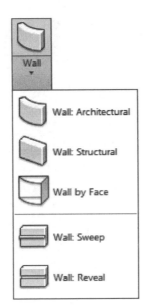

Initiating the Wall command is accomplished by simply clicking the top portion of the Wall tool on the Architecture tab.

quick steps

Wall: Architectural

1. Ribbon
   - Draw panel
   - *Default:* straight segment
2. Options Bar
   - Level (3D Views)
   - Height
   - Location Line
   - Chain
   - Offset
   - Radius
   - Join Status
3. Type Selector
   - Wall Type
     - Basic Wall
     - Curtain Wall
     - Stacked Wall
4. Properties
   - Base Constraint
   - Base Offset
   - Cross-Section
     - Vertical
     - Slanted
   - Angle From Vertical
     - Enter angle when Cross-Section is set to "slanted"
5. In-Canvas
   - Spacebar
   - Snaps
   - Temporary Dimensions

Wall: Ribbon

When the Wall tool is selected, the Draw panel appears on the
Ribbon. Here is what each option is used for:

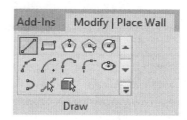

- **Line**

  Draws a straight wall segment.

- **Rectangle**

  Draws four separate walls with two clicks.

- **Inscribed Polygon**

  Draws polygon inside of specified radius; number of sides specified on Options Bar.

- **Circumscribed Polygon**

  Draws polygon outside of specified radius; number of sides specified on Options Bar.

- **Circle**

  Draws a circle using two arcs

- **Start-End-Radius Arc**

  Pick the two endpoints of the wall and then specify the radius graphically or numerically.

- **Center-ends Arc**

  Select the center point and one end point; the arc is now constrained to the specified radius—click to locate the other wall endpoint.

- **Tangent End Arc**

  The first endpoint is tangent to the adjacent wall.

- **Fillet Arc**

  Adds an arc between two non-parallel walls. Pick two walls and then graphically select radius or enter Radius on Options Bar. This option changes the length of original walls.

- **Ellipse**

  Draws an ellipse. Specify the distance of the major and minor axis of the ellipse.

- **Partial Ellipse**

  Draws ½ of an ellipse. Specify the distance of the major and minor axis of the ellipse.

- **Pick Lines**

  Create walls based on lines seen on the screen; e.g. Model/Details lines, lines for model elements, lines within CAD links. *TIP:* Hover, Tab, Click will select a chain of walls.

- **Pick Faces**

  Creates a wall based on the face of a mass element. This is the only way to create a sloped wall using a normal wall type (i.e., assembly).

Most of these options have an animated tooltip, which appears when the cursor is placed over it, to help better understand what each one does.

The above list, for the Draw panel, is common to several commands and will not be repeated again.

Architecture

Wall: **Options Bar**

The options, on the Options Bar, vary based on the Draw tool selected on the Ribbon. For walls, the two settings to pay close attention to are Height and Location Line.

Based on the image to the left, here is what each option does:

1. **Level**

This option is only visible while in a 3D view. It controls the location of the wall's bottom (Base Constraint). In a floor plan view, the wall's bottom is automatically associated with the view's Associated Level—which is a read-only parameter listed in the Properties Pallet for each plan view.

2. **Height/Depth**

Determines whether the wall is drawn up (height) or down (depth) relative to the current view. This is not a setting that is saved with the wall.

Architects typically model walls with the Height setting, and structural engineers use Depth. Structural drawings show the floor framing and the walls supporting that framing, which are below the floor.

3. **Top Constraint**

Pick a level to parametrically constrain the top of the wall—the wall height will automatically change if the levels are adjusted. Select **Unconstrained** to set a specific height—the wall height will not change if levels are adjusted.

4. **Unconnected Height**

When the Top Constraint is set to Unconnected (item #2), this field is active; enter the desired height of the wall.

5. **Location Line**

Determines how the points picked in the model relate to the construction of the wall. The three "core" options relate to the user-defined structural portion of the wall. The remaining three options relate to the overall thickness of the wall.

Location Line:	Wall Centerline ▼	☑
	Wall Centerline	
	Core Centerline	
	Finish Face: Exterior	
	Finish Face: Interior	
	Core Face: Exterior	
	Core Face: Interior	

6. **Chain**

When Chain is selected, the endpoint of a wall is also the start point for the next wall segment—thus, less clicking is required.

7. **Offset**

In conjunction with the Location Line (above) this value determines the location of the wall relative to the points picked on the screen. Example: this option might be used to create the second corridor wall by clicking points on the previously modeled wall. **Note:** The **offset** option is not available when creating a wall using the **ellipse** tool from the **Draw panel**.

8. **Radius**

   Adds an arc-shaped wall at the intersection as walls are being drawn.

9. **Join Status**

   Select Allow or Disallow; Allow is the default. When set to Disallow, <u>both</u> ends of the wall do not clean up with other walls (see image below). Another option is to select a wall and then right-click on one of the two end grips to control the wall join setting for just <u>one</u> end.

   Wall joins are dependent on each Layer's **Function**, **Material** and **Core Boundary** position (FYI: when a Layer is within the Core Boundary it is considered a structural element of the wall). To see these settings, Select a wall → Edit Type → Edit Structure.

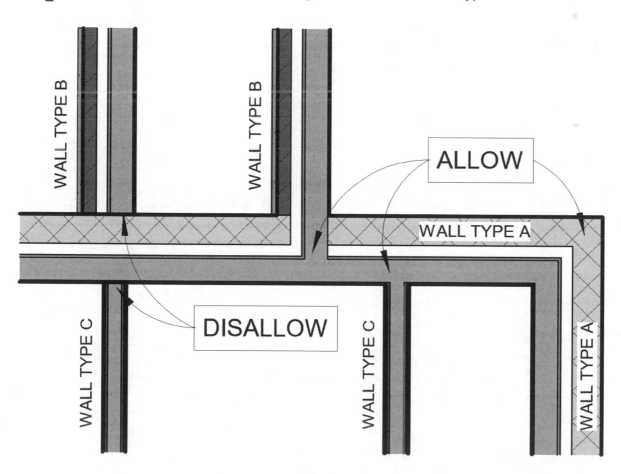

Wall: Type Selector

When the wall tool is active, there are three types of walls listed in the type selector:

- **Basic Wall**

  The most often used wall. It consists of one or more Layers (i.e., individual materials within the wall assembly—e.g. Gypsum Board) and continuous/horizontal Sweeps and/or Reveals.

- **Curtain Wall**

  A wall divided by horizontal and vertical Mullions and filled with individual Curtain Panels. The Mullions can vary in size and the Curtain Panels can represent various types of glass, spandrel panels, doors or a basic wall type. There are special corner mullions which work between two walls to make a more realistic looking corner condition.

- **Stacked Wall**

  A Stacked Wall consists of two or more Basic Walls stacked on top of each other. This is sometimes used to model more complicated exterior walls. The height of one wall is fixed and the other variable. Right-click a Stacked Wall and select **Break Up** to separate the two walls into normal Basic Walls.

Wall types can only exist within a project. If a wall type is not in the current project, one must be made by Duplicating another wall type, or by Copy/Pasting from another project. For Copy/Paste to work, both projects must be open in the same session of Revit.

> Rather than have every possible wall type needed in your firm template, create "container" projects that just have wall types. Users can open from a shared network location and Copy/Paste the desired wall type into their current project. This "additive" method reduces the possibility of selecting the wrong wall type from the Type Selector.

Walls are listed alpha-numerically in the Project Browser. Consider naming wall types so they sort by function and construction. A thoughtful naming convention also aids in applying Filters to views. Note that the wall type list can be searched via the search box at the top of the list.

The **most recently used wall types** are listed at the bottom of the list. This is helpful on large projects with a long list of walls to scroll through.

Wall: Properties

Here we are specifically talking about Instance Properties available while creating new walls—via the Properties Palette. For example, the **Cross Section** property will allow you to create a wall with a slanting face instead of a vertical wall (see Figure 3.1-0).

Most of the values which typically need to be adjusted are located on the Options Bar. Caution should be used when adjusting any other parameters during initial placement. Making an adjustment will become the default for future walls. This is not always a problem, just something to be aware of. For example, adjusting the Base Offset to 8'-0" to model a bulkhead will make all future walls have the same Base Offset until it is manually changed back. This can be confusing several days later when adding a wall to the floor plan and it is not showing up. Another example is editing the Description during placement—this would be better to edit after the wall was placed than during placement.

Custom wall parameters can be added via **Manage → Project Parameters → Add**. Here a name is provided, and one or more Revit categories selected (e.g. Walls). This custom information can appear in schedules but not tags. A Shared Parameter must be used if the information must also appear in a tag.

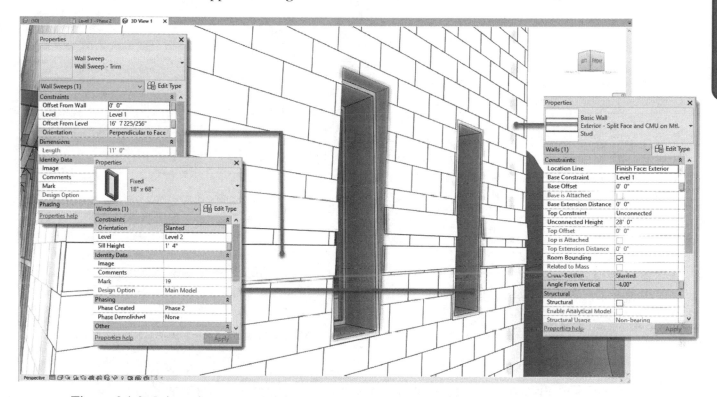

Figure 3.1-0 Orientation properties for windows and sweeps on slanted walls

Wall: In-Canvas

After clicking the first point, pressing the Space Bar will flip the wall about the Location Line. When a wall is selected, the Flip icon always appears on what Revit considers the exterior side of the wall.

Also, when a wall is selected, right-clicking on an end grip provides the **Disallow Join** via a pop-up menu. Selecting this will prevent the wall from cleaning up with any other wall. This is helpful between new and existing construction or with some curtain wall configurations. Once Disallow Join has been selected, an **Allow Join** icon will appear when the wall is selected—clicking it will toggle the feature off.

Wall: Structural

A structural wall is the same as the architectural walls with two different default settings. A structural wall also has an Analytical definition.

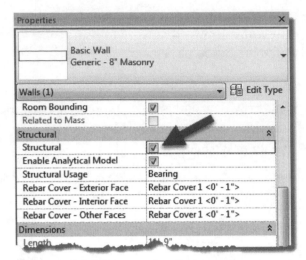

> A structural wall is meant to designate a wall which supports a load, such as a floor, roof, beam or joist. This setting affects visibility in some views and supports external structural analysis software workflows.

Figure 3.1-1 Structural wall properties

When the structural wall tool is selected, the Options Bar is automatically set to **Depth** rather than Height.

Also, the **Structural** parameter is checked by default. This option enables and reveals a few more structural parameters as shown to the right.

> Note that any wall can be made a structural wall by simply checking the Structural option in the Properties Palette.

For stacked walls not initially created with the Walls: Structural tool; the structural usage can only be set by selecting each wall sub-component and selecting the structural option in Properties. To select a sub-component, i.e. one of the two walls in the stacked wall, hover your cursor over the wall and tap the Tab key until the desired wall highlights and then click to select it.

When a view's **Discipline** parameter is set to Structural, only "Structural" walls will appear in the view. All other walls, i.e., "non-bearing," are automatically hidden in structural views.

Wall by Face

The **Wall by Face** tool is used to attach a Revit wall type to the face, i.e., a single side of a previously created Mass element. This command is repeated here for convenience, but more closely relates to the massing tools found on the **Massing & Site** tab. Thus, please reference *Chapter 9 – Massing & Site* for detailed coverage of this feature.

Wall: Sweep

A wall sweep is an extrusion of a profile along a wall; it can be horizontal or vertical. This is similar to a horizontal extrusion which can be added to the Structure of a wall type (but is only horizontal and cannot wrap wall ends).

Figure 3.1-2 shows three wall sweeps, each hosted by the same wall in this example. The wall/ceiling cove trim is hosted on the face of the wall. The Wall copping is hosted by the top edge of the wall and has a profile which places it over the wall. Finally, the masonry sill cuts into the wall.

Sweep: **Ribbon**

The two Placement options are **Horizontal** and **Vertical**. Once a surface is selected, continue selecting adjacent surfaces to place additional segments at the same elevation. Interesting corners will automatically miter.

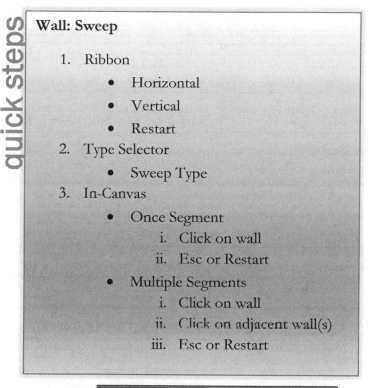

quick steps

Wall: Sweep

1. Ribbon
    - Horizontal
    - Vertical
    - Restart
2. Type Selector
    - Sweep Type
3. In-Canvas
    - Once Segment
        i. Click on wall
        ii. Esc or Restart
    - Multiple Segments
        i. Click on wall
        ii. Click on adjacent wall(s)
        iii. Esc or Restart

Architecture

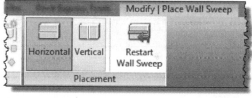

Clicking **Restart Wall Sweep** allows a new elevation to be started and will not miter with previously created segments.

Sweep: Type Selector

The defined wall sweep Types appear in the Type Selector. It is helpful to understand that loading a profile does not automatically create a new wall sweep Type. A profile defines the shape of a wall sweep, but there are other properties needed, which are not directly associated with the profile Family—e.g. Phase, Material, Cost.

To create a new Type, simply duplicate an existing Wall Sweep, select the desired profile and adjust other properties as needed.

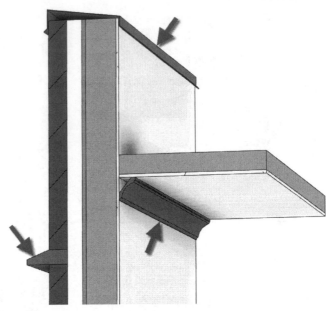

Figure 3.1-2 Wall sweeps

> Profiles can be parametric and have multiple Types loaded. For example, the wall copping, in the previous image, could have multiple types to accommodate different wall thicknesses. This would require an equal number of wall sweeps, each with a different profile/type selected.

Sweep: In-Canvas

Sweeps can only be placed in 3D views or an elevation/section view perpendicular to a wall's surface (not perspective or plan views). Wall Sweeps **cannot be angled**, only horizontal or vertical—and angled extrusion would need to be accomplished with an In-Place family.

While selected, the **end grips** can be dragged to adjust the length of the sweep. To delete a sweep, simply select it and delete.

The Type Properties control whether a sweep **cuts a wall** it overlaps, or if it is **cut by inserts** such as doors and windows.

> Selecting a previously created Wall Sweep reveals a **Modify Return** tool, on the Ribbon, allowing the profile to return on itself at any angle (via the Options Bar).

Wall Sweeps can sometimes create extra linework in plan views. Wall Sweeps can be adjusted or turned off in a view's Visibility Graphics/Overrides.

Sweeps may be **Aligned and Locked** to adjacent geometry. In this example the cove trim is hosted by the ceiling and will move with it. However, if the ceiling moves, the trim will be out of place. The Align tool can be used to keep the trim in the correct position. The Align tool only works with the lines within the profile, not any Reference Planes.

Wall: Reveal

A reveal is very similar to a Sweep. The main difference being the shape (i.e., profile) of a reveal is a void that is only seen by its impact on other elements.

Notice, in Figure 3.1-3, that sweeps are not cut by reveals.

Reveals do not have a **material**. The material of the surface being cut is used.

Sweep: Ribbon

The options here are identical to the Wall Sweep tool just covered.

Sweep: Type Selector

The options here are identical to the Wall Sweep tool just covered.

Sweep: In-Canvas

Both horizontal vertical reveals can be seen in the face brick in Figure 3.1-3. Also, one vertical reveal had its length adjusted after initial placement; simply select the reveal and drag one of the end grips. It is not editable, but the total **length** is listed for reference in the Properties Palette.

Sweeps and reveals can be dimensioned, but not tagged or used with keynotes.

quick steps

Wall: Reveal

1. Ribbon
   - Horizontal
   - Vertical
   - Restart
2. Type Selector
   - Sweep Type
3. In-Canvas
   - Once Segment
     i. Click on wall
     ii. Esc or Restart
   - Multiple Segments
     i. Click on wall
     ii. Click on adjacent wall(s)
     iii. Esc or Restart

Architecture

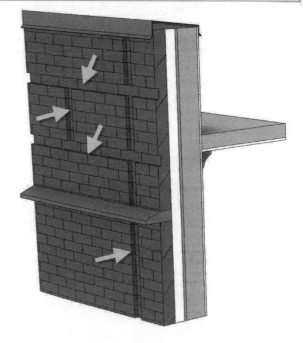

Figure 3.1-3 Wall reveals

Door

Door content can vary significantly within the industry – rarely is the Out Of The Box (OOTB) content used. The goal here is to focus on the command rather than the content itself.

Door: Ribbon

The **Load Family** tool is here for convenience when the desired family cannot be found in the Type Selector. However, trying to load a non-door family will result in an error.

The **Model In Place** option will start that tool, defaulting to the Door category.

Tag on Placement, when active, will place a Door Tag with each door. This is often not desired while initially placing doors, as they are not needed during the early design phase of a project.

quick steps

Door

1. Ribbon
   - Load Family
   - Model In-Place
   - Tag on Placement
2. Type Selector
   - Select Door Type
3. Options Bar
   - Tag options
4. Properties
   - Sill/Head Height
   - See Warning
5. In-Canvas
   - Hover over wall
   - Swing Direction
     i. Hover cursor near swing side of wall
   - Hinge Side
     i. Spacebar to flip

Door: Type Selector

The doors loaded in the current project are listed in the **Type Selector**. It is ideal to have a consistent naming convention, so doors are sorted and easy to find.

Most door families are parametric for door width and height. If the desired size is not listed in the Type Selector, select the door type (for the desired family) and then click **Edit Type** in the Properties Palette. Here, the selected Type can be **duplicated** and its parameters adjusted.

Door: Options Bar

Options related to tag placement and type are available on the Options Bar when Tag on Placement is selected on the Ribbon. These options are also available in "locked" 3D views—a 3D view must be locked before tags can be placed.

Door: Properties

Important parameters to keep in mind are Level, Sill Height and Phase. Also, custom parameters are often added to track Door Type, Jamb/Head Detail, Hardware Group, etc.

Warning: changing a parameter in the Properties Palette during placement will make that change the default for future placement.

Door: In-Canvas

Doors can be quickly tagged using the Tag All command. Keep in mind, this tool will tag all doors regardless of its **Phase** setting. Sometimes, existing doors are not tagged if no work is to be done on them or they are outside the project limits.

A door must be placed in a wall. If a door is set to be **demolished**, a special wall replaces the door in future phases. This wall cannot be deleted but can be hidden with a **View Filter** if needed.

The Door tool cannot be used to place a door in **Curtain Wall**. To place a door in Curtain Wall, select a Curtain Panel and then select a special Curtain Panel door family from the Type Selector. These doors will stretch to match the current panel dimensions – which is what is reported in a door schedule. A few example families are provided with Revit and others may be found online.

While placing the door, the direction of the **swing** is based on which side of the wall your cursor is favoring. Pressing the Spacebar before placement will flip the **hand** of the door. To change the swing and/or hand of the door after creation, select it and click one of the two flip controls.

Depending on the wall Type and Structure settings, and the presence of special reference planes within a door family, it is possible to control if and how **wall materials return at each opening** as shown in the following image (Figure 3.1-4). For more on this topic, follow these links:

- http://revitoped.blogspot.com/2012/07/reference-planes-and-wall-closure.html
- https://bimchapters.blogspot.com/2017/07/revit-20181-new-content-part-2.html

Also notice, in this image, that the custom door family (all four doors are the same door family) has an adjustable frame. It can match the thickness of the wall or have a specific depth and offset from the exterior face of the wall. Developing custom content like this is important to fully leverage the power of Revit.

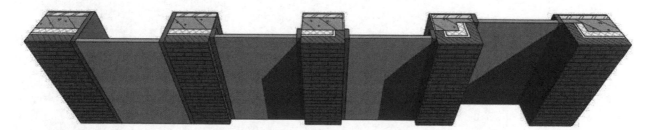

Figure 3.1-4 Custom door family with adjustable frame and special reference planes for material returns

If multiple walls are Joined together, a door will cut all of them—not just the selected host wall.

Window

Windows function similar to doors. Refer to the previous section for additional related details.

The Width and Height parameters are typically **Type** parameters and the Sill Height is an **Instance** parameter as it can vary by instance (i.e. each location).

For additional reading on how the opening in a window cuts the host wall see this author's article at *Cadalyst Magazine's* website: http://www.cadalyst.com/aec/editing-windows-autodesk-revit-13356

Component

If an element does not have its own tool (e.g. Door, Window, Ceiling) then it is placed using the catch-all **Component** tool.

Component: Ribbon

Load Family and **Model In Place** are displayed for convenience. See the following In-Canvas comments on **Face-Based** content and how it affects the Ribbon options.

Window: Options Bar

Checking the **Rotate after placement** option will immediately start the Rotate tool after placement.

Window: Type Selector

This list can get confusing and long on large projects. In this list one might find trees, parking stripes, light fixtures, plumbing fixtures and more.

quick steps

Window

1. Ribbon
   - Load Family
   - Model In-Place
   - Tag on Placement
2. Options Bar
   - Rotate after placement
   - Level (in 3D views)
3. Type Selector
   - Select Door Type
4. Properties
   - Sill/Head Height
   - See Warning
5. In-Canvas
   - Hover over wall
   - Swing Direction
     i. Hover cursor near swing side of wall
   - Hinge Side
     i. Spacebar to flip

quick steps

Place a Component

1. Ribbon
   - Load Family
   - Model In-Place
2. Options Bar
   - Rotate after Placement
3. Type Selector
   - Select Family to place
4. Properties
   - Options vary
   - See Warning
5. In-Canvas
   - Spacebar to rotate
     i. Cursor location

Everything is sorted alphabetically, so it is a good idea to develop a naming convention that will group various categories together. One way to do this is to add a two-character prefix that corresponds to the Revit category the family is found in. For example:

- Casework:
  - o **CW**_Base Sink_False Drawer
  - o **CW**_Upper_One Door
- Furniture:
  - o **FN**_Chair
  - o **FN**_Sofa
- Plumbing Fixtures:
  - o **PF**_Wall Hung Toilet
  - o **PF**_Drinking Fountain
- Entourage:
  - o **EN**_Man Walking

> With the Type Selector expanded, pressing a letter on the keyboard will jump down to that section.

Component: Properties

The options vary based on the type of family selected. The **Level** parameter matches the level of the current plan view. Some content allows the element to be **Offset** relative to the level. Checking the **Moves With Nearby Elements** causes the element to move if the closet wall (when initially placed) moves.

Component: In-Canvas

Pressing the **Spacebar** will rotate the element during placement—if near an angled wall (relative to the computer screen) then the element will also rotate aligned with the wall.

When a **Face-Based** family is selected for placement, the Ribbon displays additional options as shown in Figure 3.1-5. In plan view, the default option is **Place on Vertical Face**, which is generally meant to facilitate placement on a wall

Figure 3.1-5 Ribbon options during face-based component placement

while in a plan view. If the element needs to be placed on a floor, roof or ceiling the option should be changed to **Place on Face**. Selecting **Place on Work Plane** allows elements to float in space using *named* Reference Planes. Finally, selecting Tag on Placement will add a Tag in addition to the element. The tag used is based on the element's category and can be specified in the **Loaded Tags and Symbols** dialog—which is found in the extended panel area of the Tag panel on the Annotate tab of the Ribbon.

Model In-Place

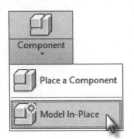

This tool is used to create **unique**, one off, custom families in the context of the model. For example, one might use this tool to create a built-in bookshelf or reception desk that ties into a column of curved wall within the model. This command is not available in Revit LT.

This tool should not be used to create something that needs to be repeated multiple times in the model—such as a chair or table. These items should be created in the family editor and loaded into the project.

When In-Place families are copied around within a model, the definition is duplicated. So, changing one does not update the others (unlike a regular Loadable Family). This would also make the project file larger compared to a loaded family which only has one definition in the project and can be copied around hundreds of times without making the file significantly larger.

quick steps

Model In-Place

1. Select Family Category
   - Controls Visibility
   - Tips:
     - Use *Generic Model* if others are not a good fit
     - Typically avoid *Mass*
2. Name
   - Provide Family Name
3. Create Tab
   - Develop Family
     - Forms
       i.   Solids
       ii. Voids
     - Component, i.e., nested family
     - MEP Connectors
     - Reference Planes/Lines
     - Dimensions/Parameters
     - Family Types
4. Properties
   - Options vary
5. In-Place Editor Mode
   - Finish Model to Keep Changes
   - Cancel Model to Discard Changes

Model In-Place: Family Category

The first step is to select the category the new element will be associated with. This selection will control visibility and inclusion in filters and schedules. This selection can be changed later as required. The options, shown in Figure 3.1-6, are built-in to Revit and cannot be modified.

Most of the Category names are self-evident as to what they are. Below are descriptions for a few that might not be obvious to some:

- Casework: Cabinets, reception desks, built-in bookshelves
- Columns: Non-structural, decorative architectural
- Entourage: People and Vehicles
- Furniture Systems: Cubicles and similar modular office furniture
- Generic Models: Catch all category
- Specialty Equipment: Marker boards, Display cases, Signage, Toilet room partitions and accessories, corner guards, lockers, mailboxes
- Structural column: An element designed by a structural engineer
- Structural framing: A beam or brace—designed by a structural engineer

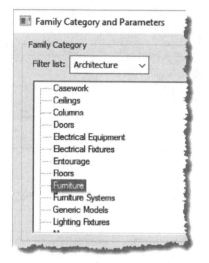

Figure 3.1-6 Specify In-Place family category

The category selection also controls whether an element can be cut in a section. Notice, in Figure 3.1-7 below, that the shaded cells under **Cut** cannot be cut. You typically would not want to see a sofa or plumbing fixture in section. If any of these elements pass through a cut plane, the entire element appears as if the section were moved just outside of the element.

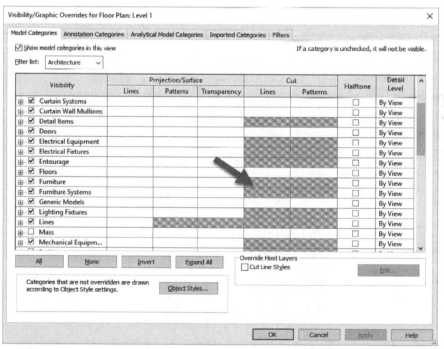

Figure 3.1-7 The elements in some categories cannot be cut – notice shared cells

Model In-Place: **Name**

After selecting the category, a unique name must be provided. Entering a name already defined in the current project will result in a warning that cannot be ignored.

Model In-Place: **Create Tab**

Once a category has been selected and a name provided, Revit switched into a special mode similar to the Family Editor—however, you are still in the project. Notice, in Figure 3.1-8, that the Architecture/Structure/Systems tabs have been replaced with a **Create** tab. This tab provides tools to create custom 3D geometry, line work and MEP connectors, all of which can be made parametric by adding dimensions and labels.

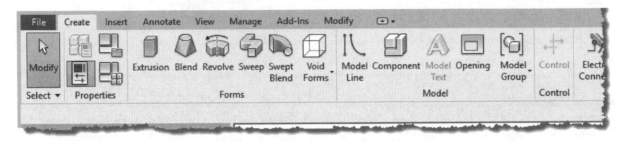

Figure 3.1-8 Create tab on Ribbon while in Model In-Place mode

The Ribbon and Options Bar changes as different tools are selected.

MEP **Connectors** allow mechanical and electrical designers to predefine locations where pipe, duct or electrical wire may be attached. Specific knowledge of these systems is required to properly define these connection points (e.g. balanced or unbalanced power connection). Some categories do not allow for connectors so these tools are grayed out on the Ribbon.

Model In-Place: **Properties**

The properties palette varies depending on the active tool. When the Extrusion Tool is selected the start and end elevation are listed as well as material and a toggle for Solid versus Void.

Custom parameters can be added by clicking on the **Family Types** button, located on the Ribbon, and then clicking the Add… parameters button.

Model In-Place: **In-Place Editor Mode**

The only way to exit the In-Place family mode is via the green check mark or red X.

> Importing a **SketchUp** or 3D **AutoCAD** file into an In-Place family will allow Revit to cut it in section.

Column: Architectural

An Architectural Column is meant to be a non-load bearing element, often representing a finish around a structural column (e.g. studs and gypsum board).

These columns do not have a Structure, i.e., layers of materials, that can be edited as walls, ceilings and floors. Rather, they take on the structure of any wall they come in contact with.

Notice in Figure 3.1-9 that the Architectural Column at the top is a simple outline. The second, or middle column, is the same family, but has taken on the properties of the stud wall that runs through it.

> Using the **Join** command on the isolated column and the adjacent stud wall would apply the wall's properties even though they do not touch.

quick steps

Column: Architectural

1. Ribbon
   - Load Family
   - Model In-Place
2. Options Bar
   - Rotate after Placement
   - Level (3D Views)
   - *Select Direction*
     - Depth (downward)
     - Height (upward)
   - *Select Top Constraint*
     - Pick a Level *or*
     - Unconnected
       i. Enter Height
3. Type Selector
   - Select Column Type
4. Properties
   - Moves with Grid
5. In-Canvas
   - Spacebar to rotate

Architecture

Architectural Columns cannot be slanted as can Structural Columns.

Architectural Column: Ribbon
The Load Family and Model In-Place commands are located here for convenience.

Architectural Column: Options Bar
Checking the **Rotate after placement** option will immediately start the Rotate tool after placement. Checking Room Bounding limits the perimeter of a Room or Space element within a given room.

When is **Height** versus **Depth** used? See the previous comments on Walls and Structural walls in this chapter. Selecting Unconnected, in the second drop-down, creates a non-parametric column whose height will not change.

Architectural Column: **Type Selector**

The most used Architectural Column families are **Rectangular Column** and **Round Column**. The Column folder has a few other options.

Keep in mind that the Architectural Column tool existed prior to the Structural Column tool—Revit Architecture came before Revit Structure. One might notice the Column folder has a **Wood Timber Column** family and the Structural Column folder has a **Timber-Column** family. The version in the structural folder should be used. The architectural version will take on properties of any walls that touch it and, as previous stated, are meant to be non-bearing.

Architectural Column: **Properties**

Checking the **Moves with Grid** option will cause the column to maintain its relative position to the nearest grid.

If the **Material** Type Property is changed, from By Category to a specific material, the column will stop taking on the adjacent wall's material—the structure (aka layers) will still appear.

Architectural Column: **In-Canvas**

Tapping the **spacebar** rotates the column 90 degrees. When near an angled wall, and while tapping the Spacebar, the column will align with the wall.

Structural Column

The Structural Column tool is repeated here for convenience. See Chapter 4 – Structural Tab for detailed coverage of this tool.

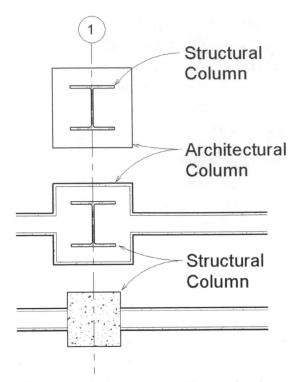

Figure 3.1-9 Comparing architectural and structural columns

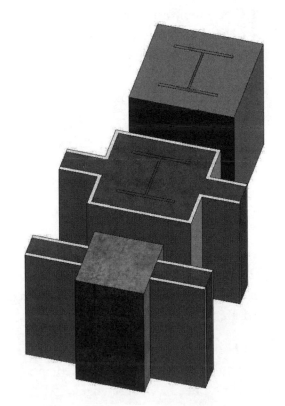

Figure 3.1-10 Comparing architectural and structural columns

Roof by Footprint

quick steps

Roof by Footprint

1. First Switch to Appropriate Plan View
   - i.e., View associated with the level the roof should be aligned with
2. Ribbon [in sketch mode]
   - Draw *Panel*
     - Boundary Line
     - Slope Arrow
     - Segment or Pick Options
   - Work Plane *Panel*
     - Set/Show Work Plane
     - Create Reference Plane
       i. Tip: Name it
     - Viewer
   - Tools *Panel*
     - Align Eaves
3. Options Bar
   - Defines Slope
     - Line defines the edge of a sloped roof plane
   - Overhang
   - Extend to core
4. Type Selector
   - Select Roof Type
5. Properties (three groups)
   - *Roofs*
     - Level settings
   - *Sketch Lines*
     - Set pitch
   - *View*
6. Green Checkmark to Finish Sketch
   - Must have enclosed footprint with no overlapping lines or gaps at intersections
7. In-Canvas
   - Select line to adjust pitch or Defines Slope Setting
   - Select Slope Arrow to adjust settings

This tool is used to draw an outline of the roof in plan view. The lines, which represent the perimeter of the roof, may define a sloped surface (see Defines Slope in Options Bar section below).

Picking this tool enters a **Sketch Mode** where all elements are grayed out and not selectable. The goal is to create a closed outline with no gaps or overlapping lines. Any secondary outlines within a larger outline define a hole.

> The only way to get out of the Roof tool is to click the **green check mark** (finish) or the **red X** (cancel) on the Ribbon.

Roof by Footprint: Ribbon

Boundary Line and Slope Arrow are the two types of line which can be drawn—next to these options are the shapes in which these lines can be drawn.

The default option is **Boundary Line**, which defines the perimeter of the roof in plan view.

Figures 3.1-12a and b show a few uses for **Slope Arrows**. The example on the left shows two Slope Arrows over one sketch line (with Defines Slope turned off). The middle example shows two Slope Arrows over two sketch lines. Finally, the right example shows a single Slope Arrow defining the slope of an entire roof.

When the Slope Arrow is selected, the Properties Palette allows the input of either a specific **height** (from head to tail of the arrow) or a **slope** (i.e., pitch).

The default sketch line option is **Pick Wall**—simply click on a wall and Revit adds a sketch line along its length. **Pick Lines** allows any line on the screen to be selected. This line could be the edge of some other element or from an AutoCAD DWG file. These two options, plus all line and arc options, can define slope. The two ellipse options and spline cannot define a slope.

Some sketches and modifications to existing roofs will cause the eaves not to align vertically. The **Align Eaves** tool will correct this problem—which requires the overhang or roof slope to adjust.

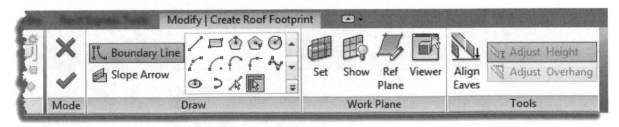

Roof by Footprint: Options Bar

To see how **Defines Slopes** impacts a roof, see Figures 3.1-11a and b. The left example has all lines set as Defines Slope—the middle has two and the right has none; notice the small triangle near the Defines Slope lines. The 3D view shows the results: Hip, Gable and Flat.

The **Overhang** is from the exterior face of the wall unless **Extend to wall core** is selected—in which case the overhang is from the exterior side of the structural core of the wall as defined in the wall Structure. Using this option will maintain the offset if the wall is repositioned later.

Selecting the **Pick Wall** option on the Ribbon changes the options on the Options Bar. In this case Overhang is replaced with **Offset**, which is essentially the same thing. There is also a **Chain** and **Radius** option.

Roof by Footprint: Type Selector

Select a roof type from this list, similar to walls and floors. This can be changed later. The bottom, or bearing, of the roof is fixed. If the thickness is changed, due to changing or editing the roof type, the top of the roof moves, not the bottom.

To create a new roof type:

- Start the **Roof by Footprint** command
- **Type Selector:** Select a roof type closest to the new type needed
- Click **Edit Type**
- Click **Duplicate**
- **Enter a name** for the new roof
- Click **Edit** for the Structure parameter
- Edit the **Layers** (i.e., materials) for the roof assembly

If trusses or rafters are to be modeled accurately, separate from the roof, then the roof type should not include a layer for them. The roof type will essentially be just the sheathing.

Roof by Footprint: Properties

While in a roof sketch, the Properties Palette shows three different things depending on what is selected.

- **If nothing is selected**
  Overall roof properties are displayed: slope, rafter cut, base level
- **If a sketch line is selected**
  Sketch line properties shown: defines slope, overhang, extends into core
- **If a slope arrow is selected**
  Slope properties displayed: specify slope vs. height, slope

The **Slope** (aka pitch) of a roof is set by the main roof properties and applies to all sketch lines (i.e., Boundary Lines) that define a slope. With a sketch line selected, the Slope of one roof plane can be changed separately from the main roof slope.

Roof by Footprint: In-Canvas

All of the basic Modify tools work in edit mode. Use Trim to clean up corners.

Architecture

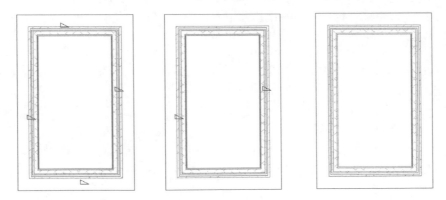

Figure 3.1-11a
Comparing various combinations of defines slope – plan view

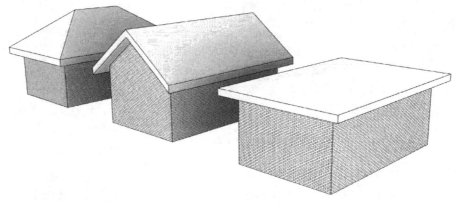

Figure 3.1-11b
Comparing various combinations of defines slope – perspective view

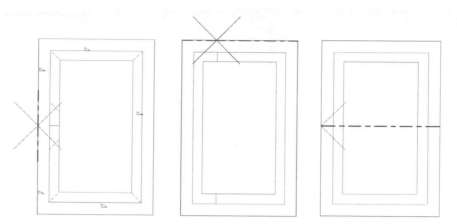

Figure 3.1-12a

Exploring various uses for slope arrows

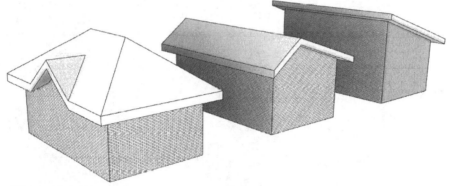

Figure 3.1-12b

Exploring various uses for slope arrows

Roof by Extrusion

Some roof forms cannot be sketched in a plan view, such as a barrel vault roof, so Revit provides the Roof by Extension tool. Using this tool, the top edge of the roof form is sketched in a non-plan view, e.g. elevation, section, 3D (see Figure 3.1-13a).

The result is a roof, based on the construction of the selected roof type, extruded perpendicular to the sketch plane (Figure 3.1-13b).

Roof by Extrusion: **Specify Work Plane**

Because this sketch is not in a plan view, Revit needs to know where in the model the sketch lines are; think depth while in an elevation view. Thus, when starting this command, Revit will prompt for a **Work Plane**.

The default option is to select a plane of some other element in the model—such as an exterior wall. Another option is to create a **Named Reference Plane** in plan view prior to using this command. These will appear in a drop-down list (Figure 3.1-13), which is often easier than trying to select a face of another element in the model.

> To create a named reference plane, sketch a Reference Plan, select it and then provide a name in the Properties Palette.

quick steps

Roof by Extrusion

1. First Switch to Appropriate View
    *   Typically, in a section or elevation view
    *   Creating a name *Reference Plan* in plan may be helpful
2. Specify Work Plane
    *   This is the plane the sketch lines will be created on
    *   Previously created named Reference Planes can be selected from a drop-down list
3. Specify Roof Reference Level and Offset
4. Ribbon [in sketch mode]
    *   Draw *Panel*
        *   Segment or Pick Options
    *   Work Plane *Panel*
        *   Set/Show Work Plane
        *   Create Reference Plane
        *   Viewer
5. Options Bar
    *   Chain
    *   Offset
    *   Radius
6. Type Selector
    *   Select Roof Type
7. Properties
    *   Extrusion End
    *   Reference Level settings
8. Green Checkmark to Finish Sketch
    *   Sketch top edge of roof - NOT enclosed outline

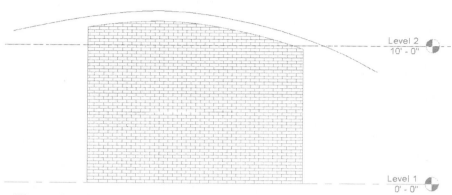

Figure 3.1-13a

Roof by extrusion sketch example – defines the top edge of roof

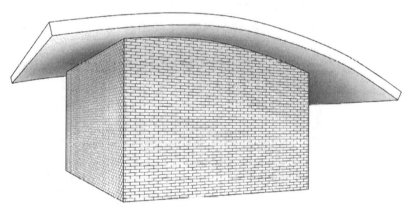

Figure 3.1-13b

Roof by extrusion result

Roof by Extrusion: **Roof Reference Level and Offset**

The next dialog (Figure 3.1-14) provides the selection of a **Level** and an **Offset**. When the selected Level is adjusted, the roof will move with it. Changing the Offset initially has no effect as the roof is modeled exactly where the sketch line was created. However, changing the Offset after initial creation of the roof will move the element vertically.

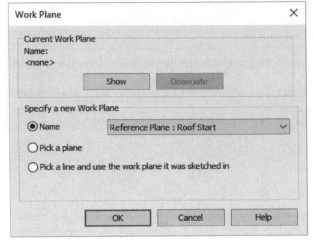

Figure 3.1-14 Work Plane options

Roof by Extrusion: **Ribbon**

The Ribbon consists of the basic **Draw** tools and options to **Set** and **Show** the work plane. The Spline tool can be used with this command.

Roof by Extrusion: **Options Bar**

The Options Bar has limited options for this tool. The **Offset** value will pick points the specified distance from the actual cursor location.

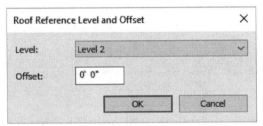

Figure 3.1-15 Level/Offset options

Roof by Extrusion: **Type Selector**

Select the desired roof construction from the Type Selector. The type selected can be easily changed later.

Roof by Extrusion: **Properties**

Various roof properties, including Room Bounding and Rafter Cut, are available. Also, the Extrusion End determines how long the roof extrusion is.

Roof by Extrusion: **Finish Sketch**

Once the top edge of the roof has been defined with one or more lines, click the green check mark to finish the command and create the roof element in the model. If changes are not needed or you need to cancel and start over, click the red X.

Roof by Face

The **Roof by Face** tool is used to attach a Revit roof type to the face, i.e., a single side, of a previously created Mass element. This command is repeated here for convenience, but more closely relates to the massing tools found on the **Massing & Site** tab. Thus, please reference *Chapter 9 – Massing & Site* for detailed coverage of this feature.

Roof: Soffit

The Roof Soffit tool is similar to the floor and ceiling tools—an outline is sketched to represent the extents of the plane. In this case, that plane represents a soffit under a roof overhang (Figure 3.1-16).

Roof: Soffit: **Ribbon**

The Ribbon contains typical sketch tools plus a toggle for **Boundary Line** or **Slope Arrow**. One unique option is the **Pick Roof Edge** edge draw option—when a roof is selected, the outer perimeter is automatically sketched. The inner perimeter still needs to be added or the soffit will extend through the entire building.

Roof: Soffit: **Options Bar**

Offset and **Extend into wall (to core)** allow for more accuracy during initial placement of sketch lines.

Roof: Soffit: **Type Selector**

Select the desired roof soffit construction. **Click Edit Type → Duplicate → Edit Structure** to create a new type.

Roof: Soffit: **Properties**

Specify the **Level** and **Height Offset From Level** to position the soffit vertically. To determine the proper offset, a section may have to be opened and the measure tool employed.

Roof: Soffit: **Finish Sketch**

Once the boundaries of the soffit have been defined, click the **green check mark**

quick steps

Architecture

Roof: Soffit

1. First Switch to Appropriate View
   - Typically, a plan view
2. Ribbon [in sketch mode]
   - Draw *Panel*
     - Sketch enclosed boundary or add slope arrow.
     - Segment or Pick Options
   - Work Plane *Panel*
     - Set/Show Work Plane
     - Create Reference Plane
     - Viewer
3. Options Bar
   - Chain
   - Offset
   - Extend into wall (to core)
4. Type Selector
   - Select Roof Soffit Type
   - Reference Level and Offset
5. Properties (two groups)
   - *Roof Soffits:*
     - Extrusion End
     - Reference Level settings
   - *Sketch Lines:*
     - Defines Slope
     - Defines Constant Slope
   - View
6. Green Checkmark to Finish Sketch
   - Objective is to sketch enclosed outline in a plan or 3D view

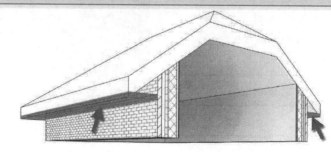

Figure 3.1-16 Roof soffit selected (section box applied)

to finish the command and create the roof soffit element in the model. If changes are not needed or you need to cancel and start over, click the **red X**.

Roof: Fascia

A Roof Fascia is an extruded element, defined by a 2D profile, representing a trim board, along the edge of a Roof element.

A Fascia can be placed on another Fascia, thus allowing a more complex condition to be created.

Roof: Fascia: **Ribbon**

The **Restart Fascia** tool allows a new section to be added which will not be joined or mitered to a previous section, which is a little easier than canceling and reselecting the command.

Roof: Fascia: **Type Selector**

The defined Roof Fascia Types appear in the Type Selector. It is

quick steps

Roof: Fascia

1. Ribbon
   - Restart Fascia
2. Type Selector
   - Select Roof Fascia Type
3. Properties
   - Offsets
   - Angle
4. In-Canvas
   - Select the edge of a Roof, Soffit or another Fascia element to add a Fascia
   - Select an adjacent edge to create a continuous/smooth transition
   - Or, Select Restart to add a new segment without restarting the tool

helpful to understand that loading a profile does not automatically create a new roof fascia Type. A profile defines the shape of a roof fascia, but there are other properties needed which are not directly associated with the profile Family—e.g. Phase, Material, Cost.

Roof: Fascia: **Properties**

Use the **Offset** and **Angle** parameters to reposition the profile relative to the edge selected.

Roof: Fascia: **In-Canvas**

Select either the top or bottom edge of a Roof element. Continuing to select multiple adjacent edges, during the same active command, will miter the corners/transitions

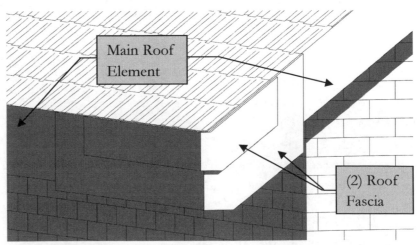

Figure 3.1-17 Roof and Roof Fascia example

(Figure 3.1-17). Selecting a fascia reveals end-grips which allow the ends of the fascia to be repositioned as shown here. Also, this example shows two built-up Fascia elements. Another option would be to use a Profile with this same overall shape.

The selected edge can be straight or curved… however, an edge created with Spline cannot be used.

Roof: Gutter

A Roof Gutter is an extruded element, defined by a 2D profile, representing a rain gutter system, along the edge of a Roof element.

Roof: Gutter: **Ribbon**

The **Restart Fascia** tool allows a new section to be added which will not be joined or mitered to a previous section, which is a little easier than canceling and reselecting the command.

Roof: Gutter: **Type Selector**

The defined Roof Gutter Types appear in the Type Selector. It is helpful to understand that loading a profile does not automatically create a new roof gutter Type. A profile defines the shape of a roof gutter, but there are other properties needed which are not directly associated with the profile Family—e.g. Phase, Material, Cost.

Roof: Gutter: **Properties**

Use the **Offset** and **Angle** parameters to reposition the profile relative to the edge selected.

---

quick steps

Roof: Gutter

1. Ribbon
   - Restart Fascia
2. Type Selector
   - Select Roof Gutter Type
3. Properties
   - Offsets
   - Angle
4. In-Canvas
   - Select the top edge of a Roof, soffit or Fascia element to add a Fascia
   - Select an adjacent edge to create a continuous/smooth transition
   - Or, Select Restart to add a new segment without restarting the tool

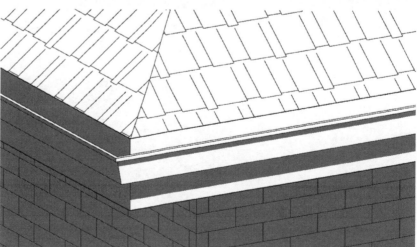

Figure 3.1-18 Roof and Roof Gutter example

Roof: Gutter: In-Canvas

Select either the top or bottom edge of a Roof or Fascia element. Continuing to select multiple adjacent edges, during the same active command, will miter the corners/transitions (Figure 3.1-18). Selecting a gutter reveals end-grips which allow the ends to be repositioned as shown here.

> The end of a gutter is not capped off and Revit does not have a downspout (aka rain leader) tool. These elements need to be modeled In-Place or with custom families.

The 2D profile family used for this gutter example is shown to the right (Figure 3.1-19).

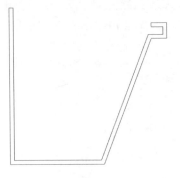

Figure 3.1-19 Profile used for Roof Gutter example

Ceiling

This tool places a ceiling assembly (i.e., Type) at the specified height in a space. Ceilings can host lights, air terminals (aka diffusers) and more.

> If a ceiling is deleted, all host elements are also deleted without warning.

Ceiling: Ribbon

The default **Automatic Ceiling** option will search for an enclosed boundary **at the specified height**. If an enclosed area is found, a ceiling can be created by clicking in the desired room.

Selecting **Sketch Ceiling** allows the perimeter of the ceiling to be manually specified. This is helpful if the ceiling is held away from the wall or an isolated "cloud" type ceiling is desired. Once this option is selected, the typical Draw tools appear on the Ribbon.

Ceilings can be sloped in one of three ways:

- By adding a **Slope Arrow**, similar to the roof tool
- Setting one sketch line to **Defines Slope**, similar to the roof tool
- By setting two parallel sketch lines to **Defines Constant Height** and specifying **Offset From Base** for each.

> A single ceiling element can only slope in one direction.

Ceiling: Type Selector

Revit lists two ceilings types: **Compound Ceilings** and **Basic Ceilings**. The Basic Ceiling option should be avoided as it does not have any thickness and cannot have certain elements hosted to it.

---

quick steps

Ceiling

1. Ribbon
   - Automatic Ceiling
     - Finds *Room Bounding* elements at the specified ceiling height
   - Sketch Ceiling
     - Enter sketch mode with draw tools available
2. Options Bar (only for Sketch Ceiling)
   - Chain
   - Offset
   - Radius
3. Type Selector
   - Select Ceiling Type
4. Properties
   - Verify Level
   - Ceiling Height
5. In-Canvas
   - TIP: Work in Ceiling View so results are visible
   - Select ceiling grid to Move and Rotate pattern
   - Select ceiling edge to delete
   - Esc key or Modify tool to cancel

Architecture

The compound Ceiling tool can have multiple structural Layers similar to floor and roof elements.

Ceiling: Properties

- **If nothing is selected**
  Overall ceiling properties are displayed
- **If a sketch line is selected**
  Sketch line properties shown: Offset, extend into wall (to core)
- **If a slope arrow is selected**
  Slope properties displayed; specify slope vs. height, slope

Make sure the ceiling is relative to the correct level and set the **Height Offset From Level** prior to clicking and placing the ceiling. If the height is not correct prior to placement, the ceiling may go under or over soffits, bulkheads or adjacent walls. Use the **Offset From Wall** parameter to maintain a fixed distance from an adjacent, parallel, wall.

Ceiling: In-Canvas

Place bulkheads and soffits first if possible. All elements at the perimeter of the ceiling must be set to **Room Bounding** in the Properties Palette.

See this author's article **Soffit and Bulkhead Modeling in Revit** on AECbytes website: http://www.aecbytes.com/tipsandtricks/2012/issue64-revit.html

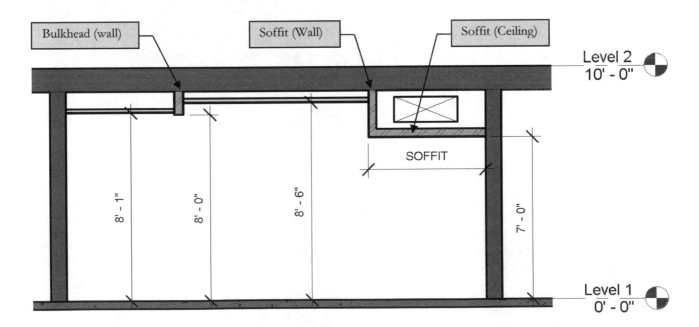

Figure 3.1-20 Section showing bulkhead and soffit conditions

Floor: Architectural

The floor tool is used to create a floor element relative to the level associated with the current plan view. The Floor may consist of one or more layers of materials. Sometimes the floor's overall thickness includes space for the structure (e.g. joists or beams). However, the Floor tool is not able to show any actual structural elements—they must be modeled separately with the Structural Framing tool.

Floor: Ribbon

Use the **Boundary Line** draw tool and sketch the perimeter of the floor in a plan or 3D view. The default method is using **Pick Walls**, which automatically constrains the sketch lines to the selected walls (FYI: see Options Bar section below). Use any of the line tools to manually define the edge.

Use the **Slope Arrow** feature to define a slope for the entire floor element. When the arrow is selected, use the Properties Palette to specify the desired slope relative to the current level (and offset).

The first line sketched defines the **Span Direction** (which is the two lines at the midpoint). This is used to determine which way the flutes run when metal deck is specified. Use the Span Direction tool to select a different sketch line, which will change the direction of the flutes (which typically run parallel to the

quick steps

Floor: Architectural

1. Ribbon [in sketch mode]
   - Draw *Panel*
     - Sketch enclosed boundary or add slope arrow.
     - Segment or Pick Options
     - Span Direction
   - Work Plane *Panel*
     - Set/Show Work Plane
     - Create Reference Plane
     - Viewer
2. Options Bar (only for Sketch Ceiling)
   - Offset
   - Extend into wall (to core)
3. Type Selector
   - Select Floor Type
4. Properties
   - Level & Offset
   - Room Bounding
   - See Warning
5. In-Canvas
   - First segment defines *Span Direction* by default
   - FYI: Top of floor assembly aligns with level
6. Green Checkmark to Finish Sketch
   - Objective is to sketch enclosed outline in a plan or 3D view

Architecture

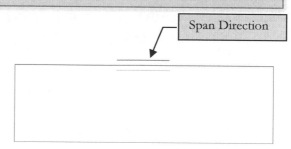

Figure 3.1-21
Floor sketch with span direction

supports; i.e., joists). Flutes in a floor element will be visible in section views when the detail level is set to Medium or Fine.

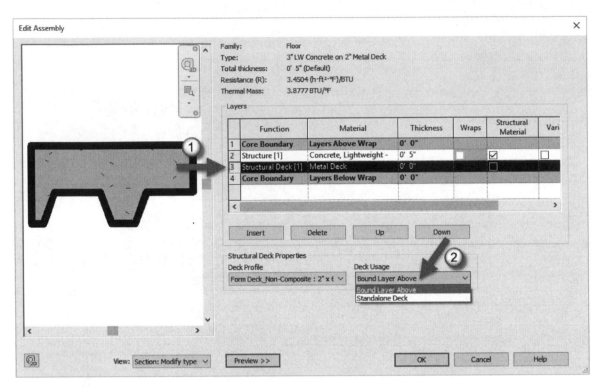

Figure 3.1-22 Floor structure with metal deck – orientation tied to Span Direction

The **Set** tool, in the Work Plane panel, can be used to specify a different level for the floor element about to be created.

Floor: Options Bar

When the Pick Walls draw option is selected, the Options Bar offers Offset and Extend into wall (to Core). The "core" is the structural portion of the wall as defined in the Edit Structure part of the walls' properties. The options vary slightly based on the selected draw tool.

Floor: Type Selector

Select from the floor types defined in the current project. Click **Edit Type** and then **Duplicate** to create new floor assemblies (i.e., Types).

Floor: Properties

Use properties to set Phase, Room Bounding, etc.

- **If nothing is selected**
  Overall floor properties are displayed

- **If a sketch line is selected**
  Sketch line properties shown: Offset, extend into wall (to core)
- **If a slope arrow is selected**
  Slope properties displayed; specify slope vs. height, slope

Floor: In-Canvas

The goal is to sketch the perimeter of the floor. Any areas sketched within the largest perimeter will define a hole in the floor. The edge of the slab is always flat. Use the separate Slab Edge tool to add a slope or a thickened edge.

Floor: Structural

See Chapter 4 – Structure Tab for coverage of this feature.

Floor by Face

See Chapter 9 – Massing & Site for coverage of this feature.

Floor: Slab Edge

The Floor element has a simple square edge. If something else is needed, you can use the Slab Edge tool to accomplish it.

The Slab Edge tool basically adds to the floor material. This addition can be above, below, to the side or combination.

This tool involves extruding a **Profile** along one of the edges of a floor. A profile is selected in the Slab Edge type properties. Continuing to select adjacent slab edges will create a continuous, mitered transition.

A profile is a simple enclosed 2D sketch created in the Family Editor. This can be started via **File Tab →** **New → Family → Profile** (template file).

Slab Edge: Ribbon

After selecting a floor edge to place a slab edge, the default option is to only select adjacent edges to create a continuation of the first segment. Clicking the Restart Slab Edge tool basically restarts the command.

Slab Edge: Type Selector

Select from the defined Slab Edge types. Simply loading a profile does not create a new Slab Edge type. To create a new type, click Edit Type and then Duplicate.

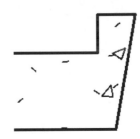

Figure 3.1-23
Slab edge added to floor

Slab Edge: Properties

The vertical and horizontal offsets allow the position of the Slab Edge to be adjusted. Additionally, the angle can be modified and the rebar cover can be adjusted.

Slab Edge: In-Canvas

The top or bottom edge of the floor can be selected. The Slab Edge should automatically join with the host floor. If it does not, use the Join command on the Modify tab.

Curtain System (by face)

The **Curtain System** tool is used to attach a Revit Curtain System type to the face, i.e., a single side, of a previously created Mass element. This command is repeated here for convenience but more closely relates to the massing tools found on the **Massing & Site** tab. Thus, please reference *Chapter 9 – Massing & Site* for detailed coverage of this feature.

TIP: The most common way to create a curtainwall is by using the Wall tool and then selecting a curtain wall type from the Type Selector.

Curtain Grid

The Curtain Grid tool is used to divide a Curtain Panel, horizontally or vertically, within a Curtain Wall. Curtain Grids also host Mullions.

Curtain Grid: **Ribbon**

With the Curtain Grid tool active, the division is across the entire wall by default—**All Segments** is selected on the Ribbon. To limit the division to one area, i.e., between two other Curtain Grids, select the **One Segment** option.

Another option is to use **All Except Picked**, which will add the

quick steps

> **Curtain Grid**
>
> 1. Ribbon
>   - Placement Options
>   - All Segments
>   - One Segment
>   - All Except Picked
> 2. In-Canvas
>   - Typically work in elevation, section or 3D view.
>   - Select the edge of a curtain wall to add a Curtain Grid.
>   - Select vertical edge to add horizontal curtain grid.
>   - Select Horizontal edge to add vertical curtain grid.

Curtain Grid across the entire wall but omit specific sections. To use this option, the location of the division is specified and then the section(s) to be omitted is selected. Finally, the **Restart Curtain Grid** button is selected to finish the command.

Curtain Grid: **In-Canvas**

Moving the cursor across the sides of the wall will create a horizontal division. The cursor positioned along the top or bottom of the wall will place a vertical division.

Angled Curtain Grids cannot be added directly.

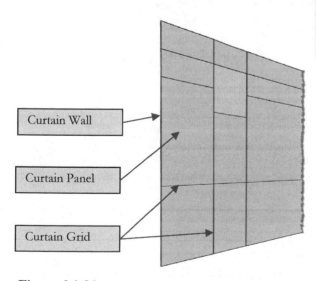

Figure 3.1-24
Curtain Grids added to Curtain Wall

Curtain Wall

Curtain Panel

Curtain Grid

Mullion

A Mullion represents the aluminum framing in a curtain wall system—which is actually the structural part of the wall. A Curtain Grid is required prior to placing a Mullion, similar to a wall being required prior to placing a door.

Mullions are centered on all Curtain Grids and aligned with the edge of the wall itself.

Mullion: **Ribbon**

The **Grid Line** option will place a Mullion along the full length of a Curtain Grid (except where the grid has been excluded). The **Grid Line Segment** option will only add a Mullion between two perpendicular Curtain Grid lines. The last option, **All Grid Lines**, will place the selected Mullion at all visible Curtain Grids and edges. If all the mullions are the same, then the last option is the most efficient, even if a segment or two needs to be deleted to accommodate door openings.

quick steps

Mullion

1. Ribbon
   - Placement Options
     - Curtain Grid
     - Grid Line Segment
     - All Grid Lines
2. Type Selector
   - Select Mullion Type
3. Type Properties
   - Angle
   - Offset
   - Profile
   - Position
   - Thickness
   - Width
4. In-Canvas
   - Typically work in elevation, section or 3D view.
   - Select Curtain Grid or Perimeter of curtain wall to place mullion.

Mullion: **Type Selector**

Select the desired mullion type from the list. To create a new option, select Edit Type and then Duplicate.

Mullion: **Type Properties**

These settings will adjust how the mullion looks in the model. The **Width** is divided into two parameters to control the amount on each side of the Curtain Grid—however, the full combined width is within the boundaries of the wall at the edge locations. A Profile can be specified if a custom shape is desired, such as a fin.

Mullion: **In-Canvas**

Mullions are typically placed in elevation, section or 3D view so the full extent of the wall can be seen. Corners do not automatically clean up as there are several variations of corner framing. See the Modify chapter for more information on adjusting the corner once the curtain wall has been placed.

Different size mullions can be added to the same Curtain Wall. In the image below, the verticals are 11" deep as they are the structural element withstanding lateral forces. The shorter horizontal elements are only 5" deep. Also, notice Mullions have been excluded in favor of a glazed butt joint above and below the colored glass.

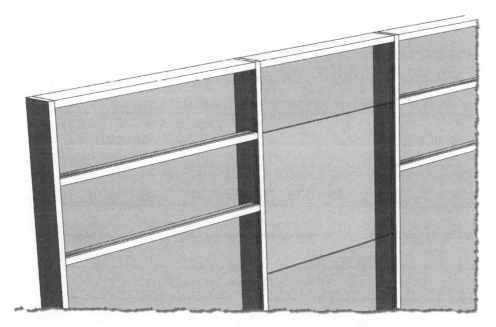

Figure 3.1-25 Curtain wall with various mullions added and excluded

For more on Curtain Walls, see this **BIM Chapters** blog post:
https://bimchapters.blogspot.com/2017/09/custom-curtainwall-and-types.html

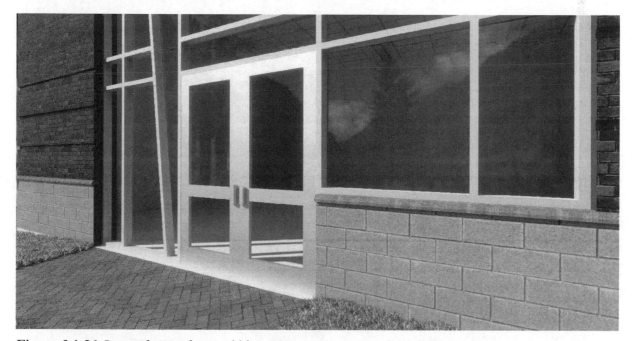

Figure 3.1-26 Image from referenced blog post on custom curtain walls in Revit

3.2 Architecture tab: Circulation panel

The circulation panel contains several tools related to vertical movement within a building: stairs, ramps and railings.

Railing: Sketch Path

The Railing tool is used to place a railing in the model. Railings are added at floor edges in atriums, at ramps and stairs. Railings can also be hosted to a topographic surface (e.g. fencing).

The Stair and Ramp tools have the option of placing a railing automatically.

Railings have a special Handrail and Top Rail which can be selected separately. These special sub-elements have their own type properties.

Railing: Sketch Path: **Ribbon**

Use the **Draw** tools to sketch one or more lines to define the path of the railing in the model. The sketch lines must form a continuous, uninterrupted path.

The **Pick Lines** option can be used to create a sketch line based on the edge of another model element previously created (e.g. a floor edge).

Use the **Pick New Host** option to select a stair or ramp to host the railing. By default the railing is hosted by the level associated with the current view.

quick steps

> **Railing: Sketch Path**
>
> 1. Ribbon [in sketch mode]
>   - Draw *Panel*
>   - Sketch connected path of railing
>   - Segment or Pick Options
>   - Work Plane *Panel*
>   - Set/Show Work Plane
>   - Create Reference Plane
>   - Viewer
>   - Tools *Panel*
>   - Pick New Host
>   - Edit Joins
>   - Options *Panel*
>   - Preview
> 2. Options Bar
>   - Chain
>   - Offset
>   - Radius
> 3. Type Selector
>   - Select Railing Type
> 4. Properties
>   - Base Level
>   - Base Offset
>   - Tread/Stringer Offset
> 5. In-Canvas
>   - Typically work in plan view
>   - Sketch connected path
>   - Path cannot close back on itself

The way in which railings come together at various angles can vary. Revit has two type parameters for railings, **Tangent Joins** and **Angled Joins**. Additionally, on the Ribbon, while in the railing tool, use the **Edit Joins** to modify the default solution.

Use the **Preview** option to see the full 3D railing while sketching. This option may be toggled on and off. This can be helpful to verify the offset or position of the railing in the current context of the model.

In the image below, Figure 3.2-1, the railing at the edge of the floor was added separately from the stair railing. It is possible to extend the stair railing, but in this case the railing needs to be modified at the edge of the floor condition. A steel tube shape was added to the bottom to continue the look of the stair stringer—for the stair; the stringer is part of the stair, not the railing.

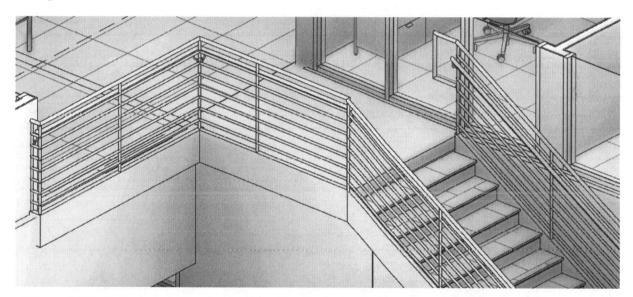

Figure 3.2-1 Railing on stair and separate railing (w/stringer element) at level 2 floor edge

Railing: Sketch Path: Options Bar

Use **Chain** to sketch multiple connected lines more efficiently. Use **Offset** to change the position of the sketch lines relative to the points picked on the screen. Negative numbers can be used, but simply pressing the space bar will flip the orientation while sketching. Use **Radius** to automatically "round off" corners while sketching.

Railing: Sketch Path: Type Selector

Select from the available railing types loaded in the project. Additional types can be created via **Edit Type → Duplicate**. To use railings from other projects, use one of these methods:

- **Copy/Paste:**
  Open current project and project containing desired railing type (both projects must be opened in the same session of Revit). Select a railing in the model and press **Ctrl + C**.

Switch to current project and press **Ctrl + V**. Click to place the new element. The pasted element can then be deleted, but the new railing type definition will remain.

- **Transfer Project Standards:**
  Open current project and project containing desired railing type (both projects must be opened in the same session of Revit). In the current project, select Manage → Transfer Project Standards. Click Check None and then ONLY check Railing Types. Click OK and then Click New Only, which will only import railing types that do not already exist in the current project.

Like Walls, Ceilings, Floors, etc., Railings are **System Families**. This means they can only exist within a project, not separately in an RFA file (like Loadable Families; e.g. Doors, Furniture, Light Fixtures). This is why the previous steps mentioned are required to access content in another project.

> In your browser, search for "**Revit Sample Stairs and Railings Files**" to see several complex examples which can be used directly or studied for tips on how to develop another railing design.

Railing: Sketch Path: **Properties**
If not hosted by a Stair or Ramp, the Base Offset can be used to adjust the vertical position of the railing.

Railing: Sketch Path: **In-Canvas**
Selecting a sketch line, while still in sketch mode, presents additional options on the Options Bar: **Slope** and **Height Correction**. The Slope option allows the control of when the railing continues to slope or flatten out past the edge of the stair or ramp. The Height Correction controls how the railing transitions between runs and landings.

The direction the railing is drawn, e.g. from left to right, determines the orientation; click the arrows while in sketch mode to swap start and end definition (Figure 3.2-2a & 2b). Selecting the railing after it is created reveals a flip icon to change the orientation if needed.

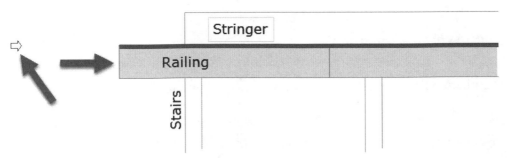

Figure 3.2-2a Toggle sketch orientation icon while in railing sketch mode; start point defined

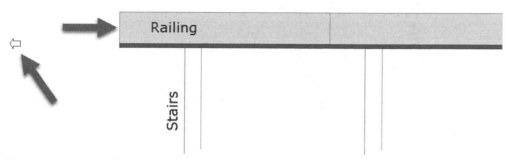

Figure 3.2-2b Toggle sketch orientation icon while in railing sketch mode; endpoint defined

Railings which are part of a multistory stair are grouped with the stairs and may be edited as a group.

Railing: Pick Host

The Stair and Ramp tools have the option of placing a railing automatically. However, these are separate elements from the Stair/Ramp and can be deleted. If deleted, use the **Place on Host** option to quickly recreate the railings as if they were created with the stair originally.

Railing: Pick Host: **Ribbon**

Toggle between **Tread** or **Stringer** prior to selecting the stair or ramp. On commercial projects, the railing is most often over the stringer.

Railing: Pick Host: **Type Selector**

Prior to selecting the stair or ramp, select the railing type desired.

Railing: Pick Host: **In-Canvas**

Click to select a stair or ramp in the model to place the railings. Seeing as this tool is meant to quickly replace the railings added during

quick steps

Raining: Pick Host

1. Ribbon
   - Position *Panel*
     - Treads
     - Stringer
2. Type Selector
   - Select Railing Type
3. In-Canvas
   - Select a previously created stair or ramp
   - Stair or ramp cannot have any hosted railings

the original creation of the stair/ramp, it cannot have any railings hosted to it—not even on just one side. Oddly, for Multistory stairs, this command works even if there is an existing railing on both sides of the stair.

Ramp

Use this tool to add a sloped walking surface to a model. This tool has several limitations and in most cases it is better to add a floor with a slope arrow.

Ramp: Ribbon

The **Run** option on the Ribbon allows the ramp to be created by picking the middle of the top and bottom of the ramp. The **Boundary** option allows the ramp to be defined by the two sides of the ramp; the top and bottom need to be closed off by adding **Riser** lines. *FYI:* the terminology here does not make sense for a ramp. The Boundary option should be used if one edge of the ramp needs to follow an angled wall.

Use the **line/arc** option to add a straight or curved ramp segment.

Railings are added by default. Use the **Railing** tool to change the railing type or exclude it.

Ramp: Type Selector

Select the desired ramp type from the list.

Ramp: Properties

By default, the ramp is set to go from the current floor to the next level above. Adjust the **Width** as needed—this only applies when the ramp is created with the Run option.

Ramp: Type Properties

quick steps

Ramp

1. Ribbon [in sketch mode]
   - Draw *Panel*
     - Select Run (default)
     - Line or Arc segment
   - Work Plane *Panel*
     - Set/Show Work Plane
     - Create Reference Plane
     - Viewer
   - Tools *Panel*
     - Railing (optional)
2. Type Selector
   - Select Ramp Type
3. Properties
   - Base/Top Level settings
   - Label settings
   - Width (3'-0" default)
4. Type Properties
   - Shape
   - Thickness
   - Ramp Material
   - Max. Length
   - Ramp Max. Slope (1/x)
5. In-Canvas
   - Work in lowest level view relative to ramp settings

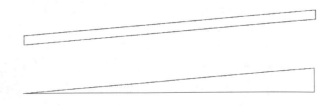

Figure 3.2-3 Ramp's Shape property; Thick vs. Solid

The **Shape** parameter toggles between a solid wedge versus a constant thickness along the slope (Figure 3.2-3). Setting the **Max. Length** and **Max. Slope** help to keep the ramp code compliant—adjust these values to match local requirements.

Ramp: **In-Canvas**

Add the ramp in the lower of the two levels defined for this element.

Stair

Stair allows the designer to pick two points in the model and instantly see a run of 3D stairs added between them.

Stairs are a **System Family** and can only exist within a project. Thus, a library of custom options must exist in a project template or copy/pasted from another project file.

Stair by Component: **Ribbon**

The **Run** option is used to sketch a run of stairs. Pick two points on the screen to define the location of the bottom (first pick) and the top (second pick).

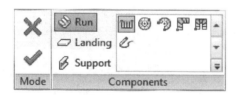

The sketch options shown in the image above allow various shapes (in plan view) to be drawn. Note that the L-shape and U-shape options would rarely be used as the riser layout created with this option does not meet any US building codes—apparently, these are for existing conditions or use in other countries. To create an L-shape or U-shape stair one would simply use the straight run (default) option, sketching the two runs manually (Revit will create the intermediate landing automatically).

quick steps

Architecture

Stair

1. Ribbon [in sketch mode]
   - Components *Panel*
     - Select from:
       i. Run (default)
       ii. Landing
       iii. Support
     - Stair shape (in plan)
   - Work Plane *Panel*
     - Set/Show Work Plane
     - Create Reference Plane
     - Viewer
   - Tools *Panel*
     - Railing (optional)
2. Options Bar
   - Location Line
   - Offset
   - Width
   - Automatic Landing
3. Type Selector
   - Select Stair Type
4. Properties
   - Base/Top Level settings
   - Actual Tread Depth
5. Type Properties (via Edit Type)
   - Maximum Riser Height
   - Minimum Tread Depth
   - Several "type" options
6. In-Canvas
   - Pick to start stair run
   - Number of risers created shown as cursor is moved

Stair: Options Bar

Similar to adding walls, the **Location Line** option allows the designer to predetermine which of the points picked on screen correspond to the width of the stair in plan view. The **Offset** option allows the final location line to be displaced from the points picked—for example, one could pick points directly along a wall with the resulting stair being offset from the wall a specified distance. The **Actual Run Width** is the width of the stair. For stairs with support (aka stringers), when the Type properties *Right Support* and *Left Support* are set to "Open," the width is between the stringers (this is the default in the Revit templates).

Stair: Type Selector

There are three fundamental types of stairs:

- Assembled Stair
- Cast-In-Place Stair
- Precast Stair.

The **Assembled Stair** type is a stair, commercial or residential, constructed with stringers, risers and treads. The **Cast-In-Place** (CIP) type would be a solid concrete stair used outside the building (i.e., site work) or in buildings with a CIP concrete

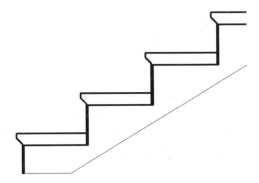

Figure 3.2-4 Assembled Stair type

structure. The **Precast Stair** type is composed of interlocking precast concrete components. Each type has slightly different properties, so switching from one system type to another, via the Type Selector, does not work well. It is possible, and often necessary, to have more than one Type associated with each of these three categories.

A project must have at least one type for each of these categories. Trying to delete the last type in any of the three main stair types will result in an error message indicating the last type cannot be deleted.

Stair: Properties

The number of risers and treads is based on the floor to floor height. Adjusting the **Base** or **Top Offset** changes that calculated result. One example of when to specify an offset would be for a stair which leads onto a roof. The roof insulation is above the structural floor/roof level. However, for the most part, it is best to have a level at each surface someone can walk on; conversely, as a best practice, don't use levels for anything else.

Building codes specify the minimum depth of a stair tread. However, they can be deeper as desired. The Actual Tread Depth value can be adjusted separately for each stair. If it is set below the **Minimum Tread Depth** specified in the **Type Properties**, a warning will be given.

Stair: Type Properties

To ensure building codes are met, we can set the **Maximum Riser Height, Minimum Tread Depth** and **Minimum Run Width**.

The component stair is based on a collection of other sub-components, each with their own types and varied settings. Here is a brief overview of each sub-component option:

Run Type	Tread/Riser thickness or profile, nosing, material
Landing Type	Same as run option, separate tread & nosing options
Right Support Type	aka Stringer; Material, Section Profile or generic width & depth
Left Support Type	aka Stringer; Material, Section Profile or generic width & depth
Cut Mark Type	Graphic indicates stair continues to floor above/below

The types listed above can be modified via **Project Browser → Families → Stairs**; right-click on a type and select **Type Properties...**

Stair: In-Canvas

Starting at the lower level, select a point to start the stair. The image below shows what we see prior to picking our second point. Notice the dashed line assures a horizontal position and the text below indicates the number of risers created and the number remaining if we clicked at that position. Creating a stair run short of the Top Level will result in a landing being automatically added if we continue to click points to draw additional stair runs, otherwise we get a warning if the stair is finished before reaching the Top level.

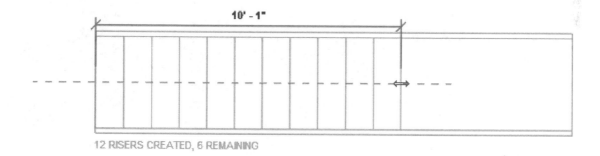

12 RISERS CREATED, 6 REMAINING

With the Stair tool, a triple (or more) switch-back stair can be drawn. This often occurs in a building with large floor-to-floor dimensions, where the stair turns back and forth multiple times before reaching the next level. To do this, simply draw short stair Runs that turn back and forth directly on top of each other. In plan the lower Runs may not be visible but will look correct in sections and 3D views.

Stair by Sketch

This command has been removed from Revit in 2021. However, when this stair type is upgraded with an old project, the steps listed here are still used to edit that stair. But new stairs of this type cannot be created.

Stair by Sketch: **Ribbon**

Use the Draw options to manually sketch the stair's risers and stringers (boundary).

Stair by Sketch: **Type Selector**

Select from predefined stair types in the current project.

Stair by Sketch: **Properties**

Specify the Base/Top level of the stair. The stair's width and annotation options are also controlled here.

Stair by Sketch: **Type Properties**

Define the geometry and rules for creating the stair.

Stair by Sketch: **In-Canvas**

When the Run option is selected, simply click two points to define a stair run. Otherwise, switch to either Boundary or Riser to manually sketch the stair in a plan view.

quick steps

Stair by Sketch

1. Ribbon [in sketch mode]
   - Draw *Panel*
     - Select from:
       i. Run (default)
       ii. Boundary
       iii. Riser
     - Stair shape (in plan)
   - Work Plane *Panel*
     - Set/Show Work Plane
     - Create Reference Plane
     - Viewer
   - Tools *Panel*
     - Railing (optional)
2. Type Selector
   - Select Stair Type
3. Properties
   - Base/Top Level settings
   - Label settings
   - Width
   - Tread Depth
4. Type Properties (via Edit Type)
   - Maximum Riser Height
5. In-Canvas
   - Pick to start stair run
   - Number of risers created shown as cursor is moved

3.3 Architecture tab: Model panel

The title of this panel, Model, may be confusing as the previous two panels, Build and Circulation, are modeled elements within the building. However, the term "Model" here is meant to contrast the 2D (and view specific) version of Text, Lines and Groups. Model Text, Lines and Groups will appear in the view they were created in <u>and</u> any other view (plan, section, elevation, 3D) they may be visible in.

Model Text

Model Text is 3D text which can appear in multiple views and renderings. This command is often used to represent text which actually exists in the final, built project. This command should not be used to add general notes and titles to the drawings.

The rendered image below shows Model Text used to create a company sign.

quick steps

Model Text

1. Edit Text dialog
   - Type in desired text
2. In-Canvas
   - Pick position in current view
   - Select/Change model text Type after placement

Model Text: **Edit Text dialog**

The **Edit Text** dialog is where the desired text is entered. The paragraph format is not editable in the model, so an **Enter** must be pressed for each new line required.

Model Text: **Type Selector**

Once placed, select the text and change the type via the Type Selector drop-down if desired. It is not possible to select the type during placement.

Model Text: **Properties**

Once placed, select the text and edit the material, depth and alignment if needed.

Model Text: **In-Canvas**

Click to place the text in the current view.

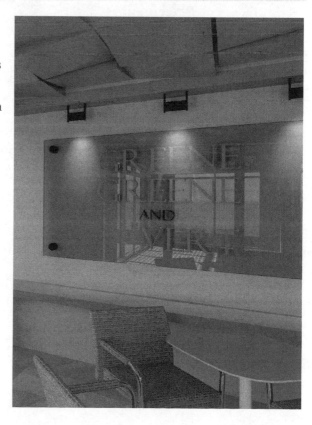

Model Line

A **Model Line** is a line that exists in 3D space, meaning it will not only appear in the elevation view it was drawn in, but also sections and 3D views of the same area. Contrast this with a **Detail Line** which only appears in the view it was created in. Keep in mind that Model lines are not 3D geometry and therefore will not appear in renderings.

Model Line: **Ribbon**

Use the **Draw** panel to control the shape of the linework. Select from the **Line Style** list to control the line's appearance.

Model Line: **Options Bar**

A Model Line is either associated with a Level, Reference Plane, or a select surface. The **Placement Plane** drop-down list on the Options Bar is used to define how the line is hosted.

quick steps

Model Line

1. Ribbon
   - Draw *panel*
     - Line Segment to Sketch
     - Pick Lines
   - Line Style *Panel*
     - Select
2. Options Bar
   - Placement Plane
   - Chain
   - Offset
   - Radius
3. Properties
   - Moves with Nearby Elements
4. In-Canvas
   - Objective: Add line segments which will appear in the 3D model.

Model Line: **Properties**

Moves with Nearby Elements really means move with nearby walls. When this option is checked, the Model Line will have an association with the closet wall, which will be maintained if the wall is moved. When the line is moved, the "nearest wall" selection is reevaluated. If another wall is now closer, that closer wall will be used.

Model Line: **In-Canvas**

Sketch lines using basic draw techniques.

Place Model Group

The **Place Model Group** command will load a collection of model elements previously grouped using the **Create Group** command (the next command to be covered).

Place Model Group: Type Selector

All Model Groups previously created with the **Create Group** command will appear here in the **Type Selector** list.

This list of groups can also be seen in the **Project Browser**, under the Groups heading.

> Model Groups can be dragged from the Project Browser into the drawing area.

Place Model Group

1. Type Selector
   - Select Model Group to Place
2. In-Canvas
   - Insertion point

Place Model Group: In-Canvas

During placement there are no special options. Note that the insertion point, at your cursor location, corresponds to the **X/Y Axis** icon pointed out in the image below.

> When a Group is selected, the **X/Y Axis** icon can be repositioned. This will affect the placement point of future instances of the Group but will not cause existing instances to move.

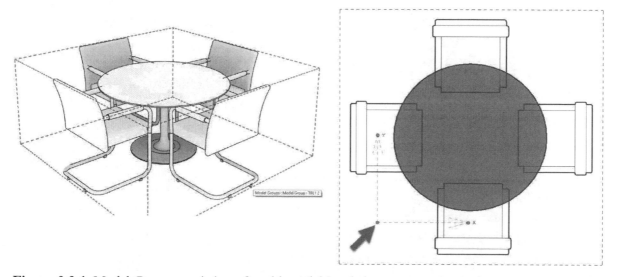

Figure 3.3-1 Model Group consisting of a table and four chairs

Create Group

With multiple model elements selected, clicking the Create Group command will create a named collection of elements. These are the benefits of groups:

- Everything in the group moves together
- The group can be placed multiple times
- Changes to one group update all instances

Elements within the group can still be tagged and scheduled. Plus, individual elements in a group can be omitted, which helps reduce the total number of groups needed.

Create Group: **Create Group dialog**

Provide a meaningful name which can be used to place additional instances of the group (Figure 3.3-2).

Check the box for **Open in Group Editor** if changes are needed right away.

Create Group: **Group Edit Mode**

If entering Group Edit mode, a floating toolbar appears (Figure 3.3-3). Click Finish or Cancel to exit Group Edit mode.

Create Group: **In-Canvas**

Click to place group in model.

Phasing can be a little tricky when working with Groups. The elements within a Group cannot have pre-assigned phase settings. Rather, when a Group is placed, the phasing is set based on the phase setting of the current view. Once placed, the Phase and Phase Demolished settings can be adjusted for each element within a group. However, if that modified Group is copied, the phase settings are not maintained.

quick steps

Create Group

1. Create Group dialog
   - *Nothing Selected first:*
     - Provide Name
     - Open in Group Editor option
   - *Model elements selected first:*
     - Provide Name
     - Group Type
2. When in Group Edit mode:
   - Floating Edit Group panel
   - Add/Remove/Attach elements to current Group
   - Finish/Cancel when done
3. In-Canvas
   - All elements outside of the current group cannot be selected until edit mode is exited

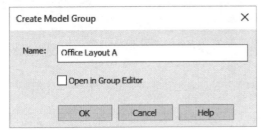

Figure 3.3-2 Create Model Group dialog

Figure 3.3-3 Group edit toolbar

Load as Group

The **Load as Group** command will take the ENTIRE current project and load its model elements as a Group into another open project.

For example, you have a Revit project with a table and chairs – which are not grouped. Selecting Load as Group will create a group of these elements within another open project. If more than one project is open, the user is prompted to select which project to load the new group into.

> In Visibility/Graphic Overrides, turn on Site→ Project Base Point to see where the insertion point for this group will be.

Load as Group: Load into Projects dialog

In a project to be loaded into another project as a group, select the **Load as Group** command. The dialog shown to the right appears (Figure 3.3-4). Select the project(s) to load the new group into.

> Use the **Load as Group** command to maintain a group in multiple projects. For example, a campus project with multiple Revit models.

Attached Detail

Groups non-model elements associated with the 3D model elements in a group. So checking this would include a view-specific 2D group with the main 3D "model group" being loaded.

Levels

If you want the Levels from the host project file to be loaded with the Model Group, check this box. When not checked, the elements within the group are re-associated with the levels in the destination project. It is bad practice to have multiple levels on top of each other, so this option is usually not checked.

quick steps

Load as Group

1. Load into Projects dialog
   - Select open family or project to load current project into as a Group.
   - Include options:
     - Attached Details
     - Levels
     - Grids
2. In other project
   - Use Place Group tool

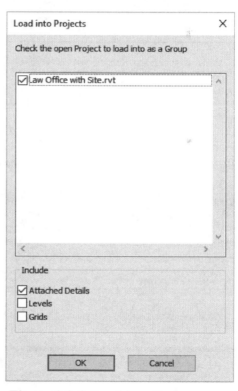

Figure 3.3-4 Load into Projects dialog

Architecture

If the host model has Grids which are not already in the destination model, check this box.

Load as Group: **In other project**

If the Group name already exists in the destination project, the Duplicate Group Names dialog will appear (Figure 3.3-5). Selecting **Yes** will update all instances of the Group in the destination project.

Once the Load as Group command is complete, the destination model is set as the current model/view. However, the Place Model Group is not active. In some cases, the **Load as Group** command is just used to push updates to multiple models so there is no need to start the placement command.

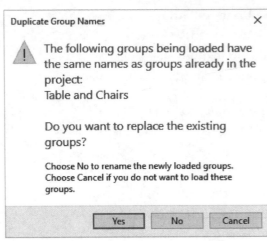

Figure 3.3-5 Duplicate Group Names

3.4 Architecture tab: Room and Area

The commands on the Room and Area panel allow the designer to track and maintain information related to each room within the building. This includes area, volume, finished and more.

Architecture

Room

A Revit **Room** element, when placed, automatically finds the perimeter of a room in the Revit model. Anytime the model is changed, the Room element will automatically update to conform to the perimeter of the room. This allows Revit to report the area (e.g. **square feet**) of each Room.

The **Volume** can also be tracked if **Areas and Volumes** is selected in the **Area and Volume Computations** dialog (covered later in this section).

All elements used to define the edge of the Room must be set to **Room Bounding**—an instance parameter associated with all system families (i.e., Walls, Floors, Ceilings and Roofs).

Click **Room** on the Ribbon to start the command.

Room: Ribbon

Clicking the **Place Rooms Automatically** sub-command will place a Room in every enclosed space in the current view. This often places more Rooms than desired, like in the plumbing chase between two toilet rooms.

quick steps

Room

1. Ribbon
   - Highlight Boundaries
   - Tag on Placement
2. Options Bar
   - Upper Limit
   - Offset
   - Tag Orientation
   - Leader
   - Room:
     - New
     - Unplaced
3. Type Selector
   - Room tag
4. In-Canvas
   - Click within enclosed area
   - Previous Rooms highlight when tool is active
   - Note on phasing
   - Esc key or Modify button to exit the Room tool
   - FYI: Rooms are only removed from the plan view using delete/erase—use a schedule to permanently delete a Room from a project.

Clicking the **Highlight Boundaries** option causes all elements set to Room Bounding to highlight.

When **Tag on Placement** is toggled on, meaning it is highlighted, a Room Tag will be placed in addition to the Room element in the current view. Note that this is an element separate from the Room element and can be deleted without affecting the Room. Also, all other views in the project will not have a Room Tag even though the Room can be seen. The Room is a 3D element, while the Room Tag is 2D and view specific.

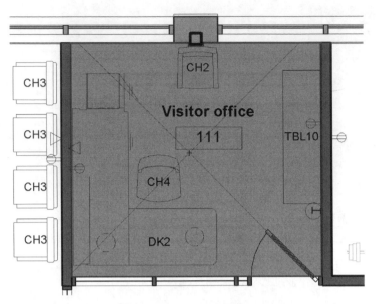

Figure 3.4-1 Selected Room element

Unfortunately, the Room element cannot be seen in 3D, only in plan and section/elevation views.

Room: **Options Bar**

The **Upper Limit** drop-down controls which Level to reference when determining the height of a room. Unlink the way Rooms work in plan view; the height of a Room is not automatic. In conjunction with the Upper Limit setting, the **Offset** value specifically controls the height of a Room. When **Areas and Volumes** is selected in the **Area and Volume Computations** dialog (covered later in this section) the height of a Room element will stop at any ceiling or Roof element it encounters. However, if a Room stops short of a ceiling or roof, the Room will NOT automatically extend up to it.

> Best Practice: Set the Room's **Upper Limit** to the Level above with an **Offset** of 0'-0". This helps to ensure the Room element will always engage the ceiling and not fall short.

The next two options, **Tag Orientation** and **Leader**, only apply when Tag on Placement is toggled on. Both of these options can be modified later.

The **Room** drop-down allows un-placed Rooms to be placed in the model, otherwise the default is to create a new Room. An un-placed room is a Room element that was created in a room schedule or the remnant of a Room deleted from the model.

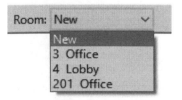

> Rooms can only be permanently deleted from a schedule.

Room: Type Selector

Use the **Type Selector** to pick the desired Room Tag to place with the Room. The selection is irrelevant if Tag on Placement is not active.

Room: In-Canvas

Click within an enclosed area to place a Room. The **point picked** defines the location of Room Tags when using the Tag All command in other views. It also determines which area the Room ends up in if new walls are added.

While in the Room command, the **previously placed Rooms highlight** in the model—never place multiple Rooms in the same area.

> The Room highlight can be turned on manually in any view via the Visibility/Graphics Override dialog and then checking the Interior Fill and Reference sub-categories under the Rooms category.

The **Phase** of a Room is hardwired based on the phase setting of the view in which the Room was placed, and not editable. Each Phase must have separate Rooms added—which must be carefully managed as this can lead to errors/conflicts.

Use the **Esc** key to exit the Room command.

One final note, **Rooms** are generally used by Architects and Interior Designers. Revit has a similar element, called **Spaces**, which is primarily for engineers.

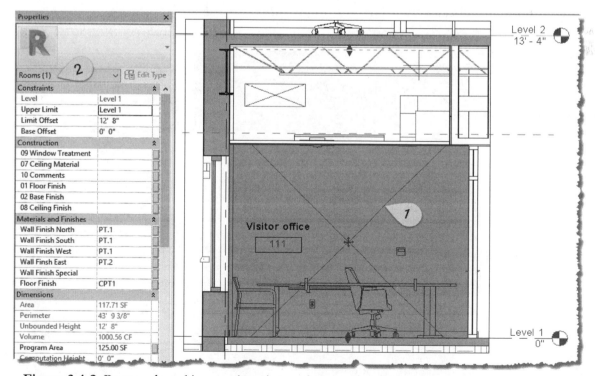

Figure 3.4-2 Room selected in a section view and properties shown

Architecture

Room Separator

When an area needs to be tracked separately, for example, an alcove or a waiting area, use a **Room Separator** line. A Room Separation line acts like a **Room Bounding** element, even though it is not a wall.

> Control the visibility of a Room Separator line in Visibility/Graphic Overrides, **Model** tab → **Lines** → **<Room Separation>**

Room Separation lines should never overlap other Room Separation lines or **Room Bounding** elements as this causes Warnings and can affect model performance.

quick steps

Room Separator

1. Ribbon
   - Draw panel
     - Line segment
     - Or Pick Lines
2. Options Bar
   - Chain
   - Offset
   - Radius
3. Properties
   - Moves With Nearby Elements
4. In-Canvas
   - Add to open areas in order to place separate Room objects (i.e., room name and number)
   - Do not add over other Room Bounding elements

The Reception area shown below is defined by <u>two</u> **Walls** and <u>two</u> **Room Separation** lines.

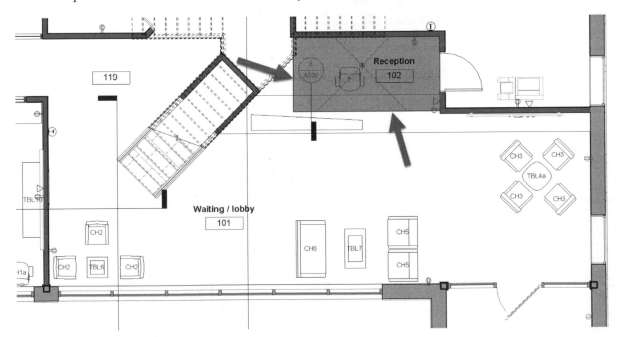

Figure 3.4-3 Room partially defined by Room Separation lines

Room Separator: **Ribbon**

Select which type of **line segment** is to be drawn. The **Pick Lines** option will add a line based on another line from a CAD file or the edge of another Revit element. Be careful not to cause Room Separation lines to overlap other Room Separation lines or Room Bounding elements.

Room Separator: **Options Bar**

The **Chain** option allows for continuous sketching. The **Offset** option will displace your picked points by the Offset value entered; negative values are allowed. Checking the **Radius** option, and then entering a value, will result in an arc being added at corners while sketching lines.

Room Separator: **Properties**

Checking **Moves With Nearby Elements** will cause the line to move with the nearby wall whenever it is moved.

Room Separator: **In-Canvas**

While using an Offset on the Options Bar, pressing the **Space Bar** will flip the offset side. Room Separation lines should not overlap walls or other Room Separation lines.

In a Revit MEP model, engineers use Revit's **Space** command, which is similar to the Room command. The Space is constrained by Room Bounding Elements and Room Separation lines contained within the linked in architectural model. Thus, an excessive use of Room Separation lines could have a negative effect on the Revit MEP model. A meeting with the MEP team should be had to verify their needs.

Room Tag

See chapter 6 – Annotation chapter for coverage of this tool.

Tag All Not Tagged

See chapter 6 – Annotation chapter for coverage of this tool.

Area Plan

An **Area Plan** is used to track square footage separately from, or in addition to, what can be done with the Room or Space elements contained in each room (assuming they have been placed in the model).

An Area Plan is similar to the Floor Plan—it is a live cut through the 3D model. However, the Area command (covered next) does not follow Room Bounding elements. Rather, an Area element will only follow Area Boundary lines. This offers a few pros and cons…

Pros:

- Specific areas may be defined such as daylight levels for LEED credit tracking.
- Can derive the gross area of a building, which includes all the exterior wall—not just to its centerline as with a Room's area.
- The area for multiple rooms, e.g. a Department, can be tracked with an Area element.

Cons:

- Depending on how Area Boundary lines are sketched, it is possible they will not update with the Revit model automatically.
- Area information is view specific and cannot appear in any other views.

Area Plan: New Area Plan dialog

Clicking to create a new **Area Plan** opens the New Area Plan dialog (Figure 3.4-4). The two options are Type and Level, which are explained next.

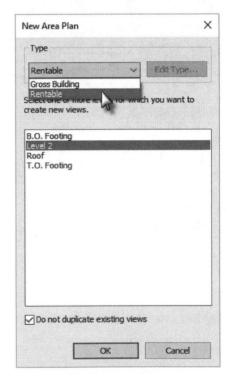

Figure 3.4-4
New Area Plan dialog

Type

The Type drop-down list provides two options by default. The Gross Building option is used for tracking all area within the extents of the building—including the walls. In contrast, the Rentable area option tracks usable area—excluding the walls.

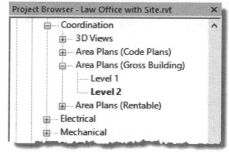

Figure 3.4-5 Area Plan Types

When multiple types exist, they are also segregated in the Project Browser as seen in Figure 3.4-5. New Types can be created via the Area and Volume Computations dialog as shown in Figure 3-4.6. All Types follow one of two rules: Gross or Rentable area.

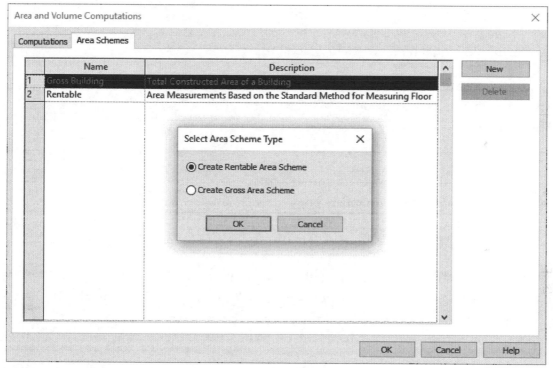

Figure 3.4-6 Area and Volume Computations dialog

Level Selection

Selecting a Level tells Revit which floor the Area Plan will be based on. When working with multiple phases, the Area Plan is created based on the Phase of the current view—but the Phase setting can be changed later if needed.

An **Area Plan** is no different than a regular **Floor Plan** with this exception: we can only place Area Boundaries and Areas in Area Plans, not floor plans. The view has the same settings (e.g. View Range, Detail Level) and the building elements can be created and modified.

> Keep in mind a Level is what we see in an elevation or section, i.e., a Level datum, and not necessarily a named floor plan view in the project.

Once a Level has an Area Plan associated with it, that Level does not appear in the list when creating a new Area Plan. Notice back in Figure 3.4-4 that Level 1 is not in the list. It is possible, however, to create multiple Area Plans based on the same level for the same Type. Simply uncheck **Do not duplicate existing views** near the bottom of the New Area Plan dialog. This will force all Levels to appear.

Creation Prompts

Once the **Type** and **Level** are selected, click **OK** to create the Area Plan.

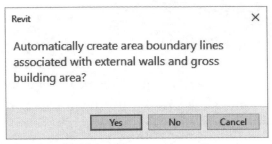

Figure 3.4-7 Area Plan creation prompt

For "gross area" plans, Revit will present the prompt shown in Figure 3.4-7. Clicking Yes will automatically create Area Boundary lines around the perimeter of the building at the exterior face.

For "rentable area" plans, Revit will offer the dialog shown in Figure 3.4-8. Clicking Yes will place Area Boundary lines at the centerline of the exterior walls.

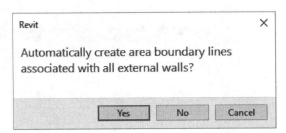

Figure 3.4-8 Area Plan creation prompt

Both options provide a quick way to get started with the process of adding Area Boundary lines. And if the goal is to quickly derive the Gross Area of a building then the only steps left would be to add an Area and Area Tag.

> Area Boundary lines created automatically or with the Pick Lines option will automatically update when the model changes. Area Boundary lines can also be Aligned and Locked manually if sketched with the other Draw tools.

Area

An **Area** is bounded by **Area Boundary** lines. The Area element is similar to the Room or Space, except it is 2D and only appears in Area Plans. Areas are not bounded by walls; thus, they can be used to derive the area of a Department which consists of several rooms.

The Area element will automatically adjust if the perimeter, consisting of Area Boundary lines, changes in any way. If there is an open gap in the perimeter, the Area element will become a 2'x2' square box until a gapless perimeter is detected.

The Area element has parameters to hold data, plus custom parameters may be created. These parameters can be seen in the **Properties Palette** when an Area is selected.

quick steps

Area

1. Ribbon
   - Tag on Placement
2. Options Bar
   - Tag Position
   - Leader
   - Area:
     - New
     - Unplaced
3. Type Selector
   - Area tag
4. In-Canvas
   - Click within enclosed area
   - Previous Areas highlight when tool is active
   - Esc key or Modify button to exit the Area tool

Architecture

> A quick way to get area, without creating an Area Plan, is to create a Filled Region. When a Filled Region is selected, its area is listed in the Properties Palette.

Area: Ribbon

When the Area command is active, the **Tag on Placement** toggle is present. When selected, an Area Tag will be placed with each Area.

Area: Options Bar

If **Tag on Placement** is selected on the Ribbon, then **Tag Orientation** and **Leader** options may be adjusted. A tag can be Horizontal, Vertical or set to Model; the Rotate tool can be used on the latter, allowing any angle. An Area Tag cannot be moved outside of an Area unless Leader is checked—otherwise the value/results are replaced with question marks.

The **Area** drop-down list, on the Options Bar, allows previously created and unplaced Areas to be placed. Areas can be created in a schedule before being placed in the

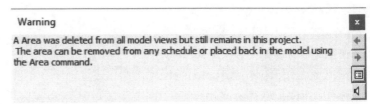

model. Also, if an Area is deleted from the model, it is not deleted from the BIM database. An Area element can only be completely deleted from a schedule.

Area: Type Selector

The **Area Tag** Type can be selected from the **Type Selector** while placing an Area—assuming **Tag on Placement** is toggled on via the Ribbon. The tag type can easily be changed later.

> Use the **Loaded Tags and Symbols** dialog to preset the default Area Tag. This is found at Annotate → Tag Extended Panel → Loaded Tags and Symbols.

Clicking **Edit Type** provides a way to change the Leader arrow symbol for each Area Tag type (Figure 3.4-9).

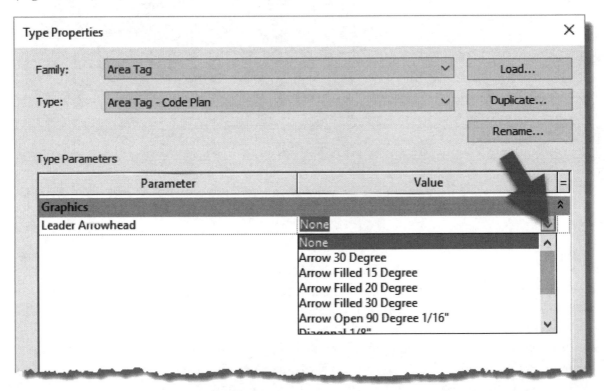

Figure 3.4-9 Type properties for Area Tag

Area: In-Canvas

When the Area command is active, all the previously created areas are highlighted so it is clear where Areas can be placed—it is considered a mistake to place two Area elements within the same area boundary; Revit will display a warning but will still allow it. However, the subsequent areas will say "Redundant" for the square footage.

The same highlights which are visible while the Area command is active can be turned on manually if desired. In the Area Plan view, type **VV** and then **check** preferred options under Areas as shown in Figure 3.4-10.

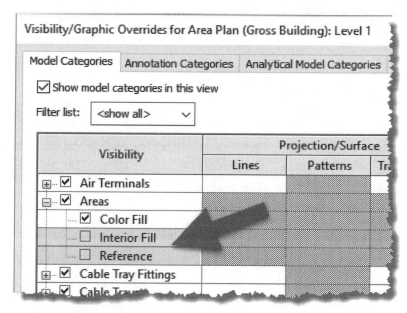

Figure 3.4-10 Graphic settings for Area Plans

Architecture

Area Boundary

The **Area Boundary** command allows lines to be sketched, in an Area Plan, which define the extents of an Area element.

The **Area Boundary** does not need to be perfectly enclosed. Small gaps are ignored when the Area element is searching for a perimeter.

Room Boundary: **Ribbon**

The typical **Draw** tools are available for sketching lines in the Area Plan. Using the default **Pick Lines** option will create a parametric relationship with any walls selected.

Room Boundary: **Options Bar**

The **Apply Area Rules** applies when using the **Pick Lines** on **walls**. When this option is checked, the position of the line, within a wall, will move if conditions change. For example, if a room is added on the side of a building, the Area Boundary line will reposition from the exterior face to the center line---and just at that room (Figure 3.1-45).

> **quick steps**
>
> **Area Boundary**
>
> 1. Ribbon
>   - Draw panel
>   - Line segment
>   - Or Pick Lines
> 2. Options Bar
>   - Apply Area Rules
> 3. Properties
>   - Moves With Nearby Elements
> 4. In-Canvas
>   - Add to open areas in order to place separate Area elements
>   - Do not add over other Room Bounding elements
>   - FYI: Areas are only 'removed' from the plan view using delete/erase—use a schedule to permanently delete an Area from a project.

Room Boundary: **Properties**

Each Boundary Line can have the **Moves With Nearby Elements** checked via the Properties Palette, which really means **moves with nearby <u>walls</u>**. This is another way to cause the Area Boundary lines to automatically update when a wall is moved within the model. The wall closest to the Area Boundary line, when created, is the only wall it will track.

Room Boundary: **In-Canvas**

Area Boundary lines can cross each other, but they should not overlap; this will lead to **Warnings** (Manage → Warnings) which can impede performance.

The Area Boundary lines can be made thicker or a different color so they are easier to see in relation to the walls. This is done via Manage → **Object Styles**.

In the image below, there is only a single Area Boundary line along the longer exterior wall. Notice how that single line breaks as needed to maintain the "Gross Area" area rules.

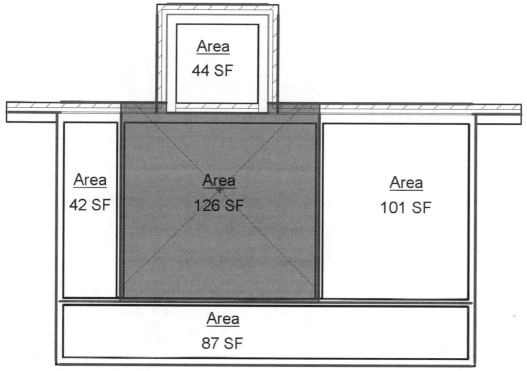

Figure 3.4-11 Area Rules for Area Boundary Lines

Use the Temporary Hide/Isolate tool to quickly see only the Area Boundary lines.

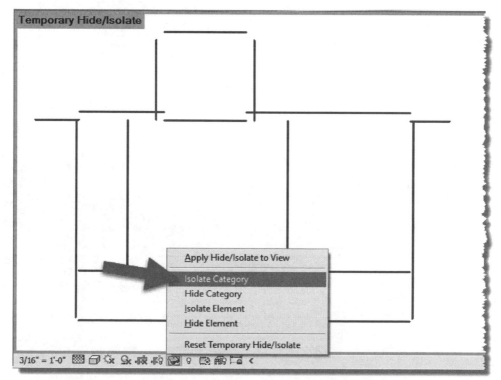

Figure 3.4-12 Isolate Area Boundary Lines using Hide/Isolate

Area Tag

See chapter 6 – Annotation chapter for coverage of this tool.

Tag All Not Tagged

See chapter 6 – Annotation chapter for coverage of this tool.

Room & Area Extended Panel

The Room & Area panel has two additional tools which are not visible by default. Any panel on the Ribbon, which has this symbol (Figure 3.4-13) has additional commands hidden on an extended panel area.

This area will remain open (Figure 3.4-14) until a command is clicked or the cursor is moved away.

Figure 3.4-13 Extended panel symbol

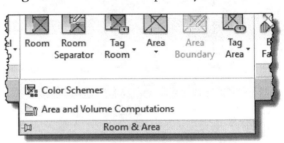

Figure 3.4-14 Extended panel open

Color Schemes

Color Scheme: **Edit Color Scheme dialog**

The **Edit Color Scheme** dialog (Figure 3.4-15) defines the colors used to fill Rooms, Spaces or Areas in a plan view; based on a selected parameter.

Color Schemes do not work in celling plan views.

quick steps

Color Schemes

1. Edit Color Scheme dialog
   - Edit Color Scheme definitions
   - Options:
     - Include elements from links

A project can have several schemes, which are defined on the left of the dialog. The default templates have a few already set up—for example, one for Spaces and two for Rooms. When one of these Schemes is selected on the left, we can see how it is defined on the right. Notice, **Scheme 1** is based on the Spaces category and looks for the Name parameter within each Space.

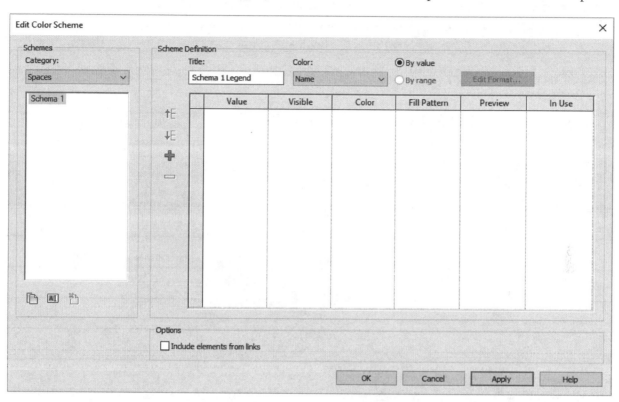

Figure 3.4-15 Edit Color Scheme dialog

In the example below (Figure 3.4-16) we see a Scheme called Department, based on the Room category. This Scheme looks for the Department parameter and assigned a color for each unique entry. As soon as the designer starts typing in Department names, this list starts to populate as shown here—there are currently four different Department names in the BIM. From here we cannot tell how many there are of each. Plus, none of this appears graphically in the model by default.

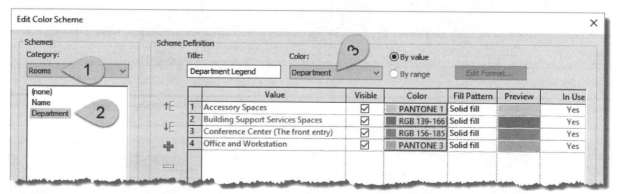

Figure 3.4-16 Edit Color Scheme definition; based on Department parameter for Rooms

Revit automatically generates a color for each unique entry. This color can be changed by clicking on the Color button. The Fill Pattern can also be changed from the default Solid Fill option.

It is possible to add rows prior to the condition existing in the model; simply click the green Plus Icon. For example, doing this in a template would allow a standard department naming convention to be applied. When a Room is selected, a designer could type what they want, but could also click the drop-down arrow and pick from the list previously defined in the Color Scheme dialog.

The image below is an example of the Color Scheme in Figure 3.4-17 applied to a floor plan. This can be activated in any floor plan by placing a **Color Fill Legend**, via the Annotation tab, or by setting a Color Scheme via the view's properties. Consider creating a separate plan view in the Project Browser so the Color Scheme can be left on. This will also help avoid the colors printing on construction document (CDs) sheets if the Color Scheme was accidently left on.

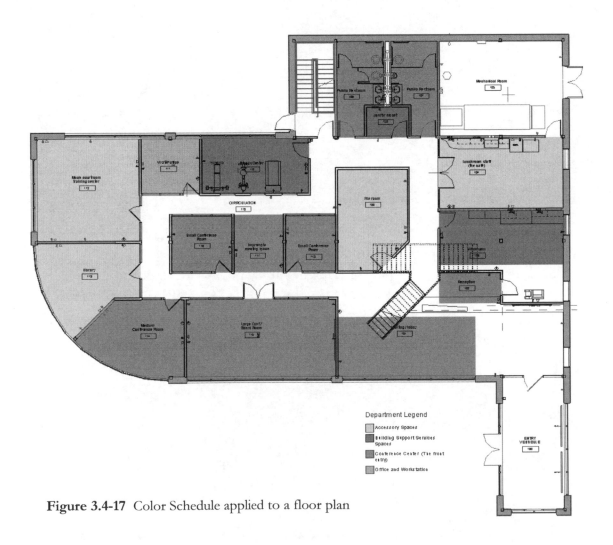

Figure 3.4-17 Color Schedule applied to a floor plan

Checking **Include elements from links** allows Rooms, Spaces or Areas in linked models to be color filled. By default, this applies to every linked model, although some models, such as structural, typically do not have Rooms so nothing would show up. However, maybe an existing model has Room elements which are not desired. In this case the **Revit Links** settings can be adjusted for each link in the view's **Visibility/Graphic Overrides** dialog. Each model, including the host model, has a **Color Fill** sub-category under Rooms (Figure 3.4-18).

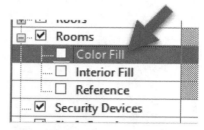

Figure 3.4-18
Color Fill sub-category

Area and Volume Computations

Area and Volume Computations: **Area and Volume Computations dialog**

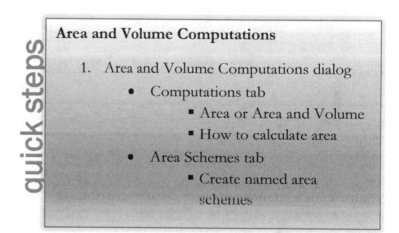

Computations Tab

The Volume Computations setting determines if Area or Area and Volume are calculated for each Room in the BIM. The default is to just compute Area.

Figure 3.4-19 shows the calculated area for a selected Room. Notice, when **Areas only** is selected, the Volume is listed as **Not Computed**.

Notice the height of a Room (or Space) is affected by the Area/Volume setting; compare Figures 3.4-20 & 21.

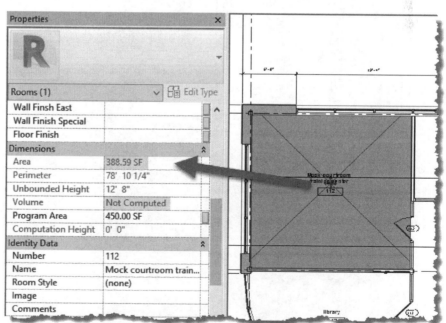

Figure 3.4-19 Area calculated, but not Volume for selected Room

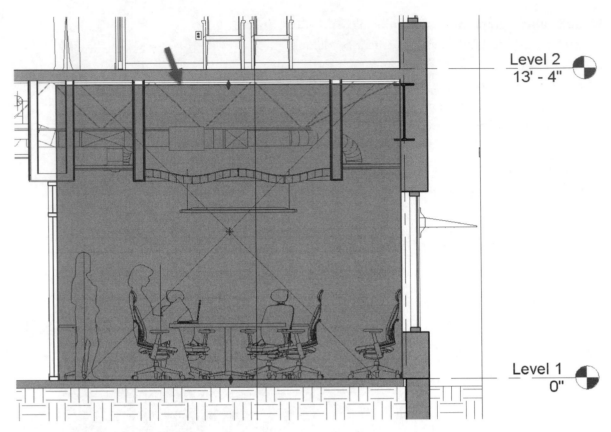

Figure 3.4-20 Room height as specified when **Areas only** is selected

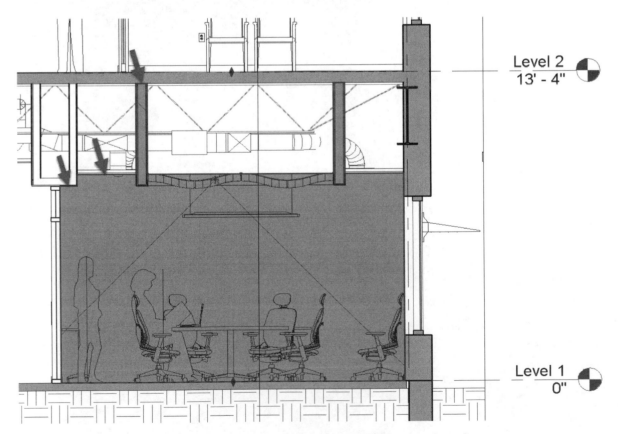

Figure 3.4-21 Room height bounded by ceiling; **Areas & Volumes** selected

When **Area and Volume** is selected, the top of the Room is limited by elements set to **Room Bounding**; this parameter is available for most system family elements like walls and ceilings. However, when the bottom of a wall is part of the upper boundary of a Room/Space element, the wall is ignored and the Room/Space extends up into the wall as seen in the previous image.

Area Schemes Tab

Somewhat hidden, in that the command on the Ribbon only suggests this dialog will contain the **Area and Volume Computations** setting, is the **Area Scheme** tab.

This dialog, shown below (Figure 3.4-22), allows the designer to create additional Area Schemes for Area Plans. The main purpose for this is to create another grouping of Area Plans in the Project Browser. When creating a new scheme, there are only two options, which relate to how areas are calculated. The only thing editable about an area scheme is the name.

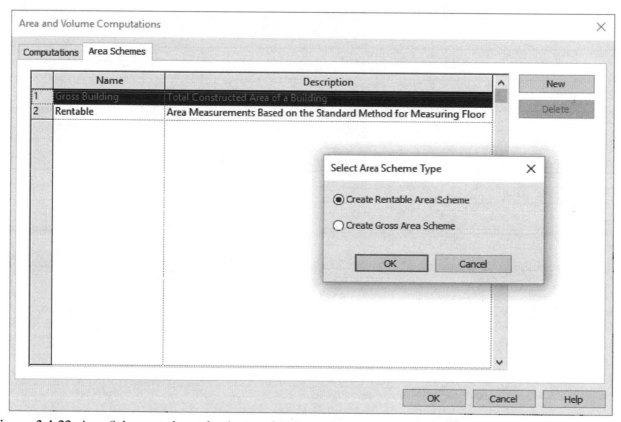

Figure 3.4-22 Area Schemes tab on the Area and Volume Computations dialog

See more on the application of Area Schemes in the Area Plans section earlier in this section.

3.5 Architecture tab: Opening panel

Revit provides a tool to show openings within Revit's **System Families** (Walls, Floors, Ceilings and Roofs). Most of these tools only work on Floors, Ceilings and Roofs; only one works on a Wall. Hover the cursor over a button to see which elements are supported via the tool tips.

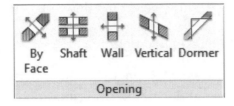

Opening by Face

Creates an opening which is perpendicular to a Floor, Roof or Ceiling as shown in the example below (Figure 3.5-1). This tool also works on structural columns and beams.

> Multiple openings can be defined on the same element, but only one at a time.

Opening by Face: **Selection**
With the command active, hover the cursor over a face until its perimeter highlights and then click to select that face.

Opening by Face: **Ribbon**
Use the **Draw** tools to define the shape of the opening's perimeter.

Opening by Face: **In-Canvas**
Sketch a single enclosed area, defining the perimeter of the opening. Click the Green Checkmark when finished.

quick steps

Opening by Face

1. Select a planar face of a Roof, Floor, Ceiling, Beam or Column
2. Ribbon [in sketch mode]
   - Line segment
3. In-Canvas
   - Sketch perimeter of opening

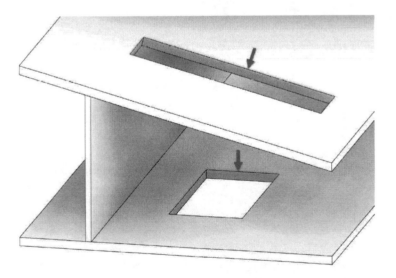

Figure 3.5-1 Opening by Face added to floor and roof

Shaft

Place a Shaft element to vertically cut through multiple floors, ceilings and Roofs. This is a good way to ensure the opening of an elevator or mechanical shaft remains in alignment.

> Add a Shaft Opening to cut through multiple floors to define an elevator shaft vertically. No need to edit each Floor's sketch to define an opening.

Shaft: Ribbon

Use the **Draw** tools to define the shape of the opening's perimeter.

Often, the opening has two crossing lines forming an "X" to help graphically identify the location of the opening in plan views. Revit offers a **Symbolic Line** option to sketch lines which will appear at each level. When this option is selected on the Ribbon the **Line Styles** panel appears as seen to the right.

quick steps

Shaft

1. Ribbon [in sketch mode]
    - Boundary Line
        - Defines the perimeter of opening
    - Symbolic Line
        - Defines lines shown within opening on each level – usually an "X" to suggest a mechanical shaft
    - Line Style
2. Options Bar
    - Chain
    - Offset
    - Radius
3. Properties
    - Base/Top Constraint
4. In-Canvas
    - Sketch perimeter of opening

Shaft: Options Bar

The Options Bar has settings related to sketching the perimeter of the opening. **Chain** allows continuous sketching, **Offset** moves the lines relative to the points picks and **Radius** allows for automatic rounding of corners.

Shaft: Properties

The Properties Palette (Figure 3.5-2) allows the design to define the top and bottom of the shaft. Any Roof, Floor or Ceiling which comes in contact with the shaft will be cut.

Shaft: In-Canvas

Typically start at the lowest level, define the perimeter with Boundary Lines and, optionally, add Symbolic Lines (Figure 3.5-3); these only appear in plan views.

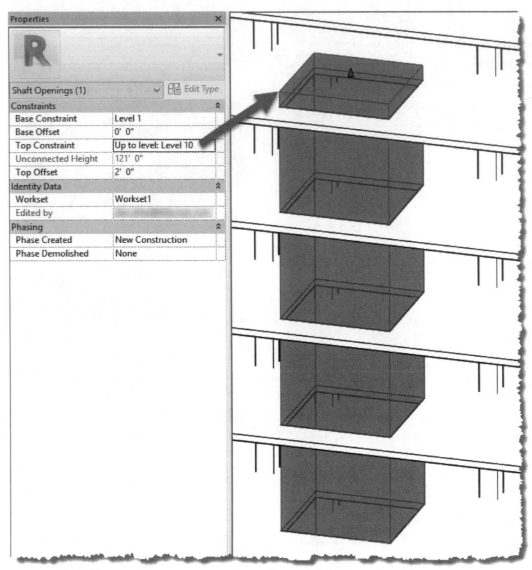

Figure 3.5-2 Shaft opening added to multi-story building

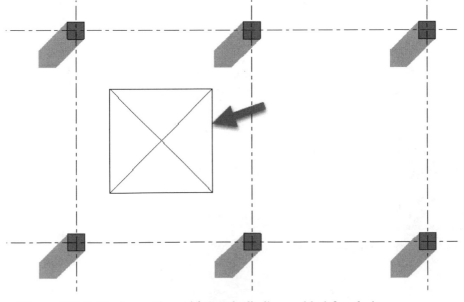

Figure 3.5-3 Shaft opening with symbolic lines added for clarity

Wall Opening

As the name implies, this tool created an opening in a wall. Drawing a continuous wall and then adding an opening is preferred over two walls that appear to have an opening. This ensures the portion of wall above the opening is accounted for, and that the entire wall stays in alignment if moved.

> Wall Opening elements are not **Room Bounding** if the opening extends all the way down to the floor. Also, if the Level's Room Calculation Height parameter is adjusted such that it intersects an opening, the Room will not be bounded. Use Room Separation Lines to define room boundaries if needed.

The Opening's **Phase** matches the host wall's Phase setting.

Wall Opening

1. In-Canvas
   - Select Wall
   - If in a plan view:
     - Pick starting point of opening
     - Pick endpoint of rectangular opening
     - Opening extends from floor level to 8'-0" high by default
   - If in an elevation or section view:
     - Pick two diagonal points to define a rectangular opening
     - This defines the width and height of the wall opening
   - Click temporary dimension to adjust opening size

Wall Opening: In-Canvas

The creation of the Wall Opening varies slightly depending on the current view type: floor plan or elevation, section or 3D.

- If in a plan view:
  - Pick starting point of opening
  - Pick endpoint of rectangular opening
  - Opening extends from floor level to 8'-0" high by default
- If in an elevation or section view:
  - Pick two diagonal points to define a rectangular opening
  - This defines the width and height of the wall opening
- Click temporary dimension to adjust opening size

The next two images, Figures 3.5-4 & 5, show a selected Wall Opening in a wall. Notice the temporary dimensions and the available parameters via the Properties Palette.

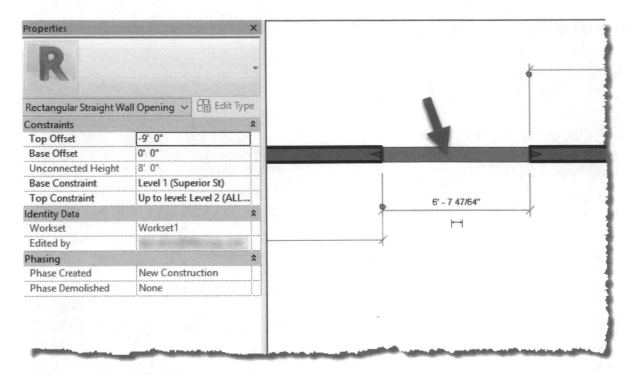

Figure 3.5-4 Wall Opening selected in a floor plan view

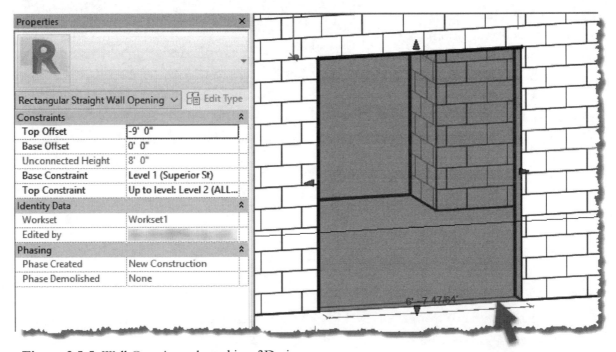

Figure 3.5-5 Wall Opening selected in a 3D view

Wall Openings can only be rectangular. If a non-rectangular opening is required, use the **Edit Profile** option on the Ribbon when the wall is selected (while in a non-plan view).

Vertical Opening

A Vertical Opening only cuts through the selected element. As the name implies, the opening is vertical—that is, perpendicular to the floor level (Figure 3.5-6).

This tool only works on Floors, Ceilings, Roof and Soffit elements.

Vertical Opening: Selection

Once the **Vertical Opening** tool is activated, simply select the model element to impose an opening upon.

quick steps

Vertical Opening

1. Select a Floor, Ceiling or Roof
2. Ribbon [in sketch mode]
   - Line Segment
3. Options Bar
   - Chain
   - Offset
   - Radius
4. In-Canvas
   - Sketch perimeter of vertical opening

Vertical Opening: Ribbon

The Ribbon has the typical **Draw** tools which are used to sketch a closed-loop perimeter of the desired opening. A plan view is often the easiest way to accurately position and sketch the opening. Click the **Green Checkmark** (Finish Edit Model) to complete the command or the **Red Checkmark** (Cancel Edit Mode) to cancel it.

Vertical Opening: Options Bar

The Options Bar offers the **Chain** toggle to allow for continuous sketching. Adjusting the **Offset** value will affect the position of the sketched lines relative to the points picked on the screen. Use **Radius** to automatically round off corners while sketching.

Vertical Opening: In-Canvas

Use the Draw tools to sketch a closed-loop perimeter of the desired vertical opening.

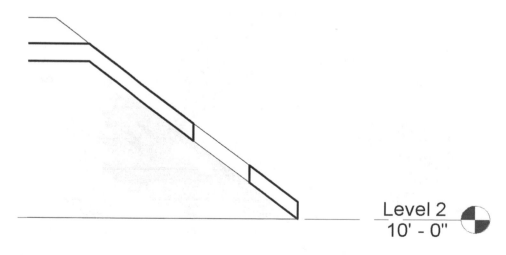

Level 2
10' - 0"

Figure 3.5-6 Vertical Opening added to sloped roof element

Architecture

Dormer Opening

This tool cuts an opening in a main roof element, below a dormer roof. The "exploded" perspective view shows a resulting opening after using this tool.

Dormer Opening: **Prerequisite**

This tool has very specific prerequisites before it can be used:

1. Main roof created
2. Dormer roof created
3. Walls below dormer roof created

Dormer Opening: **Selection**

Once the Dormer Opening tool has been started, select the main roof to be cut.

Dormer Opening: **In-Canvas**

The goal of the remaining steps is to define the perimeter of the opening – but the lines do not have to be a closed loop.

To pick the inside of the dormer walls, try setting the view's display style to **Wireframe**.

> The opening will not automatically adjust if the dormer walls or roof is modified.

All opening elements can be selected and deleted. An opening's **Phase** is automatically set to match the host element. This cannot be changed.

quick steps

Dormer Opening

1. Prerequisite:
   - Must have the main roof, dormer roof and dormer walls created
   - Temporarily make dormer roof transparent so dormer walls can be selected in roof plan view
2. Select main roof (to be cut)
3. In-Canvas
   - Select dormer roof
   - Select dormer walls

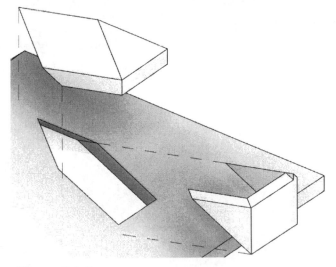

Figure 3.5-7
Dormer Opening added to sloped roof element

3.6 Architecture tab: Datum panel

Levels and Grids are annotation instruments used to manage vertical and horizontal, respectively, relationships within the Revit model. Seeing as nearly all model elements are associated with a level in one way or another, it is important to understand them.

Level

Levels are horizontal reference planes which typically define a surface a person can walk on. Graphically, Levels have a start and end point, but technically they are infinite.

Levels are only visible, and can only be created, in elevation or section views. Also, Levels only appear in views they intersect.

> **Delcting a Level** will delete all elements associated with it – and Revit does not provide any warnings, so delete Levels with care!

Level: Ribbon

In an Elevation or Section view, use the **Line** option to define a Level by picking the start and endpoint, and use the **Pick Line** option to define the extents of a Level based on a line.

Level: Options Bar

The Options Bar offers the ability to create plan views while creating levels. Checking **Make Plan View**

---

quick steps

Level

1. Must be in elevation or section view
2. Ribbon
   - Sketch or Pick Line
3. Options Bar
   - Make Plan View
   - Plan View Types…
   - Offset
4. Type Selector
   - Select Level Type
5. Properties
   - Story Above
   - Computation Height
   - Scope Box
   - Structural
   - Building Story
6. Type Properties
   - Elevation Base
7. In-Canvas
   - Pick two points to define level
     - First point is ended without symbol
   - When ends align during placement, the new Level will adjust in length with adjacent Levels.

---

will create floor plan views in the Project Browser (See Figure 3.6-1). Clicking the **Plan View Types…** opens the dialog shown in Figure 3.6-1—clicking a view type name will unselect it.

Adjusting the **Offset** value will affect the position of the Level line relative to the points picked on the screen.

> If a Level does not have any views associated with it, the target symbol is black (Figure 3.6-2). If it has associated views, it will be blue. Double-clicking on a blue target, Revit will open the appropriate plan view.

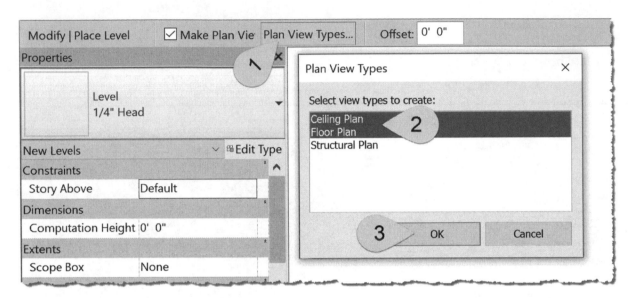

Figure 3.6-1 Create plan views

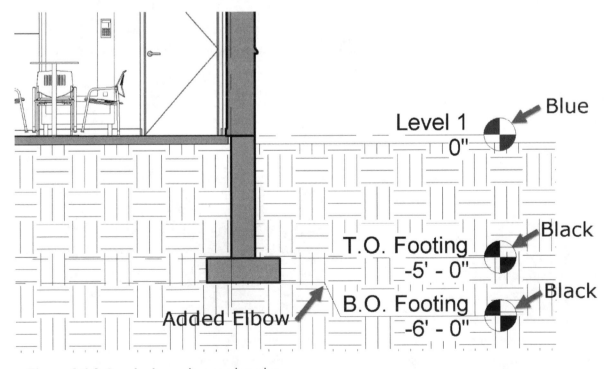

Figure 3.6-2 Levels shown in a section view

Level: Type Selector

Use the Type Selector to specify the Level type to be created; the type can be changed later if needed. Additional types can be created; for example, one shows the project elevation while a copy might show the civil or sea level elevation.

Level: Properties

There are several parameters related to Levels which are important to understand (Figure 3.6-3).

Story Above

Use the **Story Above** drop-down list to explicitly set the Building Story above. This setting has little impact in the Revit environment as it relates more to IFC compliance— IFC stands for Industry Foundation Class, and is a generic format used to share content between programs.

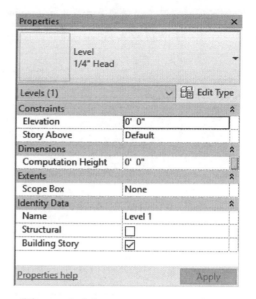

Figure 3.6-3

Level datum Type Properties

Computation Height

The **Computation Height** parameter determines how the perimeter of a space is defined— specifically, at what elevation. The default is 0'-0" and does not typically need to be changed. However, if a space has sloped walls, adjusting this value to something higher can average out the area (square footage). Adjusting this value, to something above the default, can sometimes help resolve breaks in the model where calculations are reporting an error. Of course, the modeling issue should be investigated and corrected, but this setting can help with the troubleshooting process.

The next two images show a situation where a wall is sloped outward, away from the center of the room. With the Level 1 **Computation Height** set to 0'-0" the floor Area is correct, but the Volume is wrong (Figure 3.6-4). The next image shows the Level 1 datum has its instance parameter adjusted to 4'-6" which gets the Volume closer to accurate but now the floor Area is wrong (Figure 3.6-5).

Thus, unless there is a specific need to track volumes of spaces, when sloped walls exist in the model the Computation Height should not be changed from the default 0'-0".

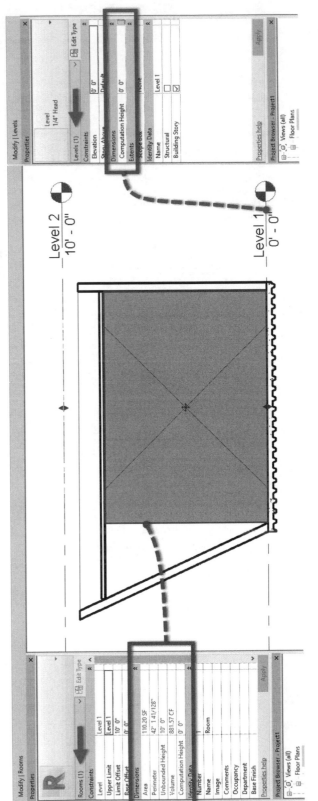

Figure 3.6-4 Level Computation Height; 0'-0"

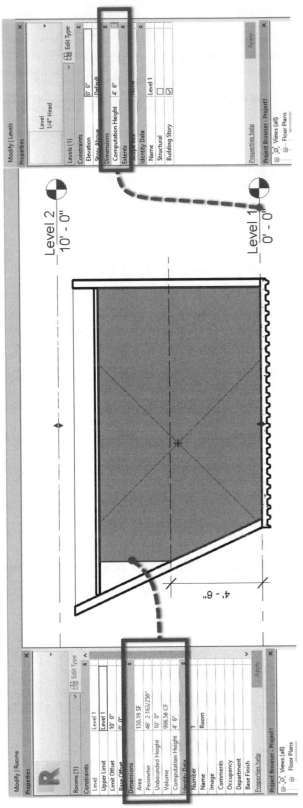

Figure 3.6-5 Level Computation Height; 4'-6"

Scope Box

This drop-down list allows the selection of previously created **Scope Boxes**. When a Level is associated with a Scope Box, a per view setting, the extents of the Level will conform to the extents of the Scope Box and will automatically update.

Name

The name of the Level defines the text which appears on the drawing near the target symbol.

> Use caution when renaming plan views, as this may prompt to change the Level name to match. Sometimes the two, view name and level name, should not match. For example, changing a view name from Level 1 to Level 1 Existing (Superior Street) is not what the Level name should be changed to. Select No to the prompt in this case.

Structural

Check this box to indicate the Level defines a structural reference, e.g. top of steel bearing, rather than the more architectural focused top of finished floor.

Building Story

Check this box when the Level defines a major building story in the building. This is generally the main use of Levels, so this is checked by default. Avoid using Levels to define things like top of window or ceiling heights—this will complicate the use of tools such as the Stair tool which automatically sets the top of the stair run to the next level above.

Level: Type Properties

Most of the Type Properties relate to the graphic representation of the Level, which are self-explanatory. However, the Elevation Base parameter is one we will cover in detail.

Elevation Base

This parameter has two options as described here:

Project Base Point: Selecting this option displays the level relative to the default Revit model origin: 0'-0", 0'-0". 0'-0" (XYZ). Thus, Level 1 is 0'-0" by default.

Survey Point: Selecting this option allows the Level elevation to match the Civil engineer's numbers, which are typically based on sea level.

The Project Base Point and Survey Point can both be seen in any model view (plan, elevation, section, 3D). By default, the two points are both at 0,0,0. When a Civil Engineer or Surveyor provides a geo-referenced CAD site model (AutoCAD Civil 3D for example), that file is linked into Revit, repositioned and rotated to align with the Revit model. Then the Survey Point is moved to align with the CAD file's origin and also repositioned vertically. Figure 3.6-6 shows a site plan view with both points turned on (i.e., visible). The Survey Point has been moved in the X, Y and Z directions.

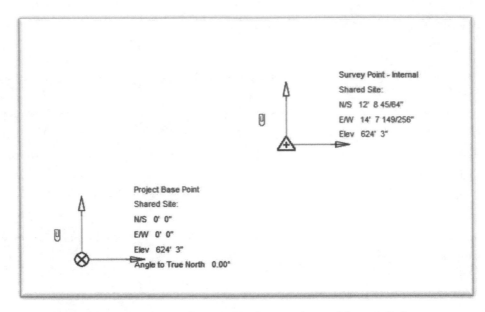

Figure 3.6-6 Site Plan showing Project Base Point and Survey Point

In an elevation or section, the default Levels report the Project Base Point elevation. If this Level Type is duplicated and the new Level's Type Properties are changed so Elevation Base is set to Survey Point, the elevation displayed is relative to the Survey Point. Notice in Figure 3.6-7 that both elevation types can appear in the same view.

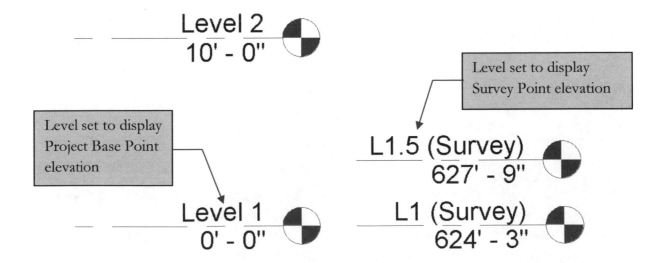

Figure 3.6-7 Elevation/Section showing levels with different Elevation Base settings

Some design firms prefer to have Level 1 at 100'-0" rather than 0'-0" so lower levels and basements are not negative numbers. To do this, open an elevation or section view, make the Project Base Point visible. (Figure 3.6-8).

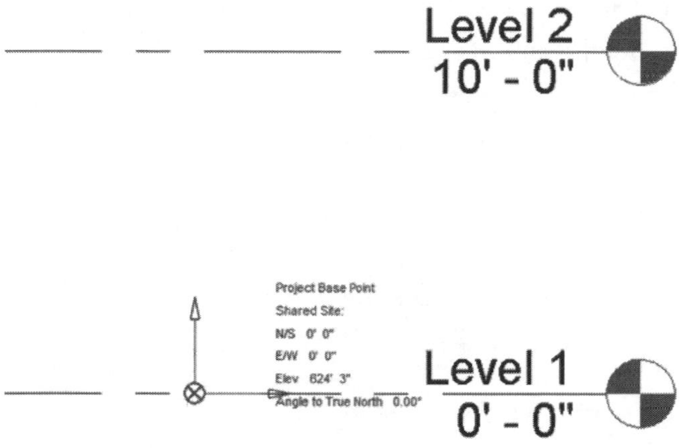

Figure 3.6-8 Adjusting Project Base Point elevation value

Now select the Project Base Point and move it downward 100' 0". This will update the elevation value for all Levels with the Elevation Base set to Project Base Point.

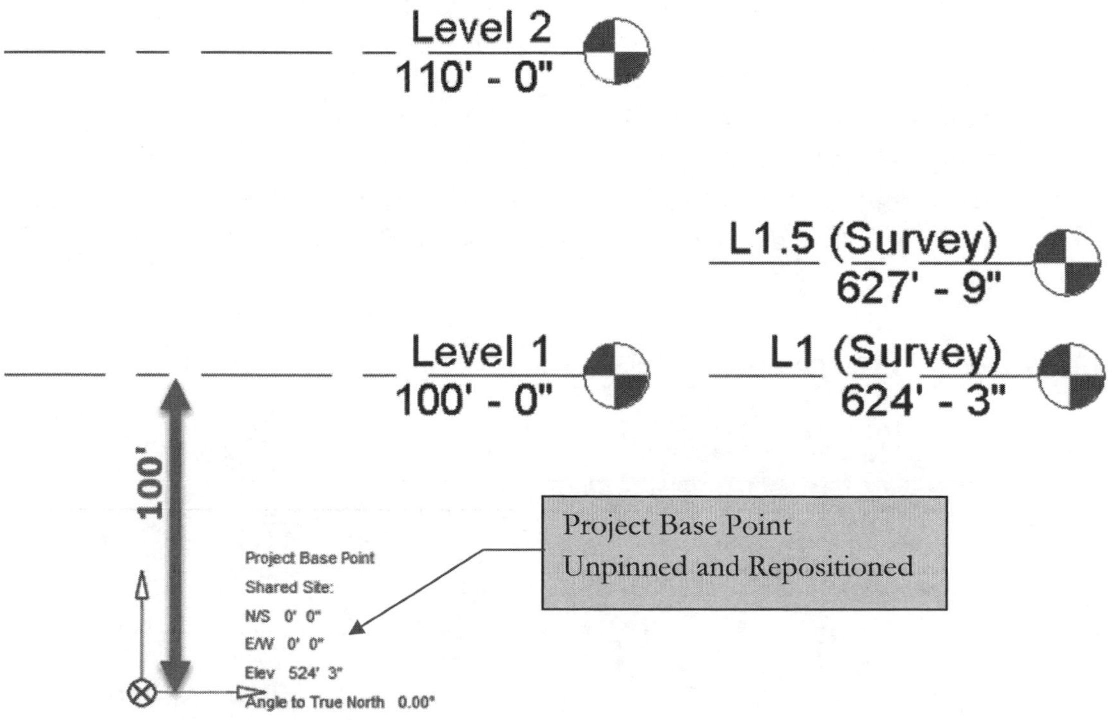

Figure 3.6-9 Adjusted Project Base Point elevation value

Level: In-Canvas

Levels must be placed in an elevation or section view. Picking start and end points that align with adjacent levels will cause them to automatically move together as any level is adjusted.

A Level can be placed approximately and then modified to the specific elevation required. Levels are also visible in 3D views. Care must be taken in adjusting the extents of a level in a 3D view. If you want to adjust the position of level indicators in a 3D view it is a best practice to first turn on a **Section Box** and then adjust the section box to control the position of the level indicators.

Do not draw a level on top, or aligned with, another level. This is bad practice and correcting this issue can be challenging when several Revit elements are associated with each level. Revit does not provide a warning when Levels are placed at the same elevation. See the next section on how to find a Level if it is not graphically showing up in a view.

If a Level is not visible in a view, there are a few simple steps one can take to find the Level datum. First, to ensure the Level exists in the project, create a schedule listing all Levels. Second, follow these steps in an elevation or section view until the Levels appears:

1. Turn off far clipping
2. Turn off cropping
3. Flip the section to look in the opposite direction

Once the Level has been located, adjust its extents so that its graphical extents intersect the original view in question. This often requires the use of a perpendicular section view.

> For existing buildings with multiple levels, which may only be inches apart from each other, there should be a Level and a floor plan view for each. Only one floor plan view is placed on a sheet in the construction documents set. However, the other views are used when placing elements like doors, windows, ductwork and furniture to ensure they are all associated with the correct level.

Anatomy of a selected Level datum:

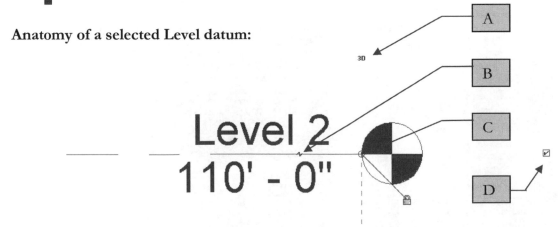

A. Toggle **3D/2D**, where 2D adjusts the length of Level in current view only.

B. **Add Elbow** adds a jog which allows the text and target to be moved up/down.

C. **Adjustment grip** repositioned the end of the Level line; a padlock symbol means other Levels are aligned and will move as well. Click the padlock to adjust the current level. The grip is solid if 2D and hollow if 3D (see 'item A' above).

D. The **checkbox** controls text and target visibility on each end of the Level.

Grid

Grids are used to coordinate and define the location of major structural elements within a building, such as foundations, columns and beams. They also become a communication tool during the design and construction phase of a project.

Grid: **Ribbon**

Grids are typically placed in a floor plan view.

On the Ribbon, select the type of **line segment** to represent the Grid line. Use the Pick Lines option to create grid lines based on previously drawn lines or linked CAD linework.

The Multi-Segment tool allows a series of sketched connected lines to define a single grid line element. This allows for jogging a grid line in plan.

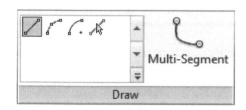

quick steps

Grid

1. Ribbon
   - Line Segment
   - or Pick Line
   - Multi-Segment
2. Options Bar
   - Offset
3. Type Selector
   - Select Grid Type
4. Properties
   - Scope Box
5. In-Canvas
   - Pick two points to define Grid
     - First point is ended without symbol – this will be the default for all views
   - When ends align during placement, the new Grid will adjust in length with adjacent Grids.

Architecture

For elevation and section views, Grids only appear if they intersect the cut plane of the view and are perpendicular to the view. Thus, Grids curved in plan never show up in elevation/section views.

Grid: Options Bar

While the Grid command is active, the **Offset** option, on the Options Bar, simply allows the actual Grid placement to be repositioned (i.e., offset) based on the points picked on the screen.

Grid: Type Selector

Multiple Grid Types can be created and used in the same project. For example, the grids for new construction can have a circle on the end, while existing grids are represented by a hexagon. Grids can also be defined to have a gap in the center of the line, if it is not desirable for the line to continue through the entire floor plan drawing.

With the Grid command active, select the desired Type from the **Type Selector**.

Grid: Properties

The only property available while creating Grids is to assign a **Scope Box**. A Scope Box is essentially a way of cropping the **model** and then applying that definition to multiple **views**. Changes to the Scope Box can then update hundreds of views.

> A Scope Box is defined in a plan view (usually the lowest one) using the Scope Box command on the View tab. A simple rectangular area is defined and a height specified.

The image to the right shows four grids associated with a Scope Box as seen in a plan view. Selecting the Scope Box and adjusting an edge position will cause the Grids to automatically follow. Any views whose properties are set to match this Scope Box will also update. Thus, a sheet containing this view would also update.

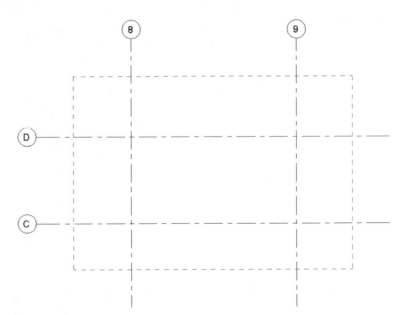

Figure 3.6-10 Grids associated with Scope Box

Grid: In-Canvas

There are two things to keep in mind while creating grids.

The **first** is to pick the Grid endpoints in the correct order. If you pick backwards, meaning the grid bubble is on the opposite side from what you wanted, you should click Undo and pick the points in the opposite order. It is possible to toggle the grid head so that the bubble is on the correct side without using Undo, but that change is an override and only applies to the current view. Thus, all other views, including the consultant models, will be backwards.

The **second** is to align the endpoints of the Grid being created with the endpoints of previously drawn grids. When this is done, they will be constrained and move with each other. Simply watch for the green dashed line before clicking each endpoint.

Curved Grid lines can have a **center point** turned on (Grid 2 in Figure 3.1-73). This point can be selected and moved using Snaps. Dimensions cannot snap to this point directly so other grid lines or model/detail lines would be required to dimension this location point.

Column Location Mark: Structural Columns can track their location relative to grid lines as seen in Figure 3.6-11.

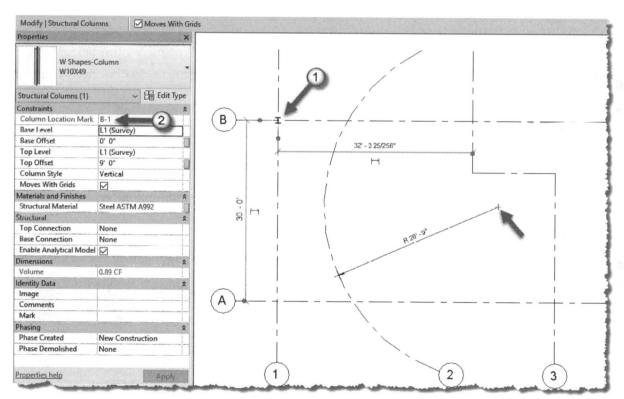

Figure 3.6-11 Grid options: straight, curved and multi-segment

If a Grid is not visible in an elevation or section view, there are a few simple steps one can take to find the Level datum. First, to ensure the Level exists in the project, create a schedule listing all Grids (and add a column for Scope Boxes). Second, follow these steps in an elevation or section view until the Levels appears:

1. Set Scope Box, under Extents, to none.
2. Turn off far clipping.
3. Turn off cropping.

4. Flip the section to look in the opposite direction.

> Grids at an angle to the projection plane of the view and curved grids will never display.

Once the Grid has been located, adjust its extents so that its graphical extents intersect the original view in question. This often requires the use of a perpendicular section view.

Another common issue is that a Grid line is adjusted in an elevation view so the line does not extend through the drawing. However, simply dragging the endpoint upward changes the physical extents of the line such that it does not engage the floor plan cut plane, which means this grid will not appear in a plan view. The proper method to adjust the elevation view is to select the Grid, click the 3D icon, and toggle it to say 2D, on the end to be adjusted. Now dragging the endpoint only affects the current view.

3.7 Architecture tab: Work Plane panel

The tools on the Work Plane panel allow the designer to define and manage 2D planes to be worked on within the context of 3D space. Revit's fundamental philosophy of breaking the creation of complex 3D geometry down into simple 2D tasks is part of what makes the platform so compelling to the industry.

Set Work Plane

The current Work Plane is where Model and Detail lines are created. Also, when using Model In-Place, any extrusions, for example, are sketched and based on the current work plane.

For Plan and 3D Views the Work Plane is set by default to match the associated Level. Elevation and Section views do not have a Work Plane set initially.

Use the Set tool to explicitly specify a Work Plane for any model view.

quick steps

Set Work Plane

1. Set Work Plane dialog
   - Current Work Plane
     - Show
     - Dissociate
   - Specify a new Work Plane
     - Name
     - Pick a plane
     - Pick a line and use the work plane it was sketched in

Set Work Plane: Set Work Plane dialog

When clicking the **Set** tool, the Work Plane dialog for the current view appears (Figure 3.7-1).

All plan views are associated with a level as seen in the Properties Palette—this setting cannot be changed. Notice, for plan views, that the Work Plane is initially set to match the Level 1 as pointed out.

There are three options for setting a work plane:

Name:

The easiest way to set the work plane is to select a **Named** reference. This can be any Level, Grid or named Reference Plane (see the upcoming section on Reference Planes for more on naming a reference plane). Figure 3.7-2 shows the various options available in a project. Note that Grids and Named reference planes only appear if they exist in the project.

Selected reference cannot be perfectly perpendicular to the view, otherwise you should be working in another view. Thus, in a plan view you would select levels or named reference planes.

Architecture

In elevations/sections grids and reference planes can be used. In 3D views, any of the options are valid.

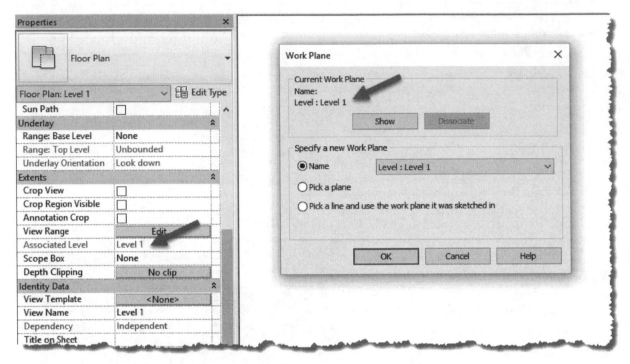

Figure 3.7-1 Set Work Plan dialog and Level 1 view properties shown

Pick a plane:

It is also possible to set a work plane by clicking on the surface of a 3D element in the model.

This option can be tricky. By default, to select a face one needs to position the cursor near the edge of the desired face. However, Revit sometimes will highlight the adjacent surface rather than the one facing you. The desired surface should have its perimeter highlighted prior to clicking.

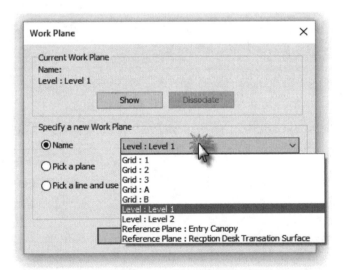

Figure 3.7-2 Selecting work plane by name

When using the "pick a plane" option it may be helpful to toggle on **Select Elements by Face** in the lower right of the Revit UI.

Pick a line and use the work plane it was drawn in:

This option is a quick way to get back to a previously used setting. For example, model lines added to a sloped roof. Clicking one of those lines will cause the work plane used to create them to be reestablished.

The **Show** button is the same as what is covered in the very next section. The **Disassociate** button only applies to a few elements which can have their work plane changed after placement. For example, when selecting **Roof by Extrusion** element, clicking the **Edit Work Plane** button on the Ribbon allows the Disassociate button to be selected. This permits the element to be moved irrespective of the original work plane.

Show Work Plane

The Show work plane tool allows us to visualize the current work plane in a given view.

Show Work Plane: **Toggle**

In any view except sheets and schedules, click the **Show** button to see the Work Plane. Some views, such as elevations and sections, do not have a work plane set by default. In this case, when the Show command is selected, a prompt is displayed which offers to activate the **Set** command.

When a work plane is visible we see a grid that we can snap to. The work plane can also be selected and its size adjusted via the four edge grips. When selected, the Properties palette allows the grid spacing to be adjusted (Figure 3.7-3).

In plan and 3D views the default work plane is Level 1 and appears as shown in Figure 3.7-4.

When the Set command has been used to change the default position of the work plane, or set its initial position, the Show command provides visual confirmation and reference of the work area as shown in Figure 3.7-5.

The work plane will never print and its visibility is only controlled by the Show button on the Ribbon. There is no visibility setting in the Object Styles or Visibility/Graphic Overrides dialog. However, it is possible to select the work plane, right-click, and Hide Element or Override Element. Hiding by element should be avoided to minimize confusion, but the color could be changed if desired. Again, the work plane does not print.

quick steps

Show Work Plane

1. Show Work Plane toggle
    *   Shaded grid appears in view to articulate the current work plane

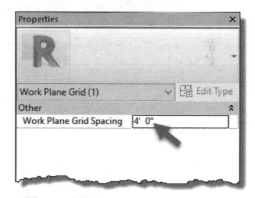

Figure 3.7-3
Work plane grid spacing

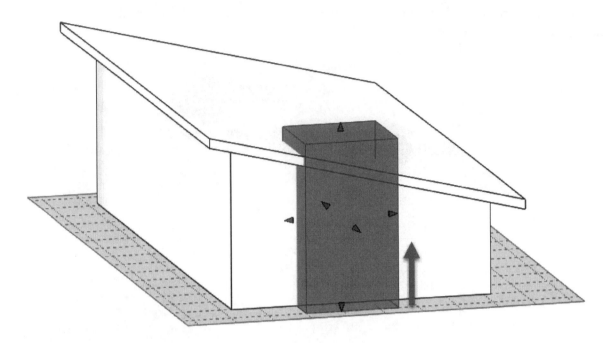

Figure 3.7-4 Work plane shown aligned with Level 1; note extrusion result

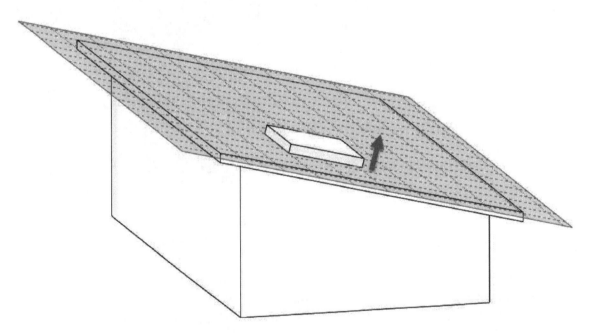

Figure 3.7-5 Work plane shown aligned with sloped roof plan; note extrusion result

Reference Plane

Reference Planes are an essential part of how complex 3D geometry creation tasks are broken down into simple 2D tasks. They also help to maintain relationships and alignments within the model.

Graphically, Reference Planes have extents—related to the two points picked in the model to define it. However, when sketching on one it is really an infinite plane.

Reference Planes appear in all views they pass through if perpendicular to that view.

quick steps

> **Reference Plane**
>
> 1. Ribbon
>   - Sketch or Pick Line
>   - Line Style
> 2. Options Bar
>   - Offset
> 3. Properties
>   - Scope Box
>   - Name
>   - Subcategory
> 4. In-Canvas
>   - Pick two points to define extents of Reference Plane

Reference Plane: Ribbon

When the **Reference Plane** command is active, the Ribbon offers two ways to define the single straight line which graphically defines it: **Line** and **Pick Lines** as shown to the right.

The **Line Style** drop-down does not actually list line styles as it normally does. But, in the case of Reference Planes, it lists any subcategories created.

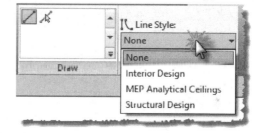

Reference Plane: Options Bar

While the Reference Plane command is active, the **Offset** option, on the Options Bar, simply allows the actual Reference Plane placement to be repositioned (i.e., offset) based on the points picked on the screen.

Reference Plane: Properties

A **Scope Box** is essentially a way of cropping the **model** and then applying that definition to multiple **views**. Changes to the Scope Box can then update hundreds of views. Grids and Reference Planes both work essentially the same way in the context of Scope Boxes. Please see the previous section on Grids for more information.

After a Reference Plane is created it can have a **Name** given—which must be unique (even if on separate subcategories). A name is only required if it will define a work plane or be used to create an extrusion from within an In-Place family. Simply select a Reference Plane and enter a name In-Canvas or via the Properties palette (Figure 3.7-6).

Architecture

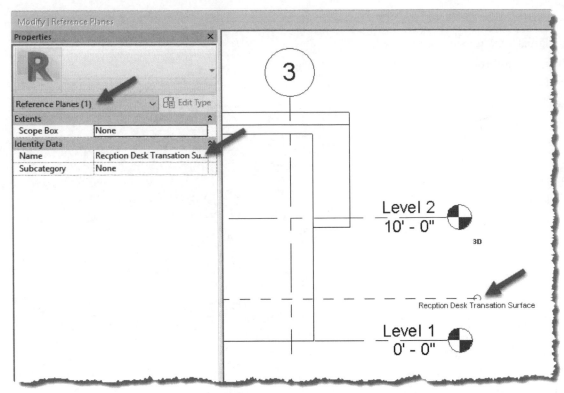

Figure 3.7-6 Reference plane added and then named "Reception Desk Transaction Surface"

Subcategories are a relatively new option which allows Reference Planes to appear graphically different as well as control visibility separately.

New subcategories can be created in the **Object Styles** dialog as shown in Figure 3.7-7.

The **Visibility/Graphic Overrides** dialog allows each subcategory to be hidden independently (Figure 3.7-8).

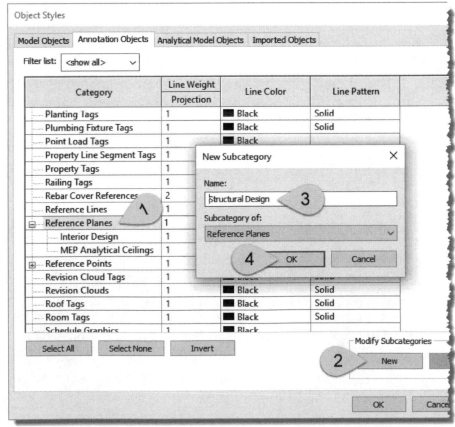

Figure 3.7-7
Object Styles dialog; creating a new subcategory for reference planes

Reference Plane: In-Canvas

Reference Planes can be created in any plan or elevation/section view. They cannot be created or even seen in a 3D or Camera view.

It is important to draw a Reference Plane in the correct direction. Notice in the image below, the Reference Place drawn from

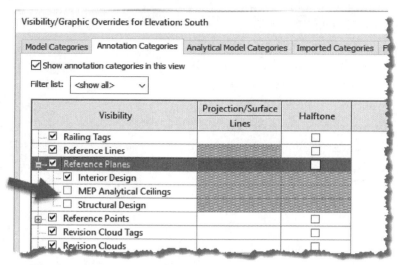

Figure 3.7-8 Visibility/Graphic Overrides dialog

left to right (L-R) produced incorrect results when the face based light fixture was placed in a ceiling plan view (Figure 3.7-9).

Figure 3.7-9 The direction a reference plane is drawn in affects how elements are hosted

Reference planes print. They can be turned off in the Visibility/Graphic Overrides dialog per view. Another option is to ensure they never print via the **Hide ref/work planes** option in the print dialog (Figure 3.7-10).

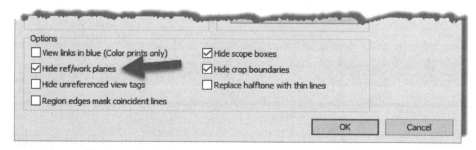

Figure 3.7-10 Plot dialog visibility options

> Reference Planes, along with the **Cut** tool on the Modify tab, can be used to trim columns and beams (in plan and elev/sect views).

Work Plane Viewer

This command opens a floating
window which can be repositioned
outside of the Revit window—even
on a second monitor. This view is
meant to provide a simplified
perpendicular view of elements
hosted by the current work plane.

Work Plane Viewer: **Toggle**

Clicking the Viewer command
opens the Workplane Viewer
window. In the image below, the

Work Plane Viewer

1. Work Plane Viewer toggle
   - Floating window to view and edit
     elements associated with the
     current Work Plane

current view is an elevation and the work plane has been set to the sloped roof plane. The
viewer shows the "flattened" view of the extruded element being modified.

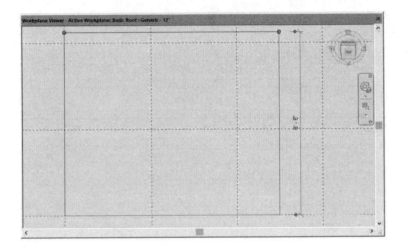

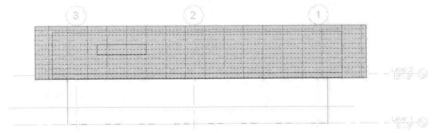

Figure 3.7-11 Workplane Viewer open

Chapter 4
Structure Tab

The Structure tab contains tools to model or document the structural elements of a building, such as columns, beams, footings, etc. Some of the tools on this tab have already been covered in the previous chapter and will not be repeated; a note is added to refer to the previous chapter in each case. Additionally, given the platform/architectural focus of this book, the reinforcement tools, the Steel tab, and the Precast tab will not be covered.

4.1 Structure tab: Structure panel

The Structural panel contains the fundamental structural element commands. These tools are used by architects and structural engineers to model existing conditions and new construction. The icon in the lower right opens the **Structural Settings** dialog, which is covered in the Manage tab chapter.

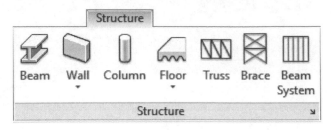

It is often helpful to model structural elements in the order they are built, which directly corresponds to how things are supported... for example, a beam is supported by a column or structural wall. However, footings and foundations are often modeled last, unless modeling existing conditions.

It is often best to add **Grids** first and then **Columns** and/or **Structural Walls**. Once the columns/walls are in place, **Beams** will automatically connect and clean up with them.

One more thing to point out: architectural plans are a view of the model from the **floor up**. Structural plans are a view of the model from the **floor down**. The beams and columns in a structural plan are what support the floor. Thus, the View Range settings are quite different between an architectural floor plan view and a structural floor plan view.

Beam

A Beam is any **horizontal load-bearing** member. For example, a wide flange beam, a bar joist, a wood joist, a precast plank, etc.

Beam: Ribbon

The **Load Family** option provides a convenient way to bring in additional

quick steps

Beam

1. Ribbon
   - Load Family
   - Draw *panel*
     i. *Default:* straight segment
   - Multiple: On Grids
   - Tag on Placement
2. Options Bar
   - Placement Plane
   - Structural Usage
   - 3D Snapping
   - Chain
3. Type Selector
   - Select Beam type
4. Properties
   - Several geometric position parameters
   - See warning
5. In-Canvas
   - Pick two points to place beam

content without needing to cancel the current command and switch to the Insert tab.

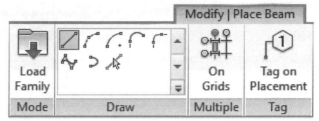

The **Draw** panel offers the typical sketch options used to define the shape of the beam in plan view. The **Pick Lines** option provides a fast way to create beam lengths based on other line work in the model, including lines from lined CAD files.

The **On Grids** option is used to place multiple beams based on selected grid lines; a beam is added along a grid line, between two intersecting grids. This feature requires columns at each grid intersection.

The **Tag on Placement** toggle causes Revit to place a beam tag with each beam created. Tags can be deleted at any time and they can also be added later using the Tag by Category command.

> During preliminary design tags are often left off the drawings to avoid clutter while the design is being developed. Also, in many cases, existing structural elements are never tagged unless they are in close proximity to an area being remodeled or expanded.

Beam: Options Bar

The **Placement Plane** drop-down defines the vertical position of the beam. In a plan view, the default setting corresponds to the level associated with the current view. All Levels and named horizontal reference planes appear in this list. In a 3D view, this list also shows grid lines to help constrain the placement of beams.

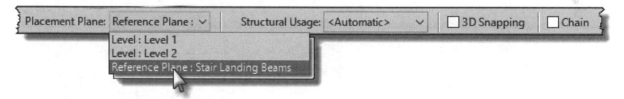

Structural Usage is a way to categorize different types of support elements, such as in schedules or to control visibility. It is also relevant information when the structural model is exported to a structural analysis software such as Autodesk Robot Structural Analysis, RISA 3D or Bentley STAAD/RAM. Best practice is to leave this set to "Automatic."

When working in 3D views and a sloped beam is desired, clicking the **3D Snapping** is required. Otherwise, Revit is constrained to the work plane.

Chain is a common option in Revit which allows continuous sequential picks without the need to re-pick the previous point to define the start point of the next element.

Structure

Beam: **Type Selector**

Available beam types, which have been previously loaded in the current project, are listed in the **Type Selector**. If a required size/shape is not listed, use the **Load Family** command.

> Do not try to browse to the Windows folder containing the beam family files as this will bypass the **Type Catalog** feature which limits the number of sizes/shapes loaded for each family.

Beam: **Properties**

By default, beams are placed with the top, which is the bearing surface for floors and joists, aligned with the level as shown in Figure 4.1-1 below. This often needs to be repositioned downward as the top of the floor is what needs to align with the Level (the main level is often referred to as "finished floor"). This vertical adjustment can be achieved by editing the beam's **Start Level Offset** and **End Level Offset** values (e.g. -0'-4").

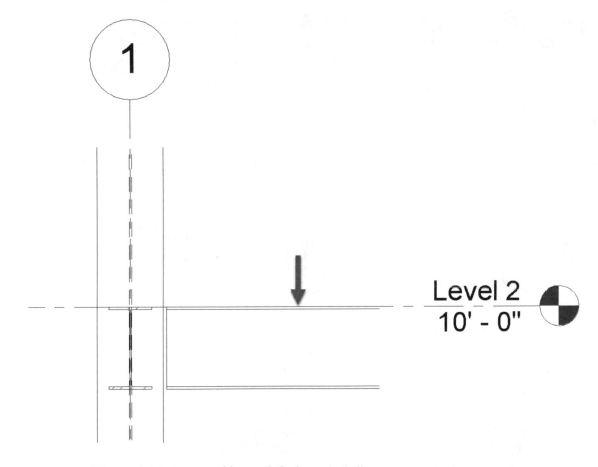

Figure 4.1-1 Structural beam default vertical alignment

One exception is if additional Levels are added to manage "top of steel." In this case, the special Level should have **Building Story** un-checked, and **Structural** checked in the Level's properties.

One challenge for architects who want to create **photo-realistic renderings** with exposed structural is that the material setting is an instance parameter for every beam and column. The default Appearance Asset is Stainless Steel with a Satin finish. This looks great, but is an unrealistic, not to mention expensive, option. The easiest way to adjust the material is to just edit the one already assigned to the beams and columns rather than trying to apply a new/different material. Changing the Appearance Asset will not affect the Physical Asset, which is important for structural analysis.

Warning: If 3D Snapping is checked, a beam could be sloped but not apparent in a plan view. It is helpful to have a 3D view which only shows the structural model to quickly visually validate the structural model.

Beam: In-Canvas

With the Beam command active, click at the intersection of Grids or Columns or on Structural Walls.

> If a View's **Discipline** parameter is set to **Structural**, only Structural walls will appear. All other (non-structural) walls are hidden—there is no way to make these walls show up without changing the Discipline setting.

The **View Control Bar** at the bottom left of the document window has a toggle called **Show Analytical Model**. This is helpful in 3D views to ensure all structural elements are properly connected. This also corresponds to the Analytical Model Categories tab within the Visibility/Graphic Overrides dialog (Figure 4.1-2).

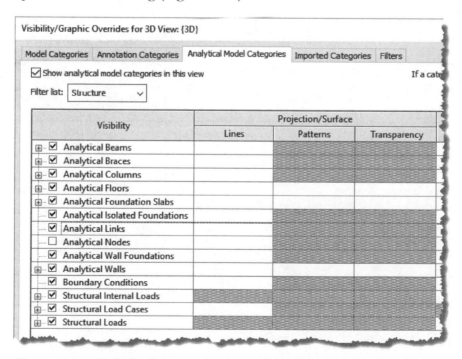

Figure 4.1-2 Visibility/Graphic Overrides dialog; Analytical tab

Structure

Wall: Structural

Note: See Chapter 3 – Architecture Tab for detailed coverage of the Wall tool.

A structural wall is any wall, as determined by a structural engineer, which supports a load from another element (such as a floor, column or roof) or acts as a shear wall. A Structural wall is an Architectural wall with the *Structural* parameter checked; notice in *Object Styles* and *Visibility/Graphic Overrides* there is no Structural Wall category.

For a Wall, when the Structural parameter is checked, several additional structural parameters become available (Figure 4.1-3).

There are two main things one should understand about Structural Walls:

- When a view's Discipline parameter is set to Structural, only structural walls are visible. All non-structural walls are automatically hidden.
- When using external structural analysis tools (e.g. Autodesk Robot or RISA Technologies) only Structural walls are exported and considered.

The **Structural Bearing** parameter has three options to choose from:

- Bearing
- Shear
- Structural Combined

While creating a structural wall, changing the **Properties Filter** to New Analytical Walls (Fig. 4.1-4) reveals additional parameters related to structural design.

Wall: Architectural

Note: See Chapter 3 – Architecture Tab for detailed coverage of this tool.

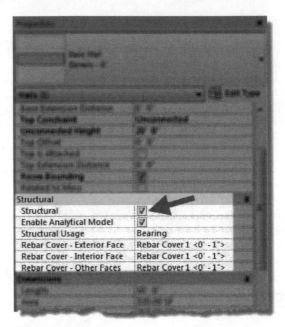

Figure 4.1-3
Structural wall setting and parameters

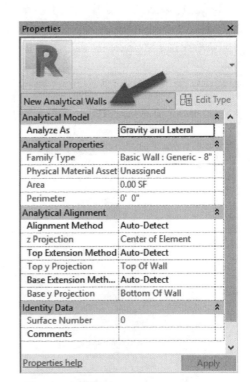

Figure 4.1-4
Analytical Wall Properties

Wall: Sweep

Note: See Chapter 3 – Architecture Tab for detailed coverage of this tool.

Wall: Reveal

Note: See Chapter 3 – Architecture Tab for detailed coverage of this tool.

Structural Column

A Structural Column is a **vertical load-bearing element** which works in conjunction with other structural elements to define, geometrically and analytically, a building's structural system.

Do not confuse Architectural Columns, found on the Architecture tab, with this command. Those columns are meant to represent aesthetic shapes and column covers (i.e., interior finishes)—but do not work with the structural analytical features.

Structural Column: Ribbon

The **Load Family** option provides a convenient way to bring in additional content without needing to cancel the current command and switch to the Insert tab.

> Be sure to select structural, not architectural columns when loading this way.

The default Placement option is **Vertical Column**. This simply means the column is vertical to the floor level. Switching to **Slanted Column** provides the means to model an angled column—via additional options on the Options Bar.

---

quick steps

Structural Column

1. Ribbon
   - Load Family
   - Vertical Column
   - Slanted Column
   - *Insert multiple columns*
     - At Grids
     - At Columns
   - Tag on Placement
2. Options Bar
   - Rotate after Placement
   - Level (3D Views)
   - *Select Direction*
     - Depth (downward)
     - Height (upward)
   - *Select Top Constraint*
     - Pick a Level *or*
     - Unconnected
       i. Enter Height
3. Type Selector
   - Select Column Type
4. Properties
   - See Warning
5. In-Canvas
   - Spacebar to rotate
   - FYI: May not see column until placed or due to view settings

Structure

The **At Grids** option is used to place multiple columns based on selected grid lines; a column is added at the intersection of all selected grids. Columns are not added at grid intersections which already contain a column.

The **At Column** option provides a way to place a Structural Column at the center of an Architectural Column. This option is likely not used very often as the structural engineer usually places columns at Grid intersections

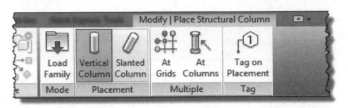

and cannot assume the architectural column (aka an interior finish) is placed accurately enough.

The **Tag on Placement** toggle causes Revit to place a beam tag with each beam created. Tags can be deleted at any time and they can also be added later using the Tag by Category command.

> During preliminary design tags are often left off the drawings to avoid clutter while the design is being developed. Also, in many cases, existing structural elements are never tagged unless they are in close proximity to an area being remodeled or expanded.

Structural Column: **Options Bar**

When the **Rotate after Placement** is checked, Revit prompts for a rotation angle after clicking a position in the model to place the column. It is also possible to simply press the **Space Bar** to rotate the column in 90 degree increments, or relative to nearby angled elements.

Structural Columns, like Structural Walls, are placed from the floor level down—thus, **Depth** is the default setting on the Options Bar. This can be changed to **Height** if desired.

The Depth/Height of a Structural Column can have a specific height (i.e., **Unconnected**) or be parametrically associated with a Level in the model. This can be changed after placement as well. When Unconnected is selected, a length value is entered in the textbox.

When **Slanted Column** is selected on the Ribbon, the Options Bar changes as shown below. These options allow a sloped column to be defined in a plan view by picking two points. The first point is the bottom of the column and the second corresponds to the top. Both the Level and Offset can be set for each pick to define the vertical position for the top/bottom.

The **3D Snapping** option, which appears when Slanted Column is selected on the Ribbon, will snap to geometry in the model rather than the values entered for the two picks.

Structural Column: **Type Selector**

Available Structural Columns types, which have been previously loaded in the current project, are listed in the **Type Selector**. If a required size/shape is not listed, use the **Load Family** command. Architectural Columns are not listed in the Type Selector.

> Although technically the same thing, a W12x36 beam family is not the same as a W12x36 column family. Both must be loaded into a project if that size is required for both beams and columns.

Structural Column: **Properties**

Structural Columns have two sets of properties associated with them. The main properties are presented by default (Figure 4.1-5). The second set of properties is found by changing the Properties Filter to New Analytical Columns (Figure 4.1-6).

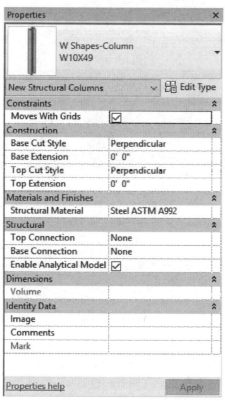

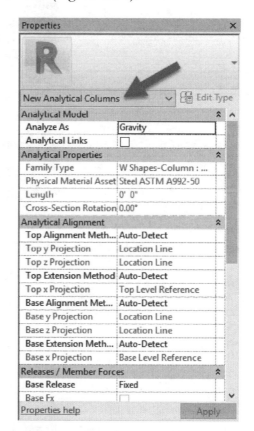

Figure 4.1-5
Structural Column Properties

Figure 4.1-6
Analytical Column Properties

When **Moves with Grids** is selected, the column will move when placed directly on a grid. This requires the column to have a valid **Column Location Mark** (an Instance Property after placement)—which requires two grids to intersect. If the grids are adjusted so they no longer intersect, the columns will stop moving with the grids.

Structure

Structural Column: In-Canvas

While placing a Structural Column, press the **Space Bar** to rotate it before picking a placement point. Ensure the **View Range** is set properly so columns and beams appear in the view. Structural plans typically have the Detail Level set to **Coarse**.

Floor: Structural

Note: See Chapter 3 – Architecture Tab for detailed coverage of the Floor tool.

A structural floor is any floor, as determined by a structural engineer, which supports a load. A Structural floor is an Architectural floor with the *Structural* parameter checked; notice in *Object Styles* and *Visibility/Graphic Overrides* there is no Structural Floor category.

For a Floor, when the Structural parameter is checked, several additional structural parameters become available (Figure 4.1-7).

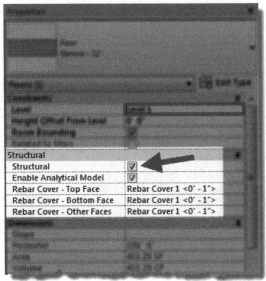

> When a floor is supported by the ground, that is—a slab on grade, use a **Structural Foundation Slab** (see this tool later in this chapter). When the floor is being supported by other building elements, use a **Structural Floor**.

Figure 4.1-7 Structural floor properties

When using external structural analysis tools (e.g. Autodesk Robot or RISA Technologies) only Structural floors are exported and considered.

Most floors are structural floors. However, floors may be used architecturally to represent a floor finish; for example, a 1/8" thick floor positioned on top of the structural floor (which may be coming from a linked structural Revit model). This allows each room to have a separate finish apart from the main structural floor element (which would not be editable if contained within a link). **TIP:** These finish floors should be set to non-Room Bounding.

Because most floors are structural, all floors appear in a view, even when the view's Discipline is set to Structural. Thus, a view Filter may need to be employed to hide any non-structural floors.

When using the Floor: Structural tool, Revit will automatically add a **Span Direction Symbol** to the current view. This can be deleted without loss of information. It may also be manually added to a view using the **Span Direction** tool on the Annotate tab.

Floor: Architectural

Note: See Chapter 3 – Architecture Tab for detailed coverage of this tool.

Floor: Slab Edge

Note: See Chapter 3 – Architecture Tab for detailed coverage of this tool.

Truss

The Truss command is used to create a structural floor or roof truss in a building model. A Truss is a unique family which defines a repeating pattern of three fundamental components of a truss: **Top Chord**, **Web** and **Bottom Chord**. The structural shape associated with each of these components can be changed to any structural framing type loaded in the current project (Figure 4.1-8).

Truss: **Ribbon**

The **Load Family** option provides a convenient way to bring in additional content without needing to cancel the current command and switch to the Insert tab.

Trusses can only be straight segments; thus, there are only two Draw options: **Line** and **Pick Lines**.

Tag on Placement allows the new truss to automatically be tagged in the current view.

quick steps

Truss

1. Ribbon
   - Load Family
   - Line Segment or Pick
   - Tag on Placement
2. Options Bar
   - Placement Plane
   - Chain
3. Type Selector
   - Select Truss Type
4. Properties
   - Set truss Height
5. Type Properties
   - Edit all "Set Framing Type" values to desired framing member
6. In-Canvas
   - Pick two points to define length of truss
   - FYI: An error may result if points are too close, as the truss design may not work

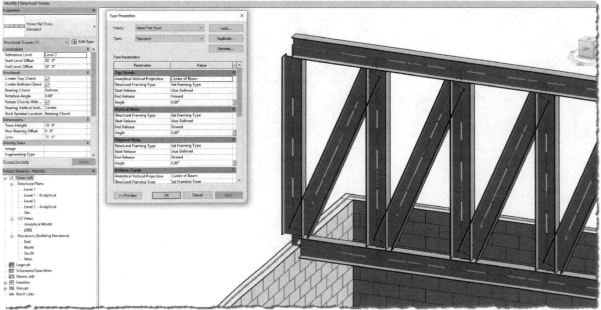

Figure 4.1-8 Structural truss and its properties

Truss: **Options Bar**

The **Placement Plane** drop-down defines the vertical position of the truss. In a plan view, the default setting corresponds to the level associated with the current view. All Levels and named horizontal reference planes appear in this list. In a 3D view, this list also shows grid lines to help constrain the placement of trusses.

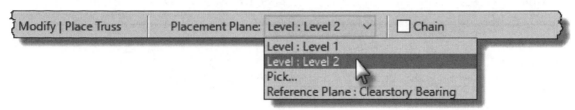

Chain is a common option in Revit which allows continuous sequential picks without the need to re-pick the previous point to define the start point of the next element.

Truss: **Type Selector**

Available truss types, which have been previously loaded in the current project, are listed in the **Type Selector**. If a required truss type is not listed, use the **Load Family** command on the Ribbon—or from the Insert tab. Several truss families are installed with Revit (Figure 4.1-9).

Figure 4.1-9 Structural truss families provided with Revit

Structure

Truss: **Properties**

In the Properties Palette (i.e., Instance Properties) set the **Bearing Chord**; options are Top and Bottom.

Truss: **Type Properties**

The Type Properties for a truss provide a means to change the structural shape and position of the Top Chord, Web and Bottom Chord components.

Truss: **In-Canvas**

Pick two points within the model to define the length of the truss. Picking points on a structural wall or column will create a parametric relationship and connect the structural analytical mode properly. Thus, if a wall or column is repositioned, the truss adjusts as well.

Brace

A Brace is a structural framing element, which has its Structural Usage set to Other, that slopes and connects to two other structural elements (columns and/or beams) as seen in Figure 4.1-10.

Brace: **Ribbon**

The **Load Family** option provides a convenient way to bring in additional content without needing to cancel the current command and switch to the Insert tab.

Braces can only be straight segments; thus, there is only one Draw option: **Line**.

Tag on Placement allows the new brace to automatically be tagged in the current view.

> **quick steps**
>
> **Brace**
>
> 1. Ribbon
>   - Load Family
>   - Tag on Placement
> 2. Options Bar (plan view)
>   - Start Level/offset
>   - End Level/offset
>   - 3D Snapping
> 3. Type Selector
>   - Select Bracing/Framing Type
> 4. Properties
>   - Several under Geometric Position
> 5. In-Canvas
>   - Pick two points to define bracing

Brace: **Options Bar**

When in a plan view, the Options Bar provides a way to define the vertical position of the **Start** and **End** of the brace. 3D Snapping is the only option in a 3D view.

Brace: Type Selector

Available brace types, any Structural Framing family loaded into the current project, are listed in the **Type Selector**. If a required truss type is not listed, use the **Load Family** command on the Ribbon—or from the Insert tab.

Brace: Properties

The Properties Pallet offers several settings in the Geometric Position section which control how the geometry appears within the model.

Brace: In-Canvas

In either a plan, elevation, section or 3D view, pick two points to define the location, length and angle of the desired bracing member.

When in a 3D view, use the 3D Snapping option to force Revit to look for points along structural members.

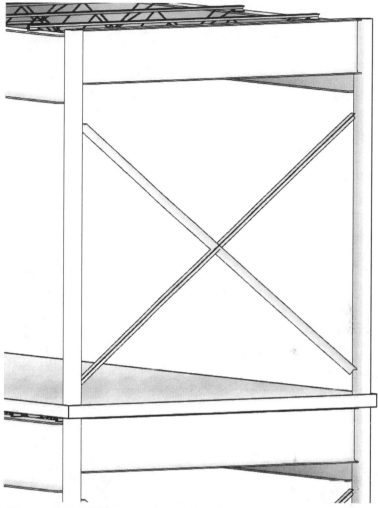

Figure 4.1-10 Structural (cross) bracing

Elevation and section views offer more visual controls to adjust the bracing once placed. For example, grips for the analytical connection point and grips for the cutback position of the actual geometry appear as well as a flip end (start vs. end) points icon.

Braces appear offset in plan views according to the Brace Symbols settings in Manage (tab) → Structural Settings → Symbolic Representation Settings (tab). **TIP:** Settings for this may need to be imported from the structural template using Transfer Project Standards.

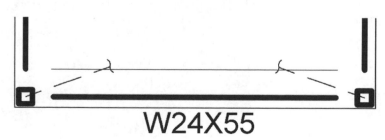

W24X55

Beam System

The Beam System command allows us to quickly place an expanse of joists in the model without requiring the individual use of the Beam command for each instance.

Beam System: **Ribbon**

The Ribbon offers Automatic and Sketch options to define the perimeter of the area to receive an array of framing members.

The default option is **Automatic Beam System**. Use this when a series of beams define an enclosed area. Otherwise switch to the **Sketch** options and use the Draw options to manually define the perimeter. A special Draw option in Sketch mode is Pick Beams, which places a sketch line at the center of a beam or structural wall.

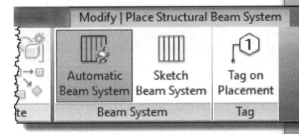

Tag on Placement allows the new brace to automatically be tagged in the current view.

Beam System: **Options Bar**

While in Automatic mode, the Options Bar offers a convenient way to set Beam Type, Justification and Layout Rule for the joists about to be placed.

quick steps

Beam System

1. Ribbon
   - _Automatic:_ If one or more enclosed beams (i.e., structural bay) are visible in the current view
     - Automatic Beam System
     - Sketch Beam System
     - Tag on Placement
   - _Sketch:_ If no enclosed beam areas are visible or don't exist
     - Boundary Line
     - Beam Direction
     - Line Segment
     - _Or_ Pick Line
     - _Or_ Pick Beam
     - Work Plane Settings
     - Reset System
2. Options Bar
   - For _Automatic_
     - Beam Type
     - Justification
     - Layout Rule
     - 3D
     - Tag Style
   - For _Sketch_
     - Chain
     - Offset
     - Radius
3. Type Selector
   - Select Type
4. Properties
   - 3D
   - Elevation
   - Layout Rule
   - Spacing Value
   - Justification
   - Beam Type

If the top of the walls sloped (using Edit Profile on the wall, for example), and the bead system should follow that slope, then ensure **Walls Define Slope** is checked. This is for either orientation relative to the wall, parallel or perpendicular framing.

When switching to Sketch mode, the Options Bar changes to this:

While in Sketch mode, to change the beam system "pattern" edit the same parameters in the Properties palette.

Beam System: Type Selector

There is not much need for Types unless tracking some metadata (e.g. Cost) as all the parameters related to the geometry are instance parameters.

Beam System: Properties

The Elevation parameter may be edited to ensure the bearing of the framing members aligns properly with the support elements.

Beam System: In-Canvas

When hovering over a beam or structural wall, for the Automatic option, the direction and spacing of framing elements will appear (before clicking) as an aid. Thus, in this case, the Span Direction is implied as seen in the two examples below (Figure 4.1-11).

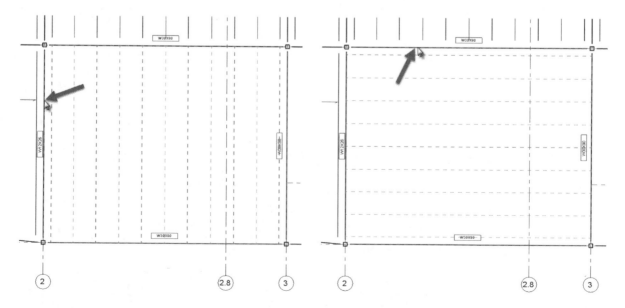

Figure 4.1-11 Defining Structural Beam System direction; hover cursor over beam

Curved beams and structural walls, in plan, can be used to define the perimeter of a beam system (Figure 4.1-12). However, they cannot be the element selected as it cannot be used to infer the Span Direction of the framing in the Beam System.

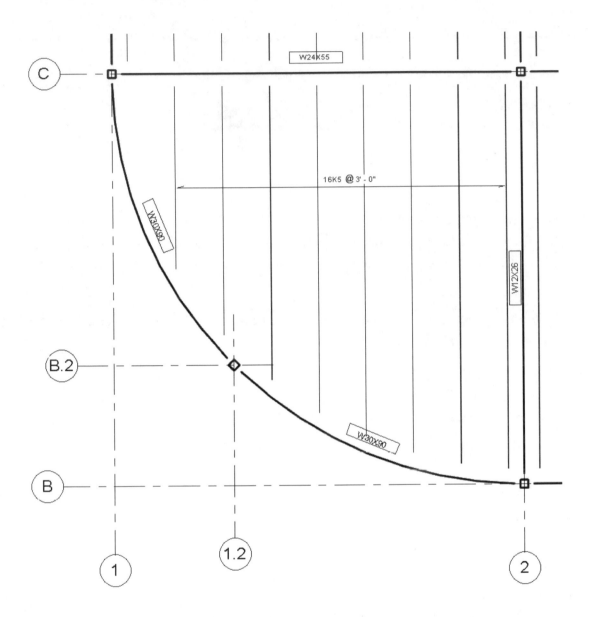

Figure 4.1-12 Structural Beam System with tag added

Structural Settings

The icon in the lower right of the Structure panel provides quick access to the **Structural Settings** dialog.

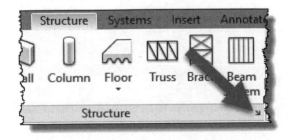

4.2 Structure tab: Connection panel

Structural Connections offers a way to show geometry to represent connections and base plates; it also offers a means to share these important decisions with structural analysis <u>and</u> detailing software.

Important Information About This Feature:
The Autodesk provided add-in **Steel Connections for Autodesk Revit 2021 64-Bit** must be installed to realize the full potential of the Structural Connections feature. Also, this feature only works with certified Revit structural families. This DOES NOT include the basic content in the Revit provided templates or the content in the main Column or Beam folders. The only steel content that works with this feature is located here:

- C:\ProgramData\Autodesk\RVT 2021\Libraries\US Imperial\Structural Framing\Steel\AISC 14.1
- C:\ProgramData\Autodesk\RVT 2021\Libraries\US Imperial\Structural Columns\Steel\AISC 14.1

Connection

Without the add-in mentioned above, Revit only places an analytical connection with no 3D geometry being created.

Connection: **Ribbon**
The Ribbon only displays a grayed out **Modify Parameters** button. This is only available after the Connection is created.

Connection: **Type Selector**
The options listed here are dependent on the connections loaded into the current project from **the Structural Connection Settings** dialog—which is accessible via the small icon in the

quick steps

Connection

1. Ribbon
   - Modify Parameters
2. Type Selector
   - Select Type
3. Properties
   - Approval Status
4. In-Canvas
   - Select structural element(s); hold Ctrl if selecting more than one.
   - Press Enter to complete command.
   - Detail Level must be set to Fine to see structural connections.

lower right corner of the Connection panel on the Ribbon. The list of *Available Connections* will be empty if the required add-in mentioned above is not installed.

Structure

Connection: **Properties**

The **Approval Status** allows the structural engineers to keep track of Connections they have reviewed and ones which still need work. The list of options is editable via the **Structural Connection Settings** dialog.

Connection: **In-Canvas**

With the Connections command active, select the necessary structural elements and then press Enter.

> The **Detail Level** must be set to **Fine** to see the connection geometry in a view.

Once a connection is placed, the **Modify Parameters** button becomes active. Clicking it reveals a dialog specific to the connection type selected. This dialog has several categories to choose from on the left, with the remainder of the dialog showing a multitude of options for the selected category.

The image below shows the basic steps after activating the Connections command:

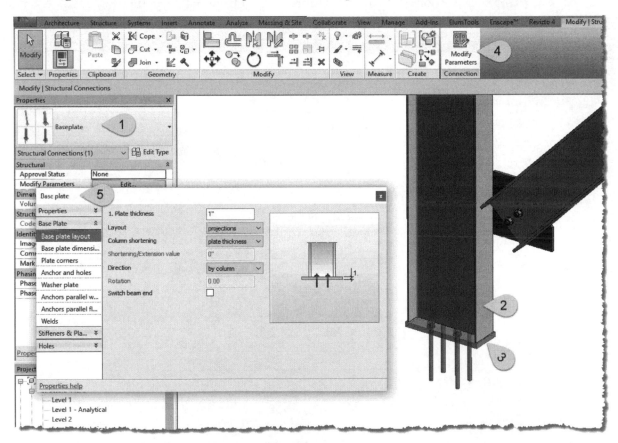

Figure 4.2-1 Structural Connections workflow

4.3 Structure tab: Foundation panel

Revit provides a **Foundation** tool (or, perhaps better described as a **Footing**) to model the building element which transfers the structural load from a wall or column to the earth. This element is often larger than the wall or column to help spread the load—this larger size is determined by a geotechnical engineer based on soil analysis at the project site.

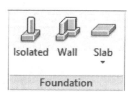

Isolated Foundation (i.e., Footing)

This command places a structural element intended to support a structural column. "Isolated" simply means it may have earth all around it and may not be in contact with any other foundations/footings. When associated with a column, the foundation automatically aligns with the bottom of the column and will maintain that relationship if modified in any way.

Isolated: **Ribbon**

The **Load Family** option provides a convenient way to bring in additional content without needing to cancel the current command and switch to the Insert tab.

If the Isolated Footing is unique and there is no family loaded in the project, we can click the **Model In-place** command to quickly switch gears and create the desired geometry directly in the project.

quick steps

Isolated Foundation

1. Ribbon
    - Multiple; at Grids
    - Multiple; at Columns
2. Options Bar
    - Rotate after Placement
    - Level (3D view only)
3. Type Selector
    - Select Type
4. Properties
    - Level/Offset
    - Structural Material
    - Enable Analytical Model
5. In-Canvas
    - Spacebar to rotate

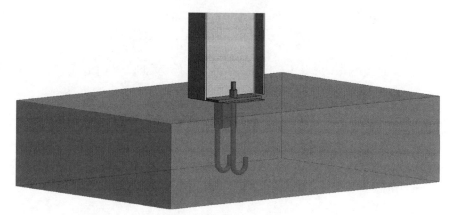

Figure 4.3-1 Isolated foundation added at structural column

Revit offers two options on the Ribbon to quickly place multiple Isolated Footings: **At Grids** and **At Columns**.

Using the **At Grids** option requires two or more Grids to be selected. A temporary representation of the Isolated Foundation appears. Clicking the **Green Checkmark** completes the task and creates the elements. Changes to the Grid locations causes the Isolated Foundations to update as seen in Figure 4.3-2; notice Grid 5 has been repositioned. The foundations also moved—even the one along the angled grid.

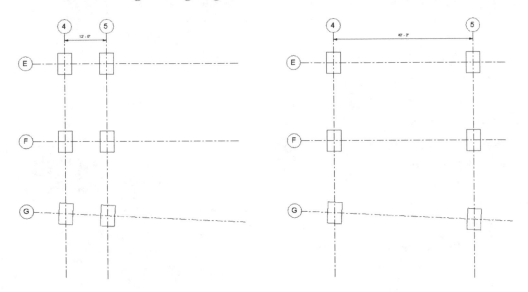

Figure 4.3-2 Isolated foundations added at grid intersections update automatically

The **At Column** option is preferred in that the Isolated Foundation becomes associated with the element it actually supports. If the column moves, the Isolated Foundation updates accordingly. Of course, this option requires that columns exist first.

Isolated: Options Bar

When the **Rotate after Placement** is checked, Revit prompts for a rotation angle after clicking a position in the model to place the element. It is also possible to simply press the **Space Bar** to rotate the element in 90 degree increments, or relative to nearby angled elements.

While in a 3D view, the Options Bar also provides an option to select the **Level** to host the placed element.

Isolated: Type Selector

Available Isolated Foundation types, which have been previously loaded in the current project, are listed in the **Type Selector**. If a required size/shape is not listed, use the **Load Family** command.

Isolated: **Properties**

If the Isolated Foundation is not associated with the bottom of a column, the **Level** needs to be selected via the Properties Palette (i.e. Instance Properties). The only options for the Level parameter are Levels that exist in the project—named Reference Planes are not accessible here. If there is a lot of variation in elevation, for a sloped site as an example, use the **Height Offset from Level** parameter to avoid adding an excessive amount of Levels.

The **Structural Material** is used to specify which Revit material is assigned to the element. This determines how the element looks in the model and renderings as well as what the physical properties are when sent to a structural analysis program.

Moves with Grids only applies when an element is placed directly on a Grid—not necessarily at a Grid intersection.

When **Enable Analytical Model** is checked, Revit understands this element should be included in any structural analysis calculations. For existing elements in a remodel or addition, which is not being modified in any direct or in-direct way, then this option can be unchecked.

Isolated: **In-Canvas**

Switch to the plan view associated with the Level you need to associate the element with. If a view or level does not exist, switch to the closest level and then specify the **Level** or **Offset** via Properties.

Spacebar to rotate the element before placement.

When using one of the Multiple options, drag a window or use Ctrl to select multiple columns/grids.

Structure

Wall Foundation (i.e., Footing)

A Wall Foundation attaches to the bottom of a wall, structural or non-structural. If the wall is adjusted or moved, the element is updated accordingly.

This command is dependent on walls already existing in the model. Also, if the host wall is deleted, the wall foundation is also deleted—just like doors and windows in the same wall.

Wall foundations join, and clean up, with Isolated Foundations as seen in Figure 4.3-3.

Wall Foundation: **Ribbon**

The Ribbon offers a Select **Multiple** option so several walls may be selected at once, rather than clicking on each one separately. After selecting multiple walls, click the Finish option—the Green Checkmark.

quick steps

Wall Foundation

1. Ribbon
   - Select Multiple
2. Type Selector
   - Select Type
3. Properties
   - Eccentricity
   - Enable Analytical Model
4. Type Properties
   - Structural Material
   - Structural Usage
     - Bearing
     - Retaining
   - Dimensions
     - Options vary depending on Structural Usage selection
5. In-Canvas
   - Select a wall—continuous footing added to its bottom

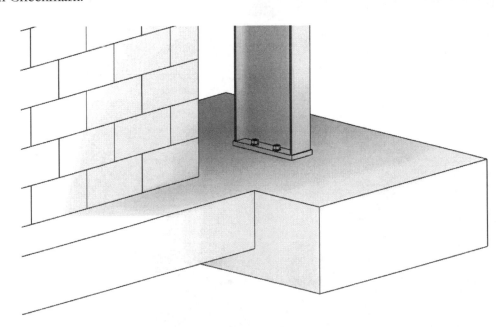

Figure 4.3-3 Wall foundation joined to isolated foundation

Wall Foundation: Type Selector

Available Wall Foundation types, which have been previously defined in the current project, are listed in the **Type Selector**. If a required size is not listed, use the **Edit Type → Duplicate** option. A Wall Foundation is a system family, meaning it can only exist within the context of a project, like walls. Thus, there are no families (rfa files) to be loaded.

Wall Foundation: Properties

By default, the Wall Foundation is centered on the overall wall thickness. Use the **Eccentricity** parameter to reposition the Wall Foundation relative to the center of the wall; both positive and negative numbers are allowed.

When **Enable Analytical Model** is checked, Revit understands this element should be included in any structural analysis calculations. For existing elements in a remodel or addition, which is not being modified in any direct or indirect way, this option can be unchecked.

Wall Foundation: Type Properties

The **Structural Material** is used to specify which Revit material is assigned to the element. This determines how the element looks in the model and renderings as well as what the physical properties are when sent to a structural analysis program.

Changing the **Structural Usage** option affects which dimensions appear as described below. The two options are **Bearing** and **Retaining**.

When the Structural Usage is set to Bearing, the Dimension options are (Fig. 4.3-4):

- Width
- Foundation Thickness
- Default End Extension Length

When the Structural Usage is set to Retaining, the Dimension options are (Fig. 4.3-5):

- Width (read only - instance)
- Toe Length
- Heel Length
- Foundation Thickness
- Default End Extension Length

Figure 4.3-6 shows a section with a "bearing" wall foundation on the left and a

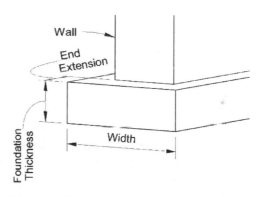

Figure 4.3-4
Wall foundation (bearing) dimensions

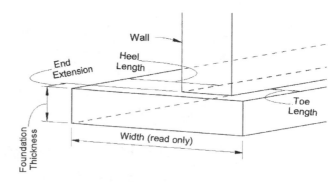

Figure 4.3-5
Wall foundation (retaining) dimensions

"retaining" example on the right. Notice, wall foundations are aligned from the top.

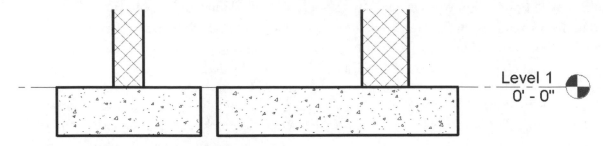

Figure 4.3-6 Wall foundations; Bearing (left) and Retaining (right)

Wall Foundation: In-Canvas

With the Wall Foundation command active, simply select individual walls, or use the Select Multiple option to select several at once.

> Once created, in a section view, use the Cut Profile command on the View tab to add a keyway as shown in the image below (Figure 4.3-7). Keep in mind this is a view specific override and does not change the 3D geometry.

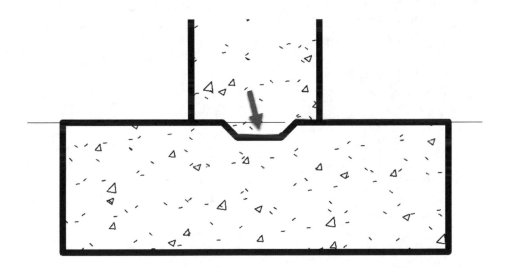

Figure 4.3-7 Cut Profile applied to Wall and Wall Foundation

Structural Foundations: Slab

The Structural Foundation Slab is primarily meant to model a slab on grade. This tool is exactly like the Floor and Structural Floor commands but with the added convenience of being sorted into a separate category, Wall Foundations.

> See Walls and Structural Walls for more information.

Slab Foundation: **Ribbon**

When sketching a slab, there are three "pick" options: **Pick Lines**, **Pick Walls** and **Pick Supports**. The Pick Supports option can be walls or beams.

Slab Foundation: **Options Bar**

When Draw is set to Pick Walls or Supports, the Extend into wall (to core) option appears. When this is checked, the slab extends to the edge of the layers, within the wall, defined as the Core, or structural, part of the wall.

Slab Foundation

1. Ribbon
   - Boundary Line
   - Slope Arrow
   - Span Direction
   - Draw (boundary options)
     - Line Segment
     - Pick a Line
     - Pick a Wall
     - Pick a Beam
   - Work Plane Options
2. Options Bar (varies based on Draw panel selection)
   - Chain
   - Offset
   - Radius
   - Extend into wall (to core)
3. Type Selector
   - Select Slab Type
4. Properties
   - Level/Offset
   - Enable Analytical Model
5. In-Canvas
   - Define perimeter of slab

Slab Foundation: **Type Selector**

Available Foundation Slab types, which have been previously defined in the current project, are listed in the **Type Selector**. If a required size is not listed, use the **Edit Type → Duplicate** and then **Edit Structure**. A Foundation Slab is a system family, meaning it can only exist within the context of a project, similar to walls. Thus, there are no families (rfa files) to be loaded.

Slab Foundation: **In-Canvas**

Sketch the perimeter of the slab—the sketch cannot have gaps or crossing lines. Any closed loops within the outer perimeter become holes in the slab.

Floor: Slab Edge

Note: See Chapter 3 – Architecture Tab for detailed coverage of this tool.

4.4 Structural tab: Reinforcement

These tools are not covered due to the architectural focus of this book.

4.5 Remaining Structural tab Panels

See Chapter 3 – Architectural tab for detailed coverage of the remaining tools on the structural tab. These tools are useful to both the architectural and structural discipline and are therefore provided on their respective tabs for convenience.

Steel & Precast tabs

The Steel and Precast tabs, as shown below, has several advanced structural tools which will not be covered in this book. Refer to Revit's **Help System** for more information on these commands and workflows.

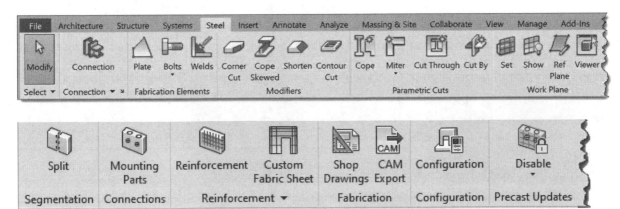

Chapter 5
Insert Tab

The Insert tab contains the tools which facilitate bringing external models, laser scans and Revit content into the current project. External files can either be **Linked** or **Imported**. Linked content will continue to update as the linked file is modified whereas an imported file has no connection to the original external data.

There are many uses for linked files. For linked Revit files, these are the most common uses:

- Separate discipline models (e.g. Architectural, Structural, MEP, Parking Ramp, Food Service, etc.) linked into each other.
- Separate buildings linked into each other, or into an overall campus model.
- Individual apartment, hotel or hospital unit/room layouts linked and copied around the main building model. **Tip:** another option here is the use of Groups

Some reasons for linking CAD data:

- Referencing an AutoCAD/Civil 3D (DWG file) site plan, showing contours, roads and sidewalks. This file may remain as part of the printed documents or be used to generate a Revit Toposurface and then removed.
- Temporarily linked CAD floor plans provided by a client can be used to aid in creating a 3D model of an existing building. The DWG might be removed if the entire building is modeled in Revit. However, for a large project, such as a hospital, if the work is contained to a small area, the DWG can remain as part of the construction documents to show context and not require the entire facility to be modeled in Revit.

Linked data does not update in real time, or even when the current project is saved. The linked data automatically updates when the project is first opened. To force an update, without closing the project and reopening it, the **Reload** options can be used in the **Manage** → **Manage Links** dialog.

For linked CAD data, such as AutoCAD DWG files, SketchUp SKP files or Microstation DGN files, a copy of the entire file is saved within the Revit project. If the external file cannot be found when the project is opened, Revit will show the saved internal version until the external file is found. Reasons for a file not being found vary; for example, the file name is changed, a folder name is changed, the file is moved or a Revit project file was sent without the external links.

Linked Revit models are not saved within the project. Each time the project is opened, the linked Revit file must be found or nothing will appear for that link. Revit will present a warning indicating links are missing.

Link or Import?

In general, the best practice is to Link external data rather than Import. Linked data is easier to manage as it appears in the Managed Links dialog. The only way to delete an imported file is to visually see it, select it in the model and then use the delete command. This can be tricky on large projects with hundreds of views and the imported file might even be hidden and not readily visible. Where the same file, if linked, can simply be selected in the Manage Links dialog and unloaded or removed.

▎Link is generally preferred over Import

Another problem with Importing CAD data is that it is possible to **Explode** CAD data when selected. This action results in several additional line types, text styles and other clutter which is not easy to clean up.

5.1 Insert Tab: Link Panel

Link Revit

The **Link Revit** command can only be used in a Project, not in a Family. Use this command to reference another Revit model into the current project.

Link
Revit

> The Revit file being linked cannot be open in the same session of Revit.

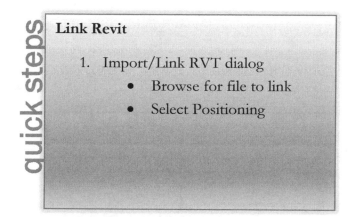

quick steps

Link Revit

1. Import/Link RVT dialog
   - Browse for file to link
   - Select Positioning

When the Link Revit command is started, the **Import/Link RVT** dialog appears (Figure 5.1-1). The first step is to browse to the desired Revit model. Make sure this is the correct location for this file and it will not move or be renamed before linking it into the current project.

When the file is selected, a small preview appears to the right. Notice in the previous image there are two backup files in addition to the main project file. If backup files exist, use care to ensure the correct/current Revit file is being linked.

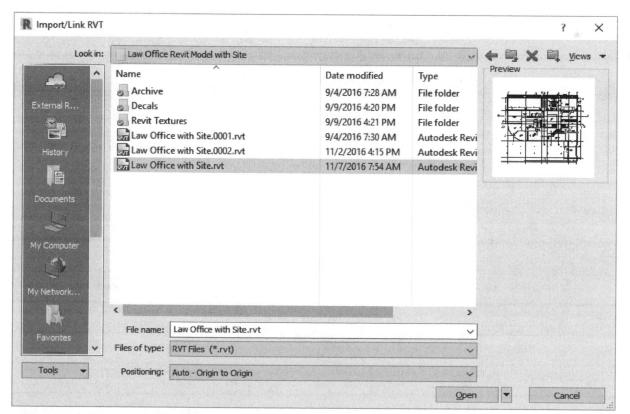

Figure 5.1-1 Import/Link RVT dialog

Insert

> The Revit file being linked should be the same version (e.g. 2021). Linking an older version will require a temporary upgrade every time the file is accessed, causing time delays. Revit will simply not allow a newer version of a Revit model to be linked into the current project.

After selecting the file, there are only two options: **Positioning** and [Open] **Specify** Worksets.

Positioning

As the name implies, the **Positioning** option determines where the linked model is placed within the current project. The default option is Auto – Origin to Origin, which aligns the internal origin of both models; this is the easiest option to use. However, if the two models were started totally independent of each other, one of the other options may be required to properly align the incoming model with the current one.

The positioning options are shown here (Figure 5.1-2).

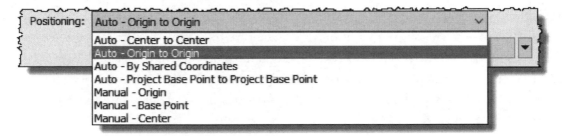

Figure 5.1-2 Import/Link RVT dialog; positioning options

Here is what each Positioning option does:

- **Auto – Center to Center**
  The link's center point is aligned with the current model center point. The center is calculated based on an imaginary bounding box around all geometry.

- **Auto – Origin to Origin**
  The default option; this is the origin of the Revit model. In a default template, this is at the center of the four exterior elevation tags. When everyone on a project agrees to do all linking with this setting, the linking process is greatly simplified.

- **Auto – By Shared Coordinates**
  If one of the files has "shared coordinates" defined, use this option to properly place the linked model. This eliminates the need to manually move and possibly rotate the model into the correct location.

- **Auto – Project Base Point to Project Base Point**
  Aligns the Project Base Point in both models. This point can be seen in a project by turning on the Site sub-category called Project Base Point.

- **Manual – Origin**

  The linked model is placed by clicking a point within the canvas. The linked model's origin is aligned with the cursor.

- **Manual – Base Point**

  The linked model is placed by clicking a point within the canvas. The linked model's Base Point is aligned with the cursor.

- **Manual – Center**

  The linked model is placed by clicking a point within the canvas. The linked model's center point is aligned with the cursor.

Specify

When worksharing (aka, collaboration) is used, and the Revit elements have been segregated into different worksets to manage system resources on large projects, the designer can control which worksets are opened while initially placing the link. This can save a lot of loading time and computer memory if there are chunks of information not needed in the current project.

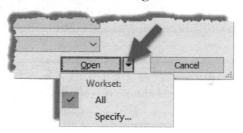

Clicking the down-arrow next to the **Open** button reveals the **Specify…** (Figure 5.1-3) option. Once Specify is selected, clicking the Open button will result in the Manage **Worksets for Link *File Name*** dialog to open (Figure 5.1-4).

Figure 5.1-3 Specify worksets option

Manage Worksets for Link Elementary School MEP17.rvt ✕

Name	Editable	Owner	Borrowers	Opened	
Architectural_Grids	No			Yes	New
Architectural_Levels	No			Yes	Delete
Architectural_Link	No			Yes	Rename
Mechanical_Grids	No			Yes	
Mechanical_Levels	No			Yes	
Mechanical_Link	No			Yes	
Structural_Grids	No			Yes	Open
Structural_Levels	No			Yes	
Structural_Link	No			Yes	Close
Workset1	No			Yes	
XY_REFERENCE	No			Yes	Editable
					Non Editable

Show:
☑ User-Created ☐ Project Standards
☐ Families ☐ Views

OK Cancel Help

Figure 5.1-4 Manage worksets for link dialog

Insert

Link IFC

The Link IFC tool is used to coordinate with other BIM platforms. The major BIM authoring tools, including Revit, have the ability to create a general file format which contains 3D geometry and element data which can then be used in other programs for coordination and collaboration. This is required because each software company develops their own proprietary format so they have freedom to implement new tools and element types.

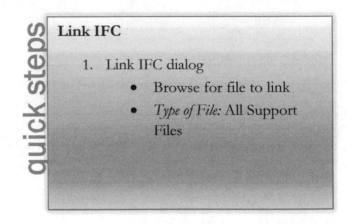

Link
IFC

quick steps

Link IFC

1. Link IFC dialog
   - Browse for file to link
   - *Type of File:* All Support Files

> IFC stands for **Industry Foundation Classes**. For more information on this open and neutral data format see: http://www.buildingsmart-tech.org/specifications/ifc-overview

To use this tool, simply **browse** to an IFC file and click **Open**. The IFC file is positioned center to center in the current model. It can be selected and repositioned as required. Once positioned, consider **pinning** the element so it is not accidentally moved.

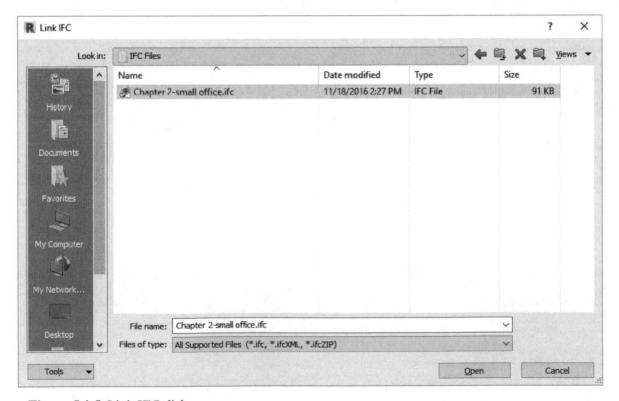

Figure 5.1-5 Link IFC dialog

Link CAD

The **Link CAD** command can only be used in a Project, not in a Family. Use this command to reference a CAD file, such as AutoCAD DWG, Microstation DGN, SketchUp 2018 SKP and more, modeled into the current project.

When this command is activated, the **Link CAD Formats** dialog appears (Figure 5.1-6).

The first step is to browse to the desired file; be sure it is in the correct location and will not be renamed before linking. Revit will not be able to find the file in the future if it has been moved to another folder or renamed.

The dialog has several options at the bottom. A brief overview of each of these settings/options will be covered next.

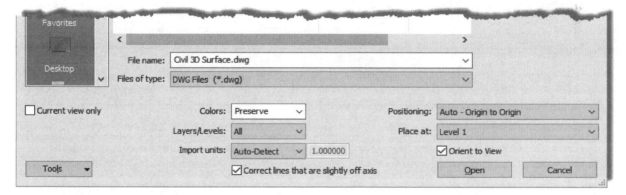

quick steps

Link CAD

1. Link CAD Formats dialog
   - Browse for file to link
   - Files of Type
     i. DWG – AutoCAD
     ii. DXF – Generic
     iii. DGN - Microstation
     iv. SAT – Generic 3D
     v. SKP – SketchUp
   - Current View Only
   - Colors
   - Layers/Levels
   - Import Units
   - Correct lines that are slightly off axis
   - Positioning
   - Place at (level)
   - Orient to View

Insert

Figure 5.1-6 Link CAD dialog

Current View Only

If this option is not selected, the CAD geometry, 2D or 3D, will appear in all views (plans, elevations, sections and 3D). If the entire CAD file is only 2D line work, the link will look like a single line in a section or elevation view.

> The visibility of the CAD file geometry can be controlled per view by typing **VV** and then selecting the **Imported Categories** tab in the Visibility/Graphic Overrides dialog. Expanding a listed link will reveal its Layers/Levels if they exist in that file.

When this option is checked, the CAD geometry will only appear in the current view. Do not use this option if the link is to be used to create a Revit toposurface.

This option is not available in 3D views.

Colors

This drop-down list offers three options:

- Invert
- Preserve
- Black and White

Preserve means the color of the elements in the CAD file will be used. Black and White changes all geometry to black lines while Invert changes lighter colors geometry to darker colors and darker to lighter.

There are various uses and reasons to select a color option. There are a few ways this feature might be used. If the CAD link will remain in the file, consider selecting **Black and White** to conform to Revit's way of displaying and printing geometry. If the CAD link is temporary, for example it is being used to create a toposurface or being traced, then use **Invert**. Use the **Preserve** if you want the colors to exactly match the CAD file—knowing that some colors may be more difficult to see because most CAD users work on a black background whereas Revit defaults to a white background in the canvas view. Seeing the colors provides a reminder that the file should be removed at some point.

> The line weights can be controlled using the **Import Line Weights** dialog. This is accessed by the small icon in the lower right of the Import panel on the Insert tab of the Ribbon. On the surface, it appears this is only for Imports, but it also applies to Links. This must be set prior to linking/importing the CAD file as changes to this dialog are not retroactive.

Layers/Levels

Revit uses **Categories** to segregate geometry in order to control graphics and visibility. In contrast, AutoCAD uses **Layers** and Microstation uses **Levels**.

This drop-down list offers three options:

- All (default selection)
- Visible
- Specify…

When **All** is selected, all Layers/Levels will be available within the Revit project. Layers/Levels currently set to not be visible within the CAD file are also turned off within Revit but can be turned on if desired.

If **Visible** is selected, only the Layers/Levels currently set to be visible within the CAD file are available. The Layers/Levels currently set to not be visible within the CAD file are not visible and are also not available to turn on either.

When Specify… is selected, once Open is clicked the Select Layers/Levels dialog is presented (Figure 5.1-7). This allows the

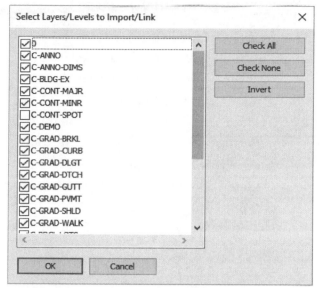

Figure 5.1-7 Select Layers/Levels to Link

visibility of Layers/Levels to be controlled before the link is placed in Revit. The Layers/Levels set to not be visible in the CAD file are unchecked by default. This requires an understanding of the Layer/Level naming convention contained within the CAD file.

> When a link is selected in Revit, click the **Query** button on the Ribbon and then click an element in the CAD file to see what Layer/Level it is on. The **Delete Layer** option will hide a Layer and remove access to control it. FYI: Revit does not actually change the linked file in any way.

Import Units

This option controls the scale of the CAD file being brought in. If Revit is using feet and inches (i.e., 12 units equal 1ft), and the CAD file is decimal feet (i.e., 1 unit equals 1ft) the file needs to be scaled up 12 times. When this option is set to **Auto-detect** things usually work out… but when the CAD file's units are not correct (possibly due to translation or export issues from the CAD program) then select the known unit (inches, feet, meter, etc.) or **Custom** and manually enter a scale factor.

> The scale of the linked file can be adjusted after placement. Select it, click Edit Type and then modify the **Scale Factor** parameter.

Correct Lines that are slightly off axis

This feature adjusts lines which are less than 0.1 degrees off the main axis. This is helpful for floor plans as CAD files can create issues when used to generate Revit elements—for example, if two walls are not perfectly parallel in Revit, it is not possible to add a dimension between them.

Given the irregular nature of site plans, this feature should be turned off to avoid changes.

Positioning

- **Auto – Center to Center**

  The link's center point is aligned with the current model center point. The center is calculated based on an imaginary bounding box around all geometry.

- **Auto – Origin to Origin**

  The origin of the Revit model and the CAD file are aligned. FYI: In a default Revit template, this is at the center of the four exterior elevation tags.

- **Auto – By Shared Coordinates**

  If shared coordinates has been previously defined between these two files, use this option to properly place the linked CAD file. This eliminates the need to manually move and possibly rotate the drawing data into the correct location.

- **Manual – Origin**

  The linked CAD file is placed by clicking a point within the canvas. The linked file's origin is aligned with the cursor.

- **Manual – Base Point**

  The linked CAD file is placed by clicking a point within the canvas. The linked file's Base Point is aligned with the cursor. This only applies to AutoCAD files which have a Base Point defined.

- **Manual – Center**

  The linked CAD file is placed by clicking a point within the canvas. The linked file's center point is aligned with the cursor.

Place at

This affects the vertical position of the CAD file within the Revit model. The default corresponds to the level associated with the current view, or, if in a 3D view, the lowest level is the only option. For plan views, the current level is the default, and only levels lower than the one selected are available to select from the list.

> Once the CAD file is placed, open an elevation or section to verify the vertical position of the file. Site drawings are often geo-referenced… meaning they are far from the origin (which relates to a benchmark in a specific state, county or city). Additionally, the geometry might be a large distance vertically relative to the origin (which relates to the ground surface position relative to sea level). This often requires the file be manually repositioned in plan (x,y adjustments) and in selection/elevation (z adjustment).

Orient to View

When this option is checked, the CAD file is aligned to the Revit model's **Project North** regardless of the current view setting. For example, if the current view is showing **True North**, then the CAD file is placed based on Project North.

That concludes the dialog options overview.

AutoCAD files have two "spaces" which can have drawing content: **Model Space** and **Paper Space**. There can only be one Paper Space. Each DWG file has at least one Paper Space but can have several. By default, Revit will only show content within Model Space. However, if Model Space is empty, Revit will show data contained in Paper Space.

The **question mark** in the upper right provides a quick link to open **Revit Help** and is focused on information related to this dialog.

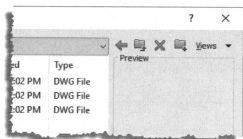

The **View** list offers a Thumbnail option to see a small preview of the CAD file selected.

Linked/Imported SKP, SAT and 3DM files also create new materials in Revit; these have a custom "Class" assigned, which can be used for sorting.

Link Topography

When collaborating with a Civil Engineer who is working with **Autodesk Civil 3D** you can link their topographical TIN surfaces into your Revit model. Once linked to your Revit model, changes made by the civil engineer will be synchronized as they make design changes. The Link Topography tool only works with TIN surface files from **Autodesk Civil 3D** files published to a **BIM 360** project. You must also have **Autodesk Desktop Connector** installed to facilitate the linking process.

Link Topography

quick steps

Link Topography

1. Publish Civil 3D file to a BIM 360 project.
2. Link Topography dialog
   - Navigate to BIM 360 project with the Civil 3D file you want to link.
   - Select the file.
3. Work with the resulting toposurface in your Revit model.
4. Reload the link from the Manage Links dialog in Revit to see published changes to the Civil 3D file.

Insert

When the Civil 3D file is linked to your Revit model, a toposurface element is created in your Revit model for each TIN surface in the linked file. If the incoming file has geolocation information included, the file is linked using the "Auto- By Shared Coordinates" option for linking. This positions the origin, or World Coordinate System (WCS) of the linked file to the shared coordinate point in Revit. Without geolocation information defined, the incoming file is linked using the "center to center" positioning option instead.

Once a Civil 3D file is linked you can work on the resulting toposurface elements in Revit as you would any other toposurface elements, adding sub-regions, building pads and site components. When the Civil 3D file is changed and reloaded the edits you have made to the toposurface will be retained while the points of the surface itself are updated. Linked topography files are updated in the same way as other linked files from the **Manage Links** dialog.

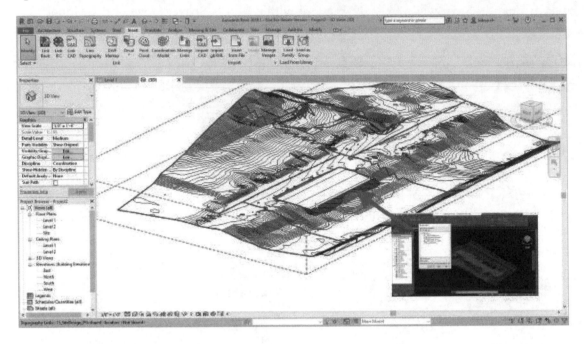

Figure 5.1-8 Civil 3D file and resulting toposurface in Revit.

Link DWF Markup

A DWF file is Autodesk's **Drawing Web Format**, which is meant to be an optimized format used to share CAD and BIM data. This format is generally not editable, so it offers the added benefit of not being able to change anything.

Autodesk Design Review allows markups to be added to sheets printed from Revit in DWFx format. A markup is a clouded area with text describing a change to the design.

quick steps

Link DWF Markup

1. Import/Link DWF File dialog
   - Browse for file to link
   - Type of File
     i. DWFx newer format
     ii. DWF older format
2. Link Markup Page to Revit Sheets dialog
   - Match DWF view (on left) with Revit View (on Right)

A current version of **Autodesk Design Review** can be downloaded from
https://www.autodesk.com/products/design-review/overview

The Link DWF Markup tool only loads the markups added to a DWF file. None of the original line work from the Revit model comes back into the project. The markups only appear on the sheets, not in any of the views on those sheets. The markup elements are pinned and cannot be unpinned. The only editable parameters are **Status** and **Notes** (Figure 5.1-9). When one of these parameters is changed, and then the Revit project is saved, the DWFx file actually is updated to record these comments. This is one of the few examples of Revit making changes to an external linked file.

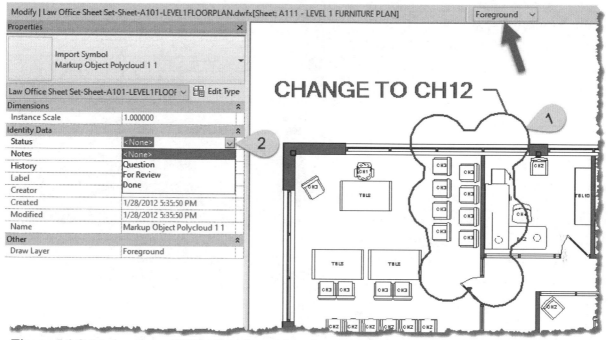

Figure 5.1-9 Markup from DWFx file selected in a sheet and Status expanded in Properties

There are two file formats for DWF files; the original format has a **.DWF** extension while the newer format has a **.DWFx** file extension. The Link DWF Markup command works with both file formats. The model shown below is from the *Interior Design using Autodesk Revit* book.

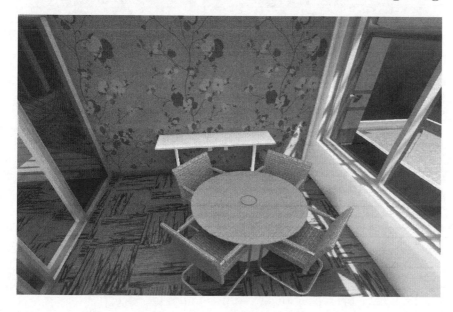

Decal

The **Decal** command is a slight departure from the other commands on the Link panel. Where the other commands are bringing in other editable design data (editable in other design software), the Decal command simply links to an image file which can then be placed one or more times within the model.

Decals can be used to represent artwork, graphics on monitors and even an entire building elevation. Two examples are pointed out in Figure 5.1-10.

The Decal image only appears when the View's Visual Style is set to **Realistic** or when the view is **rendered**. In all other cases the outline of the decal is all that appears.

If there are no Decal types defined, clicking the Decal command opens the **Decal Types** dialog (which is the next command covered in this section).

Each time the Revit model is opened, the Decal is loaded into the project, so the name and location of the image file cannot change. If the file cannot be found, no warning is offered, and the image is not displayed.

quick steps

Decal

1. Options Bar
   - Decal Dimensions (width and Height)
   - Lock Proportions
   - Reset button
2. Type Selector
   - Select Decal type
3. In-Canvas
   - Place Decal on the face of any Revit element
   - TIP: View cannot be set to wireframe as an element's face is not found

Figure 5.1-10 Decals used as art and computer screen

Decal: Options Bar

When the Decal command is started, the Options Bar presents a way to specify the size of the image before placing it in the model (Figure 5.1-11).

Decal: Type Selector

The **Type Selector** lists Decals which have been previously defined in the Decal Types dialog.

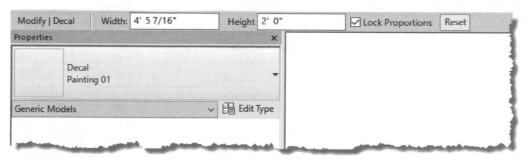

Figure 5.1-11 Options Bar and Type Selector for decal command

Decal: In-Canvas

Click on any surface to place the Decal; this works on walls, beams, doors (contained within a family) and more. The Display Style should not be set to Wireframe or Realistic. A Decal cannot be placed in wireframe mode and the image or outline is not visible until placed when in realistic mode.

The same Decal can be placed in multiple locations in a Revit project.

When to use a Decal or a custom material in Revit?

Decals have defined extents which make it easy to use for artwork and similar uses. Materials are often set to be repeating and cover an entire surface. The appearance asset of a material, which is often a raster image, and a Decal are both dependent on finding the external image file. If missing in either case, Decal or Material, the graphic just does not appear as it is not saved in the project file.

Some external applications do not support Decals. In this case it is usually possible to create a custom Revit material to do the same things—it just takes a little more time to set up.

In the image to the right, the entire building is represented by a single decal. When a Decal extends off the face it is hosted on, the portion of the image not on the face is ignored. The decal appears on the higher roof area in the center of the building, but the adjacent sky in the image is omitted.

For more information on Decals in Revit, see the author's guest post on the Enscape blog: https://enscape3d.com/best-practices-using-revit-decals-enscape/

Decal Types

Before a Decal can be placed in a Revit project, one has to be defined. This is done using the Decal Types command.

Clicking this command opens the Decal Types dialog (Figure 5.1-12). Use the small icons in the lower left to create, copy, rename and delete types.

Once a new named type is created, the only other essential step is to click the "..." button and browse to a **Source** file; these are the supported file formats:

- BMP
- JPG
- PNG
- TIF

quick steps

Decal Types

1. Decal Types dialog
    - Select from list of Decals in current project
        i. View and edit selected Decal's properties
    - Buttons to Create, Copy, Rename and Delete Decals
    - Buttons to toggle preview type: List, Medium Image, Large Image

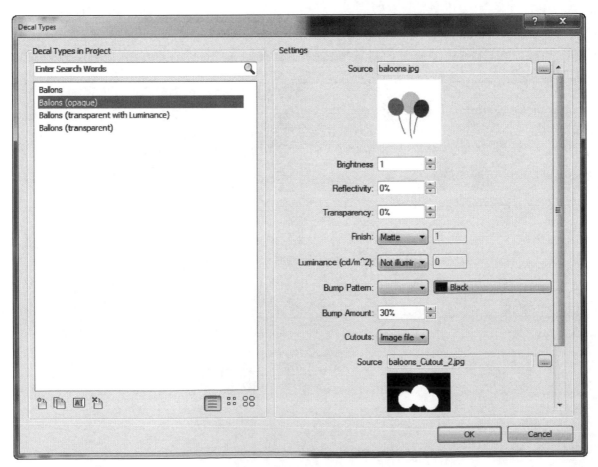

Figure 5.1-12 Decal Types dialog

The Decal Types dialog offers a number of options to control how the decal appears in the model.

The background for a decal cannot automatically be transparent (see image to the right). However, you can add a "cutout" image to get the various options shown below. Notice that any color other than white creates a shade of transparency, and solid black creates a completely transparent background.

Like texture files (e.g. jpg or png), you need to save these files in a shared location so everyone has access to them.

> Unfortunately, transparent backgrounds in PNG files will not work in a Decal.

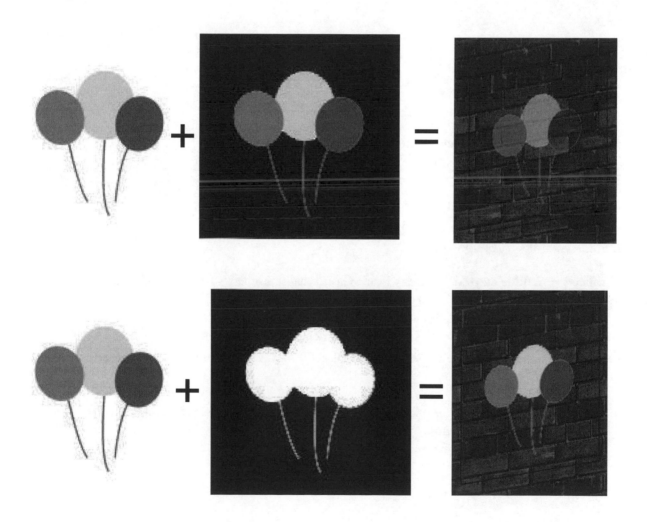

Insert

As seen in the Decal Types dialog previously shown, there are several settings for each decal type. One option is to set a Luminance value, to make the surface appear to glow. As seen in the image to the right, this option washes out the image, but may still have some useful applications.

> When you select a Decal in your model, you can adjust its size on the Options Bar or by dragging one of the corner grips. Turning off "lock proportions" allows the image to be stretched to any size.

Link Point Cloud

Use this command to link Point Cloud data into the current project.

A Point Cloud is the result of laser scanning a building—where each point represents a surface and color of the existing conditions. This can be inside or outside of a building. The Point Cloud files are usually very large, multiple GB, and contain millions or billions of points.

quick steps

Link Point Cloud

1. Link Point Cloud dialog
    - Browse for file to link
    - *Type of File:* Several standard formats supported
    - Positioning

When this command is started, the Link Point Cloud dialog appears. Refer to the Files of Type drop down near the bottom to see a list of supported file types.

Autodesk ReCap offers additional Point Cloud functionality; using this tool one can import raw data and optimize it for use in Revit (or AutoCAD). It is also possible to measure and view the scanned data in ReCap.

The three **Positioning** options are all manual, which means a point must be picked in the canvas area to place the scanned data.

Link Coordination Model

Use this command to link a **Navisworks** model into Revit. Autodesk Navisworks is a separate application used to coordinate and manage multiple model file formats.

A coordination model is extremely light weight, which allows large complex models to be referenced in without significant performance degradation.

Because a coordination model is linked, there is no import or bind option; the original file can be overwritten. When Revit is opened, all linked files are reloaded.

Link Coordination Model

1. Coordination Model dialog
    - Positioning
        i.   Origin to Origin
        ii. By Shared coordinates
    - Add
        i.   Browse for file
    - Reload From
    - Reload
    - Unload
    - Remove

Another benefit of a coordination model is that the file cannot be edited. A lot of design firms are not willing to share their native Revit models with contractors and/or manufacturers due to concerns about intellectual property and liability. A Navisworks NWD or NWC file provides a way to share the model, for coordination purposes, without the possibility of Revit content being reused or the project model being modified.

To use this command: simply click the **Add** button, **browse** for a file and click **OK**. Unlike other links, a coordination model can only be managed using the original **Coordination Model** command on the Insert tab—rather than using Manage Links. Notice, in Figure 5.1-13, that the linked models are listed along with **Reload From**, **Reload**, **Unload** and **Remove** options.

The individual elements within the coordination model are not selectable, nor can their visibility be controlled. This model has no impact, it is completely ignored, in energy modeling. Also, while the coordination model does not produce any shadows, it receives shadows generated by adjacent native Revit geometry seen in an image to follow (Figure 5.1-14).

Once a coordination model is loaded into a project, the **Visibility/Graphics Overrides** dialog will have a **Coordination Models** tab (Figure 5.1-15). Use this tab to control the visibility of the linked model, or expand to see each instance—only do this if more than one instance has been placed. The linked model can also have a transparency applied.

When selected, the coordination model has limited instance properties (Figure 5.1-16). It can have a name, which is helpful when multiple instances exist. Notice **Phase Created** and **Phase Demolished** parameters are also present; this applies to the entire link. Unfortunately, phasing within the coordination model is not supported.

Insert

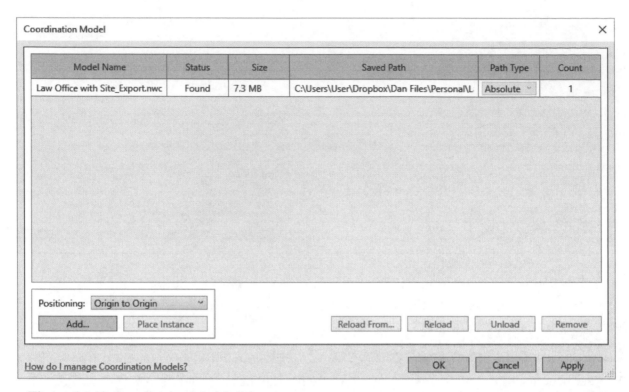

Figure 5.1-13 Coordination Model dialog

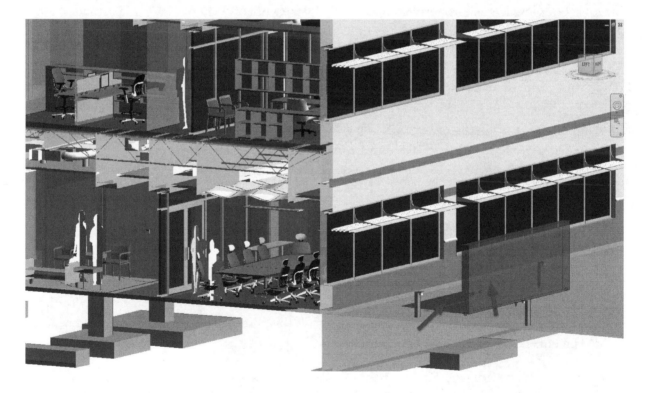

Figure 5.1-14 Coordination model (Navisworks file) linked into Revit model; native Revit wall/shadow

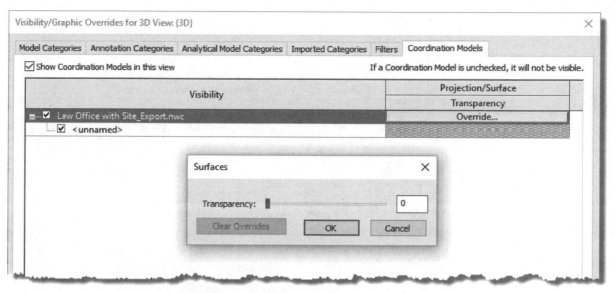

Figure 5.1-15 Visibility/Graphics Overrides dialog – Coordination Models tab

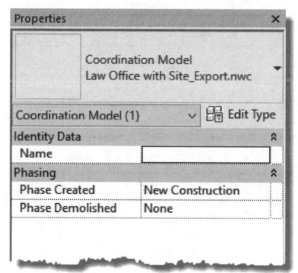

Figure 5.1-16 Instance properties for
selected Coordination Models

Link PDF/Image

These tools are used to link PDF and raster image files into Revit. When working with a multiple page PDF file, each page is processed as an individual "image". Each page needs to be imported one at a time from the multi-page PDF file. In the Import PDF dialog (Fig. 5.1-17), each page is displayed as an individual image. Select the page you wish to import and click ok to import the page from the PDF file.

Linked PDF files with vector information will allow you to snap to lines in the imported PDF image. Raster PDF files will not offer snapping points.

quick steps

Link PDF/Image

1. Link PDF/Image dialog
   - Browse for raster image file to place
   - For multi-page PDf files, select the page to import
2. In-Canvas
   - Move cursor to position scaled rectangle, which represents size of image to be placed, and then click.

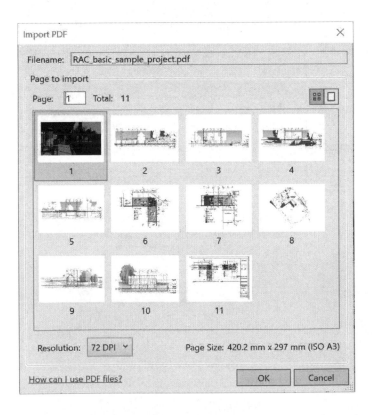

Figure 5.1-17 Import PDF dialog with multipage PDF

The imported images can be placed in all views except 3D and Schedule views. When this command is selected, simply browse for an image file and then click to place it. The size is not editable until after it is placed.

File formats supported:

- PDF
- BMP
- JPG
- JPEG
- PNG
- TIF

This feature might be used to bring in a scanned image of old hand drawn floor plans to use as a reference while modeling an existing building. When finished, select the image and delete it. Another use is to place a scanned copy of an architect or engineer's signature on a sheet for the final bid set.

This image type does not appear in 3D or renderings. For those needs a Decal or Material should be used (both of those options never save the image file within the project). Note that a transparent background in a PNG file is also transparent in Revit.

> Use a photo editing program to reduce the size and quality when possible.

Some settings such as Size (Width and Height), Draw Layer (Background and Foreground) and draw order (Arrange) may be changed after the image is placed.

Use the **Manage Links** dialog, on the Manage tab, to **Delete**, **Reload** or **Reload From** any imported image files. Reloading will update the import if the external file has changed. The Reload From allows a different image to be used, saving time not needing to delete an image and then import another, and then position and scale the image. See the Manage tab chapter for more information on the Manage Links dialog functionality.

Manage Links

This tool is repeated on this tab for convenience; see the **Manage tab** chapter for detailed coverage of this feature.

Insert

5.2 Insert Tab: Import Panel

Import CAD/PDF/Image

This tool is essentially the same as the "Link" tool of the same file type with one exception: the result is a static file with no ties to an external file. **In general, this "Import" feature should never be used!** Several negative issues can occur when a file is Imported rather than Linked. First, the file must be selected before it can be deleted from the Revit project. Selecting an imported link can be difficult if it only appears in one view and may even be hidden in that view. Additionally, for CAD files, multiple text and line styles are created in the Revit project, which are not easy to clean up.

> It should be noted that the Link option still saves an entire copy of the CAD file in the Revit project. This file will be displayed if the external source cannot be located—otherwise, the saved internal copy is updated.

See the previous **Link CAD/PDF/Image** tool for detailed coverage of this feature.

Import gbXML

This command is used to import the results from an energy analysis of the current project. Heating and Cooling loads, for example, can be saved in the project and then displayed in schedule and tags on the construction documents.

A gbXML file contains data and not geometry. See the gbXML Export section in the Application Manu chapter for more on this file format.

quick steps

Import gbXML

1. Import gbXML dialog
   - Browse for XML file
2. Select data to import from gbXML dialog
   - Select the categories to import data from, into your project

Import Line Weights

The small icon in the lower right of the Import panel opens the Import Line Weights dialog. This controls the thickness of each line based on the color assigned in the CAD file—in AutoCAD the line weight is usually based on the color of each element.

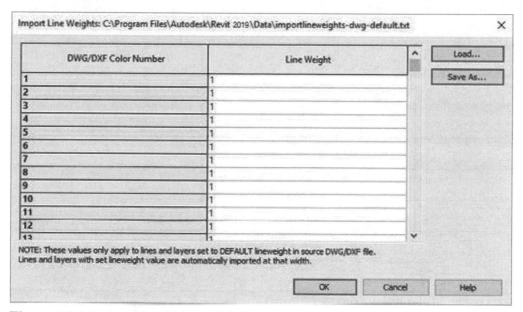

Figure 5.2-1 Import Line Weights dialog

As seen in the previous image (Figure 5.2-1), by default everything in the CAD file is mapped to Line Weight 1 in Revit. This means all lines are the same thickness. Either manually edit the Line Weight column or click Load to import one of the standards provided with Revit.

> Revit line weights are defined in **Additional Settings → Line Weights** on the **Manage** tab.

The line weight mapping is only applied when the file is first linked in. Making changes to this dialog has no effect on previous links/imports.

When creating a custom Line Weight file, use the Save As button to save a copy to be used in other Revit projects. In an office setting, this file should be saved in a shared location so everyone can access it. It is also possible to create a custom deployment, which is a preconfigured way to install Revit on new computers. The default Revit.ini file can be edited so all users immediately have a custom Import Line Weight settings file loaded. This will ensure all CAD files have the correct line weights. This is the line which can be edited within the Revit.ini file:

Insert

Line in Revit.ini file:

ImportLineweightsNameDWG=Data\importlineweights-dwg-default.txt

Location of Revit.ini file:

C:\Users*your user name here*\AppData\Roaming\Autodesk\Revit\Autodesk Revit 2021

As the Revit dialog states, lines in the CAD file which have a specific line weight assigned by element or via the Layer Manager, will automatically have that line weight assigned in Revit. This dialog is for everything whose line weight is set based on color—which is the most common method.

The image (Fig. 5.2-2) below shows the contents of a saved import line weights settings file.

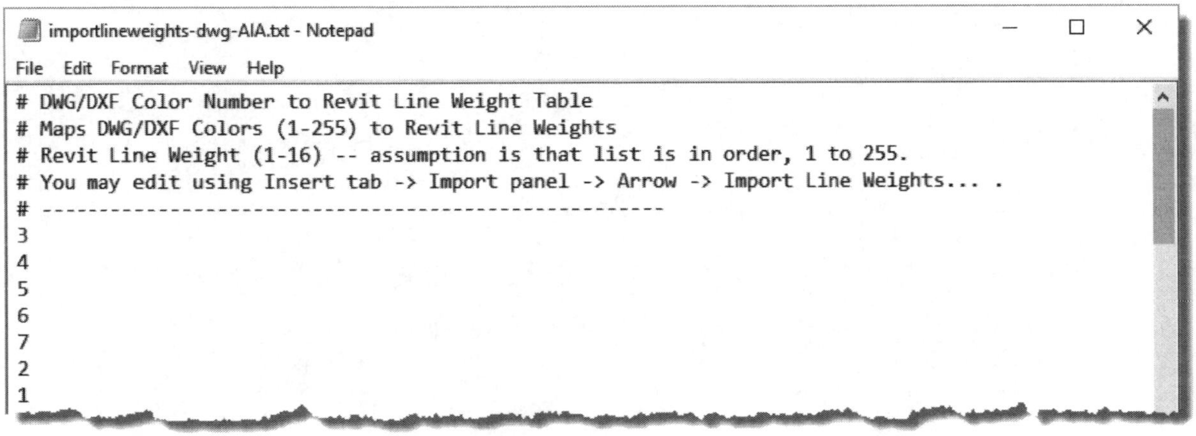

```
importlineweights-dwg-AIA.txt - Notepad                                —    □    ×

File  Edit  Format  View  Help
# DWG/DXF Color Number to Revit Line Weight Table
# Maps DWG/DXF Colors (1-255) to Revit Line Weights
# Revit Line Weight (1-16) -- assumption is that list is in order, 1 to 255.
# You may edit using Insert tab -> Import panel -> Arrow -> Import Line Weights... .
# --------------------------------------------------------
3
4
5
6
7
2
1
```

Figure 5.2-2 Example of import line weights text file contents

Import Family Types

This tool is only available while in the family editor, which is whenever a family (rfa) file is open.

Import Family Types

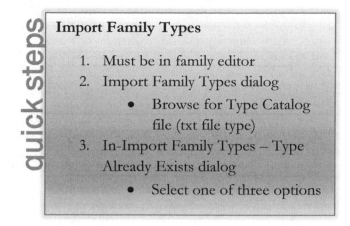

In a family, multiple **Types** can be created to save various configurations of values for the **Type Parameters**.

When this list gets too long, the types can be moved into a Type Catalog as a way of managing how many types are loaded into a project. Controlling the types loaded with a family helps keep the Type Selector manageable and also limits the selection to relevant options for a given project.

When the Types need to be loaded into a family for testing or to move away from a Type Catalog for a particular family,

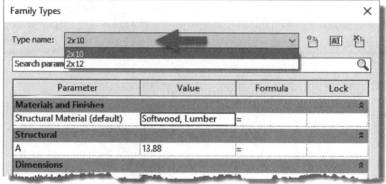

Figure 5.2-3 Original types in family

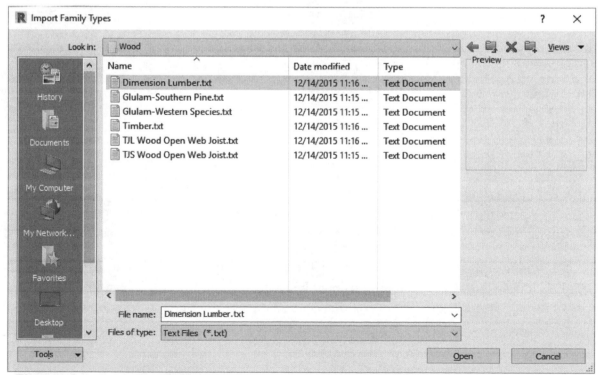

Figure 5.2-4 Import Family Types dialog

Insert

Revit offers the **Import Family Types** command.

To demonstrate how this command works, the Dimension Lumber.rfa family is opened from the Structural Framing\Wood folder. Notice, in Figure 5.2-3, that there are two types in the current family: 2x10 and 2x12.

After clicking the **Import Family Types**, browse to the Dimension Lumber.txt file located in the same folder as the family.

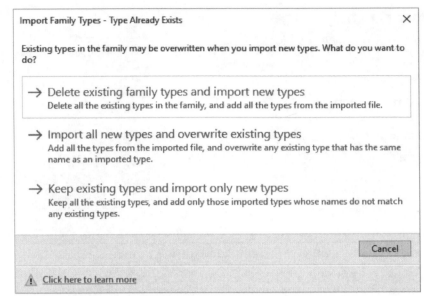

Figure 5.2-5 Import family types prompt

> A Type Catalog has the same name as the family and is located in the same folder.

After clicking Open, if there are duplicated Types—in the family and the type catalog—then Revit provides the prompt shown in Figure 5.2-5.

The final result, in this example, is several more Types loaded into the family (Figure 5.2-6).

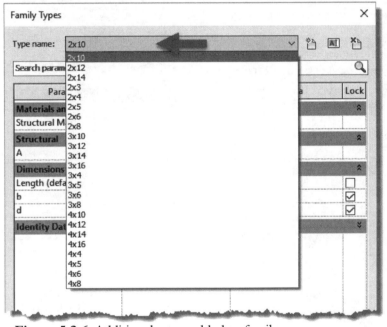

Figure 5.2-6 Additional types added to family

If the intent is to eliminate the Type Catalog, then after the import is complete and the family is saved, the Type Catalog should be renamed, moved or deleted so it is not found when this family is loaded into a project.

> Looking at this example family, when Type Catalogs are used, the family itself should only have one type and the Type could be named "**Type Catalog Missing**" rather than "2x10." This will alert the user if they load a family the wrong way (drag and drop from Windows) or the Type Catalog is missing (common with downloaded content).

5.3 Insert Tab: Load from Library Panel

Load Family

By far, the most used command on the Insert tab... Only families loaded in a project are available for placement within that project. This tool works in a project or a family. When used in a family, the loaded family becomes a nested family. In a project, it becomes available for placement within the project using the appropriate command or dragging it out of the Project Browser into the canvas area.

> Third-party developers have created add-ins for Revit to simplify the family loading process. They look similar to the Project Browser but are pointing to content on the hard drive or server. Double-clicking or drawing a listed item automatically loads the family and starts the appropriate command to place the item.

quick steps

Load Family

1. Load Family dialog
   - Browse Revit family to load; rfa file type
   - Multiple families from the same folder may be loaded at once
     i. Hold CTRL key while selecting multiple files
2. Specify Types dialog
   - This only appears if a Type Catalog exists for this family
   - Select the family on the left (if loading more than one)
   - Select one or more types to load on the right.

When the **Load Family** command is clicked, the default folder can be changed. This is done via **Application Menu → Options → File Locations → Places**. The first item in this list determines the default location when the Load Family command is activated (Figure 5.3-1).

To use this command, simply browse to a family (rfa) file location on the hard drive or network and click **Open**.

Note that **System Families**—like walls, floors, ceilings, roofs, ducts, pipes, cable tray—cannot exist outside of a project environment. Thus, they

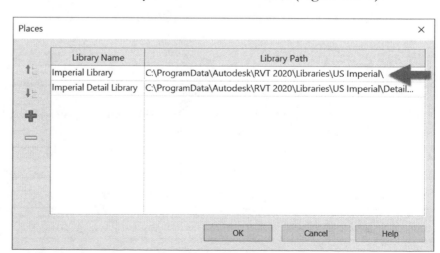

Figure 5.3-1 Options, File Locations, Places dialog

cannot be loaded into a project using the Load Family command. The same is true for **Model In-Place** families. The Load Family command only works with **Loadable Families**.

Type Catalogs

Continuing with the example family, <u>Dimension Lumber.rfa</u>, from the previous command, the process of using a **Type Catalog** in conjunction with the **Load Family** command will be shown now.

The image below shows the **Dimensional Lumber.rfa** family selected (Figure 5.3-2). Notice this dialog is being filtered to only show **RFA** and **ADSK** files by the Files of Type drop-down list at the bottom of the dialog. This Type Catalog, which is a **TXT** file, is not visible here.

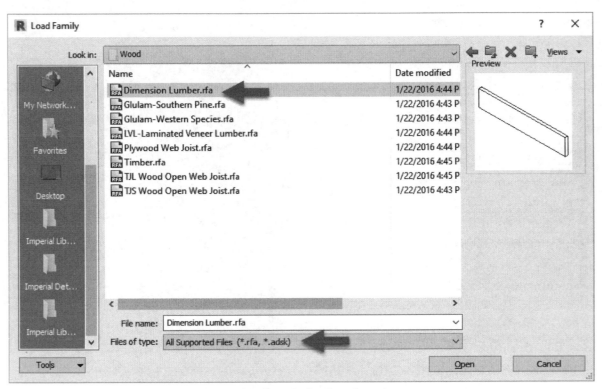

Figure 5.3-2 Load Family dialog; provided Revit family selected

If a Type Catalog exists when **Open** is selected, the **Specify Types** dialog is displayed (Figure 5.3-3). Because multiple Families can be opened at one time (from a single folder), all the Families queued up to be loaded are listed on the left. In this example, only one Family was selected. Selecting a Family listed on the left will cause its available Types to be displayed on the right side. Note that some of the families may not have a Type Catalog and this will not display anything on the right when selected. But the Specify Types dialog will only appear if one of the selected Families has a Type Catalog associated with it.

Select one or more types, holding the Ctrl key to selected multiple types, and then click **OK**.

If the family already exists, the new types will be added to that family. If any of the Types already exist, a prompt will be shown asking if the one being loaded should overwrite the values of the Type in the project.

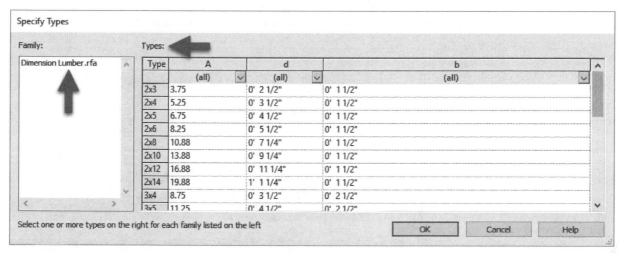

Figure 5.3-3 Type Catalog; Specify Types dialog

The image to the right shows the contents of the **Type Catalog** (Fig. 5.3-4). This file can be edited to add new types. The first row defines the parameters each value is associated with in the family.

The image below shows the **Type Catalog** file, and associated family, as seen in **Windows Explorer** (Fig. 5.3-5).

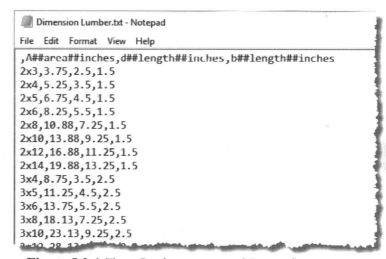

Figure 5.3-4 Type Catalog contents (Notepad)

Figure 5.3-5 Revit family and Type Catalog files in Windows Explorer

Get Autodesk Content

When you install Revit a selected base set of content is installed. This base content is enough to give you basic functionality, but there is more sample content created by Autodesk that is available, including localized content in different languages. Use the Get Autodesk Content button to be directed to an online location where additional content packs can be downloaded and installed for your use.

Load as Group

The Load as Group command takes another project file and loads it as a Group into the current project.

This command has several uses. One might be to manage the layout of a kitchen or an entire apartment unit in its own Revit project file. The file is then loaded as a group, rather than being linked, so the walls in the group clean up with walls in the main project. When a change is made to the external file, it is loaded again—Revit provides a prompt to update the group.

To describe how this command works, a new project was created and has only two model elements placed: a chair and a desk. Additionally, two furniture tags were added, two Detail Lines (the "X" on the chair) and one dimension.

quick steps

Load as Group

1. Load File as Group dialog
    - Browse Revit project file to load; rvt file type
    - Options:
        i.   Include attached details
        ii. Include levels
        iii. Include grids
2. Large file warning
    - If the file is larger than 10MB, Revit will warn the user of the potential performance issues with loading a large file.
3. Duplicate Types
    - Usually just click OK
4. In-Canvas
    - Nothing to do In-Canvas as the project is just loaded in the Project as a Group.

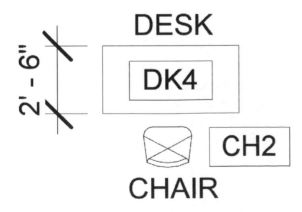

For simplicity in understanding this command, open a new project with nothing placed yet.

Select the **Load as Group** command and browse to a Revit (RVT) project file; for this example, the one shown above (chair and desk).

In addition to selecting a Revit project file there are three options at the bottom of the **Load File as Group** dialog: **Include attached details**, **Include levels** and **Include grids**.

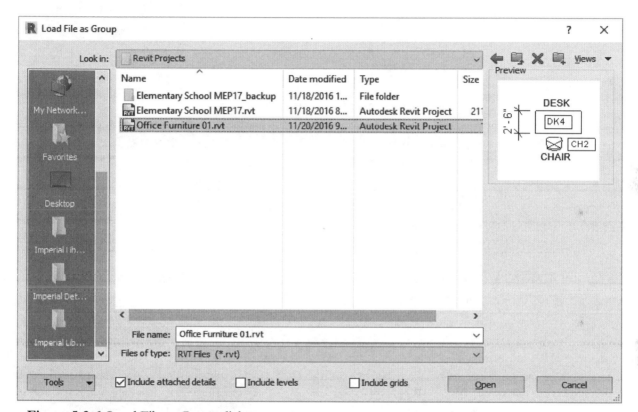

Figure 5.3-6 Load File as Group dialog

Here is what each of the options do:

Include attached details
Any non-model elements, such as Text and Detail Lines, which have been grouped in the external Revit project file can be loaded, too. Otherwise, the only thing loaded is model elements. More on this in the example to follow.

Include levels
Levels in the external Revit model are created in the current project.

This option should only be used if the external project files have unique Levels not already in the current project. Even then it might be better to create the necessary level(s) first. The challenge with using this option is the potential to load multiple levels at the same elevation. Doing so makes the model messy to adjust floor levels and use tools like Stairs. Plus, the problem is difficult to correct, and it is not easy to select elements based on the levels they are associated with. Also, if a level is deleted, all the elements associated with it, e.g. walls, furniture, etc., are deleted without warning.

When elements are brought in with this command, and Include levels is turned off, the elements automatically associate with the nearest levels, thus ensuring the proper levels already exist in the current project, and at the correct elevation, providing the ideal result.

Include grids

Grids in the external Revit model are created in the current model.

Once the external file is loaded, with **Include attached details** checked, the **Model Group** and **Detail Group** are now loaded in the project—but have not been placed yet. These two Groups can be seen in the **Project Browser** (Figure 5.3-7).

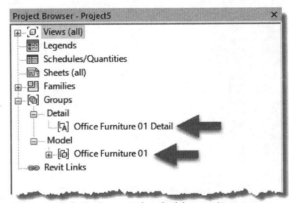

The **Model Group** can now be placed in the current project. This is done in one of two ways.

Figure 5.3-7 Groups loaded in project

Placing a Model Group:

- **Architecture** tab → **Model Group** → **Place Model Group**
- From the **Project Browser**: Right-click and select Create Instant or drag into canvas area

For our example, the result is a Model Group instance which only contains the model elements—the chair and desk (Figure 5-3-8). When it is selected in the project, the **X,Y insertion point** (or origin) appears; this can be repositioned by dragging it. The new origin will be used when placing future instances of this group.

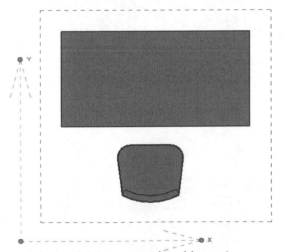

Figure 5.3-8 Group placed in project

Once an instance of the Model Group is placed, the **Detail Group** can be placed. There are multiple ways to place a Detail Group.

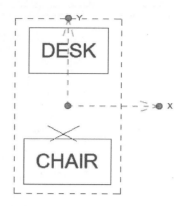

Placing a Detail Group:

- **Annotate** tab → **Detail Group** → **Place Detail Group**
- From the **Project Browser**: Right-click and select Create Instant or drag into canvas area
- Select a Model Group instance then click **Attached Detail Groups** on the Ribbon.

Figure 5.3-9 Detail group placed manually

The result of using either of the first two options is an independent instance of the Detail Group   meaning it is not associated with the Model Group (Figure 5.3-9). Also notice the two hosted elements, the furniture tags, are not present. Only the two text elements and the two detail lines appear.

The third option listed produces a different result. The image to the right, Figure 5.3-20, shows the Ribbon when the Model Group is selected.

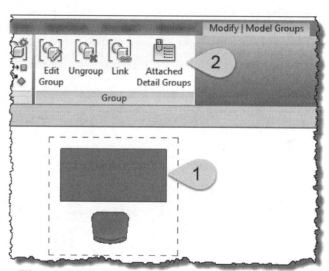

Figure 5.3-10 Model group selected

Clicking the **Attached Detail Group** automatically positions the Detail Group properly relative to the Model Group; the Detail Group cannot be moved apart from the Model Group (Figure 5.3-11). Additionally, notice the two furniture tags appear.

Creating a Detail Group in the external file is not required. If the dimension, two text elements, two detail lines and two furniture tags existed in the external file—but not part of a Detail Group—they would still be imported if the **Insert Attached Details** item is checked. They just would not be visible in the Project Browser.

When **Include Attached Details** is selected, any view, the Model Group appears in will have that view's 2D elements imported as well. For

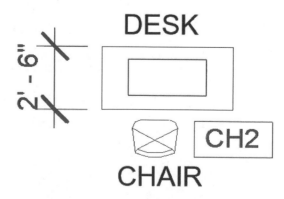

Figure 5.3-11 Group placed in project

example, in the south exterior elevation, if text and Detail Lines are added, then clicking Attached Detail Group in the south elevation of the current project, while the Model Group is selected, will result in those elements being positioned in that view.

Reloading a Model Group

When the external Revit file is updated, it can be reloaded into the current project. Revit presents the message shown here (Figure 5.3-12). Click **Yes** to replace the original version; click **No** to create a new group in the current project. Click **Cancel** to do nothing…thus canceling the command.

The image on the left shows the original group in the current project (Figure 5.3-13). In this case we notice the desk is backwards relative to the chair. To fix this issue, open the original file and edit the desk, save the file and close it (or use the Architecture tab → Model Group → Load into

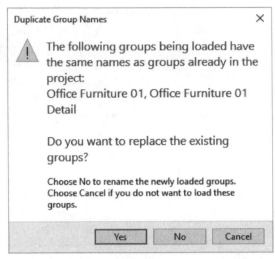

Figure 5.3-12 Duplicate group names

Open Project as Group command at this point). Back in the main Revit project, use the Load as Group command again to reload and replace the original group. The result is seen in Figure 5.3-14.

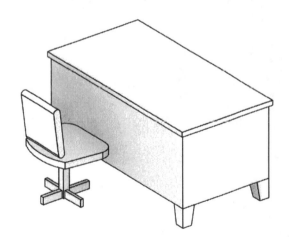

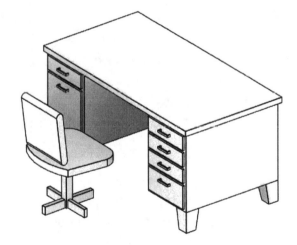

Figure 5.3-13 Original group

Figure 5.3-14 Reloaded group

Group Origin

The **origin** of the Model Group is defined relative to the Project Base Point in the external file that contains the loaded model elements. To see this in that project, type VV and expand the Site category (on the Model Categories tab) and turn on Project Base Point for the floor plan view.

Removing a Group from a Project

If a large Model Group is no longer needed, it should be deleted from the project to clean house and keep the overall file size down (not that file size is a primary indicator of Revit project file performance). Simply **right-click** on the Model Group in the **Project Browser** and select **Delete**; Delete is only available if there are no instances of the group placed in the current project. To delete all instances of the group selected, right-click and select all instances and then press the Delete key. If an instance cannot be found to select, place a new copy and then follow the steps just mentioned. The Manage tab → Purge unused command will also remove the group from the project… if unused of course (meaning there are no instances placed).

Insert Views from File

Use this command to extract Views, Schedules or Sheets from another Revit project file without opening it directly. Only "2D" views, such as Drafting, Schedules and Sheets, may be inserted from another file. Views with 3D geometry, like Plans, Sections and Camera, are project specific and may not be inserted using this tool.

> When inserting Sheets and Views at the same time, from a project, the Views will already be placed on the sheet to match the layout in the selected external project.

quick steps

Insert Views from File

1. Open dialog
   - Browse for RVT file
2. Insert Views dialog
   - Select type of view from drop-down list
     i. All (default)
     ii. Drafting only
     iii. Schedules only
     iv. Sheets only
   - Check each View/Sheet to import

Insert

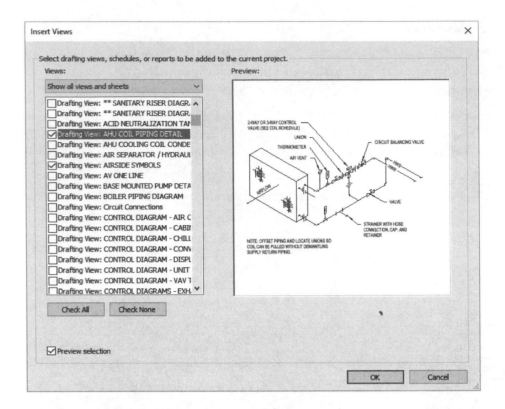

Figure 5.3-15 Insert Views dialog

When **Insert Views from File** → **Insert Views from File** is selected, the **Insert Views** dialog opens (Figure 5.3-15). In this dialog, the **Views** drop-down at the top filters the list of available views to import. Before changing the selection, click **Check None** to ensure items are not checked which are not visible.

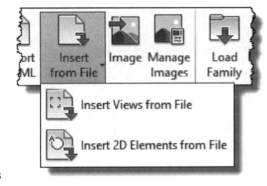

When a view is checked its preview is displayed to the right—assuming the **Preview selection** option is checked.

> Sheets with Legend views placed on them cannot be inserted. Also, if the file is not the same version of Revit a temporary upgrade will be required.

Clicking OK will import all the checked views, sheets and/or schedules into the current project. One of the views will be set as current on the screen. The items may not appear in the expected location in the Project Browser if either file is using a custom filter or view type naming.

When any text types, fill patterns or materials already exist in the current project Revit will provide a **Duplicate Types** prompt as shown here. The current project settings take precedence over anything being imported. If the type does not exist it will be created as defined in the original document.

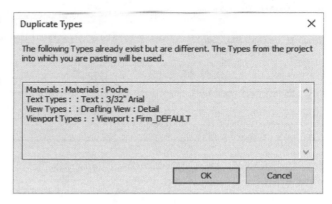

This command can be used to facilitate a detail library within a design firm. Revit project files with a firm's standard details can be placed in a central location and then loaded into projects as needed. Maintaining a detail library can save a designer, and their firm, an enormous amount of time.

Insert 2D Elements from File

This tool is a quick way to insert all the 2D elements in a view from another Revit project. The 2D elements are placed in the current view. If the current view is a model view, such as Plan, Ceiling, Section, etc., any dimensions which cannot find something to reference will be deleted.

> What exactly are "2D" elements? Anything on the Annotate tab is considered 2D and view specific; Text and Detail Line, for example.

quick steps

Insert 2D Elements from File

1. Open dialog
   - Browse for RVT or ADSK file format
2. Insert Views dialog
   - Select a single view to copy all of its 2D elements to the current view in the current project

Insert

When this command is selected, the **Insert 2D Elements** dialog appears (Figure 5.3-16).

In this example, a newer version of the project file needs 2D elements added back in from an older, archived, project file. The model shown behind the dialog does not have dimensions on

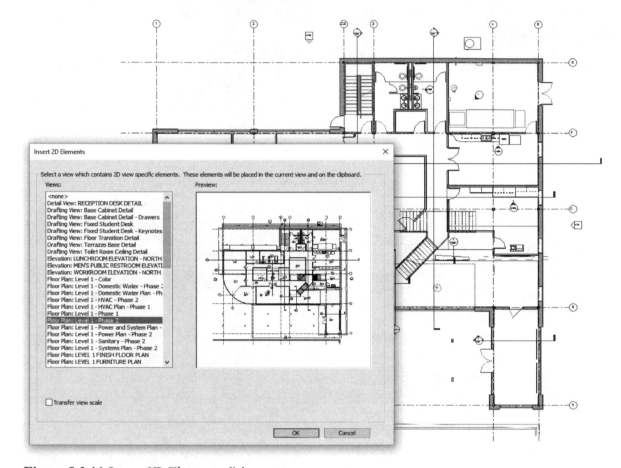

Figure 5.3-16 Insert 2D Elements dialog

the grids, for example. Notice the preview in the dialog does have several dimensions between the grids on the north side of the building.

Clicking **OK** closes the dialog and allows a point to be picked on the screen to place the new 2D elements. One thing that is a bit tricky is the initial preview is in the exact same location as the original file, but there is no corresponding point to pick… making it difficult to accurately place the elements in the exact same location (which is needed for the hosted elements, like dimensions, to find their new hosts).

Once a point is clicked, a unique opportunity to **Edit Pasted Elements** is found on the Ribbon, in addition to **Finish** and **Cancel** buttons (Figure 5.3-17).

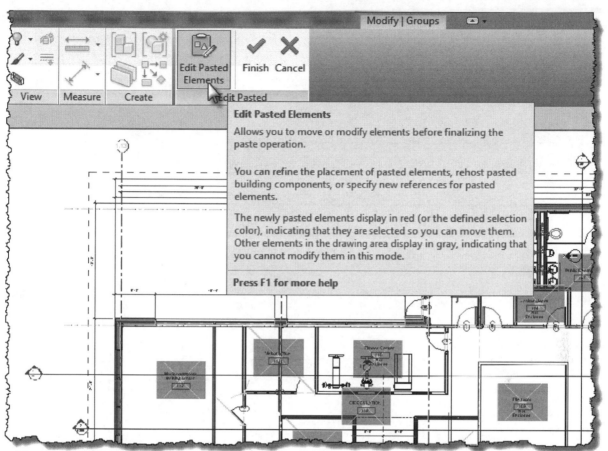

Figure 5.3-17 Insert 2D Elements command in progress

Clicking Edit Pasted Elements opens a special edit mode where elements can be repositioned before finishing the paste command. While in this mode, the floating **Edit Pasted** panel is displayed as shown here.

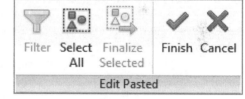

When **Finish** is selected, any elements which cannot find a host are deleted. A prompt similar to this one is provided by Revit. In this example, 8 dimensions will not be copied (Fig. 5.3-18).

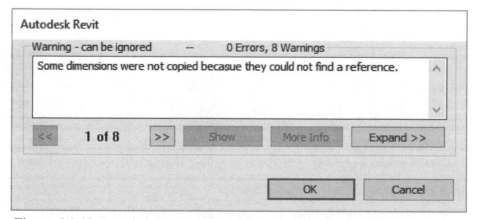

Figure 5.3-18 Pasted elements will be deleted because host not found

Notes:

## Chapter 6

Annotate Tab

The Annotate tab can be summarized as containing **all 2D and View Specific commands**. Any command used from this tab will only appear in the view it was created in; the only exception being dependent views. Notes added in a Code Plan will not appear in the Furniture Plans, and dimensions added in an Architectural Plan will not appear in an Electrical Power Plan.

Most of these commands can be used in most views including sheets, legends, 3D views, model views and drafting views. None of these commands can be used in schedule views.

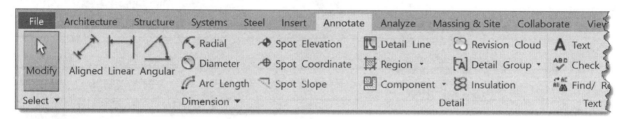

6.1 Annotate tab: Dimension Panel

All the commands available for dimensioning the model can be found here. This image also shows the **expanded panel area** which provides access to modify and create (via Duplicate) types for each dimension command.

This first section will review each of these dimension commands.

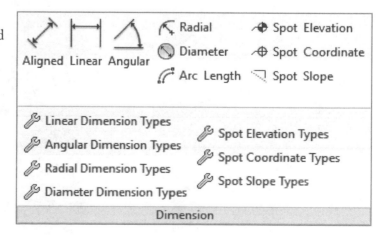

Dimension: Aligned

The **Aligned** dimension tool is the most used of the multiple dimensioning tools. Its purpose is to graphically indicate the distance between two elements or points. The Aligned dimension is dependent on two picked elements or points. Like a door or window being dependent on a wall to exist, if a referenced element is deleted, the dimension is deleted as well. Keep in mind, the deleted dimension may not be in the same view where the model element is being deleted.

The resulting dimension is based on how the two reference picks are made. In the image below, the "Aligned" dimension on the right was created by selecting directly on the angled line on the <u>left</u> and then tapping **Tab** to pick a point on the <u>right</u>. The other "aligned" dimension used **Tab** to pick points for <u>both</u> picks.

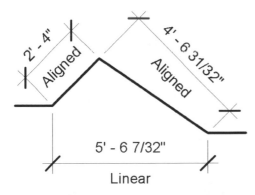

For the dimension on the right, notice the witness line is parallel (or aligned, as the name implies) with the selected element. By contrast, the **Linear** dimension is always horizontal or vertical; this command is covered next.

---

quick steps

Dimension: Aligned

1. Ribbon
   - Option to switch/toggle to another dimension tool
2. Options Bar
   - 'Dimension to' setting
     - i. Wall Centerlines
     - ii. Wall Faces
     - iii. Center of Core
     - iv. Faces of Core
   - Pick
     - i. Individual References
     - ii. Entire Walls
   - Options button
     - i. Available when Pick: Entire Walls is selected
     - ii. Select what to automatically dimension to on a selected wall
3. Type Selector
   - Select dimension type
4. Type Properties
   - Units Format; rounding
5. In-Canvas
   - Tap the Tab key to toggle between layers in a wall
   - Picking multiple elements will create a dimension string
   - Picking same element a second time will remove the reference
   - Click the padlock icon after placing a dimension to lock it
   - Click the "EQ" icon after placing a dimension string to equally space the dimension elements

Aligned: **Ribbon**

Once the Aligned command is started, the Ribbon provides an easy way to switch to another dimension command. Maybe the wrong command was selected, or a significant dimensioning effort requires various dimensions; after adding a dimension, when another command is needed it can be easily selected from the Ribbon.

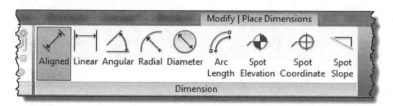

Aligned: **Options Bar**

The first drop-down list relates to how walls are to be dimensioned. The options are **Wall Centerline**, **Wall Faces**, **Center of Core** and **Faces of core**.

When set to **Wall Centerline**, clicking on the wall will result in a witness line being placed at the geometric center of the wall (not including any sweeps). The **Wall Faces** option will favor faces (or the construction "Layers" within the wall). This latter option requires greater attention as it is not just looking for outer faces; it is even possible to dimension to sweeps contained within the wall. The two core options, **Center of Core** and **Faces of Core**, relate to what is considered the structural Layer, or Core, of the wall as pointed out in Figure 6.1-1.

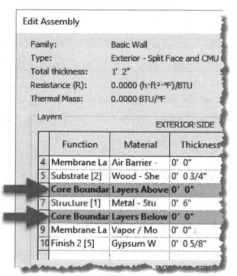

Figure 6.1-1 Wall core defined

The **Pick** drop-down list offers two options: **Individual References** and **Entire Walls**. The default is picking individual references. The Entire Walls selected enables the adjacent Options button which controls how an entire wall is dimensioned. An example is shown in the image below (Figure 6.1-2).

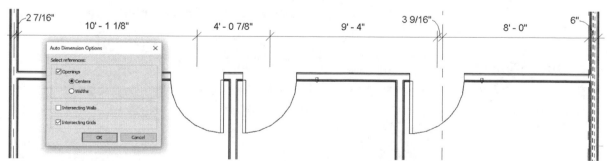

Figure 6.1-2 Aligned dimension command; pick "entire walls" result shown

Annotate

Aligned: Type Selector

The Type Selector lists the available Aligned dimension types in the current project. Types for the same dimension command are used to show the length in different units, e.g. imperial versus metric, as well as the ability to graphically distinguish between them.

> Dimensions are NOT listed in the **Manage** (tab) → **Object Styles** → **Annotation Objects** (tab) because everything graphically is defined in the type settings. The **Object Styles** dialog is the project-wide setting for most elements, unless defined in a family or overridden in a view.

Aligned: Type Properties

There are several options in Type Properties that control how the dimension looks; many are self-explanatory. The one setting to be keenly aware of is **Units Format**. This setting controls the value shown and any rounding applied (Figure 6.1-3).

When **Use project settings** is selected, the **Project Units setting, from Manage** (tab) → **Project Units** → **Common** (Discipline) → **Length** is used.

When this is un-checked, the units/format can be set for just this dimension type.

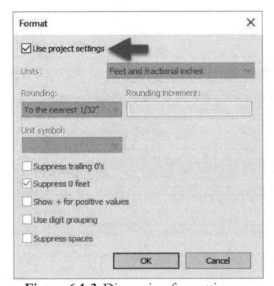

Figure 6.1-3 Dimension formatting

Aligned: In-Canvas

There are several things, while using the dimension tool, that affect the result:

- Tap the **Tab** key to toggle between layers in a wall
- Tap the **Tab** key multiple times, near an endpoint, to select a point
- Picking multiple elements will create a **dimension string**
- Picking the same element a second time will remove the reference
- Click the **padlock** icon after placing a dimension to lock it
- Click the **EQ** icon after placing a dimension string to equally space the dimension elements

> **Dimension visibility** is controlled on the Annotation tab in the Visibility/Graphic Overrides dialog for the current view. Additionally, a dimension will be hidden if the referenced element is hidden (e.g. if the doors are hidden, any dimensions to doors will also be hidden) even if the Dimension category is turned on.

Dimension: Linear

Use this tool to create horizontal or vertical dimensions relative to the current view. This command does not work on Grids or Reference Lines.

Unlike the **Aligned** command, the **Linear** dimension will never get modified in such a way that the dimension is no longer orthogonal with the view.

Linear: Ribbon

Once the Linear command is started, the Ribbon provides an easy way to switch to another dimension command.

Linear: Type Selector

The Type Selector lists the available Linear dimension types in the current project. Types for the same dimension command are used to show the length in different units, e.g. imperial versus metric, as well as the ability to graphically distinguish between them.

Aligned: Type Properties

There are several options in Type Properties that control how the dimension looks; many are self-explanatory. The one setting to be keenly aware of is **Units Format**. This setting controls the value shown and any rounding applied (Figure 6.1-3).

When **Use project settings** is selected, the **Project Units setting, from Manage** (tab) → **Project Units** → **Common** (Discipline) → **Length** is used.

When this is un-checked, the units/format can be set for just this dimension type.

Linear: In-Canvas

Hover over a point to select it. If the desired point is not highlighted, tap the **Tab** key to cycle through the valid options directly below your cursor. While pressing the Tab key, do not move the cursor. When the cursor is moved, the cycle starts over.

After picking two points, the direction the cursor is moved determines if the dimension will be horizontal or vertical. Additionally, pressing the **Spacebar** will toggle between the two options—once this option is used, the cursor position can no longer be used to change between horizontal and vertical.

---

quick steps

Dimension: Linear

1. Ribbon
   - Option to switch/toggle to another dimension tool
2. Type Selector
   - Select dimension type
3. Type Properties
   - Units Format; rounding
4. In-Canvas
   - Tap the Tab key to highlight and then select a point; e.g. a line endpoint or intersection
   - After picking two points, the direction you move your mouse determines if the dimension is vertical or horizontal

Annotate

Dimension: Angular

Use this command when an angular dimension is needed in a view. The Angular dimension is limited from 0 to 180 degrees.

The image below shows several angular dimension examples. These are applied to simple line work but work the same for model elements.

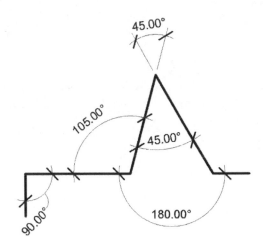

Angular: **Ribbon**

Once the Angular command is started, the Ribbon provides an easy way to switch to another dimension command.

Angular: **Options Bar**

See the Aligned command for information on the selection options available in the drop-down list.

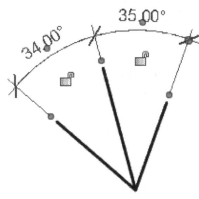

Angular: **Type Selector**

The Type Selector lists the available Angular dimension types in the current project. Types for the same dimension command are used to show the length in different units, e.g. imperial versus metric, as well as the ability to graphically distinguish between them.

Angular: **In-Canvas**

Points cannot be selected with this command. Model elements, lines, grids and reference lines can all be used with this command. Picking multiple references, which pass through a common point, will create a **dimension string**; this offers the **Equality** option as shown above.

DImension: Radial

Use this command to define the Radius of a curved element in a view. The image below shows two Types: one with the "Center Marks" turned on and the other with it off, a prefix added, a different arrow and the length of the dimension line adjusted.

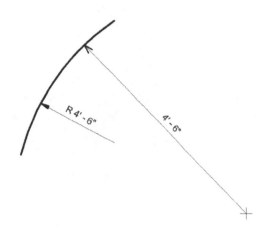

In addition to a **Center Mark** option for the Angular dimension, when a curved line or wall is selected, it also has an option to turn on a visible Center Mark. When the elements Center Mark is visible, it can be dimensioned to, and even repositioned (manually or parametrically) with other dimensions.

quick steps

Dimension: Radial

1. Ribbon
   - Option to switch/toggle to another dimension tool
2. Options Bar
   - Dimension to setting
     i. Wall Centerlines
     ii. Wall Faces
     iii. Center of Core
     iv. Faces of Core
3. Type Selector
   - Select dimension type
4. In-Canvas
   - First pick: Select curved line
   - Second pick: Locate dimension line
   - Drag center point grip to shorten radial line – this will not change the dimension

Annotate

Radial: Ribbon

Once the Radial command is started, the Ribbon provides an easy way to switch to another dimension command.

Radial: Options Bar

See the Aligned command for information on the selection options available in the drop-down list.

Radial: Type Selector

The Type Selector lists the available Radial dimension types in the current project. Types for the same dimension command are used to show the length in different units, e.g. imperial versus metric, as well as the ability to graphically distinguish between them as shown above.

Radial: In-Canvas

With the command active, the first pick is to select curved line or wall. The second pick is to position dimension line. Drag center point grip to shorten radial line – this will not change the dimension as in the example above.

Dimension: Diameter

Use this command to add a Diameter dimension to the current view.

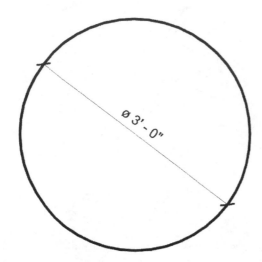

Diameter: **Ribbon**

Once the Diameter command is started, the Ribbon provides an easy way to switch to another dimension command.

Diameter: **Options Bar**

See the Aligned command for information on the selection options available in the drop-down list.

Diameter: **Type Selector**

The Type Selector lists the available Diameter dimension types in the current project. Types for the same dimension command are used to show the length in different units, e.g. imperial versus metric, as well as the ability to graphically distinguish between them. Notice the example above has a prefix and tick marks for arrowheads specified.

Diameter: **In-Canvas**

Simply select a curved or circular reference and then pick again to position it in the view.

> Even though Radial and Diameter are separate commands, when selected they are interchangeable via the Type Selector.

---

quick steps

Dimension: Diameter

1. Ribbon
   - Option to switch/toggle to another dimension tool
2. Options Bar
   - Dimension to setting
     i. Wall Centerlines
     ii. Wall Faces
     iii. Center of Core
     iv. Faces of Core
3. Type Selector
   - Select dimension type
4. In-Canvas
   - First pick: Select curved line – usually a circle
   - Second pick: Locate dimension line

Dimension: Arc Length

This command is used to measure the length along a curved reference.

This command is a little unusual. The result is a dimension element that references three elements in the view and is tied to the same type properties as those used by the Linear dimension command.

The example below consists of two lines and a wall; however, it could be any combination of these elements. If any of these elements are moved or deleted, the **Arc Length** dimension will change or be deleted accordingly.

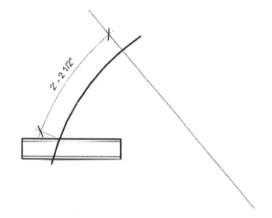

quick steps

Dimension: Arc Length

1. Ribbon
   - Option to switch/toggle to another dimension tool
2. Options Bar
   - Dimension to setting
     i. Wall Centerlines
     ii. Wall Faces
     iii. Center of Core
     iv. Faces of Core
3. Type Selector
   - Select dimension type
4. In-Canvas
   - First pick: Select curved line
   - Second Pick: intersecting reference along curved line
   - Third Pick: Second intersecting reference along curved line
   - Optional: continue selecting reference lines
   - Final pick: locate dimension line

Annotate

Arc Length: Ribbon

Once the Diameter command is started, the Ribbon provides an easy way to switch to another dimension command.

Arc Length: Options Bar

See the Aligned command for information on the selection options available in the drop-down list.

Arc Length: Type Selector

The Type Selector lists the available Linear dimension types in the current project. This is not a typo; the Arc length command uses the same Types as the Linear dimension command.

Arc Length: In-Canvas

This command has prerequisites and requires specific steps to place an Arc Length dimension.

First, the curved surface must have two elements/lines (just lines in a view, or lines within a sketched element like a floor or ceiling) which intersect this curve.

Second, with the Arc Length command active, select the curved reference, and then the two intersecting elements; select the elements directly, not where they intersect the curved reference. Finally, pick a position for the dimension line relative to the picked elements.

Continuing to pick intersecting elements creates a dimension string, rather than separate dimension elements. However, for the Arc Length element, a selected dimension string <u>does not</u> offer the **Equality** option.

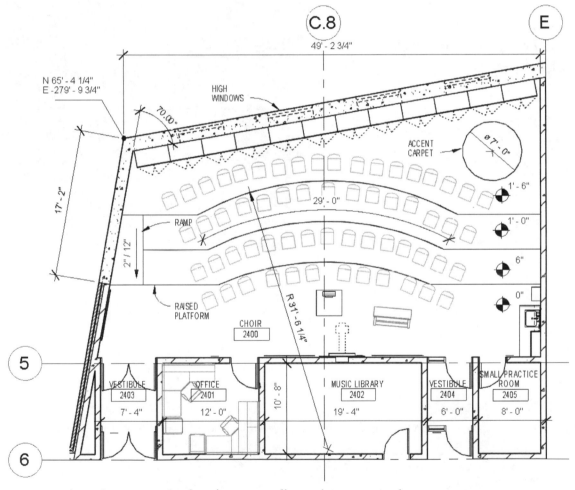

A project example showing every dimension type used

Dimension: Spot Elevation

A spot elevation can be used to document the elevation of a surface relative to the Project Base Point, the Survey Point (if defined) or Relative to a Level in the project.

The example below shows several **Spot Elevations** added to a reflected ceiling plan. If any of these ceilings have their elevation changed, the Spot Elevations will update automatically. Plus, when the ceiling is selected, the elevation can be modified via the Spot Elevation— this actually changes the elevation of the ceiling.

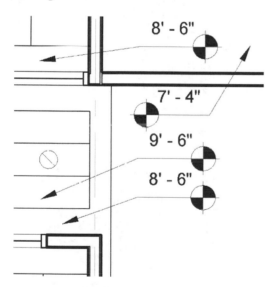

This command can also be used on other surfaces like topography, duct, pipe, structural framing and more.

quick steps

Dimension: Spot Elevation

1. Ribbon
   - Option to switch/toggle to another dimension tool
2. Options Bar
   - Leader
   - Shoulder (for leader)
   - Relative Base
     i. Type dependent – see notes in this section
   - Display Elevations
     i. Actual (selected) Elevation
     ii. Top Elevation
     iii. Bottom Elevation
     iv. Top & Bottom Elevations
3. Type Selector
   - Select dimension type
4. In-Canvas
   - View should not be in Wireframe so Revit can "see" the 3D faces
   - When desired elevation is seen, click to place
   - TIP: Use Spot Elevation to quickly verify elevations of ceilings (in RCP views), stoops and grade (in site plan views), etc. Just start tool and move cursor around, press the ESC key when done – no need to actually place a Spot Elevation.

Annotate

Use this command to discover an elevation and then press Esc to cancel the command if a permanent dimension is not required.

Spot Elevation: Ribbon

Once the Diameter command is started, the Ribbon provides an easy way to switch to another dimension command.

Spot Elevation: **Options Bar**

When this command is active, the Options Bar has two checkboxes related to a leader and two drop-down lists related to what elevation value is displayed.

The three examples below represent the various Leader options—note that when **Leader** is not checked, **Shoulder** is grayed out.

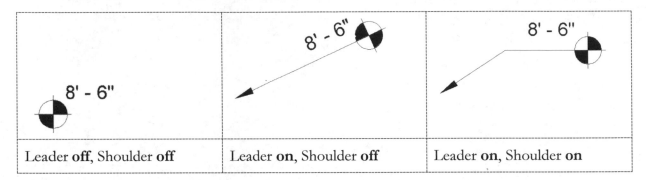

| Leader **off**, Shoulder **off** | Leader **on**, Shoulder **off** | Leader **on**, Shoulder **on** |

The **Relative** option is grayed out unless the Type Parameter **Elevation Origin** is set to "Relative"—and then this drop-down lists all the **Levels** in the current project. Picking a level above will make the value negative, and a level below will result in positive numbers indicating the distance from the selected level.

The **Display Elevations** list has four options:

- Actual (Selected) Elevation
- Top Elevation
- Bottom Elevation
- Top & Bottom Elevations

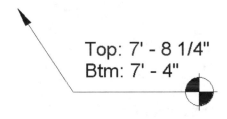

When selecting something with a thickness, such as a floor, beam, ceiling, etc., it is possible to specify if the elevation should reference the Top, Bottom, Both Top and Bottom, or just the face selected. As shown to the right, when the Spot Elevation is listing a Top and Bottom elevation, that condition can add a custom **prefix** to help distinguish what the two values represent—this is set in the Type Properties.

Spot Elevation: **Type Selector**

The Type Selector lists the available Spot Elevation dimension types in the current project.

Spot Elevation: **In-Canvas**

Ensure the view is <u>not</u> set to **Wireframe**, as faces cannot be detected by the Spot Elevation command. When the command is active, hover over a surface and click to place the point, and then, depending on the leader options, click 1-2 more points to locate the leader, text & target.

Dimension: Spot Coordinate

A **Spot Coordinate** defines a point in space relative to the **Project Base Point** in the current Revit project (Figure 6.1-4). If the Project Base Point position is adjusted, the Spot Coordinates will automatically update (Figure 6.1-5).

This tool could be used to locate items on the site, such as trees or a fence corner, or to define an irregular edge of a wall or floor finish.

quick steps

Dimension: Spot Coordinate

1. Ribbon
   - Option to switch/toggle to another dimension tool
2. Options Bar
   - Leader
   - Shoulder (for leader)
3. Type Selector
   - Select dimension type
4. In-Canvas
   - Click a location to place the Spot Coordinate
   - Three clicks total when Leader and Shoulder are checked

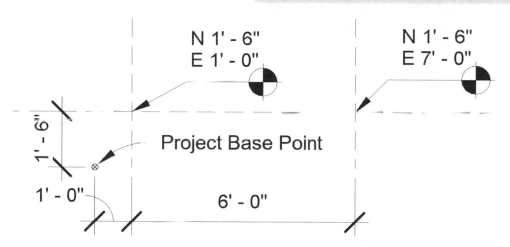

Figure 6.1-4 Spot Coordinate with Project Base Point displayed

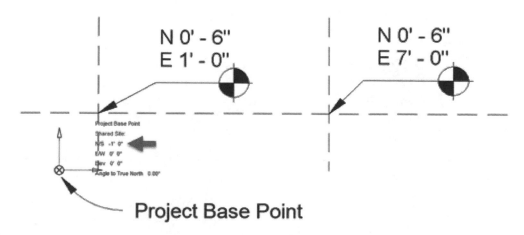

Figure 6.1-5 Spot Coordinate with modified Project Base Point position

Annotate

In addition to a distance in the north direction, and the east direction, from the Project Base Point, an elevation can also be displayed. This means the Spot Coordinate must be hosted on an element (e.g. floors, walls, toposurfaces and boundary lines) and not just floating in empty space.

Spot Coordinate: **Ribbon**
Once the Diameter command is started, the Ribbon provides an easy way to switch to another dimension command.

Spot Coordinate: **Options Bar**
See the Spot Elevation command for information on the **Leader** and **Shoulder** options shown on the Options Bar while the Spot Coordinate command is active.

Spot Coordinate: **Type Selector**
The Type Selector lists the available Spot Coordinate dimension types in the current project.

Spot Coordinate: **In-Canvas**
Ensure the view is <u>not</u> set to **Wireframe**, as faces cannot be detected by the Spot Coordinate command—although walls and lines can be selected. When the command is active, hover over a surface, wall or line and click to place the point, and then, depending on the leader options, click 1-2 more points to locate the leader, text & target.

For all dimension commands, the length value can be modified by clicking directly on the text—which opens the Dimension Text dialog.

Here, text can be added as a prefix, suffix, above and below the dimension value. To change the dimension value itself, click the Replace with Text option. Revit prevents faking dimensions by entering another length value—however, adding a simple period to the end will trick Revit and allow it.

When a dimension is selected, some allow it to be locked so the length value cannot be changed. This will cause other geometry to move rather than allowing the dimension to change.

Dimension Text ×

Note: this tool replaces or appends dimensions values with text and has no effect on model geometry.

Dimension Value
- ◉ Use Actual Value — 2' - 7 1/2"
- ○ Replace With Text
- ○ Show Label in View

Text Fields
Above:
Prefix: | Value: 2' - 7 1/2" | Suffix:
Below:

Segment Dimension Leader Visibility: By Element

OK | Cancel | Apply

Dimension: Spot Slope

The **Spot Slope** dimension command lists the slope of the selected surface of a Floor, Roof, Beam, Piping, etc. But it does not work on a Ramp element.

> Placing a Spot Elevation on an adjacent *Floor* element and then dragging it to a *Ramp* will cause the Spot Elevation to work properly on a ramp.

Figure 6.1-6 shows the Spot Slope in a plan view, and Figure 6.1-7 shows it in an elevation view. The appearance of the Spot Slope can be changed in elevation—which will be covered in a moment.

Spot Slope: Ribbon

Once the Diameter command is started, the Ribbon provides an easy way to switch to another dimension command.

Spot Slope: Options Bar

In a plan or 3D view, the only option is **Offset from Reference**. When picking the boundary line of an element (e.g. floor) this determines how far away the Spot Slope line is placed from it. In a section/elevation view, the Slope Representation setting is active. This changes the graphic from an **Arrow** (basically, a line) to a **Triangle**.

Spot Slope: Type Selector

The Type Selector lists the available Spot Slope dimension types in the current project.

Dimension: Spot Slope

1. Ribbon
   - Option to switch/toggle to another dimension tool
2. Options Bar
   - Slope Representation (elevation and section views only)
     i. Arrow
     ii. Triangle
   - Offset from Reference
     i. Default: 1/16"
3. Type Selector
   - Select dimension type
4. In-Canvas
   - Plan view: Single click on sloped surface to place
   - Elev/Sect view: First click to select edge of sloped surface, and second click to select which side [of sloped edge] to place graphic.

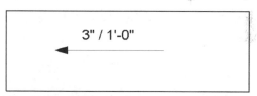

Revit Roof Element

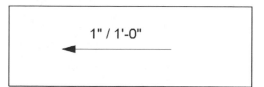

Revit Floor Element

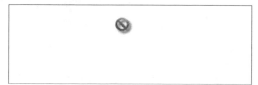

Revit Ramp Element

Figure 6.1-6 Spot Slope on Roof and floor, but not Ramp

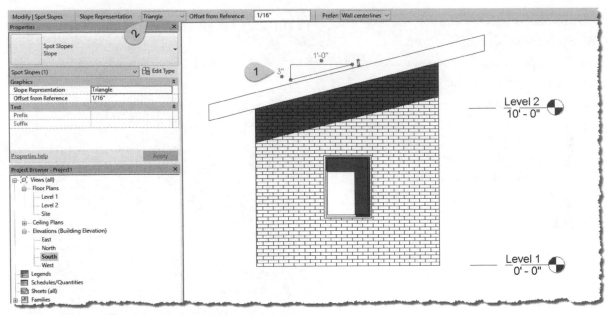

Figure 6.1-7 Spot Slope on sloped roof in an elevation view

Spot Slope: **Type Properties**

It is possible to change the way in which the slope is described via Type Properties. If the current type is used elsewhere in the project, consider using Duplicate to make a new type; the other instances should not change. In Type Properties, edit Units Format to see the various Units available (Figure 6.1-8).

Spot Slope: **In-Canvas**

With the Spot Slope command active, click on a surface in a plan view, or on a boundary/edge in a plan/elevation/section view. When picking an edge or boundary line, a second pick is required to select which side of the line to place the graphic.

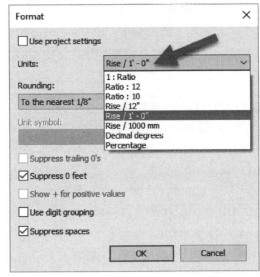

Figure 6.1-8 Spot Slope format options

Dimension Panel Extension: Dimension Types

The extended panel portion of the Dimension panel provides access to the Type Properties for each of the dimension commands. Click on the panel title to expand the panel as shown here.

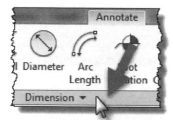

Each of these commands open the same as when the related Dimension command is active and the Edit Type button is clicked in Properties; they both open the Type Properties dialog (Figure 6.1-9).

From the Type Properties dialog new dimension styles are created using the Duplicate button and providing a name. The only way to delete a dimension type is via the **Manage** (tab) → **Purge Unused** command (Fig. 6.1-10). Be sure to click Check None and then select the specific things to remove. Only types not used in the project can be purged/deleted from a project.

> To make a dimension style not used, and qualify to be purged, place one of the types somewhere in the project, select it, and then right-click and pick Select All Instances in Entire Project. Then switch to another type.

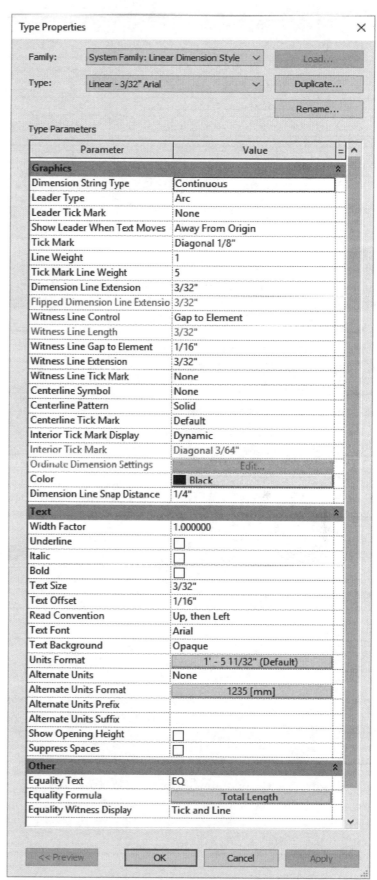

Figure 6.1-9 Linear dimension type properties

Each of these "Types" commands control how a dimension type appears and specifics about the precision and units of the value displayed. Most of these parameters are self-explanatory or with a little testing can be easily understood. Thus, these dialogs will not be covered in-depth. Figure 6.1-11 defines some of the major terms used to describe the parts of a dimension element.

In the example below, **Alternate Units** is turned on. This provides a way to show two units; for example, imperial and metric. Some federal government projects in the United States require both feet and inches and metric dimensions on the construction documents (CDs). This can also be used on large projects with international design teams collaborating.

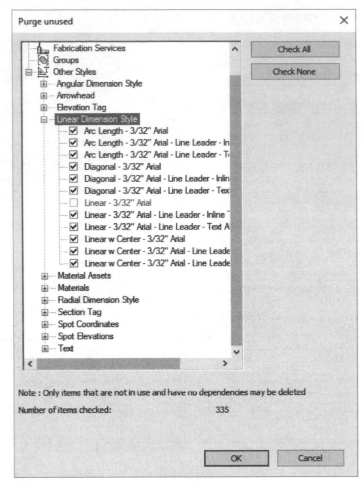

Figure 6.1-10 Purge unused dialog

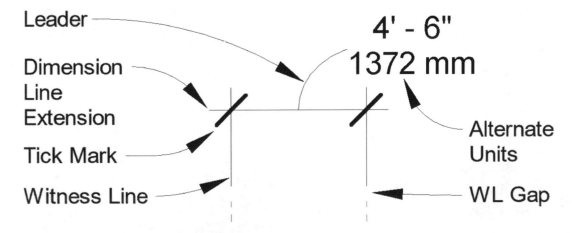

Figure 6.1-11 Anatomy of a dimension element

Linear Dimension Types

Used to access Type Properties for the following dimensions commands:

- Angular
- Linear
- Arc Length

Angular Dimension Types

Used to access Type Properties for Angular dimensions.

Radial Dimension Types

Used to access Type Properties for Radial dimensions.

Diameter Dimension Types

Used to access Type Properties for Diameter dimensions.

Spot Elevation Types

Used to access Type Properties for Spot Elevation dimensions.

Spot Coordinate Types

Used to access Type Properties for Spot Coordinate dimensions.

Spot Slope Types

Used to access Type Properties for Spot Slope dimensions.

Annotate

6.2 Annotate tab: Detail panel

The Detail panel, on the Annotate tab, provides 2D and view specific drawing tools. These tools can be used in model views, e.g. Plans, elevations, sections, 3D views, and they can also be used in Drafting Views and Legend Views. Some of these tools can also be used on Sheets.

In model views, these tools should be used as little as possible as they are not 3D and will not show up in other views or in linked consultant models. Additionally, these 2D embellishments will not schedule with other building elements they are meant to represent.

In Drafting Views, these tools offer a way to create static typical details which contain a wealth of embodied knowledge within the notes and dimensions added as shown in the example below. These 2D details are not affected by the constant change and development of the model. Of course, this creates a disconnect between the two and can create costly or timely conflicts during construction. Thus, care must be used when using these tools.

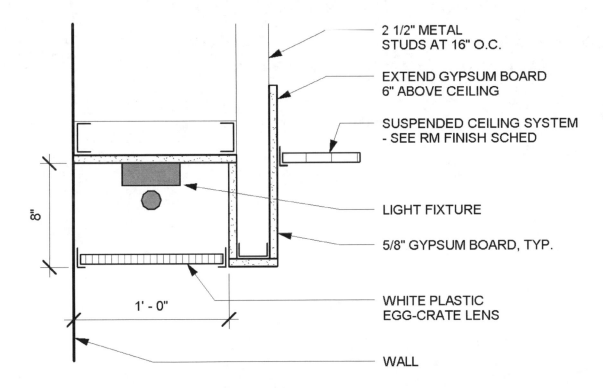

Typical detail created in a Drafting View using the tools on the Annotate tab.

Detail Line

Detail Lines are view specific lines used to add 2D detail to the model. Not everything can be modeled three-dimensionally. In some cases, over modeling creates performance issues or detracts from the readability of the documents; therefore, 2D detail lines are added to a given view to convey additional information, for example, a line representing the wall base in an interior elevation would be shown using Detail Lines as it is not practical to model all the wall base in a building, and seeing this line is likely only important in this view. Contrast this with **Model Lines** which are lines in 3D space; i.e., those lines would show up in all views of the same area (sections, 3D, Camera).

Detail Line: Ribbon

When the Detail Line command is activated, the Ribbon provides the typical **Draw** tools and the **Line Style** selector (Figure 6.2-1).

In the **Draw** panel select which type of line to sketch. The Pick Lines option allows a new line to be created by picking other lines. These "other" lines can be part of some other element, e.g. a floor edge, or even from a linked CAD file.

The **Line Style** selection determines appearance of the line in terms of line width and dashes. This setting can be changed later if required.

quick steps

Detail Line

1. Ribbon
   - Select type of line segment to draw
   - Select Line Style
2. Options Bar
   - Chain
   - Offset
   - Radius
3. In-Canvas
   - TIP: SZ is a keyboard shortcut to close a loop after sketching two or more line segments

Annotate

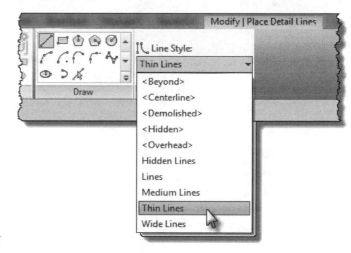

Figure 6.2-1 Detail lines Ribbon options

Detail Line: Options Bar

This is what the Options Bar looks like when this command is active:

The **Chain** option allows multiple picks in the canvas without needing to pick the same point twice, once for the end of a line and again for the beginning of the next line.

If an **Offset** value is entered, the points picked in the canvas will be repositioned, i.e., offset the specified amount. After picking the first point, the line will appear offset to one side. Before picking subsequent points, if the line is not on the desired side, press the **Spacebar** to flip to the other side.

If the **Radius** box is checked, the **Radius Value** textbox becomes editable. When a value is entered, a radius is automatically added at all corners (Figure 6.2-2) until this option is unchecked.

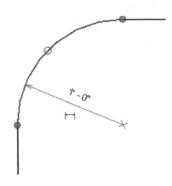

Figure 6.2-2 Arc added via Radius setting on Options Bar setting

Detail Line: **In-Canvas**

Draw lines using Snaps; see Manage (tab) → Snaps for control and the keyboard shortcuts. After picking the first point, move the mouse in the desired direction and angle, and then type a length for accurate lines.

Object Snap Symbols:

If you hold your cursor still for a moment while a snap symbol is displayed, a tooltip will appear on the screen. However, when you become familiar with the snap symbols you can pick sooner.

The TAB key cycles through the available snaps near your cursor.

The keyboard shortcut turns off the other snaps for one pick. For example, if you type SE (snap endpoint) on the keyboard while in the Wall command, Revit will only look for an endpoint for the next pick.

Finally, typing SO (snaps off) turns all snaps off for one pick.

Symbol	Position	Keyboard Shortcut
✕	Intersection	SI
☐	Endpoint	SE
△	Midpoint	SM
○	Center	SC
✕	Nearest	SN
⌐	Perpendicular	SP
⊙	Tangent	ST

Filled Region

This command created a view-specific 2D pattern within a sketched boundary. The fill pattern scale can be tied to the View Scale (**Drafting** pattern) or to specific dimensions (**Model** pattern). The boundary can consist of one or more **Line Styles**, including Invisible.

Model-based patterns can be repositioned within the boundary. Simply hover over one of the pattern lines and tap the Tab key, on the keyboard, until the line highlights—and then click to select it. It can then be modified using Move, Rotate or pressing one of the arrow keys on the keyboard.

When the Filled Region boundary is sketched over the boundary of another model element, Revit will offer padlocks to lock the alignment. To align with detail lines, use the Align command on the Modify tab.

quick steps

Filled Region

1. Ribbon [in sketch mode]
   - Select type of line segment to draw
   - Select Line Style
2. Options Bar
   - Chain
   - Offset
   - Radius
3. Type Selector
   - Select fill pattern type to use
4. Type Properties
   - Fill Pattern
   - Background
     i. Opaque
     ii. Transparent
5. In-Canvas
   - One of the options for Line Style is <Invisible>, making the perimeter of the filled hidden

Annotate

In the image below, the wall tile is a **Model** pattern and the blocking pattern is a **Drafting** pattern. If the view scale is changed, the blocking pattern will change but the tile pattern will not as it is meant to represent a real-world building material.

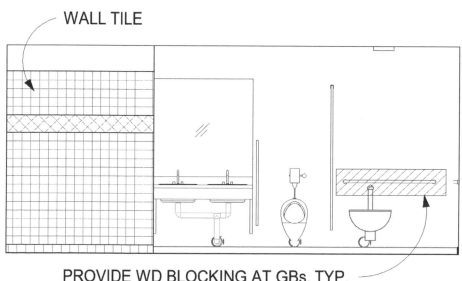

WALL TILE

PROVIDE WD BLOCKING AT GBs, TYP.

The fill patterns used by this command are the same options available within Revit materials, which appear automatically in model views when an element with that material assigned is seen in elevation or cut in section.

▌ Fill pattern elements report the **area** within its boundary in the Properties Palette.

Filled Region: Ribbon

When the Filled Region command is activated, the Ribbon provides tools related to defining the boundary of the area to be filled.

In the **Draw** panel select which type of line to sketch. The Pick Lines option allows a new line to be created by picking other lines. These "other" lines can be part of some other element, e.g. a floor edge, or even from a linked CAD file.

The **Line Style** selection determines appearance of the line in terms of line width and dashes. This setting can be changed later if required.

Filled Region: Options Bar

The **Chain** option allows multiple picks in the canvas without needing to pick the same point twice, once for the end of a line and again for the beginning of the next line.

If an **Offset** value is entered, the points picks in the canvas will be repositioned, i.e., offset the specified amount. After picking the first point, the line will appear offset to one side. Before picking subsequent points, if the line is not on the desired side, press the **Spacebar** to flip to the other side.

If the Line Style needs to vary for each line, turn off Chain so the Line Style option is editable after the creation of each line. Otherwise, press the Esc key to cancel the current "chain" of lines to activate the Line Style drop-down.

If the **Radius** box is checked, the **Radius Value** textbox becomes editable. When a value is entered, a radius is automatically added at all corners (Figure 6.2-2) until this option is unchecked.

Filled Region: Type Selector

The Drafting and Model patterns defined in the current project appear in the Type Selector.

Filled Region: Type Properties

The Type Properties control the appearance of each fill type.

In the Graphics section, as seen in Figure 6.2-3, the **Fill Pattern** determines which pattern fills the boundary sketched. Clicking in the Value box exposes the icon to open the Fill Patterns dialog. This dialog is also directly accessible via **Manage → Additional Settings → Fill Patterns**.

A **Filled Region** can be defined using two **fill patterns**, one for the **background** and one for the **foreground**. This allows you to create complex filled regions. For example, you might set a background to be a solid gray fill and the foreground to use a "concrete" fill pattern.

The **Masking** option defines if an element located behind the filled region will be visible through the fill pattern or not. Often, the same pattern is duplicated just to change this setting. A good Filled Region "type" naming convention is required to know which one to use.

The **Line Weight** and **Color** settings only control the Fill Pattern for this type.

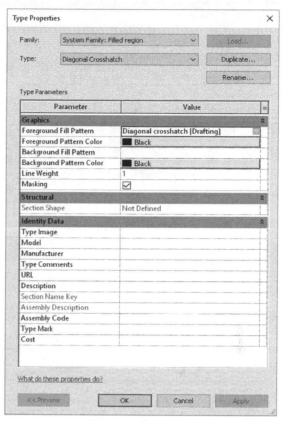

Figure 6.2-3 Filled Region Type Properties

> The line thickness of a Model pattern defined within a system family, such as a floor or wall, is controlled, universally, by Model Line Weight **#1**. This is found at **Manage** (tab) **→ Additional Settings → Line Weights → Model Line Weights** (tab).

Filled Region: In-Canvas

Create a closed loop outline of an area to be filled with the selected pattern. Creating closed loops within a larger closed loop defines a void where the pattern is omitted.

> Drafting patterns cannot be manually rotated. If a rotated pattern is required, a new pattern needs to be created or imported in the project.

Use the <Invisible> Line Style to make the perimeter not visible when the command is complete. When the cursor hovers over the Filled Region element, the invisible lines will highlight.

When creating a new fill pattern, any AutoCAD PAT file can be used via the Custom → Import option. Note that the Import Scale setting is a onetime option. If the scale turns out to be wrong, the Fill Pattern definition must be deleted and then recreated.

Annotate

Masking Region

This command works just like the Filled Region command except its main purpose is to "whiteout" or hide things.

There are some things a Masking Region cannot hide, such as text, dimensions and tags. In those cases, they should just be deleted if not needed.

A Masking Region can be larger than a view's cropping window, but the Masking Region element is cropped like model elements.

Masking Region: **Ribbon**

See the previous command, Filled Region, for Ribbon options while the Masking Region command is active.

Masking Region: **Options Bar**

See the previous command, Filled Region, for Options Bar options while the Masking Region command is active.

Masking Region: **In-Canvas**

See the previous command, Filled Region, for In-Canvas options while the Masking Region command is active.

quick steps

Masking Region

1. Ribbon [in sketch mode]
   - Select type of line segment to draw
   - Select Line Style
2. Options Bar
   - Chain
   - Offset
   - Radius
3. In-Canvas
   - One of the options for Line Style is <Invisible>, making the perimeter of the filled hidden

Detail Component

The Detail Component command organizes all the view-specific 2D symbols used to create details. These items represent things like beams, gypsum board, door frames, metal studs, insulation, metal deck and more.

The Detail Component adds a **Detail Item** to a model view or a drafting view; they cannot be added to sheet or schedule views. When used in a model view, it is often to embellish the 3D model with additional detail. For example, showing studs at 16 inches on center, as studs are rarely modeled in Revit. However, Detail Items will not automatically move if the model is changed in the area Detail Items have been placed, so care must be exercised when these elements are used in a model view.

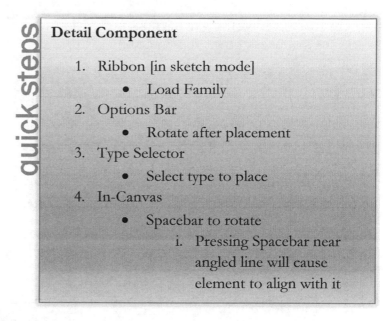

quick steps

Detail Component

1. Ribbon [in sketch mode]
    *   Load Family
2. Options Bar
    *   Rotate after placement
3. Type Selector
    *   Select type to place
4. In-Canvas
    *   Spacebar to rotate
        i. Pressing Spacebar near angled line will cause element to align with it

The image below (Figure 6.2-4) shows a Detail View made up of mostly Detail Items.

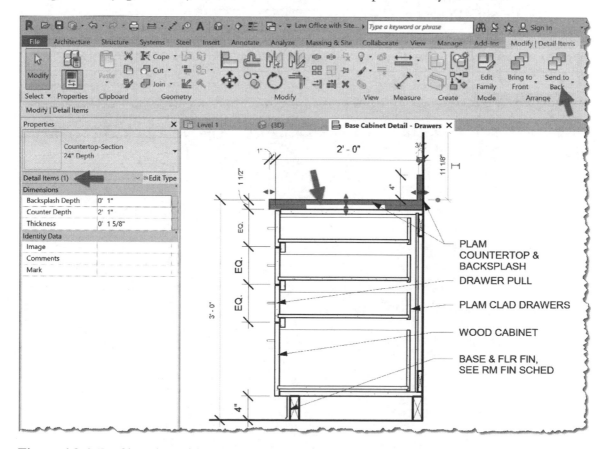

Figure 6.2-4 Drafting view with detail item selected

Detail Item families can be parametric and have multiple types. A wood stud Detail Item family can have types for 2x4, 2x6, 2x8, etc. They can also have visibility controls and utilize formulas like model-based Revit families.

Detail Component: Ribbon

When the Detail Component is active, the only option on the Ribbon is **Load Family**. This is a quick way to load additional Detail Items into the current project. When this button is clicked, the Load Family dialog appears. Click the **Imperial Detail Items** quick link on the left (Figure 6.2-5) to access the Detail Items library provided with Revit.

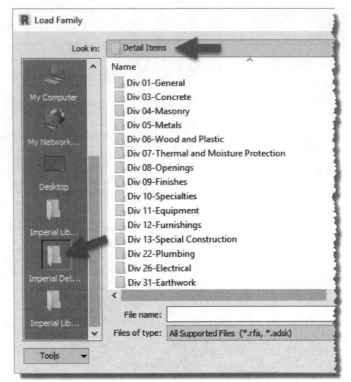

Figure 6.2-5 Load Family dialog; Imperial Detail Items quick link selected

Detail Component: Options Bar

When the **Rotate after Placement** box is checked, Revit will prompt to rotate the element immediately after a point is picked for placement. It is also possible to press the Spacebar to rotate the element prior to placement.

Detail Component: Type Selector

The Type Selector lists the available Detail Items in the current project. Click **Load Families** on the Ribbon to add more options.

Detail Component: In-Canvas

Use Snaps to accurately place the item. Press the Spacebar to rotate prior to placement.

When loading a Detail Item family (Figure 6.2-6), **Type Catalogs** are used if there are several types. For example, the steel wide flange beam has hundreds of types, which would make the type selector too long. Figure 6.2-7 shows the Type Catalog that appears while loading the "AISC Wide Flange Shapes-Section" family.

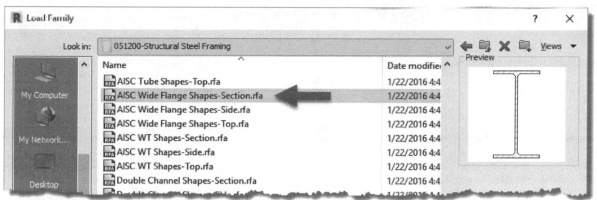

Figure 6.2-6 Detail item, steel beam in section, selected

Type	W	A	d	bf	tw	tf	k
(all)	(all)	(all)	(all)	(all)	(all)	(all)	(all)
W16X45	45.0	0.09 SF	1' 4 13/128"	0' 7 5/128"	0' 0 11/32"	0' 0 145/256"	0' 1 1/4"
W16X40	40.0	0.08 SF	1' 4"	0' 7"	0' 0 39/128"	0' 0 129/256"	0' 1 3/16"
W16X36	36.0	0.07 SF	1' 3 115/128"	0' 6 253/256"	0' 0 19/64"	0' 0 55/128"	0' 1 1/8"
W16X31	31.0	0.06 SF	1' 3 115/128"	0' 5 17/32"	0' 0 35/128"	0' 0 113/256"	0' 1 1/8"
W16X26	26.0	0.05 SF	1' 3 179/256"	0' 5 1/2"	0' 0 1/4"	0' 0 11/32"	0' 1 1/16"
W14X808 - Obsolete	808	1.65 SF	1' 10 215/256"	1' 6 143/256"	0' 3 189/256"	0' 5 31/256"	0' 6 7/16"
W14X730	730	1.49 SF	1' 10 51/128"	1' 5 115/128"	0' 3 9/128"	0' 4 233/256"	0' 6 3/16"
W14X665	665	1.36 SF	1' 9 77/128"	1' 5 179/256"	0' 2 53/64"	0' 4 133/256"	0' 5 13/16"
W14X605	605	1.24 SF	1' 8 115/128"	1' 5 51/128"	0' 2 77/128"	0' 4 41/256"	0' 5 7/16"
W14X550	550	1.13 SF	1' 8 51/256"	1' 5 51/256"	0' 2 97/256"	0' 3 105/128"	0' 5 1/8"
W14X500	500	1.02 SF	1' 7 77/128"	1' 5"	0' 2 49/256"	0' 3 1/2"	0' 4 13/16"
W14X455	455	0.93 SF	1' 7"	1' 4 205/256"	0' 2 5/256"	0' 3 27/128"	0' 4 1/2"
W14X426	426	0.87 SF	1' 6 179/256"	1' 4 179/256"	0' 1 225/256"	0' 3 5/128"	0' 4 5/16"
W14X398	398	0.81 SF	1' 6 77/256"	1' 4 77/128"	0' 1 197/256"	0' 2 109/128"	0' 4 1/8"

Specify Types — Family: AISC Wide Flange Shapes-Se — Select one or more types on the right for each family listed on the left.

Figure 6.2-7 Detail item with type catalog being loaded; use Ctrl to select multiple

Loaded Detail Items can be managed in the **Project Browser** (Figure 6.2-8). Expand the Families node and then also expand Detail Items.

Right-click on the Family name to delete, save to a file or open for editing. Caution: when a family is opened for editing, clicking Save will save back to the original location it was loaded from.

Right-clicking on a Type name offers delete (just that type), rename, duplicate and Type Properties.

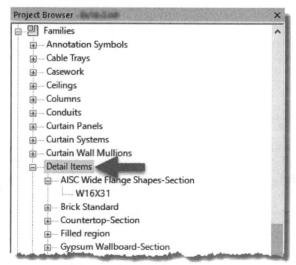

Figure 6.2-8 Detail Item listed in the Project Browser

Repeating Detail Component

The Repeating Detail Component command provides a built-in way to quickly array a Detail Item family.

In the image below (Figure 6.2-9) the brick on the left is an individual Detail Item family. To the right is a Repeating Detail Component being placed.

Placing a Repeating Detail Component only requires two picks: the first and the endpoint of the array. Keep in mind this

quick steps

Repeating Detail Component

1. Ribbon [in sketch mode]
    - Sketch line or pick a line to define length and direction
2. Options Bar
    - Offset
3. Type Selector
    - Select repeating detail family type
4. In-Canvas
    - Draw or pick line

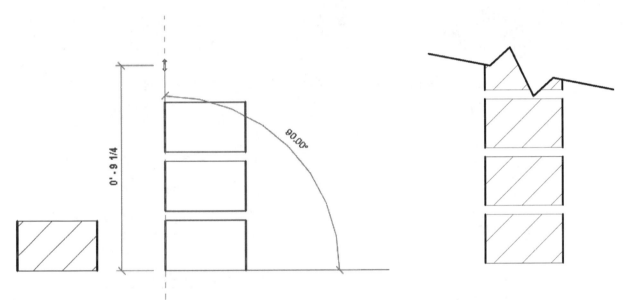

Figure 6.2-9 Detail Item on left, Repeating Detail Item being placed on right

Figure 6.2-10
Break Line family

array cannot be ungrouped or exploded. Additionally, the array always ends with a full Detail Item – the last item is never clipped relative to the second pick point. To deal with this, Revit provides a Detail Item family called **Break Line**, found in the **Div 01-General** folder. This family includes a Marking Region to cover a portion of the drawing below it.

> Repeating Detail Components cannot be tagged or keynoted. Only individual Detail Components can be tagged or keynoted.

Repeating Detail Component: **Ribbon**

The Ribbon offers two Draw options: Line and Pick Line. The line sketched defines the path (length and angle) of the array. The Pick Lines option uses a visible line on the screen, from the boundary of another element or from a linked CAD file, to define the start and end point of the array.

Repeating Detail Component: **Options Bar**

If an **Offset** value is entered, the points picked in the canvas will be repositioned, i.e., offset the specified amount. After picking the first point, the line will appear offset to one side. Before picking subsequent points, if the line is not on the desired side, press the **Spacebar** to flip to the other side.

Repeating Detail Component: **Type Selector**

The Type Selector lists the available Repeating Detail Items in the current project. Before new ones can be created, the appropriate Detail Item family must be loaded.

Repeating Detail Component: **Type Properties**

Clicking **Edit Type**, while this command is active, opens the **Type Properties** dialog (Figure 6.2-11).

This is where the Detail Item family is selected and the spacing is defined.

Repeating Detail Component: **In-Canvas**

Pick two points to define the length and angle of the repeating detail component. Press the **Spacebar** to flip the element to the other side.

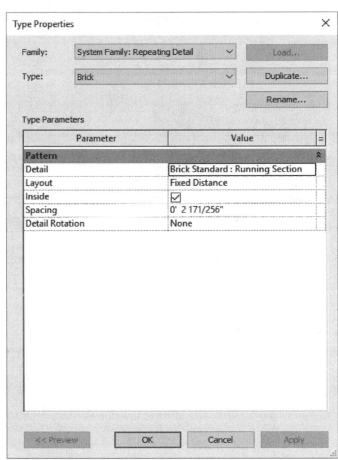

Figure 6.2-11 Type Properties dialog

Annotate

Legend Component

The Legend Component command is used in Legend views to describe elements used in the model, similar to a legend on a map.

Items to appear in the schedule do not have to be placed in the model yet but must be loaded. Families that would normally require a host, e.g., a door requires a wall, can be placed apart from a host.

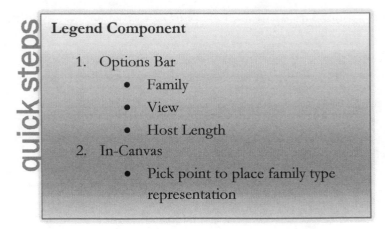

Legend Component

1. Options Bar
   - Family
   - View
   - Host Length
2. In-Canvas
   - Pick point to place family type representation

Items placed in a legend view do not get counted or appear in schedules. However, this does make the family appear as "in use" and thus that family cannot be purged. When a family is selected in the model view (plan, section, etc.) and Select all instances in entire project is employed, only the elements in the model will be selected, not the legend. Plus, when that same family is selected in a Legend view, and select all instances in entire project is used, the current family is unselected and only the instances in the actual model are selected.

The image below (Figure 6.2-12) shows the same door placed twice in a legend view: one for plan view representation and the other for elevation. A window family was placed the same way. The dimension command was used to annotate these items. Additionally, two wall families were placed and text added.

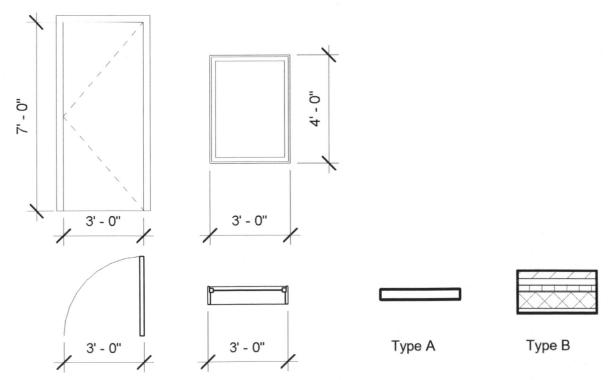

Figure 6.2-12 Legend view with one door, one window type and two wall types placed

Legend Component: **Options Bar**

The Options Bar is the only method for defining what is to be placed and what view is needed.

The **Family** drop-down lists loaded, and supported, families in the current project. These items would normally be selected from the Type Selector for other commands.

The **View** option determines how the family will appear. Depending on the category, the view options vary.

View options for the main architectural categories:

- Ceilings, Floors, Roofs   Section
- Curtain Panel, Furniture   Plan and Elevations (Front, Back, Right, Left)
- Doors, Windows   Plan and Elevations (Front and Back)
- Walls   Plan, Section

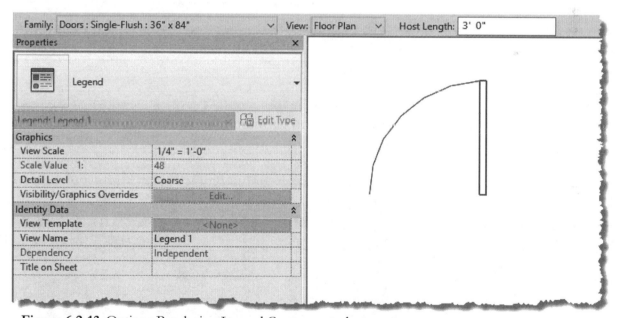

Figure 6.2-13 Options Bar during Legend Component placement

The **Host Length** determines how long a wall is. For doors and windows, if the value is set to the same width as the type, no sample host wall will be shown. If the value is greater than the width of the element placed, a generic sample wall appears and is centered on the element.

Legend Component: **In-Canvas**

Click to place a Legend Component. These elements can be moved but not rotated or mirrored.

Revision Cloud

When errors are found in the model or the drawings, a clouded area is added to highlight the issue.

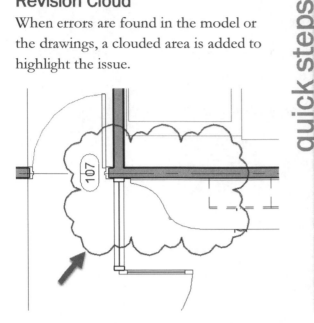

Revision Cloud

1. Ribbon [in sketch mode]
   - Select line segment to sketch outline with
2. Options Bar
   - Chain
   - Offset
   - Radius
3. Properties
   - Select Revision
4. In-Canvas
   - Pick points to define revision area

This can be done for internal comments to members of the design team, or as official communication to the contractor during construction via Change Orders, Proposal Request or Architectural Supplemental Information (ASI).

Revit provides three tools to facilitate this need: **Revision Cloud**, **Tag by Category** and **Revisions**. The latter is found on the View tab and is a dialog box used to create and manage revisions in the current project.

Revision Cloud: **Ribbon**

When the Revision Cloud command is started, Revit enters sketch mode. The only way to exit sketch mode is to click the green checkmark (finish) or the red "X."

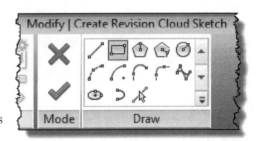

The Draw panel offers several sketch tools to define the location of the clouded area. The rectangle option is a quick way of clouding an area by clicking just two [diagonal] points. The Pick Lines option creates a sketch line based on some other line visible on screen. This line can be the edge of another element or from a linked CAD file.

Revision Cloud: **Options Bar**

The **Chain** option allows multiple picks in the canvas without needing to pick the same point twice, once for the end of a line and again for the beginning of the next line.

If an **Offset** value is entered, the points picked in the canvas will be repositioned, i.e., offset the specified amount. After picking the first point, the line will appear offset to one side. Before picking subsequent points, if the line is not on the desired side, press the **Spacebar** to flip to the other side.

If the **Radius** box is checked, the **Radius Value** textbox becomes editable. When a value is entered, a radius is automatically added at all corners until this option is unchecked.

Revision Cloud: **Properties**

When using **Revision Clouds** to formally track changes to the design, the proper Revision must be selected from the Properties palette during placement.

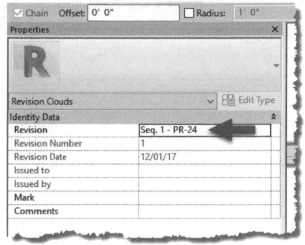

Figure 6.2-14 Options Bar and Properties

> If the wrong Revision is assigned to a Revision Cloud, select it and change the Revision selection via the Options Bar or Properties Palette.

Revision Cloud: **In-Canvas**

With the Revision Cloud command active, the Draw option selected and the Revision set in Properties, pick points to define the cloud path/location. The sketched lines do <u>not</u> have to form an enclosed area – the sketch could be a single line or several that intersect each other.

> The size of the arcs in the cloud can be adjusted in **View** (tab) → **Revisions** dialog.

The line weight of the Revision Cloud can be controlled in **Manage** (tab) → **Object Styles** → **Annotation Objects** (tab) → **Revision Clouds**. It can also be controlled per view in the **Visibility/Graphic Overrides** dialog.

For information on the Revisions dialog, shown below, see the View tab chapter.

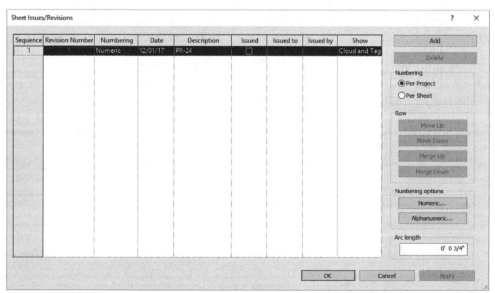

Figure 6.2-15 Revisions dialog

Annotate

Place Detail Group

The **Place Detail Group** command is used to place previously created Detail Groups in model views, drafting views, legend views and, in some cases, on sheets.

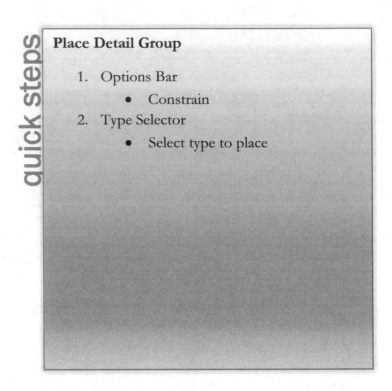

A Detail Group is a collection of 2D lines, text, tags and/or filled regions. Grouping these elements makes it easy to repeat common features in the design. Additionally, if any Detail Group is modified, they will all update.

▌ A Detail Group cannot contain any Model elements.

Both Model Groups and Detail Groups have the ability to **Exclude** elements from individual instances of a group. Just tap Tab over an element to highlight and then click to select. Once selected, right-click and pick Exclude. This feature reduces the number of overall groups required when there are minor exclusions needed. When a Group has elements excluded, the Restore All Excluded button appears on the Ribbon when the group is selected.

Attached Detail Groups

When creating a Group with Model and Detail items selected something unique happens. Revit prompts to create two groups: a **Model Group** and an **Attached Detail Group** at the same time (Figure 6.1-16). What's unique is that the Attached Detail Group has a special association to the Model Group. Anytime an instance of the Model Group is selected, the Ribbon has an Attached Detail Groups button visible. Clicking this and then selecting the Attached Detail Group will perfectly position it on the Model Group. The Attached Detail Group can also include tags.

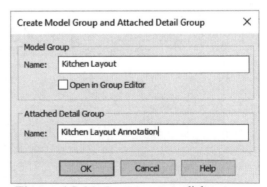

Figure 6.2-16 Create groups dialog

Place Detail Group: Options Bar

When placing multiple instances of a Detail Group, the **Constrain** option becomes available and limits the movement angle relative to previous instances.

Place Detail Group: **Type Selector**

Select from a list of available Detail Groups.

> Revit does not provide any Detail
> Groups with the software. This feature is
> meant for custom uses per project.

Place Detail Group: **In-Canvas**

Click to place an instance of a Detail Group. The element's insertion point is based on the X/Y icon—which is visible when an instance of the group is selected. This x/y icon can be repositioned. When the x/y icon is moved all future instances will be positioned accordingly, but previously placed groups will not move.

All Model, Detail and Attached Detail Groups are organized and managed in the **Project Browser** (Figure 6.2-17). It is important to have a good naming convention to minimize

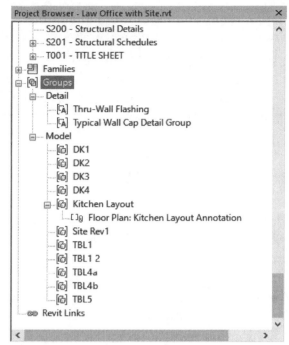

Figure 6.2-17 Groups in Project Browser

confusion amongst the design team. Model and Detail Groups can be dragged from the Project Browser into a view to initiate placement. However, Attached Detail Groups cannot be dragged from the Project Browser.

Create Group

See the Modify tab chapter for detailed coverage of this tool.

Annotate

Insulation

The Insulation command is used to represent acoustic or batt insulation (but not rigid insulation) in floor plan and section views. In the example below, acoustic insulation is represented in a plan view using the Insulation command.

The per-view visibility control is on the Model Categories tab, under Lines: **Insulation Batting Lines**. Project wide control is in the Line Styles dialog found on the Manage tag.

Insulation: **Ribbon**

The Draw panel offers the **Line** and **Pick Lines** options for sketching the start and endpoints of each Insulation segment.

quick steps

Insulation

1. Ribbon (in sketch mode)
   - Select sketch of pick line
2. Options Bar
   - Width
     i. Default 3 ½"
   - Chain
   - Offset
   - To
     i. Center
     ii. Near Side
     iii. Far Side
3. Properties
   - Insulation Bulge to Width Ratio 1/x
     i. Default: 2

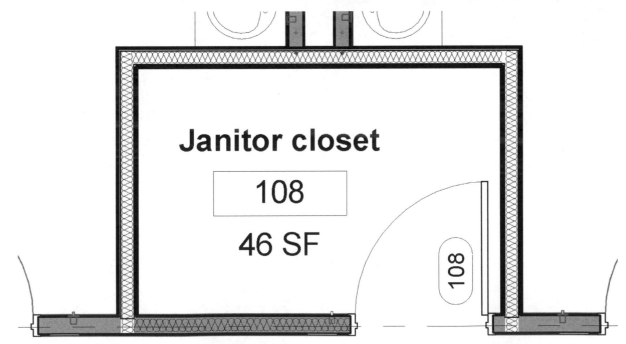

Figure 6.2-18 Insulation symbol added to floor plan view

Insulation: **Options Bar**

The **Width** controls the overall width of the insulation. This is typically adjusted to match the size of the studs in a wall; for example, 1.5" in a 2x4 wall, or 3.625" in typical commercial construction with metal studs.

The **Chain** option allows multiple picks in the canvas without needing to pick the same point twice, once for the end of a line and again for the beginning of the next line.

If an **Offset** value is entered, the points picks in the canvas will be repositioned, i.e., offset the specified amount. After picking the first point, the line will appear offset to one side. Before picking subsequent points, if the line is not on the desired side, press the **Spacebar** to flip to the other side.

The Insulation command is unique when it comes to the Offset option on the Options Bar. To help properly position the isolation symbol within a stud space, the drop-down list offers three options: **to center, to near side** and **to far side**. The image to the right shows how the insulation is positioned when picking the back side of the brick.

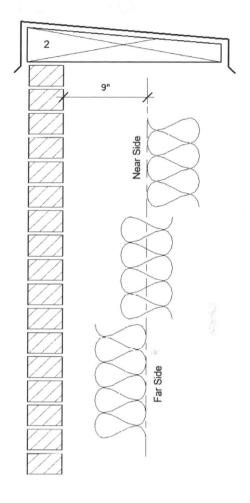

Figure 6.2-19 Options Bar with Insulation command active

Insulation: **Properties**

The Properties palette repeats the [Insulation] Width parameter and, in addition, offers the **Insulation Bulge to Width Ratio (1/x)** parameter—the graphic to the right explains.

Insulation: **In-Canvas**

Simply pick two points (Line) or select another line (Pick Lines).

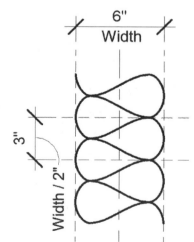

6.3 Annotate tab: Text panel

This panel contains tools to add, spell check and find/replace text within a given view.

Text

The first thing to know about the text tool is that it should be used as little as possible! Rather, live tags, keynotes and dimensions are preferred over static text to ensure the information presented is correct. Text will not update or move when something in the model changes—especially if the text is not visible or in the view where the model is being changed.

The Text tool can be started from the **Quick Access Toolbar**, the **Annotate** tab or by typing **TX** on the keyboard.

Text: Type Selector

The first thing to do is consider the current text type in the Type Selector (Figure 6.3-1). The name of a text type should, at a minimum, contain the size and font. The size is the **actual size on the printed page** regardless of the drawing scale.

There are additional steps that are optional, but at this point you could click, or click and drag, within the drawing window and start typing.

quick steps

Text

1. Ribbon
   - Various format options – see notes in this section
   - Check Spelling
   - Find/Replace
2. Type Selector
   - Select text type
3. In-Canvas
   - Single click and start typing or click and drag to define width and then start typing
   - Enter create a new line
   - Click away from text to finish the text tool

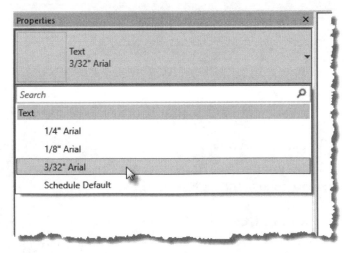

Figure 6.3-1 Type selector with text command active

Here is the difference between **click** and **click and drag** options when starting text:

- **Click**: Defines the starting point for new text. No automatic return to next line while typing. An Enter must be pressed to start a new line; this is called a "hard return" in the word processing world.

- **Click and Drag**: This defines a windowed area where text will be entered. The height is not really important. The width determines when a line automatically returns to the next line while typing. Using this method allows paragraphs to be easily adjusted by dragging one of the corners of the selected textbox.

As long as "hard returns" are not used, the textbox width and number of rows can be adjusted at any time in the future. To do this, the text must be selected and then the round grip on the right (see Figure 6.3-1) can be repositioned via click and drag with the left mouse button.

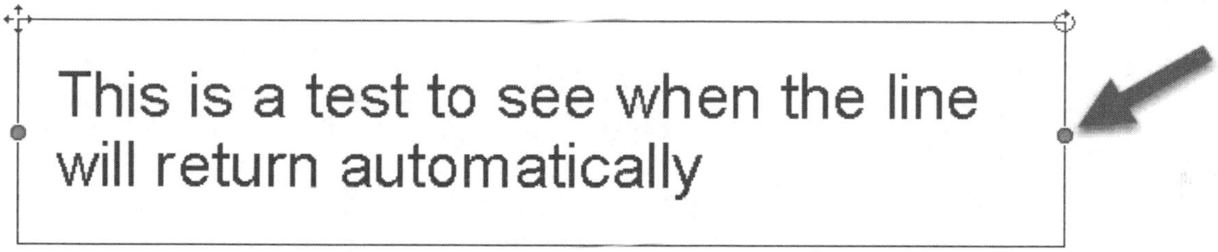

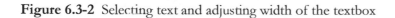

Figure 6.3-2 Selecting text and adjusting width of the textbox

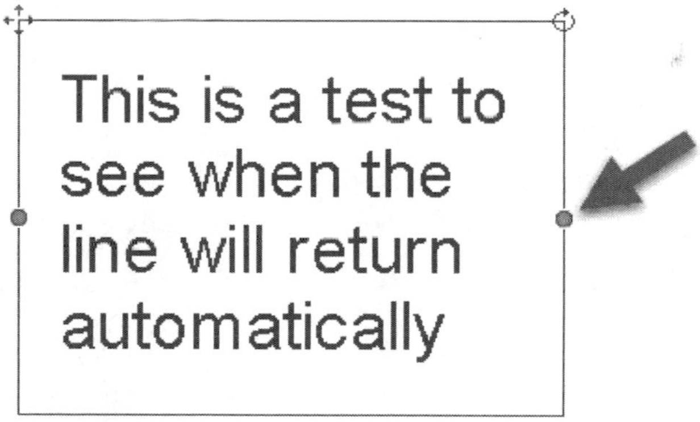

The image to the right, Figure 6.3-3, shows the result of adjusting the width of the textbox—the text element went from two lines to four.

If an Enter is pressed at the end of each line of text, when originally typed, the text will not automatically adjust as just described. An example of this can be seen in the following two images: Figure 6.3-4 & 5. Note that in some cases this is desirable to ensure the formatting of text is not changed.

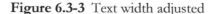

Figure 6.3-3 Text width adjusted

Keep in mind that this text tool is strictly 2D and view specific. If the same text is required in multiple views, the text either needs to be retyped or Copy/Pasted. However, for general notes that might appear on all floor plans sheets, for example, a Legend View can be utilized.

Revit has a separate tool, on the Architecture tab, called **Model Text** for instances when 3D text is needed within the model.

Annotate

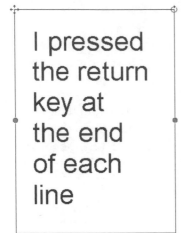

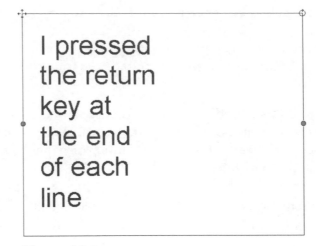

Figure 6.3-4
Text with hard returns

Figure 6.3-5
Text does not adjust when textbox width is modified

While editing text, right-click and hover over **Symbols** to see a list of symbols which can be inserted into the text (see image to right). Additional special characters are added to text via the **Character Map** function in **Windows**.

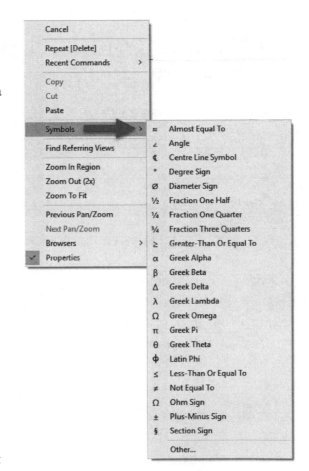

Text: Ribbon

During text placement there are two major variations of the Ribbon. First, during placement there are overall leader and formatting options. Second, while in text edit mode, there are more detailed font, paragraph and bulleting options.

Formatting Text

There are a number of formatting options which can be applied to text. These adjustments can be applied while initially creating the text or at any time later.

The formatting options are mainly found on the Ribbon while in the Text command or when text is selected.

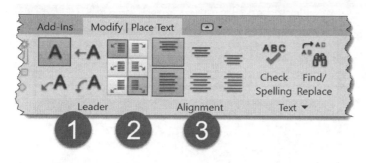

The formatting options identified in the image are:

1. Leader options
2. Leader position options
3. Text Justification

Figure 6.3-6
Formatting options on the Ribbon while creating text

Leader Options

A leader is a line which extends from the text, with an arrow on the end, used to point at something in the drawing.

The default option, when using the text command, is no leader. This can be seen as the highlighted option in the upper left.

The remaining three options determine the graphical appearance of the leader: one segment, two segment or curved as seen in Figure 6.3-7. Often, a design firm will standardize on one of these three options for a consistent look.

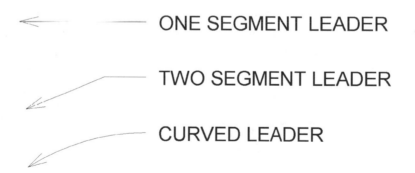

ONE SEGMENT LEADER

TWO SEGMENT LEADER

CURVED LEADER

Figure 6.3-7 Leader formatting options

Here are the steps to include a leader with text:

- Start the **Text** command
- Select text type via **Type Selector**
- Select **Leader** option
- Specify **leader location**
- **Type** text
- Click **Close,** or click away from text, to finish

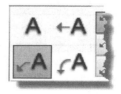

Leader Options

Annotate

Once the text with leader is created, it can be selected and modified later if needed. In the image below, Figure 6.3-8, notice the two **circle grips** associated with the leader, at the arrow and the change in direction of the line (the other two circle grips are for the text box as previously described in this section). These two grips can be repositioned by clicking and dragging the left mouse button.

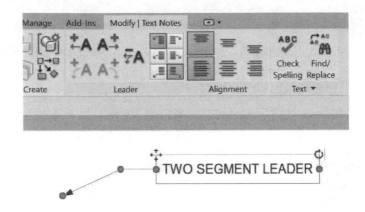

Figure 6.3-8 Text with leader selected; notice leader grips and Ribbon options

When text is selected the Ribbon displays slightly different options for leaders as seen in the image above. It is possible to have multiple leaders (i.e., arrows) coming off the text—denoted by the green "plus" symbol. In the next image, Figure 6.3-9, an additional leader was added to the left and one was also added to the right. It is not possible to have both curved and straight leader lines for the same text element. In this example, the curved leader options are grayed out as seen in Figure 6.3-8.

Figure 6.3-9 Multiple leaders added

Back in Figure 6.3-8, also notice that leaders can also be removed—even to the point where the text does not have any leaders. The one catch with the **remove leader** option is that they can only be removed in the order added.

The **Leader Arrowhead** can be changed graphically (i.e., solid dot, loop leader, etc.). This will be covered in the section on Managing Text Types as this setting is in the Type Properties for the text itself.

Good to know...

Text can be placed in **any view** and on **sheets**. The only exception is the Text command does not work in schedule views.

To place text in a 3D view, the view must be Locked via the **Save Orientation and Lock View** command on the View Control Bar at the bottom. Also, use the **Set Work Plane** before placing text to control how text looks.

Text can also be in a **Group**. When the group only has elements from the Annotate tab, it is a Detail Group. When the group also has model elements, the text is in something called Attached Detail Group. When a Model Group is placed, selecting it gives the option adding the Attached Detail Group.

Leader Position Options

These six toggles control the position of the leader relative to the text as seen in the two images to the right (Figures 6.3-10 and 11). These options also appear in the Properties Palette when text is selected, called **Left Attachment** and **Right Attachment**.

Similar to leader type, a design firm will often select a standard that everyone is expected to follow so construction documents look consistent.

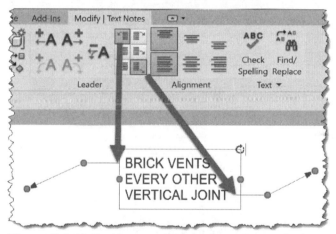

Figure 6.3-10 Leader position toggles – Example A

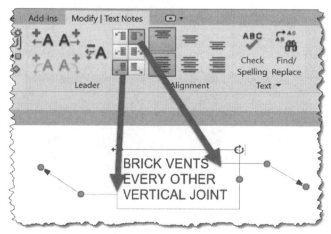

Figure 6.3-11 Leader position toggles – Example B

Text Justification

When text is selected there are six options for horizontal and vertical justification on the Ribbon: Top, Middle, Bottom, Left, Center and Right. The results for the three vertical options can be seen in the three images for Figure 6.3-12.

These options also appear in the Properties Palette when text is selected, called **Vertical Align** and **Horizontal Align**. Keep in mind that all options in the Property Palette are instance parameters—meaning they only apply to the instance(s) selected. Thus, each text entity in Revit can have different settings.

Text Formatting

The next section to cover is the options to make text Bold, Italic or be Underlined.

These options do not appear in the Properties Palette because they can be applied to individual words (or even individual fonts). In the example below, the word "Brick" is bold, "other" is italicized and "vertical" is underlined.

If all the text is selected and set to one of these three options, an edit made in the future will also have these settings.

> Notice the formatting options are different when editing the text, compared to when the text element is just selected.

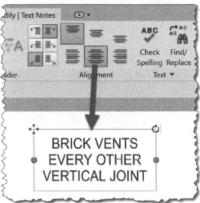

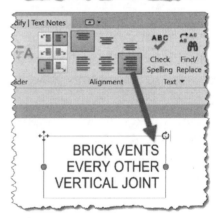

Figure 6.3-12
Text justification options

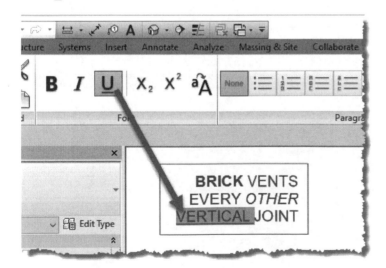

Figure 6.3-13 Text formatting

Text Formatting

The next section to cover is the **List** options on the Paragraph panel. This feature is only available while editing the contents of the text element; to do this, select the text and then click on the text to enter edit mode.

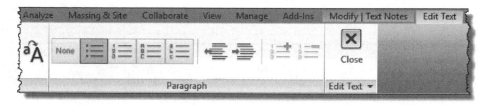

• 4" FACE BRICK	1. 4" FACE BRICK
• 1" AIR SPACE	2. 1" AIR SPACE
• 3" RIGID INSULATION	3. 3" RIGID INSULATION
• 8" CONCRETE MASONRY UNIT (CMU)	4. 8" CONCRETE MASONRY UNIT (CMU)
• 3 5/8" MTL STUDS AT 16" O.C.	5. 3 5/8" MTL STUDS AT 16" O.C.
• 5/8" GYP BD	6. 5/8" GYP BD
a. 4" FACE BRICK	A. 4" FACE BRICK
b. 1" AIR SPACE	B. 1" AIR SPACE
c. 3" RIGID INSULATION	C. 3" RIGID INSULATION
d. 8" CONCRETE MASONRY UNIT (CMU)	D. 8" CONCRETE MASONRY UNIT (CMU)
e. 3 5/8" MTL STUDS AT 16" O.C.	E. 3 5/8" MTL STUDS AT 16" O.C.
f. 5/8" GYP BD	F. 5/8" GYP BD

Figure 6.3-14 Four options to define a line within text

This is one case where you must press Enter to force the following text to a new line and automatically generate a list (i.e., bullet, letter or number).

Clicking the **Increase Indent** tool will indent the list as shown below (Figure 6.3-15). To undo this later, click in that row and select **Decrease Indent**. There does not appear to be a way to change what the indented listed value is. Using the backspace key and then indenting again allows an indent without a number/letter.

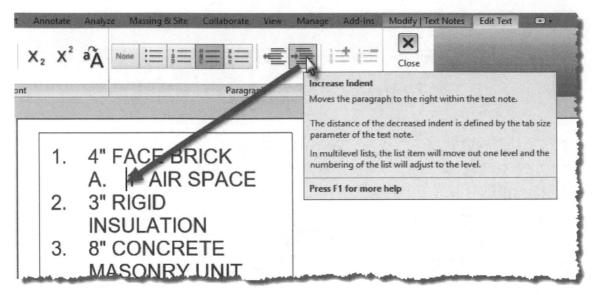

Figure 6.3-15 Indenting within a list

Clicking within the first row, clicking on the **plus** or **minus** icons will let you change the starting number/letter of the list.

The text formatting options also allow for **subscript** and **superscript** as shown in the example below (Figure 6.3-16).

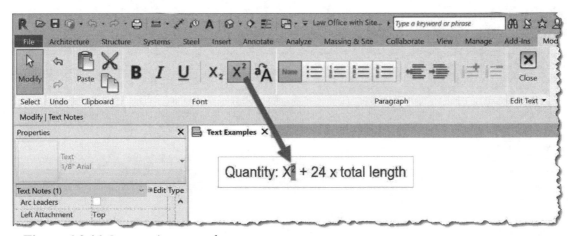

Figure 6.3-16 Superscript example

Also, notice the **All Caps** icon; clicking this icon will change selected text to all upper case. When this feature is used, Revit remembers the original formatting—thus, toggling off the App Caps feature later will restore the original formatting.

Spell Check

Revit has a tool which allows the spelling to be checked. Keep in mind this only works on text created with the Text command and only for the current view. Revit cannot check the spelling of text in keynotes, tags or families. Neither can it check the entire project.

The Spell Check command can be found on the Annotate tab or on the Ribbon when text is selected.

Spell Check: Check Spelling Dialog

When **Spell Check** is selected, the dialog to the right appears if there are any misspellings found (Figure 6.3-17).

When Revit finds a word not in the dictionary it will provide a list of possible correct words. Often the first suggestion is the right one. If not, select from the list.

Clicking the **Change** button will correct the highlighted word. Clicking **Change All** will change all of the words with this same misspelling in the current view.

Sometimes Revit will flag a word that is not misspelled. This might be a company

quick steps

Spell Check

1. Check Spelling Dialog
   - Verify "Change to" is correct
   - Click the **Change** or **Change All** button
   - Click **Ignore** if a change is not required
   - Click **Add** if the word should be added to the dictionary
   - Click **Close** to exit the dialog

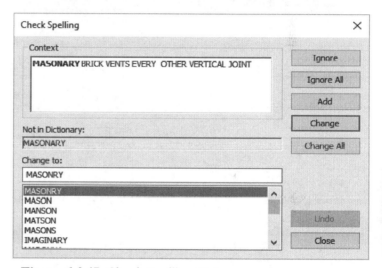

Figure 6.3-17 Check Spelling dialog

name, your name, a product name or an industry abbreviation. In this case one might select the **Add** option to add the word to the custom dictionary so you don't have to deal with this every time you run spell check.

The image below shows the settings related to the Spell Check engine—found in **Application Menu → Options**. The options are self-explanatory. In an office, consider placing the custom dictionary on the server and point all users to it.

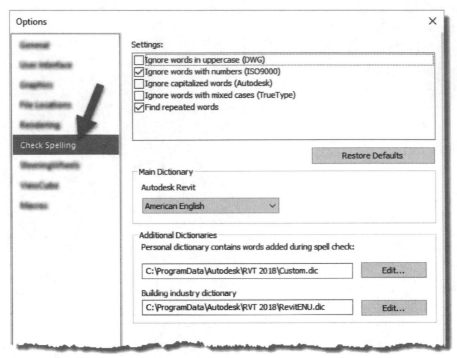

Figure 6.3-18 Check Spelling in Options dialog

Don't totally rely on Spell Check. A word may be spelled correctly but still be the wrong word. For example:

- Fill the **whole** with concrete and trowel level and smooth.
- Fill the **hole** with concrete and trowel level and smooth.

In this example the word "whole" is wrong but spelled correctly. Also keep in mind that Revit does not have a grammar check system like the popular word processing systems.

Find/Replace

Revit has a tool which allows words to be found or replaced within a view. Keep in mind this only works on text created with the Text command. Revit cannot find or replace text in keynotes, tags or families. Unlike the spelling tool, this tool can search the entire project.

Find/Replace Dialog

When this tool is selected the dialog to the right appears (Figure 6.3-19).

In this example, the current view is being searched for a brand name, **Sheetrock**, so it can be replaced with the generic industry standard term, **Gypsum Board**.

Selecting **Entire project** and then clicking **Find All** tells Revit to list all matches in the middle section of the dialog. For each row, you can click to select and see the context the found word(s) is used in.

When items are found, the **Replace** or **Replace All** buttons can be used to swap out the text in one location or all. When clicking Replace, only the selected row is replaced.

This tool can be used to just find something, and not replace it. For example, on a large project with hundreds of views in the Project Browser, using the Find/Replace to search for the details with the word "roof drain" can significantly speed up the process of locating the desired drawing.

quick steps

Find/Replace

1. Find/Replace Dialog
   - Find: Type text to be replaced
   - Replace with: Type text used to replace old text
   - Define scope
     i. Current selection
     ii. Current view
     iii. Entire Project
     iv. Match Case
     v. Match whole word
   - Select Replace or Replace All to proceed with change
   - Click Close to finish or cancel at any time

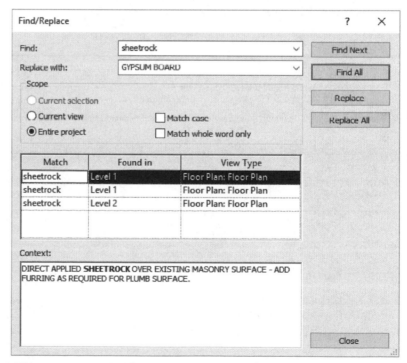

Figure 6.3-19 Find/Replace dialog

Annotate

Replacing a Text Type

In addition to replacing content within a text element, there is a way to replace the text type as well. For example, some imported details use a different font, and you want everything to match and be consistent. This is not really associated with the Find/Replace tool, but it is important to know how to accomplish this task.

Replacing a Text Type within a view or project:

- **Select** one text element within the project
- **Right-click** (Fig. 6.3-19)
- Pick **Select All Instances** →
  - o Visible in View
  - o In Entire Project
- Select a different type from the **Type Selector**

This procedure will replace all the text in either the current view or the entire project. Even text which has been hidden with the "Hide in View" right-click option will be changed.

When a specific text type is selected, the selection count, in the lower right corner of the Revit window, will indicate the total number of elements selected (Figure 6.3-21). This can be used as a quick double check before replacing the text type. For example, if the intent was to just replace a few rogue text instances but the count was several hundred, this would be a clue that some other view uses this text type and perhaps should not be changed as it was created by someone else on the project. This can be especially true if multiple disciplines are working in the same project.

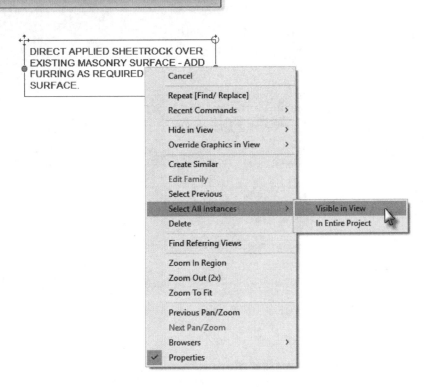

Figure 6.3-20 Select all instances via right-click

Figure 6.3-21 Total element selected count

Text Types

When using the text tool, the options listed in the Type Selector are the result of **Text Types** defined in the current project. Most design firms will have all the Text Types they need defined within their template.

There are two ways to access the text type properties. One is to start the Text command and then click **Edit Type** in the Properties Palette. The other is to click the arrow within the Text panel on the Annotate tab as pointed out in the image below.

Text Types: Type Properties Dialog

The **Type Properties**, as shown in the example below (Figure 6.3-22), are fairly self-explanatory. Following is a brief description of each.

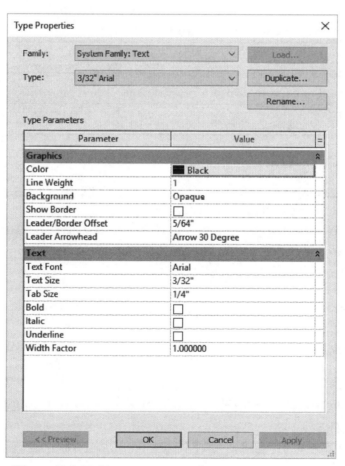

Figure 6.3-22 Text type properties

Command	What it does...
Color	This can affect printing, so it is often set to black.
Line Weight	This is only for the leader.
Background	*Toggle:* Opaque or Transparent
Show Border	*Check box:* Show or Hide
Leader/Border Offset	Space between text and board and leader – helpful when Background is set to Opaque.
Leader Arrowhead	Select from a list of predefined arrow types
Text Font	Select from a list of installed fonts on your computer
Text Size	Size of text on the printed page
Tab Size	Size of space when Tab is pressed (size on printed paper)
Bold	Default setting – can be changed while in edit mode
Italic	Default setting – can be changed while in edit mode
Underline	Default setting – can be changed while in edit mode
Width Factor	Adjusts the overall width of a line of text

Colors

The color applies to the text and the leader. The color is often set to black. If any other color is used, this can affect printing. For example, any color becomes a shade of gray when printed to a black and white printer—similar to printing a document from Microsoft Word where green text is a darker shade of gray than yellow text. In the Print dialog there is an option to print all color as Black Lines which can make colored text black. However, this also overrides gray lines and fill patterns.

Custom Fonts

Be careful using custom fonts installed on your computer as others who do not have those fonts will likely not see the formatting the same as intended. Custom fonts can come from installing other software such as Adobe InDesign. In fact, Autodesk also installs several custom fonts which are supposed to match some of the special SHX fonts which come with AutoCAD.

Custom Fonts

It is not possible to create custom arrowheads. However, the list of arrowheads is based on styles defined here: **Manage → Additional Settings → Arrowheads**. This provides many options in how these items look.

Width Factor

Some firms will use a Width Factor like 0.75, 0.85 or something similar to squish the text to fit more information on a sheet. Getting any narrower than this makes the text hard to read. This option actually changes the proportions of each letter, not just the space between them.

Misc.

Note that Text does not have a phase setting. Thus, the phase filters and overrides do not apply to text. It is sometimes desired to have text noting existing elements, such as ductwork, to be a shade of gray rather than solid—black being reserved for things that are new.

6.4 Annotate tab: Tag panel

The Tag panel offers various ways to visually expose information contained within building elements. Many tags have the ability to change the parameter they display when edited. However, deleting a tag has no effect on the information it represents. Tags are view specific but can be placed in multiple views without concern about a conflict as they all report the same data.

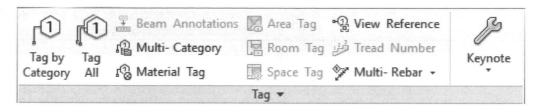

Tag by Category

This tool, often just referred to as the Tag command, is as significant as the Wall tool in the world of Revit—which is why it is also located on the *Quick Access Toolbar*. This tool is used to graphically display information contained within model elements. For example, a tag can be used to display a door's number in a floor plan, or a window type in an elevation view. If the door or window is changed, via the Type Selector, the tag automatically updates. The tag itself holds no information; it just reports information from the element it is tagging.

Each category in Revit, i.e., Door, Window, Wall, Furniture, Column, etc., must have its own tag family. Revit allows the user to specify which tag is used, by default, for each category. The tag may be changed at any time via the Type Selector. A single element may be tagged more than once. You might tag the same door in the main floor plan and an enlarged floor plan. Another example might be tagging the same element, in the same

quick steps

Tag by Category

1. Options Bar
   - Tag orientation
     i. Horizontal
     ii. Vertical
   - Tags… (button)
   - Leader
   - Leader type
     i. Attached End
     ii. Free End
   - Leader Length
2. In-Canvas
   - Hover cursor over model elements, click to place tag
   - 3D Views must be locked before placing tags
   - Notice the specific element you are about to tag is listed on the status bar in the lower left of the screen

Annotate

view, with more than one tag; a light fixture may have three different tags: circuit number, fixture type and switch system.

The image below has several tags added to a floor plan view. All the listed information is coming from the properties of the elements which have been tagged. For example, the "M1" tag within the diamond shape is listing the wall type (i.e., *Type Mark*). Because the wall tag(s) is/are listing a *Type Parameter*, all wall instances of that type will report the same value, i.e., "M1." The door number "1" is reporting the element's **Mark** value, which is an *Instance Parameter*. Therefore, each door instance may have a different number. Notice some tags have the *Leader* option turned on. This is especially helpful if the tag is outside the room. Any 3D element visible in a view may be tagged, even the floor. In this case the *Floor Tag* is actually reporting the *Type Name* listed in the *Type Selector*. The leader can be modified to have an arrow, a dot or nothing.

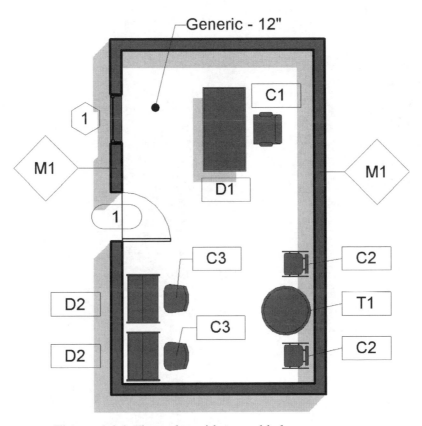

Figure 6.4-1 Floor plan with tags added

All of these tags were placed using the same tool: **Tag by Category**.

The section / interior elevation below (Figure 6.4-2) shows many of the same elements tagged as were tagged in the floor plan view above. If the wall's *Type Mark* is changed, the wall *Tag* will be instantly updated in all views: plans, elevations, sections, and schedules.

Some tags can actually report multiple parameter values found within an element. For example, the **Ceiling Tag** used in this example lists the *Type Name*, the ceiling height, and has fixed text which reads "A.F.F." Many tags can be selected and directly edited, which actually changes the values within the element. Changing the 8'-0" ceiling height will cause the ceiling position to change vertically.

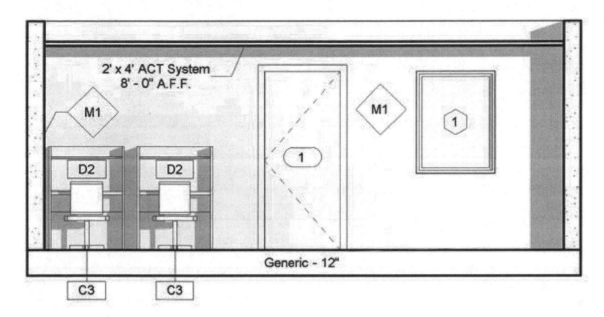

Figure 6.4-2 Section / Interior Elevation with tags added

Tag by Category: **Options Bar**

The image below shows the settings available on the *Options Bar* when the **Tag** command is active.

- The tag can only be *Horizontal* or *Vertical*.
  - o Option: check 'Rotate with Component' in family editor for a tag family
- Clicking the **Tags…** button allows you to specify which tags are used when multiple tags for the same category are loaded in the current project.
- Checking the *Leader* option draws a line between the tag and the element.
  - o *Attached End:* the leader always touches the tagged element.
  - o *Free End:* allows you to move the end of the leader.
    - ▪ *Caution: This option makes it possible to point at the wrong elements or nothing at all*
- **Length**: This determines the initial length of the leader.

Tag by Category: **In-Canvas**

Click on an element to tag it. If the Leader with Free End is selected, then pick two additional points to define the leader and tag location relative to the element being tagged.

Annotate

Tag All

This command, more formally called "Tag All Not Tagged," will add a tag to every element, in the selected category, in the current view. As the formal name implies, this tool will only tag elements that do not already have a tag in the current view.

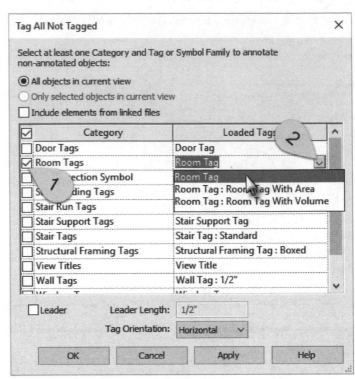

One example of using this tool would be for door tags. During the early design phase of a project, Schematic Design, the door tags are not necessary. Once the plan has formalized, the **Tag All Not Tagged** command may be used to quickly add door tags to all doors in the main floor plan view.

Tag All Not Tagged Dialog

Figure 6.4-3 Tag All dialog

The first set of options relates to what to tag in the current view. **All objects in the current view** is the first option and is self-explanatory. If elements in the model were selected when starting the Tag All command, then the option **Only selected objects in current view** is active and is the default selection. The third option, **Include elements from linked files**, can be checked if elements within a linked Revit model are to be tagged as well.

The **Categories** listed are limited based on tags loaded in the current project. If there is no window tag loaded, then "Windows" is not an option in this dialog. Multiple categories can be tagged at once by checking multiple boxes on the left – one for each category list.

Check **Leader** if each tag placed should have a leader. When this option is selected, the **Leader Length** value can be edited. This defines the leader length from the element to the tag.

The **Tag Orientation** option allows you to select the most prevalent condition: **Horizontal** or **Vertical**. Some tags will likely need to be manually rotated to maintain a clean, readable set of documents.

To complete this command:

Select a <u>row</u> (i.e., a **Category**) and then select a **Loaded Tag** from the drop-down list to the right of the selected category (if more than one tag is loaded for that category). Finally, click **Apply** and then **OK** to complete the command.

Beam Annotations

This tool places multiple beam tags, annotations and spot elevations. It is geared more towards structural design and drafting and will not be covered—other than providing an image of the dialog below, which explains how the feature works. Note: this dialog will not open unless specific tag types (structural framing) are present in the current project.

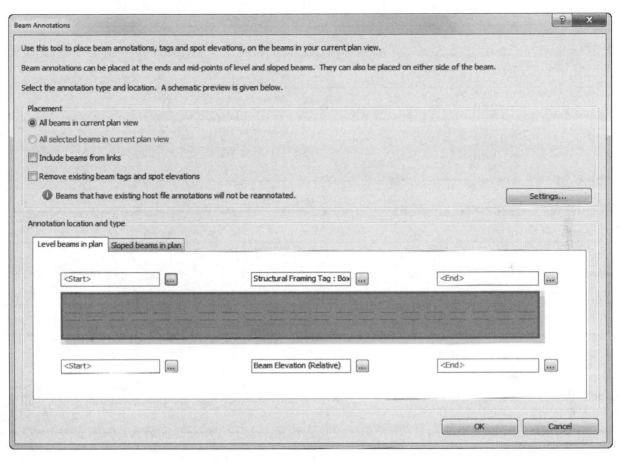

Figure 6.4-4 Beam Annotations dialog

Multi-Category Tag

The *Tag by Category* (which was covered a few pages back) is designed to use a different tag for each category. For example, a door tag, which has a distinct graphical look, is used when doors are clicked on, and a window tag is used when windows are selected. In contrast, the **Multi-Category** tag command uses one tag for any model element selected. Additionally, it is possible to use a Shared Parameter as a filter to control, or limit, which families may be tagged.

In Figure 6.4-5 the tags on top of the furniture and near the door are all "Tag by Category" tags. These are all placed with the same command, but the result is a different family used— one for furniture and one for doors in this case. The three tags pointed out are multi-category tags. These were placed with the same command and the same family is used.

Notice, for the furniture the information presented is the same, as both tags are set to show the Type Mark parameter. In contrast, the two door tags do not match. The multi-category tag is listing the Type Mark and the "by Category" tag is listing the Mark parameter.

Both tag commands have their uses. Understanding how each works is important to fully leveraging them.

Multi-Category Category Tag: **Options Bar**

The image below shows the settings available on the *Options Bar* when the **Multi-Category Tag** command is active.

---

quick steps

Multi- Category Tag

1. Options Bar
   - Tag orientation
     i. Horizontal
     ii. Vertical
   - Tags… (button)
   - Leader
   - Leader type
     i. Attached End
     ii. Free End
   - Leader Length
2. Type Selector
   - Select Multi-Category Tag type
3. In-Canvas
   - Hover cursor over model elements, click to place tag
   - 3D Views must be locked before placing tags
   - Notice the specific element you are about to tag is listed on the status bar in the lower left of the screen

- The tag can only be *Horizontal* or *Vertical*.
- Clicking the **Tags...** button allows you to specify which tags are used when multiple tags for the same category are loaded in the current project.
- Checking the *Leader* option draws a line between the tag and the element.
    - *Attached End:* the leader always touches the tagged element.
    - *Free End:* allows you to move the end of the leader.
        - *Caution: This option makes it possible to point at the wrong elements or nothing at all*
- **Length**: This determines the initial length of the leader.

Multi-Category Category Tag: **Type Selector**

Select the multi-category tag Type to use from the Type Selector. Different tags can be set to use or filter for specific parameters in a family (more on this below).

Multi-Category Category Tag: **In-Canvas**

Click on elements to tag them in the current view.

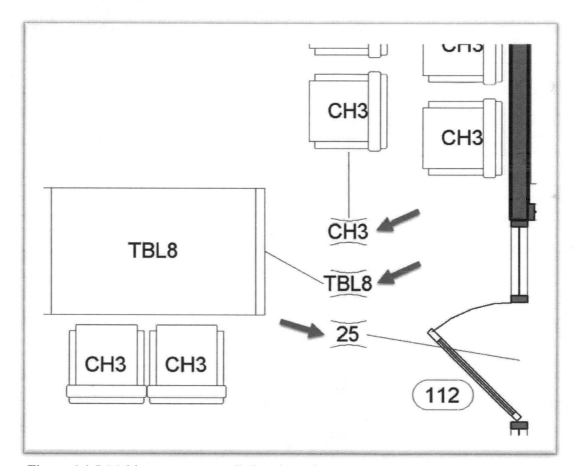

Figure 6.4-5 Multi-category tag applied to three elements

Multi-Category Category Tag: **Additional Info**

Multi-Category tags can be set up to only work on families that contain a specific **Shared Parameter**. For example, if several families have a Shared Parameter called "Salvage Tracking Number," then the multi-category tag **Label** can be modified to filter for that parameter (Figure 6.4-6). When this is done, that specific tag will ONLY tag families which contain that shared parameter. Other families do not even highlight when the tag command is active.

A Shared Parameter can be quickly added to multiple categories via **Manage** (tab) → **Project Parameters**. There, a new parameter can be created, using a Shared Parameter, and then the desired Categories selected (Figure 6.4-7).

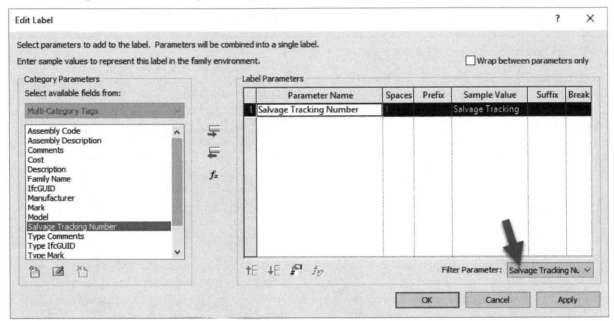

Figure 6.4-6 Family editor; Edit Label dialog for multi-category tag

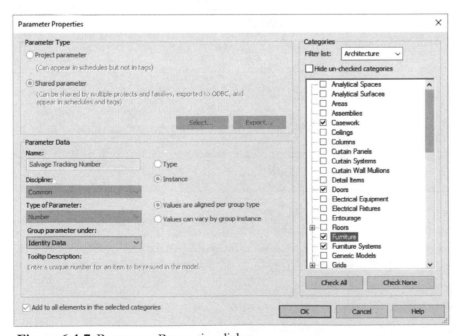

Figure 6.4-7 Parameter Properties dialog

Material Tag

All 3D surfaces within Revit can have a Material assigned – the Material Tag can report information contained within that material definition.

The Material Tag "looks" specifically at the spot picked on a surface. Thus, clicking within an area where the **Paint** tool was used to override a material, the Material Tag will report the painted material.

Tag by Category vs. Material Tag:
The Tag by Category command reports information (i.e., parameters) directly associated with a given element. In contrast, the Material Tag is reporting information indirectly associated with the given element—for example: *Material Description* (this is the default parameter the Material Tag reports). The *Material Description* may only be modified in the Material Browser (Figure 6.4-8), not in the Properties or Type Properties of a given element.

Material Tag: **Options Bar**

- The tag can only be *Horizontal* or *Vertical*.
- Clicking the **Tags...** button allows you to specify which tags are used when multiple tags for the same category are loaded in the current project.
- Checking the *Leader* option draws a line between the tag and the element.
  - ○ *Free End:* this is not an option with the material tag, thus it is grayed out.
- **Length:** This determines the initial length of the leader.

---

quick steps

Material Tag

1. Options Bar
   - Tag orientation
     - i. Horizontal
     - ii. Vertical
   - Tags... (button)
     - i. This has no apparent value within the context of this tool
   - Leader
   - Leader type
     - i. Attached End
     - ii. Free End
   - Leader Length
2. Type Selector
   - Select Material Tag type
3. In-Canvas
   - Be sure view is **not** set to Wireframe
   - Hover cursor over model elements, click to place tag
   - 3D Views must be locked before placing tags
   - Notice the specific element you are about to tag is listed on the status bar in the lower left of the screen

Annotate

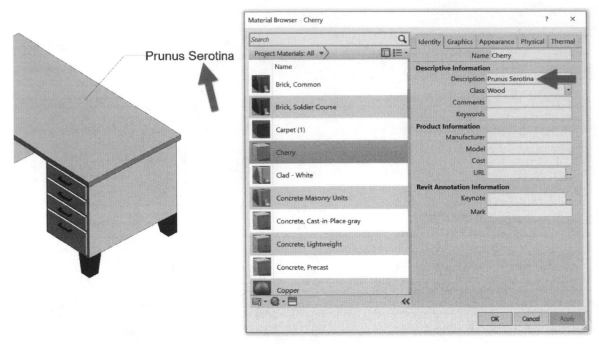

Figure 6.4-8 Material Tag and its source in the Material Browser

> The default Material Tag is looking at the **Material Description** parameter in the Material Browser (on the Identity tab). A custom tag could be made to display the Material *Name* or *Mark* if desired.

Material Tag: **Type Selector**

Select the material tag Type to use from the Type Selector. Different tags can be used in different views, or even on the same element. Two tags could be used, on the same element, to report two things; e.g. Mark and Description. It is also possible to create a custom material tag which combines two or more parameters automatically.

Material Tag: **In-Canvas**

Ensure the current view is not set to Wireframe as materials cannot be "seen" by the Material Tag command.

> Previously placed material tags are not affected by switching to wireframe view.

Some elements (families) cannot be tagged in certain views. For example, the provided "Desk" family found within the Architectural Template (default.rte) cannot be tagged with the Material Tag in a floor plan view. This is due to how the family was created. Figure 6.4-9 shows the **Family Element Visibility Settings**, in the Family Editor, for the desk surface. Notice it is set to be hidden in plan and reflected ceiling plan (RCP) views. Thus, there is no surface visible to the Material Tag command. In Figure 6.4-10, a customized version of the same desk was placed to the right and properly tagged.

There are several good reasons to use this visibility setting in a project. So, the reasons and desired functionality of each family should be considered before simply changing the visibility settings for 3D elements within the family editor.

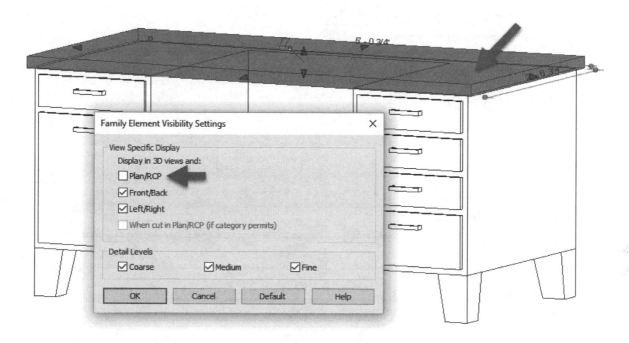

Figure 6.4-9 Family Editor: 3D extrusion set to be hidden in Plan/RCP views

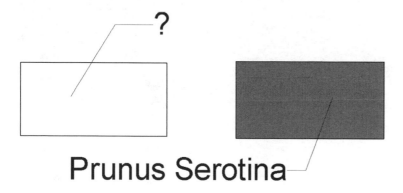

Figure 6.4-10 Project environment: original desk family on left, modified version on right

One of the challenges with using the Material Tag is that the information is disconnected from the typical way finish materials are reported in Revit. For the contract documents (CDs), the finish materials are often associated with the Room elements, and then a Room Schedule is created and placed on a sheet. The problem is that the information in the Room element has no connection to the actual materials used in a given room.

Area Tag

The Area Tag is used to tag Area elements. The name may be misleading. An Area plan is meant to track area, i.e., square footage, but an Area Tag actually reports more than just the square footage of an Area element. By Default, an Area Tag reports an area **Name** and **Area**.

The Area Tag family can be copied and/or customized to only report Area without name, or include additional information—including **custom information** by using Shared Parameters.

An Area Tag cannot report the Room Name or Number, but Room Tags can be placed separately within an Area Plan view.

Area Tag: **Ribbon**

There is only one option available during placement: **Reconcile Hosting**. This opens the Reconcile Hosting palette which lists any tags which have lost a connection to their linked elements. So, this really only applies to previously placed tags which are associated with an element in a linked model.

Area Tag: **Options Bar**

There are two options:

- The tag orientation
  - *Horizontal*
  - *Vertical*
  - *Model*
- Checking the *Leader* option draws a line between the tag and the element.

The **Model** option causes the tag to rotate if the Crop Region is rotated in a view. In contrast, Horizontal and Vertical will maintain that position relative to the screen/sheet even if the Crop Region is rotated.

---

quick steps

Area Tag

1. Ribbon
   - Reconcile Hosting
2. Options Bar
   - Tag orientation
     i. Horizontal
     ii. Vertical
   - Leader
3. Type Selector
   - Select Area Tag type
4. In-Canvas
   - Must be in an "Area Plan" which has Area elements already placed
   - Spacebar to rotate tag before placing
   - Can only place tag within boundary of area if Leader is not checked.
   - Tags can be easily aligned with adjacent areas during initial placement—watch for dashed alignment lines.

Area Tag: **Type Selector**

Select the desired area tag type from the Type Selector before placement.

Area Tag: **In-Canvas**

To use the Area Tag command, the current view must be an Area Plan. Additionally, Area elements must already exist in the model.

When the Area Tag is active, the previously placed Area elements become highlighted and the "X" appears.

> The center of the "X" corresponds to the point picked when the Area element was originally placed. Additionally, when Tag All is used, the Area Tag is placed at the "x" location by default.

The area fill and "X" can be manually turned on via **Visibility/Graphics Override** (dialog) → **Model Categories** (tab) → **Areas**, and then check **Interior Fill** and/or **Reference** sub-categories.

The image below (Figure 6.4-11) shows a common use for an Area Plan and Area Tag: to track the overall gross or net area of a floor. In this case, view filters hide the interior walls and doors and unneeded categories are turned off. The result is a simple view indicating the needed information.

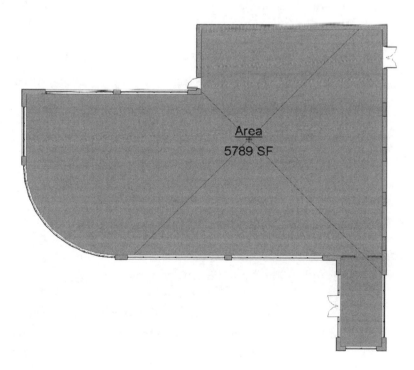

Figure 6.4-11 Area Tag added to area plan showing net square footage for level 1

Room Tag

The Room Tag command is used to tag Room elements in a plan, elevation or section view. The Room Tag reports parameters associated with Room elements, like Name, Number, Area, Department and more.

The Room Tag family can be copied and/or customized to report any of the built-in parameters as well as **custom information** by using Shared Parameters.

Room elements only appear in Plan, Elevation and Section views. Room Tags may only be added in views where the Room element appears and therefore cannot appear in 3D views. This also means that Room Tags can only be added or seen when the Room Category is turned on. View Filters, Phasing and Design Option can add an additional level of complexity when considering the visibility of a Room element in a given view—and, thus, whether or not a Room Tag may be added or seen.

> **quick steps**
>
> **Room Tag**
>
> 1. Options Bar
>   - Tag Position
>   - Leader
> 2. Type Selector
>   - Select tag type
> 3. In-Canvas
>   - Can only place tag within boundary of room if Leader is not checked.
>   - Tags can be easily aligned with adjacent rooms during initial placement—watch for dashed alignment lines.

If the Room category is turned off in a view, all Room Tags will be hidden—even though the Room Tag category may still be turned on.

Room Tag: **Options Bar**

There are two options:
- The tag orientation
  - *Horizontal*
  - *Vertical*
  - *Model*
- Checking the *Leader* option draws a line between the tag and the element.

The **Model** option causes the tag to rotate if the Crop Region is rotated in a view. In contrast, Horizontal and Vertical will maintain that position relative to the screen/sheet even if the Crop Region is rotated.

The **Leader** option must be on if the Room Tag needs to be positioned outside the room in a given view. For example, due to the view scale and/or the size of a room, the room tag may not fit within a room, so it is often positioned adjacent to the room. In this case, Revit requires the leader be on to maintain the association with a room the tag is not contained by.

Room Tag: Type Selector

Select the desired area tag type from the Type Selector before placement.

Room Tag: In-Canvas

To use the **Room Tag** command, the current view must be plan, section or elevation. Additionally, Room elements must already exist in the model.

> Room Tags can also be placed with the **Tag All** command.

When the Room Tag is active, the previously placed Room elements become highlighted and the "X" appears.

> The center of the "X" corresponds to the point picked when the Room element was originally placed. Additionally, when Tag All is used, the Room Tag is placed at the "x" location by default.

The area fill and "X" can be manually turned on via **Visibility/Graphics Override** (dialog) → **Model Categories** (tab) → **Rooms**, and then check **Interior Fill** and/or **Reference** sub-categories.

The image below (Figure 6.4-12) shows several Room Tags placed in a floor plan view.

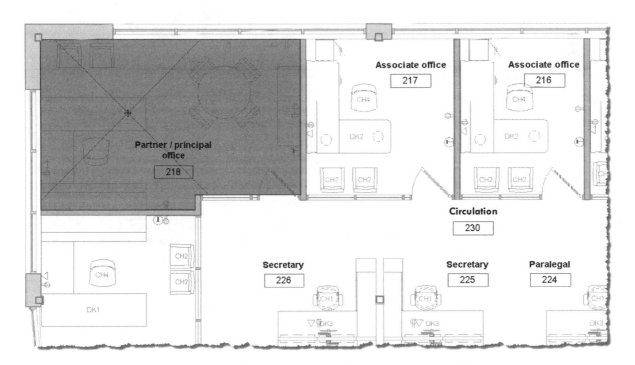

Figure 6.4-12 Room Tag added to floor plan – showing room name and number

Space Tag

Space Tags are specific to the MEP discipline and as such will not be covered in detail. See the Autodesk Help site for more information on this feature.

Spaces are functionally identical to the Room element, which is also true for Room Tags and Space Tags. The only difference is they are designed to hold engineering data such as CFM and Average Illuminance. Contrast that with a Room element which is designed to hold architectural data like wall finish and occupancy type for code compliant building egress calculations.

In the HVAC plan view below, a Room Tag and Space Tag have been added to the same room. With the Space selected, notice it has the ability to read the Room Name and Number it is aligned with in the model. The Space Name and Number can be made to match the Room Name and Number using **Analysis** (tab) → **Space Naming** command. This is a onetime adjustment between the two. If Rooms are added or changed, this command needs to be run again.

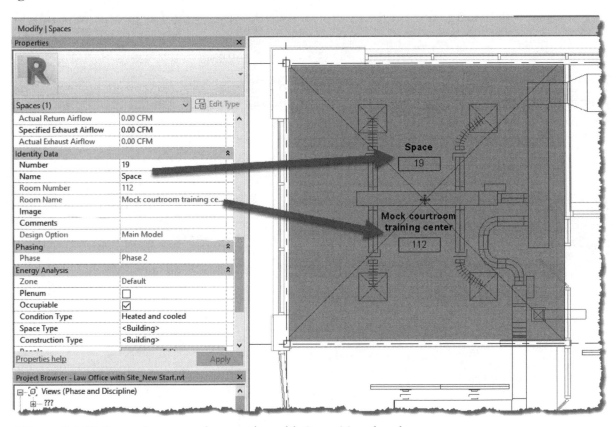

Figure 6.4-13 Space element and properties, with Space Tag placed

View Reference

The View Reference command has multiple uses for manually referencing other views in the current model. The first use is placing a reference next to a matchline, indicating the sheet to go to for a continuation of the drawing.

Another use is to create a View Reference tag that looks like an elevation tag. This special "tag" may be placed in plan views to reference more dynamic isometric views that take the place of traditional interior elevations.

Another use is to align the View Reference with a note, when the note needs to reference a detail. This is better than just typing the detail number and sheet number which would never update.

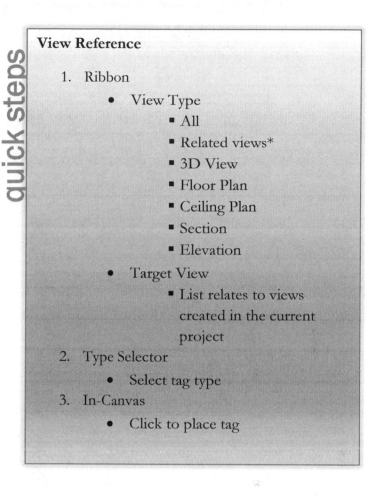

quick steps

View Reference

1. Ribbon
   - View Type
     - All
     - Related views*
     - 3D View
     - Floor Plan
     - Ceiling Plan
     - Section
     - Elevation
   - Target View
     - List relates to views created in the current project
2. Type Selector
   - Select tag type
3. In-Canvas
   - Click to place tag

The reference can be changed later if needed; the item does not need to be deleted and recreated.

View Tag: Ribbon

The Ribbon has two options: **View Type** and **Target View** (Figure 6.4-14).

The first drop-down, **View Type**, is used to filter the next drop-down. There is an "all views" option to see everything. When working in **Dependent Views**, there is also a **Related Views** option geared towards working with and labeling **matchlines (Figure 6.4-15)**.

Target View lists all views, based on the selected View Type.

Room Tag: Type Selector

Select the desired Type to be placed.

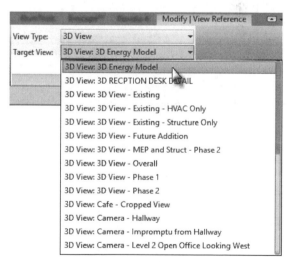

Figure 6.4-14
View Reference options for 3D views

Room Tag: **In-Canvas**

The View Reference can only be rotated after placement using the Rotate command.

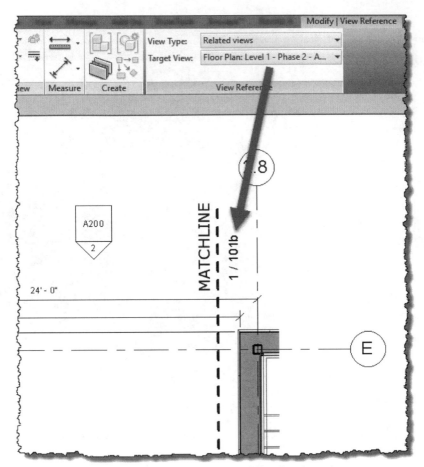

Figure 6.4-15 View Reference at Matchline in plan view

Immediately after placement, double-click on the View Reference to open the Target View and verify it is correct. Given this manual selection of the Target View, it is possible to be wrong.

> For presentation drawings use this feature to create hyperlinks to other views. This is a quick way for you, the presenter, to quickly navigate the project.

The next two images show a custom View Reference tag, which looks like an elevation tag. The View Reference, in plan view, points to drawing 4 on sheet A500 (Figure 6.4-16). The second image shows the view placed on sheet A500 (Figure 6.4-17). If the view moves to another sheet, or the sheet number changes, the View Reference will update automatically.

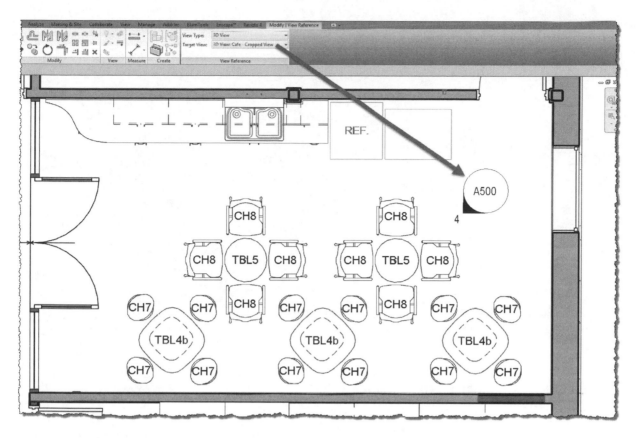

Figure 6.4-16 Custom view reference element; looks like elevation tag

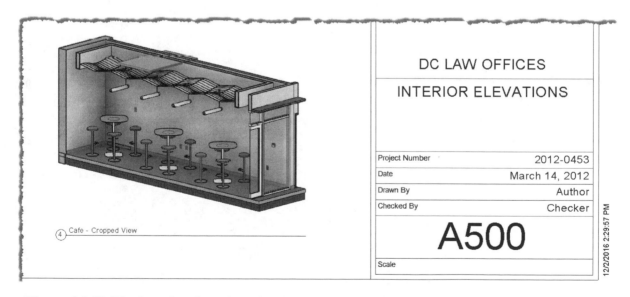

Figure 6.4-17 3D view placed on sheet A500; drawing number is 4

Tread Number

Use this tool to annotate the
number of risers or treads on a
single run of stairs.

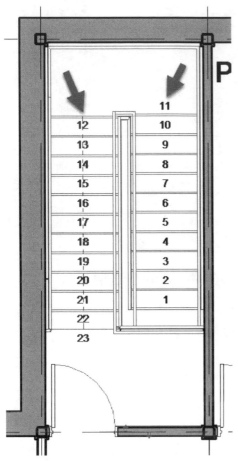

quick steps

> **Tread Number**
>
> 1. Properties
>   - Various graphics options
> 2. In-Canvas
>   - Works in Plan, Elevation or
>   Section views
>   - Click on stair to place Tread
>   Numbers
>   - During placement, a dashed line
>   appears to indicate text location

Room Tag: In-Canvas

Click on a single run of stairs in plan or section as shown
below. For the plan example, there are two runs, so this
requires two clicks—thus, two instances of the Tread
Number element.

Figure 6.4-18
Tread Number in plan view

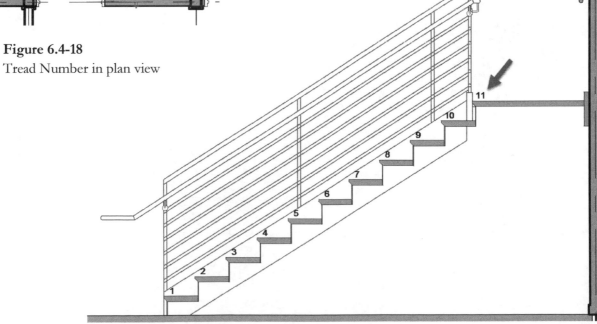

Figure 6.4-19 Tread Number in section view

Multi-Rebar

This tool is specific to the structural discipline and as such will not be covered here.

Element Keynote

Keynoting provides a consistent way to document items in a set of construction documents. The Element Keynote tool will report the Keynote value found in Type Properties for a given element. The actual keynote value corresponds to a text file defined in the Keynote Settings dialog.

quick steps

Element Keynote

1. Options Bar
   - Tag Position
   - Leader
2. Type Selector
   - Select keynote type
3. In-Canvas
   - Item to be tagged will highlight and be listed on the status bar

Element Keynote: Options Bar

- The tag can only be *Horizontal* or *Vertical*.
- Checking the ***Leader*** option draws a line between the tag and the element.
  - *Attached End:* the leader always touches the tagged element.
  - *Free End:* allows you to move the end of the leader.
- **Length**: This determines the initial length of the leader.

Element Keynote: Type Selector

Select the desired Type from the Type Selector. Note that one option is to show the keynote text rather than the number, which is more readable for most people.

Element Keynote: In-Canvas

Click items to keynote; additional clicks are required if leader is checked (Figure 6.4-20).

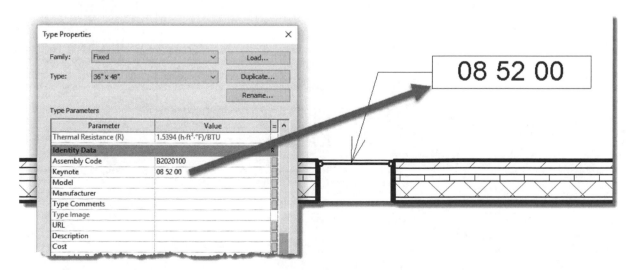

Figure 6.4-20 Element Keynote and Type Properties showing source information

Annotate

The Element Keynote tag can list the keynote number or the keynote text. The default templates have both options listed in the Type Selector.

The Keynote options are derived from the Keynotes text file; this can be one that ships with Revit (Figure 6.4-21) or a custom version.

A **Keynote Legend** can be created to track the keynotes used on a sheet (Figure 6.4-22).

The detail, shown in Figure 6.4-23, shows a detail with all key values used. It requires the Keynote Legend to know what the references mean.

The keynote command is powerful and results in less errors compared to using text; however, it is more complicated and as such is not the de facto standard in the industry due to the learning curve and some challenges managing this feature on a larger project with multiple staff working on it.

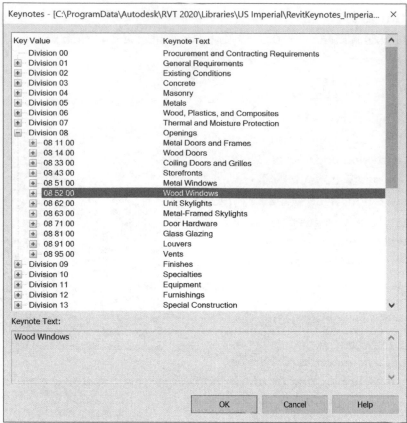

Figure 6.4-21
Keynotes dialog accessed from element Type Properties

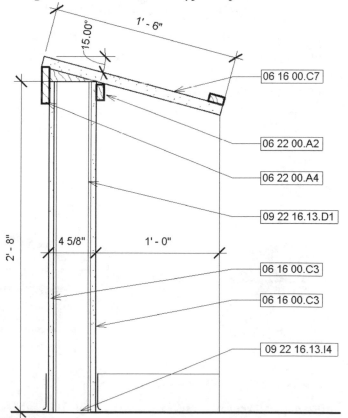

Figure 6.4-23 Detail with keynotes

Keynote Legend	
Key Value	Keynote Text
06 16 00.	1/2" Particleboard
06 16 00.	1" Particleboard
06 22 00.	1x2 Wood Trim
06 22 00.	1x4 Wood Trim
09 22 16.	3 5/8" Metal Stud Framing
09 22 16.	3 5/8" Metal Runner
09 29 00.	5/8" Gypsum Wallboard

Figure 6.4-22 Keynotes Legend

Material Keynote

Keynoting provides a consistent way to document items in a set of construction documents. The **Material Keynote** tool, similar to the Material Tag, will report the Keynote value found in **Material Browser** for a given material (Figure 6.4-24). The actual keynote value corresponds to a text file defined in the **Keynote Settings** dialog.

quick steps

Material Keynote

1. Options Bar
   - Tag Position
   - Leader
2. Type Selector
   - Select keynote type
3. In-Canvas
   - Item to be tagged will highlight and be listed on the status bar

Element Keynote: Options Bar

The tag can only be *Horizontal* or *Vertical*.
Checking the *Leader* option draws a line between the tag and the element.

Element Keynote: Type Selector

Select the desired type.

Element Keynote: In-Canvas

Ensure the view is not set to Wireframe as no materials can be detected as such.

Some materials contained within families cannot have a Material Keynote used on them. The keynote will only see the entire family rather than the material in it.

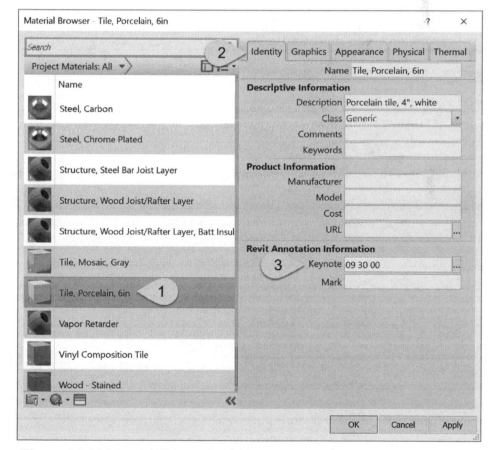

Figure 6.4-24 Material dialog; Material keynote

Annotate

User Keynote

Keynoting provides a consistent way to document items in a set of construction documents. The **User Keynote** tool still references the Keynote text file, similar to the other two keynote tools previously covered. The main difference is the keynote value displayed is selected during placement and unique to the selected element.

The User Keynote tag is unique from all other tags in Revit because it actually contains data. All other tags simply report

data contained within the element being tagged. Thus, keep in mind that this is not an Instance Property found in the Properties dialog when the element is selected. The only way to change the value displayed in a specific tag is to select the User Keynote tag itself and edit the on-screen value or the Properties palette version of the same data.

The same element can have multiple User Keynotes, displaying different keynote values, even in the same view. For example, one User Keynote might describe the electrical requirements for a reception desk and another User Keynote indicates the countertop material (different information which does not contradict).

User Keynotes can lead to inconsistent drawings due to the user selecting from a list for each instance. For example, the user intends to select countertop information and accidentally selects an adjacent item related to roof drains. Of course, this could happen with the other keynote tools (element and material keynotes), creating a larger scale problem. However, when selecting a keynote that applies to an entire project, more care is often taken in that selection.

Element Keynote: Options Bar

- The tag can only be *Horizontal* or *Vertical*.
- Checking the *Leader* option draws a line between the tag and the element.
  - o *Attached End:* the leader always touches the tagged element.
  - o *Free End:* allows you to move the end of the leader.
- **Length**: This determines the initial length of the leader.

Element Keynote: Type Selector

Select the desired type. One default option, in the templates provided by Autodesk, is Keynote Text. This shows the words rather than the numbers. Using this option would not require a

Keynote Legend and streamlines the description of elements and materials in the project. Also, if something needs to change, e.g. Sheetrock (which is a brand name) needs to change to Gypsum Board (which is a generic description), the Keynote text file can be edited—which automatically changes all notes in the project once reloaded.

Element Keynote: In-Canvas

Click on any model element to add a User Keynote – except model lines. 1-2 more clicks are required if Leader is selected on the Options Bar.

The image below (Figure 6.4-25) shows an example of how it is possible to create inconsistencies when using the User Keynote command. In this case, three User Keynotes were added to the same element—and only one is correct.

Also, the other notes on the left are Element Keynotes set to report the text rather than the number. This ensures that every countertop, for example, is described the same way everywhere in the project "COUNTERTOP WITH BACKSPLASH."

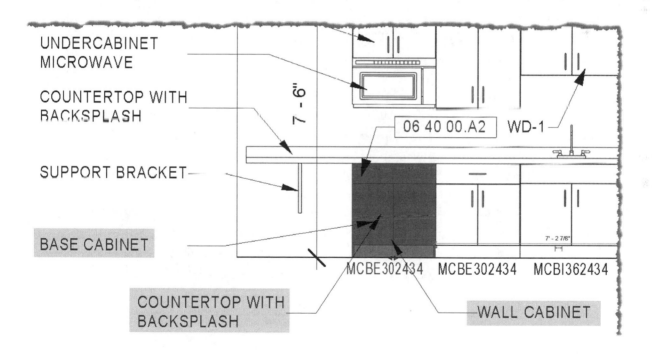

Figure 6.4-25 Multiple User Keynotes added to the same base cabinet family

Keynote Settings

This option, in the Ribbon, provides access to the Keynote Settings dialog.

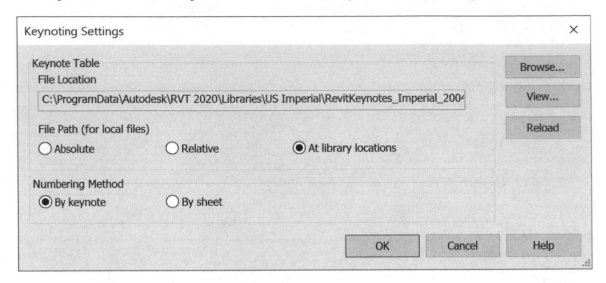

The **File Location** points to a specially formatted text file Revit reads and uses as the source for all numbers and text used by keynotes in a given project; each project can have its own unique **Keynote Table**. If the file is moved or deleted the current Keynote Table is remembered, but it cannot be modified unless the original text files are available and it is reloaded from this dialog.

Use the **Browse** button to select a relocated text file or select a new one. Changing text files can have a negative impact on previously placed keynotes. The **View** button opens the keynote list in Revit (not the text file itself). The **Reload** button is used to update the Keynote Table in Revit after changing the text file. Otherwise, Revit reloads the file every time the project is opened (similar to RVT/DWG links).

> Windows does not report TXT files as being locked when open. This means two or more people can have the TXT file open at the same time. The last person to save the file "wins"—that is, their version of the file is the one saved. Other's changes are lost. Therefore, caution must be used when multiple users need to edit the keynotes.

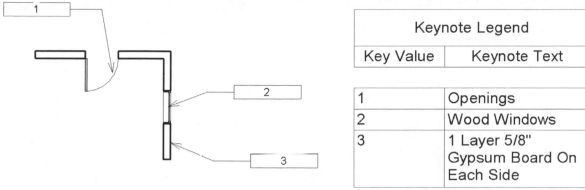

Keynote Legend	
Key Value	Keynote Text

1	Openings
2	Wood Windows
3	1 Layer 5/8" Gypsum Board On Each Side

Figure 6.4-26 Sheet view with keynotes by sheet

The **Numbering Method** has two options: **By keynote** (the default method) and **By sheet**.

The **By keynote** option keeps track of the formal number in the text file, for example "08 52 00" for Wood Windows (Figure 6.4-28).

In contrast, the **By sheet** method essentially ignores the formal number in favor of a number based on the order each element type was selected in by using the keynote command. Additionally, those numbers (by sheet) do not appear until the view is placed on a sheet (Figure 6.4-26).

```
RevitKeynotes_Imperial_2010.txt - Notepad                                    —    □    ×

File   Edit   Format   View   Help
08 51 23        Steel Windows    08 51 00
08 51 23.A1     Steel Sliding Window Sill      08 51 23
08 51 23.A2     Steel Sliding Window Jamb      08 51 23
08 51 23.A3     Steel Fixed Window Head/Jamb   08 51 23
08 51 23.A4     Steel Fixed Window Sill 08 51 23
08 51 23.A5     Steel Fixed Window Fixed Jamb  08 51 23
08 51 23.A6     Steel Fixed Window Mullion     08 51 23
08 52 00        Wood Windows    08 50 00    ⬅
08 52 00.A1     Clad Wood Double Hung Window Head        08 52 00
08 52 00.A2     Clad Wood Double Hung Window Sill        08 52 00
08 52 00.A3     Clad Wood Double Hung Window Jamb        08 52 00
08 52 00.A4     Clad Wood Double Hung Window Meeting     08 52 00
08 52 00.A5     Clad Wood Double Hung Window Head - Double Glazed    08 52 00
08 52 00.A6     Clad Wood Double Hung Window Sill - Double Glazed    08 52 00
08 52 00.A7     Clad Wood Double Hung Window Jamb - Double Glazed    08 52 00
08 52 00.A8     Clad Wood Double Hung Window Meeting - Double Glazed 08 52 00
08 52 00.B1     Clad Wood Projecting Window Head         08 52 00
08 52 00.B2     Clad Wood Projecting Window Sill         08 52 00
08 52 00.B3     Clad Wood Projecting Window Jamb         08 52 00
08 52 00.B4     Clad Wood Projecting Window Head - Double Glazed     08 52 00
08 52 00.B5     Clad Wood Projecting Window Sill - Double Glazed     08 52 00
08 52 00.B6     Clad Wood Projecting Window Jamb - Double Glazed     08 52 00
```

Figure 6.4-27 Keynote text file

Annotate

Loaded Tags and Symbols

This feature is accessed by expanding the Tags panel. Access is also provided via the Options Bar from within many of the tag tools.

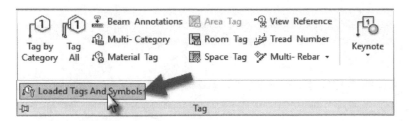

For any given category, there can be multiple tags loaded in the current project. Revit provides this dialog to predetermine which tag should be used, by default, when there is more than one to choose from. After placement, a tag can be selected and swapped out via the Type Selector if needed. Otherwise, the default option can be changed as needed here.

The **Load Family** button provides a convenient way to load a tag, i.e., Revit family, not currently available in the project. The **Filter List** option can hide irrelevant discipline categories to focus on what a specific user needs.

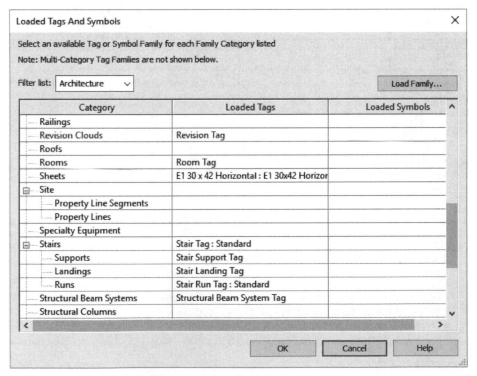

Figure 6.4-28 Loaded Tags and Symbols dialog

6.5 Annotate tab: Color Fill panel

In this section the Color Fill tools will be covered.

Duct and Pipe Legend

These tools are specific to the MEP discipline and will not be covered.

Color Fill Legend

This command adds a **Color Fill Legend** to a plan or section view. If the view does not already have an active **Color Scheme** applied, the user is also prompted to set one up for the current view.

> Color Fill cannot be applied to ceiling plan views.

quick steps

Color Fill Legend

1. Type Selector
   - Select legend type
2. In-Canvas
   - Click near floor plan or section to place legend
   - Does not work in ceiling plans

It is recommended to create new views to apply a **Color Scheme** and **Color Fill Legend** to. Doing so will avoid messing up construction drawing views and printing sheets with superfluous information.

A Color Fill Legend can be based on Rooms, Spaces or HVAC Zones. Additionally, a parameter such as Department or Name is used to determine colors.

Color Fill Legend: **Type Selector**

Select the desired type from the Type Selector. The default templates only have one type available.

Color Fill Legend: **In-Canvas**

Click somewhere in the view.

The next two images show a Color Fill Legend added to a plan view and a section view, respectively (Figures 6.5-1, 2).

Department Legend

□ Accessory Spaces

■ Building Support Services Spaces

▨ Circulation

■ Conference Center (The front entry)

▦ Office and Workstation

Annotate

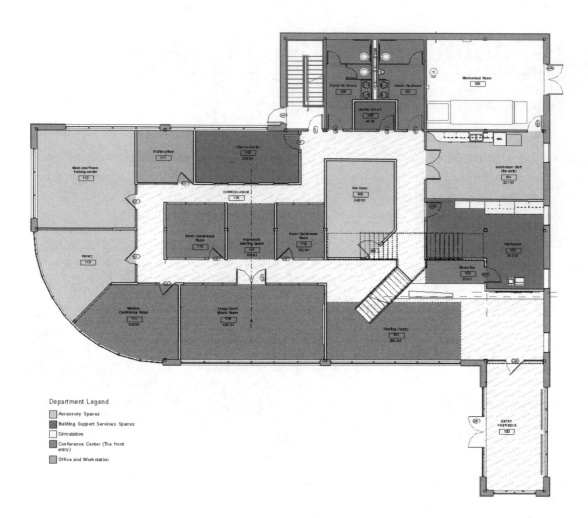

Figure 6.5-1 Color Fill Legend added to plan view

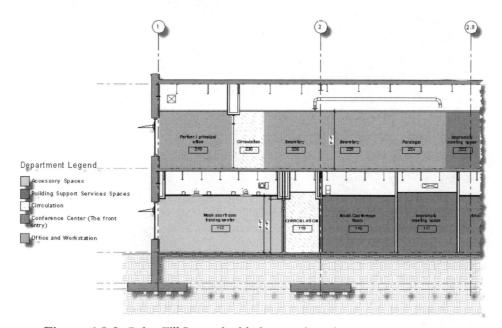

Figure 6.5-2 Color Fill Legend added to section view

If a **View Template** is applied to a view, and that View Template controls the Color Scheme, placing a Color Fill Legend will not change the Color Scheme. So, if no Color Scheme is currently applied, then the Color Fill Legend will be empty.

In some cases, a valid option might be to NOT "include" the Color Scheme in what the View Template controls as shown in the image below (Figure 6.5-3).

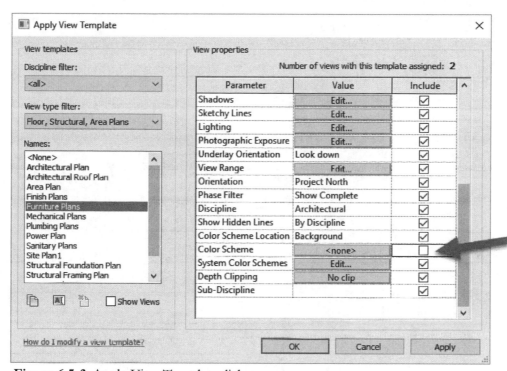

Figure 6.5-3 Apply View Template dialog

The legend itself has several **Type Properties** which control how it looks graphically. As seen in Figure 6.5-4, things like the size of the swatch, background, text and title text settings can be adjusted.

Clicking **Duplicate** allows for multiple legend types in the project. There could be one for plans and another for sections. Or, one for Rooms and another for Spaces if needed in the same project.

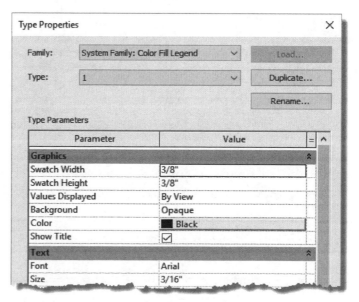

Figure 6.5-4 Color Fill Legend properties

Annotate

6.6 Annotate tab: Symbol panel

In this section, only two tools will be covered: **Symbol** and **Stair Path**. The rest are specific to the structural discipline; those are Span Direction, Beam (for beam systems), Area (rebar), Path (rebar) and Fabric (in concrete slabs).

Symbol

A Symbol is a 2D, view specific, graphic that changes size based on the current view scale.

Uses include north arrows, graphic scales and other static annotations such as the centerline symbol.

A symbol can be placed in every view type except schedules and 3D views. In **Legend** views, the type selector also lists all loaded tags in the current project. However, the tags only show the default value entered in the Family Editor.

Figure 6.6-1 shows two Symbols, north arrow and graphic scale, added near the view title on a sheet.

Tag by Category: **Ribbon**
The **Load Family** button is provided as a convenient way to load additional symbols when needed.

Tag by Category: **Options Bar**

quick steps

Symbol

1. Ribbon
   - Load Family
2. Options Bar
   - Number of Leaders
   - Rotate after placement
3. Type Selector
   - Select Symbol type to place
4. In-Canvas
   - Hover cursor over model elements, click to place tag
   - 3D Views must be locked before placing tags
   - Notice the specific element you are about to tag is listed on the status bar in the lower left of the screen

The **Number of Leaders** option provides a way to predetermine the number of leaders to add immediately after placing the symbol (Figure 6.1-2). Text is the only other element that can have multiple leaders because symbols and text are not actually associated with anything in the model. Checking **Rotate after Placement** prompts for rotation immediately after placement.

Tag by Category: **Type Selector**

Select the desired Type from the Type Selector.

Tag by Category: **In-Canvas**

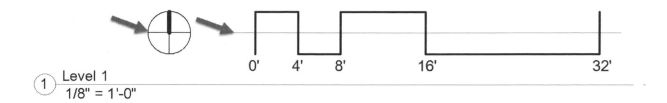

Figure 6.6-1 Two symbols added to sheet

After placement, a symbol can have leaders added or removed. Unfortunately, leaders can only be added in the reverse order they were added.

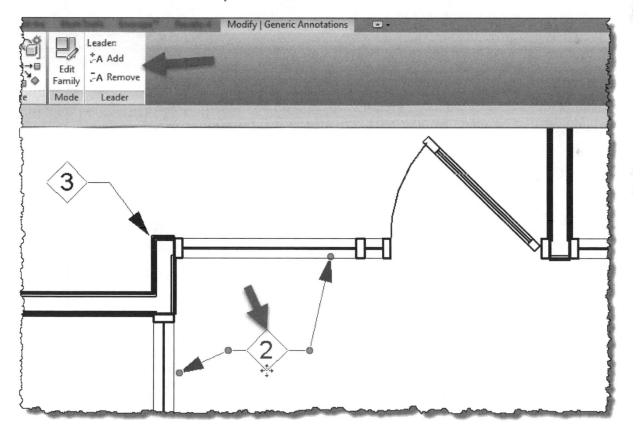

Figure 6.6-2 Two leaders added to symbol

Stair Path

The **Stair Path** annotation is added to a stair automatically when the stair is created. However, if it is deleted, or additional plan views are created, this command is used to add it to the current view.

Tag by Category: **Type Selector**

There are two fundamental types:

- Automatic Up/Down Direction
- Fixed Up Direction

Each has different Type Properties.

Tag by Category: **In-Canvas**

In a floor plan view, click on a stair to add the Stair Path graphic. After placement it can be selected and moved perpendicular to the stair run if needed.

The line weight of the line and arrow can be changed (Figure 6.6-3) in **Object Styles** (for a project-wide change) or in the views **Visibility/Graphic Overrides** dialog (affects current view only). On the Annotation tab, edit the "arrows" row(s) as shown below.

quick steps

Stair Path

1. Type Selector
   - Select stair path type to place
2. In-Canvas
   - Select a stair to add the stair path to

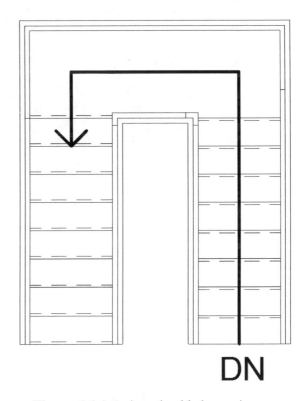

Figure 6.6-3 Stair path added to stair

Chapter 7

Analyze Tab

The Analyze tab contains tools to prepare your Building Information Model (BIM) for external analysis tools and to perform analysis directly within Revit. Many of these tools are Structural or MEP specific and will not be covered in detail—the exception being the brief overview of some of these tools in this introduction. As the AEC industry continues to develop the BIM process, more opportunities become available to do analysis and provide additional value. Revit is not inclusive of the BIM process, but represents its foundation.

Structural analysis tools highlights

The first two panels are specific to the structural discipline:

- Analytical Model; Loads and Boundary Conditions
- Analytical Model Tools

These tools allow the structural engineer to document specific loads (e.g. a heavy safe) on the structural system, and ensure the model is ready for analysis in an external application. There are multiple options for external structural analysis. Autodesk has one called Robot Structural Analysis. Another is made by RISA Technologies. Both provide an add-in for Revit to allow easy data sharing, both to and from the external application.

MEP analysis tools highlights

There are four panels on the Analyze tab which have MEP specific tools:

- Spaces & Zones
- Reports & Schedules
- Check Systems
- Color Fill (excluding the *Color Fill Legend* tool)

The **Spaces & Zones** panel contains tools which allow the MEP designers to define and maintain information about various areas within the building, similar to the architectural Room element. The **Reports & Schedules** panel contains powerful tools used by the mechanical engineer to calculate the building's overall heating and cooling loads directly within Revit. The results from this analysis are used to size key building equipment (e.g. air handling unit and air conditioning). The **Check Systems** panel provides tools to ensure the ductwork, for example, is properly connected—as this is required for accurate calculations of air flow, friction loss and more. Finally, the **Color Fill** panel has two features to graphically display information about the duct and pipe systems. For example, the duct's air flow may be displayed—showing a different color for each specified flow range.

These topics all require specific knowledge of mechanical, piping, electrical and/or structural engineering and would require a separate book to give them proper attention. Thus, these tools are not covered in depth.

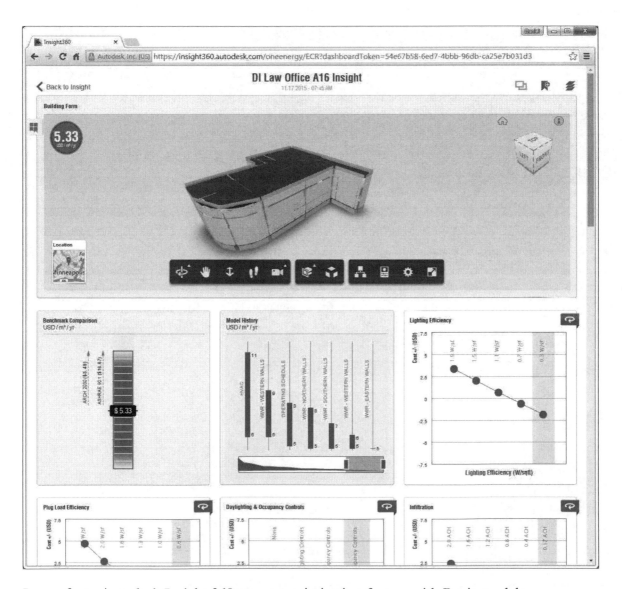

Image from Autodesk Insight 360 energy optimization feature with Revit model.

7.1 Energy Optimization panel

This panel, previously called Energy Analysis, has tools used to define, maintain, generate and view the results for the energy analysis workflow. Two of the tools on this panel, System-Zone and Systems Analysis, are designed specifically for MEP analysis and will not be covered in this book.

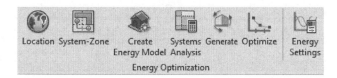

Location

This command is used to document the project's location on earth. This does not have to be set before modeling, and it can even change if needed. The project location, along with an accurate True North setting, determines how the sun casts light, which affects shade and shadows. This is also important in energy optimization, as a site's micro-climate is a significant part of the equation.

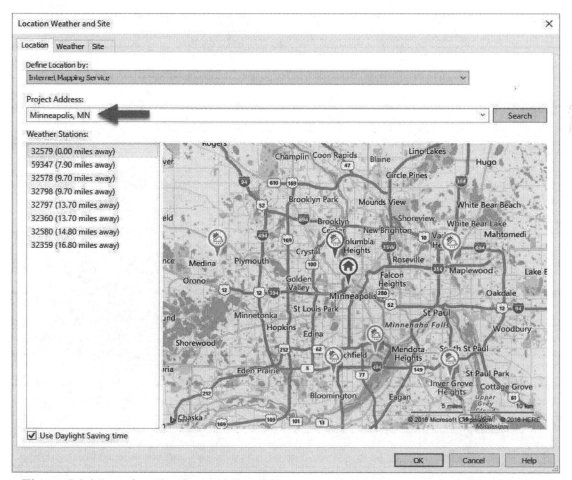

Figure 7.1-1 Location Weather and Site dialog

To use the Location command, simply type a site name or specific postal address in the **Project Address** field and click **Search** or press **Enter**. If the project is on a new site, with no address, zoom in on the map and drag the location icon to the project site. However, for daylight and energy analysis the city location should be close enough.

The Define Location by has two options: **Internet Mapping Service** and **Default City List**.

The **Internet Mapping Service** currently uses Bing to define the location on earth and provide the map graphics. Thus, an internet connection is required for this to work properly.

The **Default City List** provides a list of several cities around the world. This does not require an internet connection. When this option is selected, Revit indicates this option is best for HVAC sizing and not for energy analyses.

Once a location is selected, several **Weather Stations** near the project site are listed. The closest one is selected by default. This is usually the best option, but in some cases it should be changed based on geographical conditions, such as vertical grade or proximity to a lake or ocean.

If the municipality for the project location observes **Daylight Saving time**, then check this box.

The **Weather** tab lists specific historical weather data for the selected weather station.

The **Site** tab relates to positioning one or more instances of one Revit model in another.

Scrolling and **panning** is allowed in the map view. When zooming, an aerial view will appear.

Figure 7.1-2 Location Weather and Site dialog – aerial view results from zooming in

Create Energy Model

This command is very simple; it either creates or deletes an analytical energy model in the current project.

Even though this specific command is simple, there are some important settings to consider before using it. Mainly, the options in the **Energy Settings** dialog need to be considered. These options include phase, thermal zoning and building constructs.

When **Create Energy Model** is selected, this prompt appears (Figure 7.1-3).

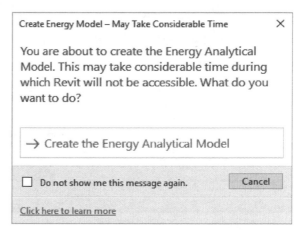

Figure 7.1-3 Create Energy Model prompt

The result is the model seen below (Figure 7.1-4) which is in a new 3D view called **3D Energy Model**. Use this view and the related analysis visibility categories to visually verify the accuracy of the model.

The energy model is static and does not change. If the Revit model changes, click **Delete Energy Model** and recreate it.

Delete
Energy Model

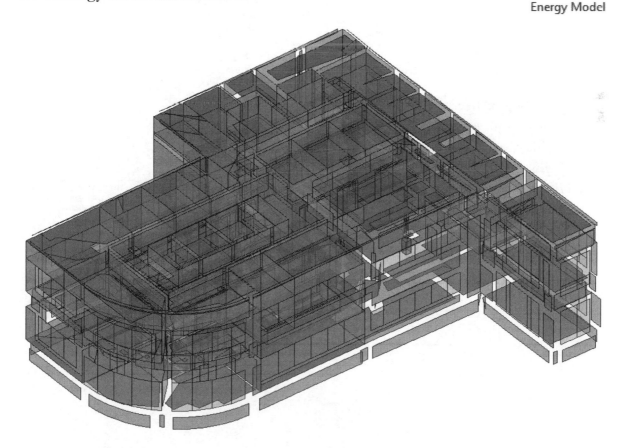

Figure 7.1-4 Energy model as seen in the 3D Energy Model view

Analyze

Generate

This command sends the energy model to **Autodesk Insight 360** to be processed in the Cloud. When the command is selected, if an Energy Model already exists, a prompt offers to update the Energy Model first or use the existing version (this does not mean "existing" phase) as seen in Figure 7.1-5.

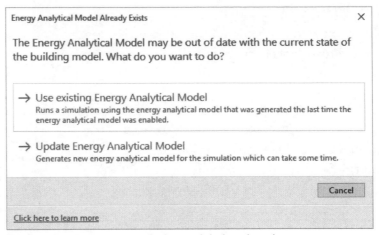

Figure 7.1-5 Energy analysis model already exists prompt

Once this command is finished, the prompt in Figure 7.1-6 is provided. The energy model has now been successfully uploaded to the Cloud. It still takes some time before the energy analysis is complete. The next command, **Optimize**, covers the process of viewing and manipulating the results.

Using this command requires an internet connection. Depending on the size of the project, this can take some time to upload.

Finally, some projects have very aggressive confidentiality clauses in their contracts between the design team and the client. Uploading your model involves sharing project information which may be confidential. Autodesk requires each user to agree to special terms related to this the first time you use this service.

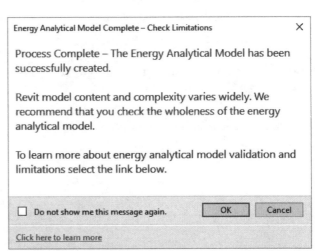

Figure 7.1-6 Energy analysis upload complete

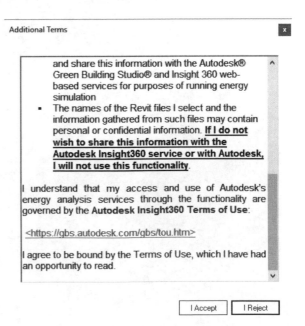

Figure 7.1-7 Additional terms

Optimize

This command opens a viewer to access **Autodesk Insight 360**. This is essentially a specialized web browser to access the Cloud-based content. This content can also be accessed from any computer using a modern browser such as Chrome or Safari. The computer does not need to have Revit installed. You just need to log in with your Autodesk account username and password.

This feature is essentially another software application with many capabilities and could be a book of its own.

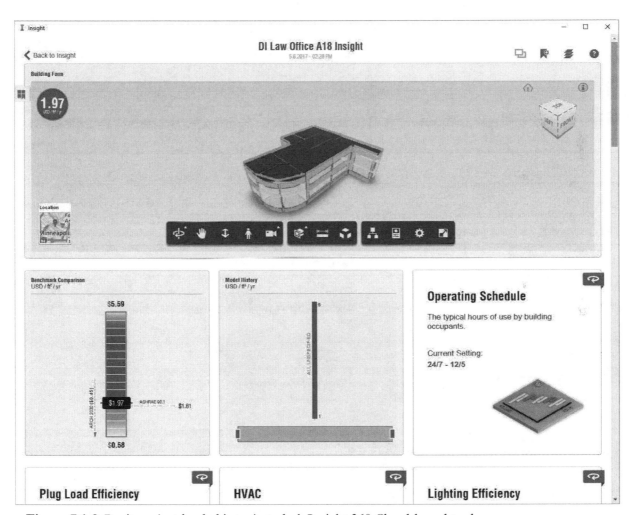

Figure 7.1-8 Revit project loaded into Autodesk Insight 360 Cloud-based tool

Energy Settings

In the **Energy Settings** dialog, which has been streamlined in Revit 2021, there is one critical setting related to running the simulation and a few more that must be considered to generate a valid Energy Analytical Model (EAM). To study the real-time impact on overall performance, all other inputs can be adjusted in the cloud later.

In Revit 2021, these important parameters have been separated from the secondary options. The Energy Settings dialog is split into two. The first dialog (Figure 7.1-9), presents the more essential parameters, while the **Other Options** "Edit…" button opens the secondary dialog where optional settings can be adjusted.

Now let's take a look at each of these settings to better understand what they do and why they are important.

Location

This feature was covered previously in this section. Refer back a few pages for details on this command.

FYI from Autodesk's help page: "Weather stations include 'actual year' virtual weather stations and typical year weather stations (TMY2 and other formats) based on 30-year averages of weather data, typically taken from airport locations."

The Energy settings essential to a valid Energy Simulation are:

- Location

The Energy settings essential to creating a valid Energy Analysis Model (EAM):

- Analysis Mode
- Ground Plane
- Project Phase
- Analytical Space Resolution
- Analytical Surface Resolution
- If using masses (separately or in conjunction with building elements)
  - Perimeter Zone Depth
  - Perimeter Zone Division

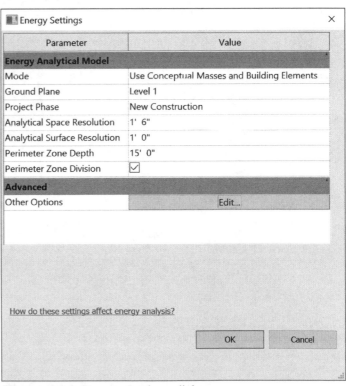

Figure 7.1-9 Energy Settings dialog

Analysis Mode

Analysis Mode determines if Revit should use Masses, Building Elements, or both to create the EAM.

These are the Analysis Mode options:

- Use Conceptual Masses (never select this option!)
- Use Building Elements
- Use Conceptual Masses and Building Elements (added in Revit 2016)

As it turns out, the "Use Conceptual Masses" only option uses an older internal algorithm, so don't use that one. Rather, use the combined option which will work on mass-only models.

Ground Plane

Properly setting the **Ground Plane** parameter ensures that spaces which occur below this level are understood to be below grade by the cloud-based Green Building Studio (GBS) calculation engine. For masses, the EAM will not generate glazing below the ground plane and a different construction can be selected for underground exterior walls. Keep in mind, when using the **Detailed Elements** option (which means use the thermal properties defined in materials) the toposurface elements are NOT used.

Project Phase

For projects with phasing—e.g., existing, new construction, phase demolished—be sure this is set correctly. The phase settings for the 3D Energy Model view have no impact on the EAM created.

EAM Creation Level of Detail

These two properties control the accuracy of the EAM when it is created:

- **Analytical Space Resolution** (default: 1'-6")
  This allows small gaps to exist in the model—both exterior and interior. Remember, Revit ignores elements in several categories such as Generic Model and In-Place families, so gaps are not uncommon.
- **Analytical Surface Resolution** (default: 1'-0")
  This setting works in conjunction with the Analytical Space Resolution setting to control the accuracy of the surface boundaries.

If portions of the Revit model are complex and not coming out right in the EAM, these values can be lowered. It is recommended that both of these values be adjusted proportionally. Lowering these values will result in a more accurate EAM but will take longer to create—the simulations will take more time to process as well. Large projects may require these values be increased.

Analyze

Automatic Thermal Zoning Settings

These two properties control the inclusion of space sub-divisions in the EAM:

- **Core Depth** (or Perimeter Zone Depth)
  The distance to measure inward from the exterior walls to define the core zone.
- **Divide Perimeter Zones** (or Perimeter Zone Division)
  Check this box to divide the perimeter into multiple zones.

These settings are used to divide large spaces into smaller subdivisions for more accurate simulations. In previous versions of Revit, these settings only work on masses, but they now also apply to large spaces defined by building elements, for example, a 200,000sf office and warehouse building. The large warehouse space would benefit from this subdivision feature.

When creating the EAM for building elements, the interior walls are typically sufficient to naturally 'zone' the model. In these situations, set the Core Depth to zero and uncheck the divide option.

Advanced Energy Settings

These "**Other Options**" can be adjusted on-the-fly in Insight 360 so they are not critical in the early stage of the design. However, when specific inputs are known they can be set here and will be used as the default, or **BIM Setting**, in Insight 360.

In the **Material Thermal Properties** section, the parameters are listed in order of priority. Schematic Types overrides Conceptual Types when used, and **Detailed Items** ignores the previous two in favor of thermal properties of model elements such as walls, doors, windows, etc.

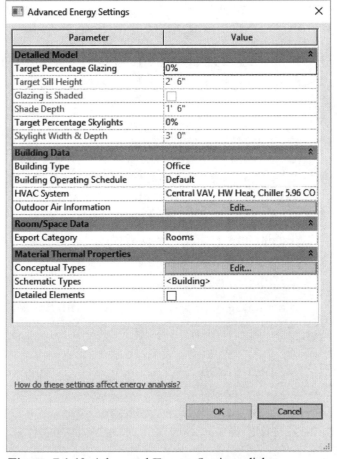

Figure 7.1-10 Advanced Energy Settings dialog

7.2 Route Analysis panel

Route analysis is a tool used to automatically calculate the distance and time between any 2 points in your model. For example, you can use this tool to determine existing distances, or something like a distance from a nursing station to a patient room.

Path of Travel

To define a path of travel you define a start and end points for the path in your model. An analysis is run by Revit that generates the shortest distance between the 2 points avoiding the categories of elements you identify as obstacles. If the design changes, path of travel lines must be updated manually.

While placing a path of travel line the contextual tab of the ribbon offers 2 settings. You can toggle the option to tag the path of travel line as you place it and you can set the

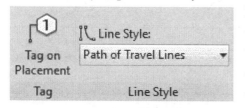

line type that will be used to draw the path of travel line.

quick steps

Path of Travel

1. Ribbon
   - Route Analysis panel
2. Contextual Tab on Ribbon
   - Tag on Placement
   - Line Type
3. Route Analysis Settings
   - Obstacle Categories
   - Analysis Zone
4. In-Canvas
   - Select Points

The path generated maintains approximately .3 meters distance away from any obstacles. This distance setting is a system setting and not able to be changed by the user. A path of travel line is limited to a single level and cannot calculate a path that includes stairs.

Path of travel lines are view specific. The lines can be both tagged and scheduled. Properties of a path of travel line are as follows:

- **Speed** – (Read only) Value for the speed of a person traveling along the path.
- **View Name** -View where the path is placed.
- **Level** – Associated level.
- **Time** – Time it will take for a person to travel the length of the path of travel.
- **Length** – Total length of all segments of the path of travel.

Analyze

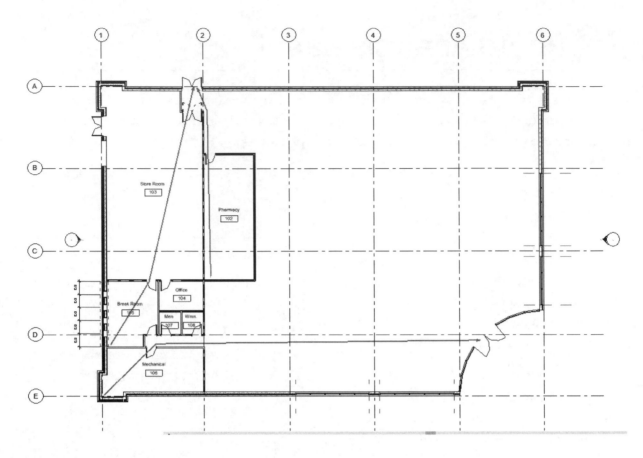

Figure 7.2-1 Floor plan with multiple Path of Travel lines placed

Reveal Obstacles

When working with a path of travel you may need to understand what is being identified as an obstacle in order to understand the path generated. The reveal obstacles turns on a mode where all elements defined as obstacles are highlighted in the view (Fig 7.2-2). Click the reveal obstacles button again to return the view to the normal mode.

Reveal Obstacles

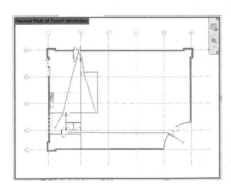

Figure 7.2-2 Floor plan with reveal objects mode on

Route Analysis Settings

The Route Analysis Settings dialog allows you to set what categories of elements are considered obstacles during the analysis. Some elements are NEVER considered as obstacles.

- Hidden Elements
- Demolished Elements
- Underlay elements

All other model categories can be excluded by selecting the check box in the settings dialog. If checked, elements in the checked category will behave as if it is turned off or hidden in the view. The analysis will not consider it as an obstacle and the path of travel line will be drawn through those elements.

The lower section of the setting dialog allows you to set a "zone" for the analysis. This analysis zone must be clear of obstacles for a path to be drawn. This setting allows you to account for small changes in something like a floor finish that might produce a line in the plan, but not be tall enough to be considered an obstacle. If your analysis fails, you may need to adjust the analysis zone.

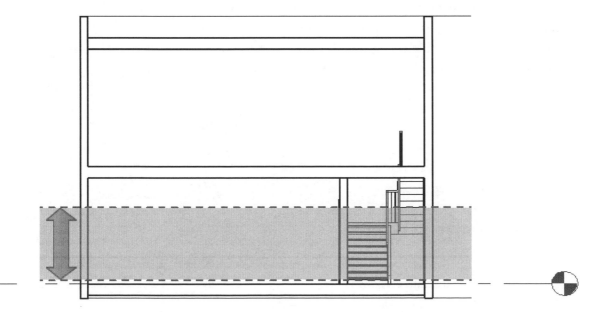

Figure 7.2-3 Analysis Zone

The Path of Travel lines may also be scheduled as shown below.

\<Path of Travel Schedule\>					
A	B	C	D	E	F
Mark	From Room	To Room	Length	Speed	Time
Level 1					
POT-01	Not available	Not available	17,199 mm	4.8 km/h	12.8 s
POT-02	Not available	Not available	10,767 mm	4.8 km/h	8.0 s
Level 2					
POT-03	Not available	Not available	13,567 mm	4.8 km/h	10.1 s
POT-04	Not available	Not available	17,338 mm	4.8 km/h	12.9 s
POT-05	Not available	Not available	6,334 mm	4.8 km/h	4.7 s

Figure 7.2-4 Path of travel schedule

7.3 Energy Analysis panel

Lighting:

This tool allows for daylighting calculations in Revit based on location, sky conditions, surface reflectance and glazing visual transmittance. Previously, this tool was hard-wired only to valid USGBC LEED® credit compliance. This newer version included with the Insight 360 tools allows for custom environment settings.

Lighting Solar

Energy Analysis

Worflow Overview:

When the command is selected, the 'Welcome' dialog presents an outline of steps, or prerequisites, required to generate accurate results (Figure 7.3-1). Click the links to read the specifics for each point.

> These tools are installed from the Autodesk App store or via Autodesk Desktop App.

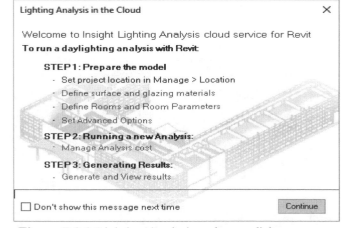

Figure 7.3-1 Lighting Analysis welcome dialog

Clicking **Continue** moves on to the Select Analysis, or project, dialog (Figure 7.3-2). Either select a previous project or **<Run New Analysis>**.

Clicking **Go** moves on to the Select Analysis Type dialog (Figure 7.3-3).

Here you select:

- Analysis type
- Floor levels in the building to analyze
- Date/Time
    - Sky model
    - Use weather data
- Illuminance Settings
- Grid Resolution

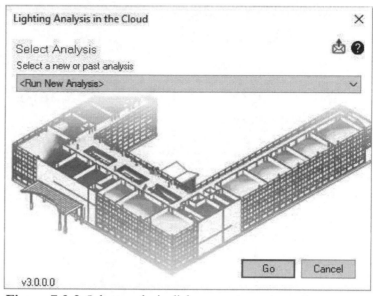

Figure 7.3-2 Select analysis dialog

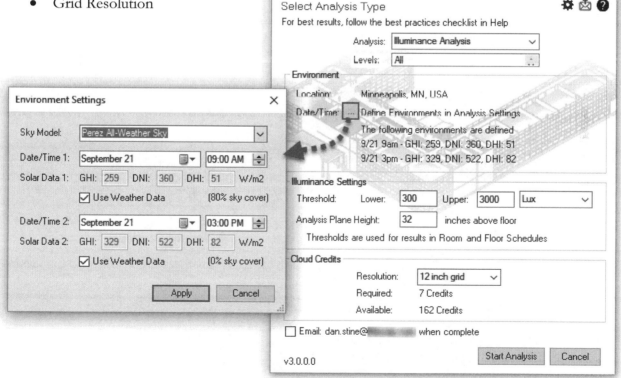

Figure 7.3-3 Select Analysis Type dialog and Environment Settings sub-dialog

Analyze

In the upper right corner of the main dialog, there are three icons:

- **Gear**: Settings (Fig. 7.3-3)
- **Envelope**: Email support
- **Question mark**: open help

Several of the dialog boxes associated with the Insight 360 add-in have a small envelope icon which can be used to send an email to Insight 360 help. This is a separate support mechanism set up by Autodesk to support the smaller sustainable design community within the Revit eco-system. Responses are quick and by very experienced people!

When the analysis is complete, the prompt shown in Figure 7.3-5 is presented. The required cloud credits are not charged until clicking one of the two **Accept** options listed.

When accepted, the data is saved into the project and available for viewing in model views and schedules.

Figure 7.3-4 Lighting Analysis Settings dialog

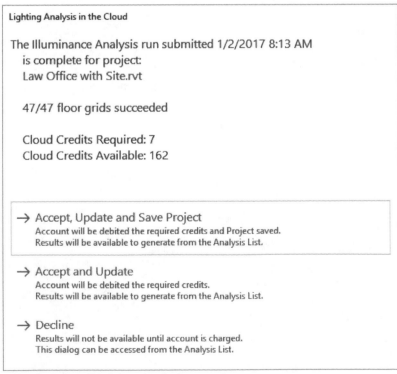

Figure 7.3-5 Lighting Analysis complete prompt

Analysis Type:

There are several types of analysis which can be performed:

- **Illuminance Analysis**
  Illuminance results visually displayed in model.
- **Daylight Autonomy (sDA preview)**
  A lower cost preview of the full LEED v4 EQc7 opt1. Analysis results will be comparable to the LEED analysis results; however, a sampling of typical hours will be used to calculate results. This is a good option for when you want to get a baseline of daylight autonomy values, reduce cloud credit consumption, or test the impact of different design options.
- **LEED 2009 IEQc8 opt 1**
  Automate settings for LEED 2009 IEQc8 opt1 criteria
- **LEED v4 EQc7 opt 1 (sDA+ASE)**
  Automate settings for LEED v4 EQc7 opt1 criteria
- **LEED v4 EQc7 opt 2**
  Automate settings for LEED v4 EQc7 opt2 criteria
- **Solar Access** (aka Daylight Access)
  Determine if a space is achieving enough daylight throughout the day

Settings Glazing Properties:

Setting the **Visual Transmittance** (T-Vis) value for glazing is essential in achieving accurate results. This must be done for every Revit material used to represent exterior glass. This is done by adjusting the Revit material's appearance asset for the following information:

- Set via the **Appearance** tab (Figure 7.3-5)
- Asset type should be **Glazing** (not Generic)
  - **Color** = Custom
  - **Custom Color** = T-Vis (per info listed below)
  - **Reflectance** and **Sheets of Glass** are irrelevant for this analysis tool
  - The chart below is from Autodesk Help | Glazing Properties for Illuminance Renderings

The **Visual Transmittance** (T-Vis) of the glass will depend not only on the **color** you define, but also on the **Thickness** of that glass and how many panes there are. You can use the table below to approximate a **T-Vis** appropriate for the actual glazing geometry modeled in Revit.

For example, if you have modeled a double pane window with geometry that is 3mm thick for each pane, and you want the **T-Vis** of the window assembly to be 70%, you would specify R50, G50, B50. Note, a "perfectly clear" piece of glass will have a **T-Vis** of about 92%.

Analyze

	R,G,B Tvis								
	Thickness	90%	80%	70%	60%	50%	40%	30%	20%	10%
single	3.0 mm	171	24	2	0	0	0	0	0	0
	4.0 mm	189	43	8	1	0	0	0	0	0
	5.0 mm	201	61	16	3	0	0	0	0	0
	6.0 mm	209	78	25	7	1	0	0	0	0
	8.0 mm	219	105	45	17	5	1	0	0	0
	10.0 mm	226	125	64	29	11	3	0	0	0
	12.7 mm	232	146	86	47	22	9	3	0	0
	25.4 mm	243	193	148	109	76	49	27	12	3
	28.6 mm	244	199	157	120	87	59	35	17	5
dual	3.0 mm	-	154	50	14	3	0	0	0	0
	4.0 mm	-	175	76	29	9	2	0	0	0
	5.0 mm	-	189	96	44	18	5	1	0	0
	6.0 mm	-	198	113	59	28	11	3	0	0
	8.0 mm	-	211	139	86	48	24	9	2	0
triple	3.0 mm	-	-	137	58	21	6	1	0	0
	4.0 mm	-	-	160	84	39	15	4	0	0
	5.0 mm	-	-	175	105	57	27	10	2	0
	6.0 mm	-	-	186	121	73	39	17	5	0
quad	3.0 mm	-	-	224	118	55	21	6	1	0
	4.0 mm	-	-	232	143	81	40	16	4	0
	5.0 mm	-	-	236	160	101	58	28	10	1
	6.0 mm	-	-	239	173	118	74	40	17	4

To use this chart:

A. Determine the **thickness** of the windowpane geometry in your Revit model and the number of panes modeled (single, dual, etc.)—this is the **Row**.

B. Determine the **Visual Transmittance** (T-Vis) value to simulate—this is the **Column**.

C. Set the **Appearance Asset Color** of the glass material in your Revit family to have RGB values that are each equal to the number in the **Cell** you have identified; high numbers are more transparent, whereas low numbers are more opaque.

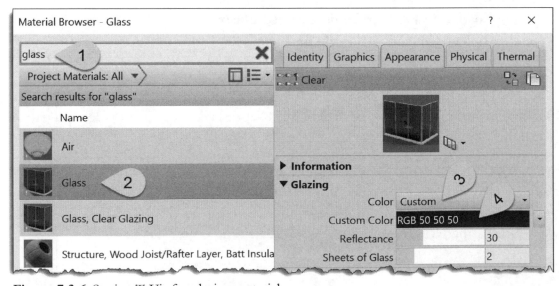

Figure 7.3-6 Setting T-Vis for glazing material

Setting Surface Reflectances

Similar to glazing, surface reflectances are set via the Appearance asset settings (Figure 7.3-6):

- Asset type should be **Generic**
- Reflectivity derived from color (RGB values), using formula: **(0.2126 R + 0.7152 G + 0.0722 B) / 255**
  - See chart to right for examples
- If you need to define a Specular Reflectivity (i.e., non-diffuse, mirror like), you need to enable Reflectivity for the Appearance Asset; see help for more information on this.
- If an image is used, the average color of the image is used in the formula stated above.

The following image, Figure 7.3-7, shows the steps required to define the reflectance for a specific material. Setting the **Reflectivity** (#4) is optional.

Reflectivity	R	G	B
100%	255	255	255
98%	250	250	250
94%	240	240	240
90%	230	230	230
86%	220	220	220
82%	210	210	210
78%	200	200	200
75%	190	190	190
71%	180	180	180
67%	170	170	170
63%	160	160	160
59%	150	150	150
55%	140	140	140
51%	130	130	130
47%	120	120	120
43%	110	110	110
39%	100	100	100
35%	90	90	90
31%	80	80	80
27%	70	70	70
24%	60	60	60
20%	50	50	50
16%	40	40	40
12%	30	30	30
8%	20	20	20
4%	10	10	10
0%	0	0	0

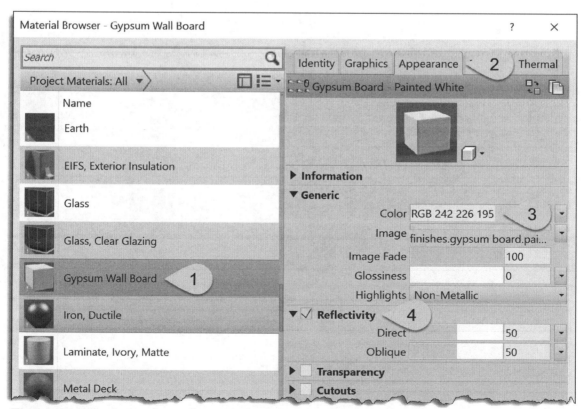

Figure 7.3-7 Setting surface reflectance for materials

Viewing the Results

- Open a 3D view; it's best if one is saved for lighting analysis specifically.
- Click the **Lighting** tool on the Analyze tab.
- Select the completed analysis (Figure 7.3-8).
- Click **Go**.

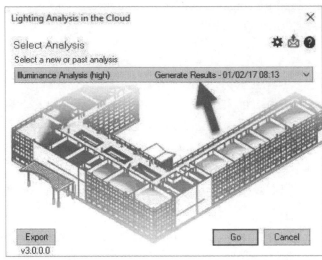

Figure 7.3-8
Lighting – Generate Results in current view

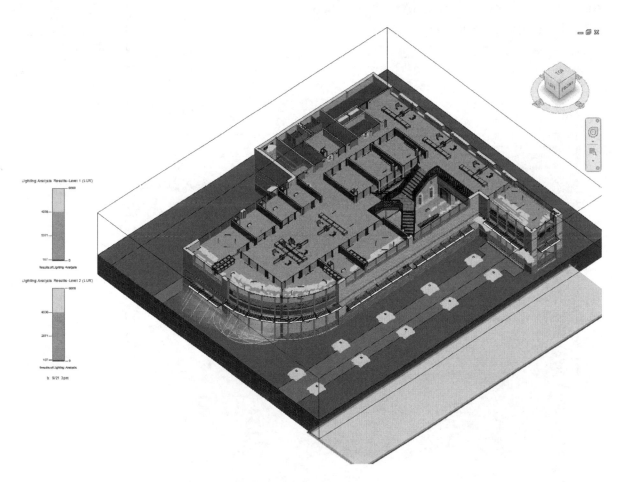

Figure 7.3-9 Lighting – View results in current view

The following Revit schedules are automatically generated when the analysis is completed. The first one is important as the rooms not required to have daylight can be identified by unchecking the **Include in Daylighting** column.

						9am threshold results						3pm threshold results					
						within threshold		above threshold		below threshold		within threshold		above threshold		below threshold	
A	B	C	D	E	F	G	H	I	J	K	L	M	N	O	P	Q	R
Level	Name	Number	Area	Include in Daylighting	Automated Shades	%	Area	%	Area	%	Area	%	Area	%	Area	%	Area
Level 2	Impromptu meeting	223	73 SF	☑	☐	100	73 SF	0	0 SF	0	0 SF	100	73 SF	0	0 SF	0	0 SF
Level 2	Partner / principal o	220	196 SF	☑	☐	96	188 SF	4	8 SF	0	0 SF	38	74 SF	62	122 SF	0	0 SF
Level 2	Partner / principal o	219	242 SF	☑	☐	100	242 SF	0	0 SF	0	0 SF	30	72 SF	70	169 SF	0	0 SF
Level 2	Partner / principal o	218	247 SF	☑	☐	100	247 SF	0	0 SF	0	0 SF	79	194 SF	21	53 SF	0	0 SF
Level 2	Partner / principal o	222	227 SF	☑	☐	86	195 SF	14	32 SF	0	0 SF	78	178 SF	22	49 SF	0	0 SF
Level 2	Associate office	221	108 SF	☑	☐	92	99 SF	8	8 SF	0	0 SF	91	98 SF	9	9 SF	0	0 SF
Level 2	Associate office	217	133 SF	☑	☐	100	133 SF	0	0 SF	0	0 SF	100	133 SF	0	0 SF	0	0 SF
Level 2	Associate office	216	109 SF	☑	☐	100	109 SF	0	0 SF	0	0 SF	100	109 SF	0	0 SF	0	0 SF
Level 2	Associate office	214	150 SF	☑	☐	100	150 SF	0	0 SF	0	0 SF	100	150 SF	0	0 SF	0	0 SF
Level 2	Associate office	213	150 SF	☑	☐	100	150 SF	0	0 SF	0	0 SF	100	150 SF	0	0 SF	0	0 SF
Level 2	Associate office	212	150 SF	☑	☐	100	150 SF	0	0 SF	0	0 SF	100	150 SF	0	0 SF	0	0 SF
Level 2	First year associat	211	91 SF	☑	☐	-1	0 SF	-1	0 SF	-1	0 SF	-1	0 SF	-1	0 SF	-1	0 SF
Level 2	Retired partners of	200	252 SF	☑	☐	67	168 SF	33	84 SF	0	0 SF	58	146 SF	42	107 SF	0	0 SF
Level 2	Paralegal	206	80 SF	☑	☐	89	72 SF	11	8 SF	0	0 SF	100	80 SF	0	0 SF	0	0 SF
Level 2	Paralegal	205	80 SF	☑	☐	82	65 SF	18	14 SF	0	0 SF	100	80 SF	0	0 SF	0	0 SF
Level 2	Paralegal	204	80 SF	☑	☐	81	65 SF	19	15 SF	0	0 SF	100	80 SF	0	0 SF	0	0 SF
Level 2	Paralegal	224	73 SF	☑	☐	100	73 SF	0	0 SF	0	0 SF	100	73 SF	0	0 SF	0	0 SF
Level 2	Paralegal	229	73 SF	☑	☐	100	73 SF	0	0 SF	0	0 SF	100	73 SF	0	0 SF	0	0 SF
Level 2	Secretary	201	67 SF	☑	☐	80	53 SF	20	13 SF	0	0 SF	100	67 SF	0	0 SF	0	0 SF
Level 2	Secretary	225	60 SF	☑	☐	99	59 SF	0	0 SF	1	1 SF	99	59 SF	0	0 SF	1	1 SF
Level 2	Secretary	228	60 SF	☑	☐	100	60 SF	0	0 SF	0	0 SF	100	60 SF	0	0 SF	0	0 SF
Level 1	Reception	102	70 SF	☑	☐	93	65 SF	0	0 SF	7	5 SF	81	57 SF	19	13 SF	0	0 SF
Level 1	Waiting / lobby	101	304 SF	☑	☐	100	303 SF	0	0 SF	0	1 SF	85	259 SF	14	44 SF	0	1 SF
Level 1	Large Conf / Board	115	426 SF	☑	☐	67	283 SF	5	23 SF	28	120 SF	82	349 SF	10	41 SF	9	36 SF
Level 1	Medium Conferenc	114	240 SF	☑	☐	94	226 SF	6	14 SF	0	0 SF	24	58 SF	76	182 SF	0	0 SF
Level 1	Small Conference	116	123 SF	☑	☐	0	0 SF	0	0 SF	100	123 SF	0	0 SF	0	0 SF	100	123 SF
Level 1	Impromptu meeting	117	106 SF	☑	☐	0	0 SF	0	0 SF	100	106 SF	0	1 SF	0	0 SF	100	105 SF
Level 1	Small Conference	118	102 SF	☑	☐	100	102 SF	0	0 SF	0	0 SF	100	102 SF	0	0 SF	0	0 SF
Level 1	Public Restroom	107	143 SF	☐	☐	-1	0 SF	-1	0 SF	-1	0 SF	-1	0 SF	-1	0 SF	-1	0 SF
Level 1	Public Restroom	106	143 SF	☐	☐	-1	0 SF	-1	0 SF	-1	0 SF	-1	0 SF	-1	0 SF	-1	0 SF
Level 1	Visitor office	111	118 SF	☑	☐	100	118 SF	0	0 SF	0	0 SF	100	118 SF	0	0 SF	0	0 SF

<_Lighting Analysis Room Schedule>
Custom Analysis Whole Building Results: Minneapolis, MN, USA
9/21 9am: 71% & 9/21 3pm: 71% of points are between 300-3000 lux
Solar Values (W/m2): 9/21 9am GHI: 259, DNI: 360, DHI: 51 & 9/21 3pm GHI: 329, DNI: 522, DHI: 82

Figure 7.3-10 Lighting – results viewed in lighting analysis room schedule

<_Lighting Analysis Floor Schedule>
Custom Analysis Whole Building Results: Minneapolis, MN, USA
9/21 9am: 71% & 9/21 3pm: 71% of points are between 300-3000 lux
Solar Values (W/m2): 9/21 9am GHI: 259, DNI: 360, DHI: 51 & 9/21 3pm GHI: 329, DNI: 522, DHI: 82

			9am threshold results						3pm threshold results					
			within threshold		above threshold		below threshold		within threshold		above threshold		below threshold	
A	B	C	D	E	F	G	H	I	J	K	L	M	N	O
Name	Floor Area Included in Daylighting	Total Floor Area	%	Area	%	Area	%	Area	%	Area	%	Area	%	Area
Level 1	7426 SF	10841 SF	52	3847 SF	6	445 SF	42	3134 SF	55	4120 SF	11	804 SF	34	2502 SF
Level 2	7937 SF	9165 SF	89	7044 SF	3	203 SF	9	690 SF	85	6742 SF	6	510 SF	9	685 SF

Figure 7.3-11 Lighting – results viewed in lighting analysis floor schedule

Analyze

Once the calculations are completed in the cloud, the following dialog is presented indicating the results and a Pass/Fail score for LEED credit analysis.

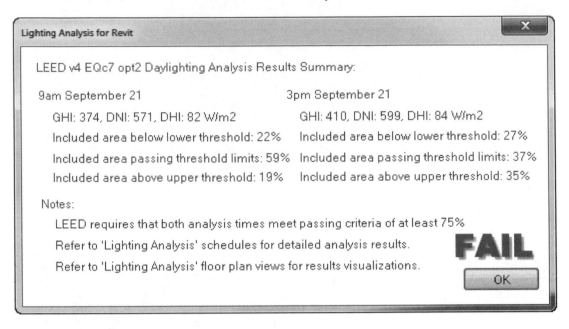

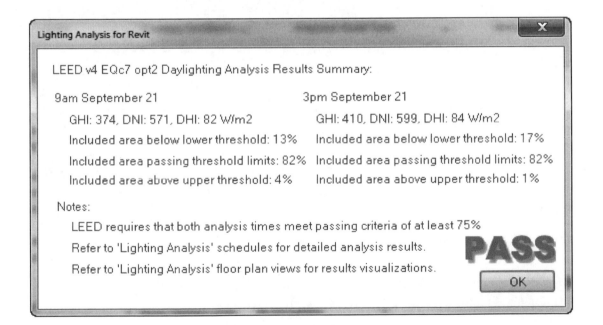

Solar:

This tool is used to analyze solar radiation on surfaces based on a building's location, orientation and form. This newer version, included with the Insight 360 tools, has been enhanced to automatically select all roof elements. Several settings can be adjusted as seen in the image below.

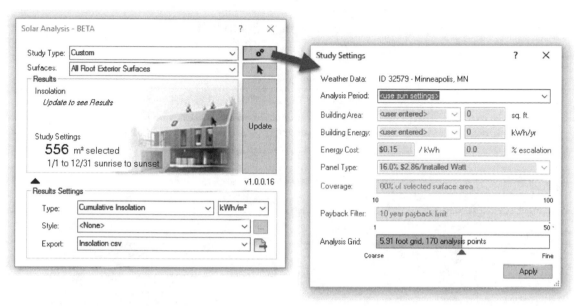

Figure 7.3-12 Solar Analysis dialogs

The following, from the Insight FAQ page, should be of interest to most:

7. Are the analysis engines Insight 360 uses validated?
    Yes, Insight 360 is proud to use validated industry leading analysis engines. Access the links below for more detailed information on our testing standards.
    - Whole building energy analysis from Revit, FormIt 360 Pro, and the Insight 360 interface is powered by Green Building Studio, a DOE-2.2 engine run in the cloud, and has been tested against ANSI/ASHRAE 140.
    - Insight 360 heating and cooling loads are calculated using EnergyPlus Cloud, and have also been tested against ANSI/ASHRAE 140.
    - The lighting and daylighting analyses use Autodesk 360 Rendering, a cloud based engine that uses bidirectional ray tracing. This engine has been validated against Radiance and real world measurements.
    - Insight 360 solar analysis uses an optimized Perez sky model and overshadowing calculation, validated with NREL provided test values. Photovoltaic panel libraries are updated often based on the range of panel types available in the US, European, and Chinese markets

Analyze

Clicking Update runs a solar analysis using the current settings. The results are displayed on the screen as shown in the image below (Figure 7.3-13). PV Production and energy offset information is also shown in the Solar Analysis dialog as shown in Figure 7.3-14.

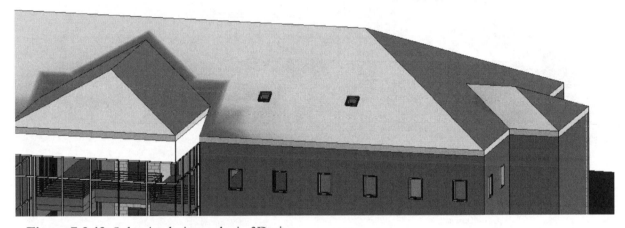

Solar Energy (kWh/m²)

1150 ┌1463
ᴜᴜᴜ ┤
        └53

Project location: Minneapolis, MN
Sun study start date time: 1/1/2010 12:00:00 AM
Sun study end date time: 12/31/2010 11:59:00 PM

Cumulative Insolation

Figure 7.3-13 Solar Analysis results in 3D view

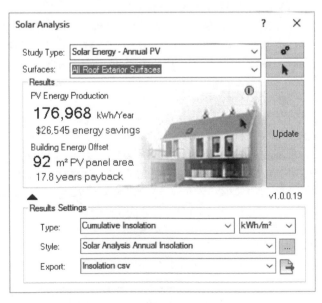

Figure 7.3-14 Solar Analysis results dialog

7.4 More on Energy Simulation

The following information is offered in addition to the specific commands outlined on the analysis tab. This information provides a broader overview of the energy simulation options and workflows within Revit and Insight 360. Some of this information will overlap the command specific items previously covered in this chapter. However, those interested in energy simulation will benefit from this detailed overview.

The built-in ability to run an energy simulation within Revit represents the most democratized opportunity for design professionals and students – period. If you or your firm have Revit, you have access to this feature. There are no add-ins or additional software costs. This is not to say it is free—Revit has to be on subscription, as the simulation itself is actually run in the Cloud using Autodesk's GBS engine. However, subscription is becoming the norm with Autodesk's new sales model.

> Energy simulations and Insight 360 do not require cloud credits at this time.

The example image below shows two scenarios being compared within Insight 360.

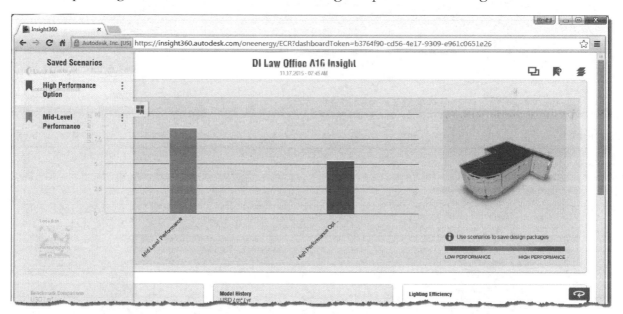

Figure 7.4-1 Comparing scenarios within Insight 360

When getting started with the energy simulation features, you may wonder where to begin given all the settings and related commands.

Analyze

Here is a basic overview of the workflow:

1. Create a Revit model using masses, building elements, or both

2. Energy Settings dialog (**Analysis** tab)
   o Set Location

   o Review *Energy Analytical Model* settings

   o All other settings are optional

3. Create Energy Model
   o Views created: 3D Energy Model, Analytical Spaces (Schedule), Analytical Surfaces (Schedule)

   o Use this tool to delete and recreate the Energy Analysis Model (EAM) anytime the Revit model changes

4. Visually Review Energy Model

5. Launch **Insight 360** for interactive project exploration

6. AIA 2030 Commitment Firms: Upload results to DDx

7. Optional Settings and Workflows

Following these steps will provide super-fast access to estimated **Energy Cost** and **Energy Use Index** (EUI) information at any phase in a project. Next, we will take a detailed look at each of these steps. Understanding 'what Revit wants' will facilitate efficient and accurate use of this tool.

The image below (Figure 7.4-2) shows the Energy Optimization tools found on the **Analyze** tab in Revit. Also shown is the **Energy Analysis** panel.

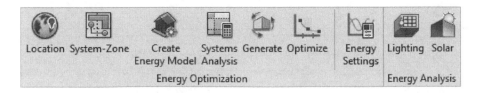

Figure 7.4-2 Energy Analysis tools

Creating the Revit Model

The obvious first step is to create the model. The ideal design process would be to start with massing, quickly studying how shape and orientation impact performance. As the project develops, Revit building elements can be added to the mix for known aspects of the design, for example, curtainwall, windows, sunshades, etc. At some point, the mass elements are abandoned in favor of a more detailed model based solely on Revit building elements.

Here are a few things to keep in mind concerning Revit model creation.

Massing

When using masses, be sure to select the mass and specify the floor levels as shown in the image below—using the **Mass Floors** tool (Figure 7.3-3). Also, use masses to define external shades such as adjacent buildings; just don't specify a mass floor for them. The Revit project must have at least one mass with mass floors assigned to create a valid Energy Analysis Model (EAM).

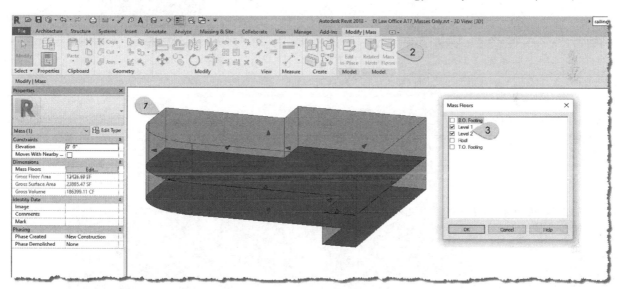

Figure 7.4-3 Masses must have 'mass floors' specified to create a valid EAM.

Combined Massing and Building Elements

The next image (Figure 7.4-4) is an example of using massing and building elements together. This example makes the glazing size and locations explicit rather than being generically based on a percentage of surface area and fixed sill height.

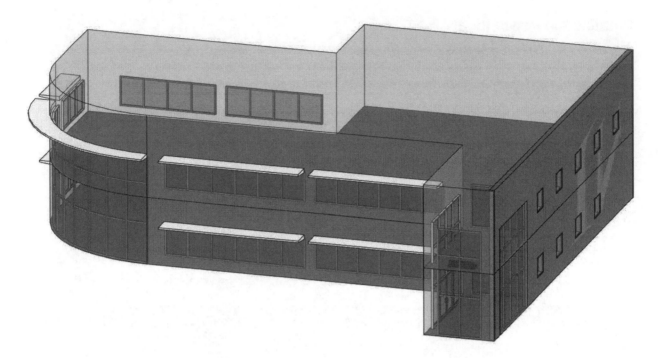

Figure 7.4-4 Masses and Building Elements used together in energy analysis

Another example might be to use masses to study future expansion in the context of an existing Revit model (Figure 7.4-5). When you get to the point in the design where you need to add masses to create sloped walls (e.g., using the **Wall by Face** tool), then it is time to stop using masses in the energy analysis. It is not possible to include some masses and exclude others.

Figure 7.4-5 Masses and Building Elements used together in energy analysis

Building Elements

Most Revit models can be used to create a decent **Energy Analytical Model** (EAM). The elements listed in the image below (Figure 7.4-6), when set to **Room Bounding**, are used in the EAM creation.

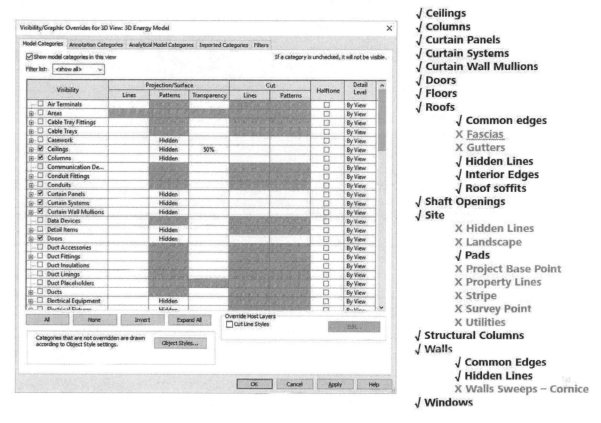

Figure 7.4-6 Building elements used in EAM creation; some sub-categories are not included

The tricky thing is dealing with aspects like sunshades as they are often modeled as Generic Model. As seen in the list above, this category is not used, as it could contain any number of irrelevant items.

The EAM will also include elements, set to **Room Bounding**, contained within **linked models**—as long as the linked model itself is set to Room Bounding in the host model—an **Edit Type** property.

Analyze

2 Energy Settings

In the **Energy Settings** dialog, which has been streamlined in Revit 2021, there is one critical setting related to running the simulation and a few more that must be considered to generate a valid EAM. To study the real-time impact on overall performance, all other inputs can be adjusted in the cloud later.

In Revit 2021, these important parameters have been separated from the secondary options. The Energy Settings dialog is now split into two. The first dialog shown, see the image to the right (Figure 7.4-7), presents the more essential parameters, while the **Other Options** "Edit…" button opens the secondary dialog where optional settings can be adjusted.

Now let's take a look at each of these settings to better understand what they do and why they are important.

Location

The **Location** setting provides **localized weather and utility data** which is vital in creating a legitimate energy simulation. There are two steps involved in accurately specifying location: **Project Address** and **Weather Station**.

Project Address specifies the project location on earth. This can be a city, a specific postal address or Lat/Long values. If the project site does not have

The Energy settings essential to a valid Energy Simulation are:

- Location

The Energy settings essential to creating a valid Energy Analysis Model (EAM):

- Analysis Mode
- Ground Plane
- Project Phase
- Analytical Space Resolution
- Analytical Surface Resolution
- If using masses (separately or in conjunction with building elements)
  - Perimeter Zone Depth
  - Perimeter Zone Division

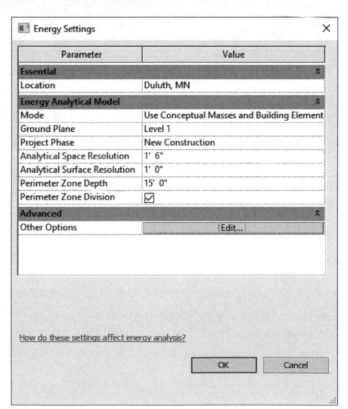

Figure 7.4-7 Energy Settings dialog

an address, enter the City name and then drag the Project Location Pin (red) to the desired location on the map. You can zoom and pan in this map view as well as make the dialog larger.

Once the geographic location has been specified, the **Weather Station** options should be evaluated. Revit will automatically select the closest option, but this may not always be the best selection. Consider the example shown in the image below (Figure 7.4-8). In this author's location, two of the closest stations have an 800 foot elevation difference. Additionally, depending on project location in this area, one of the two buoy-based weather stations may be closest—which would not be ideal (this is the largest freshwater lake in the world).

FYI from Autodesk's help page:

"Weather stations include 'actual year' virtual weather stations and typical year weather stations (TMY2 and other formats) based on 30-year averages of weather data, typically taken from airport locations."

Don't bother to change anything on the **Weather** tab as this data only relates to Revit's built-in, and older, Heating and Cooling Loads feature.

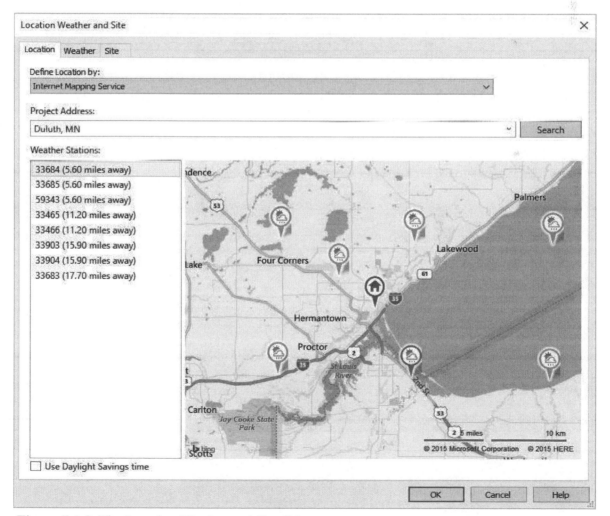

Figure 7.4-8 The Location Weather and Site dialog

Analysis Mode

Analysis Mode determines if Revit should use Masses, Building Elements, or both to create the EAM.

These are the Analysis Mode options:

- Use Conceptual Masses (never select this option!)
- Use Building Elements
- Use Conceptual Masses and Building Elements

As it turns out, the "Use Conceptual Masses" only option uses an older internal algorithm, so don't use that one. Rather, use the combined option which will work on mass-only models.

▎ Never use the Use Conceptual Masses option.

Ground Plane

Properly setting the Ground Plane parameter ensures that spaces which occur below this level are understood to be below grade by the cloud-based GBS calculation engine. For masses, the EAM will not generate glazing below the ground plane and a different construction can be selected for underground exterior walls. Keep in mind, when using Building Elements the toposurface elements are NOT used.

Project Phase

For projects with phasing—e.g., existing, new construction, phase demolished—be sure this is set correctly. The phase settings for the 3D Energy Model view have no impact on the EAM created.

EAM Creation Level of Detail

These two properties control the accuracy of the EAM when it is created:

- **Analytical Space Resolution** (default: 1'-6")
  This allows small gaps to exist in the model—both exterior and interior. Remember, Revit ignores elements in several categories such as Generic Model and In-Place families, so gaps are not uncommon.
- **Analytical Surface Resolution** (default: 1'-0")
  This setting works in conjunction with the Analytical Space Resolution setting to control the accuracy of the surface boundaries.

If portions of the Revit model are complex and not coming out right in the EAM, these values can be lowered. It is recommended that both of these values be adjusted proportionally. Lowering these values will result in a more accurate EAM but will take longer to create—the simulations will take more time to process as well. Large projects may require these values to be increased.

Automatic Thermal Zoning Settings

These two properties control the inclusion of space sub-divisions in the EAM:

- **Core Depth** (or Perimeter Zone Depth)
  The distance to measure inward from the exterior walls to define the core zone.
- **Divide Perimeter Zones** (or Perimeter Zone Division)
  Check this box to divide the perimeter into multiple zones.

These settings are used to divide large spaces into smaller subdivisions for more accurate simulations. These settings used to only work on masses, but, new to Revit 2016 R2, they now also apply to large spaces defined by building elements. For example, my firm recently worked on a 200,000sf office and warehouse building. The large warehouse space would have benefited from this subdivision feature.

When creating the EAM for building elements, the interior walls are typically sufficient to naturally 'zone' the model. In these situations, set the **Zone Depth** to zero and uncheck the **Zone Division** option.

> For Building Element-based analysis, set the Zone Depth to zero and uncheck the Zone Division option.

3

Create/Delete Energy Model

Clicking this tool creates the analytical model, resulting in three related views created in the Revit model:

- 3D Energy Model
- Analytical Spaces (schedule)
- Analytical Surfaces (schedule)

This is a newer feature added back in Revit 2016 and, as covered in the next section, allows the designer to visually validate model fidelity prior to running a simulation.

As mentioned, the analytical energy model is created from Room Bounding elements within the Revit model. The result is a simplified model consisting of surfaces somewhat analogous to a SketchUp model. By selecting surfaces and adjusting the **3D Energy Model** view's visibility, the designer can make sure there are no anomalies before starting a simulation.

The next image (Figure 7.3-9) shows the categories and sub-categories in the **Analytical Model Categories** tab of the **Visibility/Graphics Overrides** dialog, automatically turned on in the 3D Energy Model view. All other categories on this tab are related to structural analysis and can be ignored.

As more analysis tools like this become available, it is important that models are created correctly. For example, ceilings should not be used for floors or floors for countertops. If thin floors are used for finishes on top of a structural floor, they should have **Room Bounding** unchecked.

The analytical model can actually be seen in any view by adjusting the **Visibility/Graphic Overrides**. However, the 3D Energy View provides dedicated and instant access. The **Hide/Show Analytical Model** toggle on the View Control Bar, in the lower left corner of each view, will toggle the Analytical Model Categories on or off for the current view.

> This tab used to be exclusively for structural analytical visibility control

It is important to understand that the analytical surfaces do not update automatically as the model changes. When the Revit model changes, the EAM must be deleted and recreated.

A few of the Analytical Surfaces sub-categories may be confusing. Revit Windows are translated to an analytical surface called **Operable Windows** even though they may not actually be operable, and Curtainwall walls are all called **Fixed Windows**.

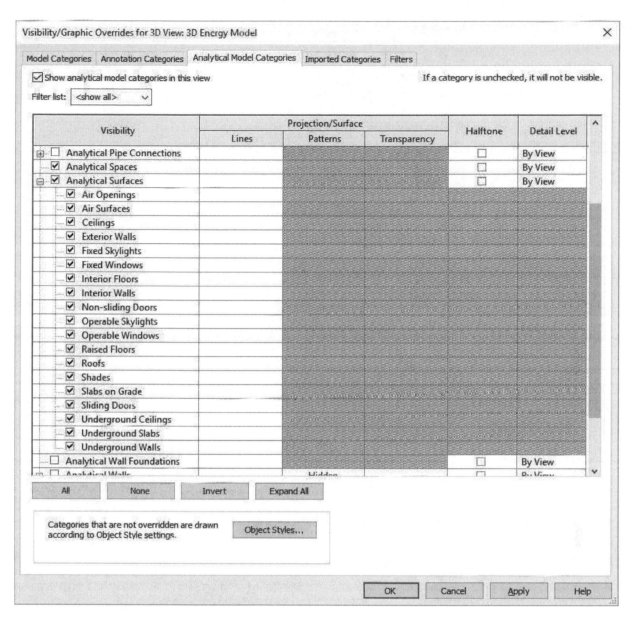

Figure 7.4-9 Analytical Spaces and Surfaces in Visibility/Graphic Overrides dialog for 3D Energy Model view

Analyze

4 Visually Review Energy Model

It is important to visually <u>validate the Energy Model</u> prior to running a simulation to ensure valid results. If there are problem areas, the Revit model needs to be adjusted and the EAM recreated. The analytical surfaces cannot be modified directly in any way.

In the **3D Energy Model** view, some of the Model Categories are turned on (and set to be partially transparent). It may be helpful to turn these off to clean up the view (Figure 7.4-10).

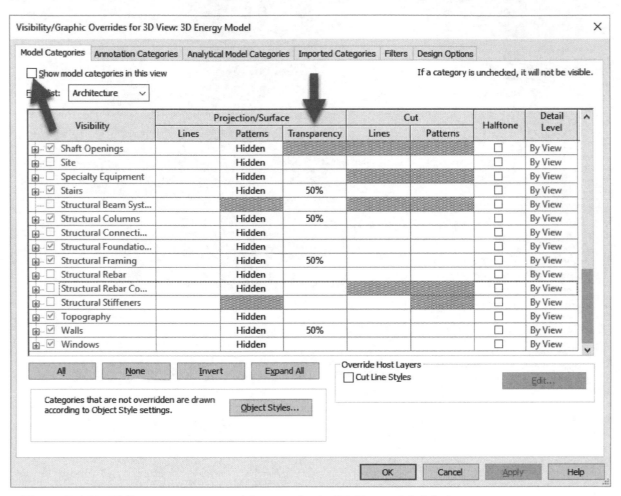

Figure 7.4-10 Hiding transparent model categories in 3D Energy Model view

Things to look for are missing spaces and excessive shade surfaces. In the 3D Energy Model view, the sub-categories can be adjusted to isolate these items, or the **Hide/Isolate** feature can also be used. The next image (Figure 7.4-11) shows just the **Analytical Spaces**: both occupiable spaces and plenum spaces (i.e., above ceilings). There are no large missing spaces within the building so all is well. Remember, the **Analytical Space Resolution** aims to simplify model complexity and, as such, some smaller spaces may be omitted.

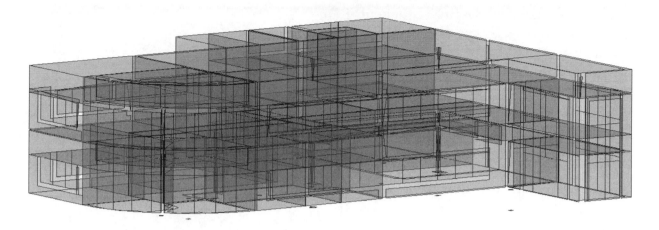

Figure 7.4-11 3D Energy Model view adjusted to show only Analytical Spaces

The next image (Fig. 7.3-12) shows just the Exterior Walls, Fixed Windows and Operable Windows' analytical surfaces. The **Function** setting (i.e., Interior vs. Exterior) for walls does not matter as the EAM algorithm automatically determines this. Notice that three interior curtainwall Walls appear due to their close proximity to the exterior.

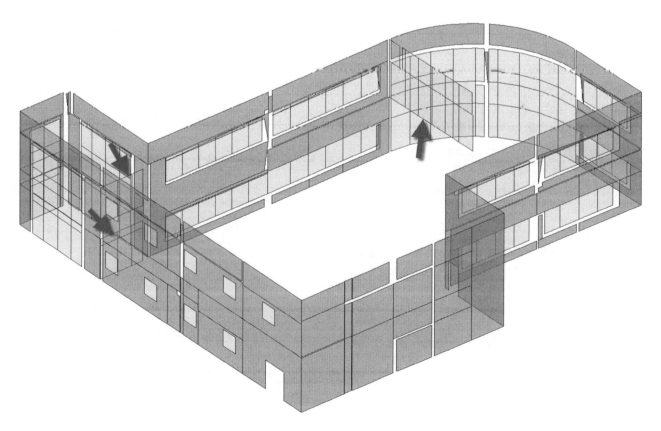

Figure 7.4-12 3D Energy Model view adjusted to show only analytical surfaces for Exterior Walls and Windows

Review the **Analytical Surfaces schedule** to verify the right mix of surface types (Figure 7.4-13). If there are only shades and/or no windows, that would be a "red flag."

<Analytical Surfaces>			
A	**B**	**C**	**D**
Area	Count	Opening Type	Surface Type
3446 SF	157	Fixed Window	
551 SF	24	Non-sliding Door	
144 SF	9	Operable Window	
9602 SF	29		Ceiling
6614 SF	98		Exterior Wall
4823 SF	30		Interior Floor
8412 SF	153		Interior Wall
5635 SF	14		Roof
2628 SF	59		Shade
5578 SF	22		Slab on Grade
10 SF	15		Underground Wall
47442 SF			

Figure 7.4-13 Analytical Surfaces Schedule

The **Analytical Spaces schedule** can be used to verify rooms and areas (Figure 7.4-14). The rows without a room name are void/shaft and plenum spaces. Splitting the screen to show both the Analytical Space schedule and the 3D Energy Model view facilitates highlighting a space in the 3D view by selecting a row in the schedule.

Due to the voids and plenum spaces, don't expect the total square footage (SF) to match the total in a typical Revit room schedule.

These two schedules are based on the Analytical Surfaces and Analytical Spaces categories. Don't get confused by the title **Analytical Spaces**—the term "Spaces" does not relate to Room versus Space elements.

If using Design Options, note that only elements in the **Main Model** and the **Primary Options**

<Analytical Spaces>			
A	**B**	**C**	**D**
Area	Count	Room Name	Volume
322 SF	1	ASSOCIATES OFFICE 106	2229 CF
322 SF	1		1159 CF
535 SF	1	PARALEGAL 208	3791 CF
539 SF	1		2181 CF
354 SF	1	CONF ROOM 104	2492 CF
157 SF	1	BREAK RM 105	1061 CF
354 SF	1		1328 CF
157 SF	1		572 CF
1449 SF	1	LAW LIBRARY 207	10015 CF
3094 SF	1		12912 CF
190 SF	1	STAIR #1	3415 CF
465 SF	1	MECH - ELEC ROOM 100	5489 CF
169 SF	1	MENS 101	1211 CF
132 SF	1	WOMENS 102	1202 CF
177 SF	1		643 CF
180 SF	1		659 CF
254 SF	1	FILE STORAGE 203	2138 CF
163 SF	1	MENS 205	1186 CF
180 SF	1	WOMENS 206	1235 CF
247 SF	1		1241 CF

Figure 7.4-14
Analytical Spaces Schedule showing rooms and areas

are used in the EAM. If you want to use **Design Options**, the desired design must be set as **Primary** before creating the EAM. This should be fine early in the design process, but later (e.g., bid alternates), changing the primary designation can mess up construction document views.

Generate (or Generate Insight)

This tool will send the energy analysis model (EAM) to the Autodesk A360 cloud for simulation. Although not required, the EAM should be created and reviewed prior to selecting this option. If this tool is selected prior to creating and validating the EAM, the **Generate Insight** command will create one and send it to the A360 cloud; however, in this case, the EAM will not be visible within Revit. If the EAM does exist, it will automatically be updated when the Generate Insight command is selected.

> Prior to selecting this command, select *Location*, specify *Energy Settings* and *Create/Validate* the energy model.

An email will be sent indicating that the analysis process has started (Figure 7.4-15).

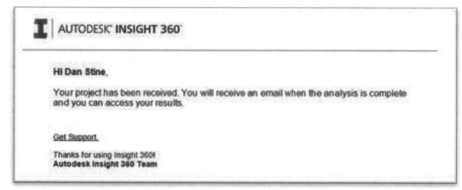

Figure 7.4-15 Email from Autodesk indicating the Insight project has been received

Insight 360 automatically varies building design inputs resulting in high and low possible annual energy costs with approximately +/- 10% accuracy. Inputs can then be adjusted, e.g., glazing properties, to see instant feedback on performance impacts.

Once the simulation is complete, another email will be sent with a link to the project (Figure 7.4-16). You must sign into Autodesk A360 to access the information.

Analyze

Figure 7.4-16 Email from Autodesk indicating the Insight project simulation is complete

Optimize (Open Insight 360)

Once an analysis has been performed, the results can be seen by clicking the link in the email or selecting the **Optimize** tool within Revit. Additionally, results from Insight can be pushed to the **AIA DDx** by firms participating in the AIA 2030 Commitment.

Let's take a closer look at what we can do with Insight 360. When the EAM is ready, simply click the **Optimize** command in Revit. Once the analysis is complete, the results can be accessed in the cloud in one of two ways: clicking the Insight360 tool within Revit or browsing to the website per the URL mentioned above (Chrome, Firefox or Safari browsers only). The browser option allows the window to be resized and will not close if Revit is closed.

The next image is the initial view of the project in Insight 360 (shown in Chrome). Right away, we see the energy cost in the upper left (red circle). This value will change as we adjust inputs. Speaking of inputs, they have all been varied across all possible values. Thus, looking at the **Benchmark Comparison** tile, we see the high and low possible cost range—this means the best and worst possible scenarios based on energy usage.

Let's take a minute to look at the User Interface (UI). There are several interesting aspects of the UI that should be understood to fully leverage this new tool. They are marked in the image below (Figure 7.4-17) and discussed subsequently.

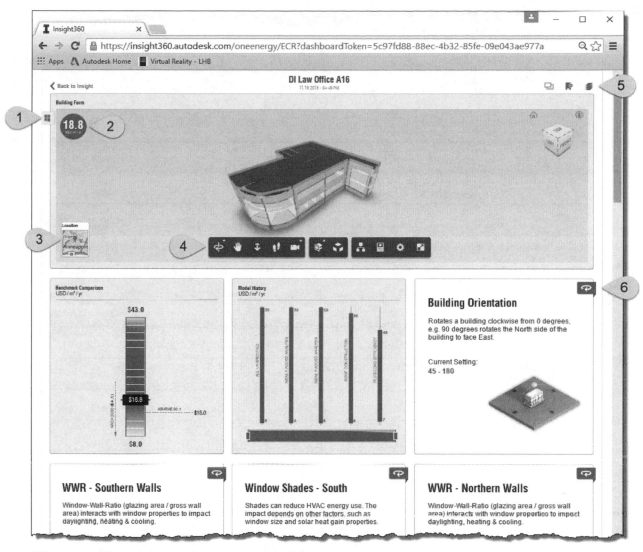

Figure 7.4-17 Initial project view in Insight 360

1. **Saved Scenarios Slide-out Panel**

   View saved scenarios in this slide-out panel. A saved scenario is a snapshot of a specific arrangement of input settings. This allows for quick comparison between specific or combined variations.

2. **Energy Cost**

   Lists the current annual energy cost per m² or ft². Adjusting the inputs will instantly change this value. You can click this graphic to toggle between energy cost **$/m²/yr** or EUI **kBtu/ft²/yr.**

3. **Location**

   Clicking here will toggle the EAM preview to a location map. Click the weather station icon to see historical weather data graphs (Figure 7.3-18).

Analyze

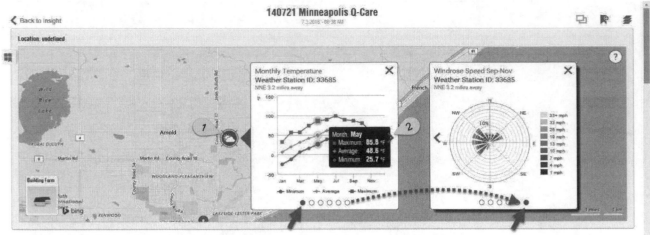

Figure 7.4-18 Weather data for local site

4. **EAM Toolbar**

   The basic navigation tools on the left are generally self-explanatory. The center tools facilitate applying a section cut to the model and exploding the elements to better visualize complex conditions in the EAM (Figure 7.4-19 & 20). The tools on the right provide access to element information and general settings. See some examples in the next few images (Figure 7.4-21).

   By default, drag with Left mouse button to orbit and right button to pan. Spin the mouse wheel to zoom.

5. **Scenario Creation and Comparison**

   Use these icons to save and compare scenarios. The **Visualize** tool provides access to solar, lighting and H/C analysis tools from within Insight 360.

6. **Input Adjustment Tiles**

   Each tile represents a specific design element. The initial view shows a generic image and the current (default) range. For example, the **Daylighting and Occupancy Controls** tile ranges from "none" to "Daylighting and Occupancy Controls" (aka worst to best). Clicking this tile allows the designer to refine the range, or even select something very specific if known. More on this in a moment.

Figure 7.4-19 EAM view shown with section cut applied

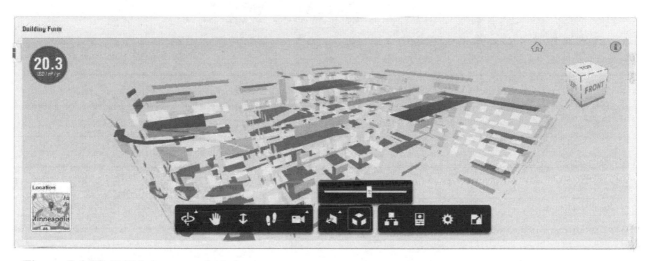

Figure 7.4-20 EAM shown exploded

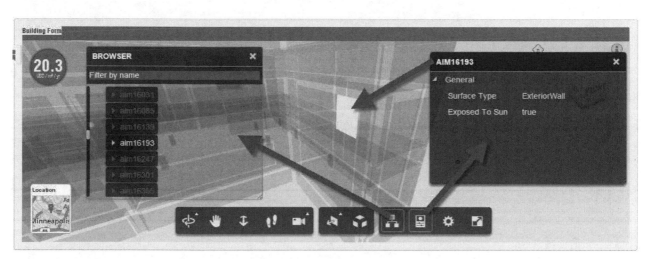

Figure 7.4-21 The Object Browser and Properties Palette provide a look "under the hood."

Analyze

Now let's take a look at the most significant feature: insight into your design and the interrelated results based on making various adjustments. This is analogous to a mixing board in a music recording studio—the combination of several adjustments produces a unique result.

First, we can see that adjusting some aspects of the design, such as orientation and location in this case, has minimal impact on overall performance (see image below – Figure 7.4-22). Clicking the **Building Orientation** tile reveals the relatively flat graph shown below. If the building orientation will not change, the range can still be adjusted to reflect this known bit of information. Note that the "0" position relates to the current orientation in the Revit model, which is also marked by a triangle.

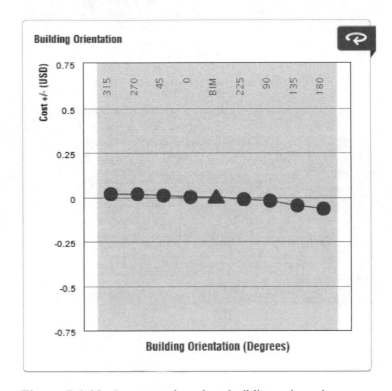

Figure 7.4-22 Cost range based on building orientation

When we contrast the **Building Orientation** with another metric such as **Lighting Efficiency**, we see a more significant opportunity to affect the overall building performance, as shown below (Figure 7.4-23).

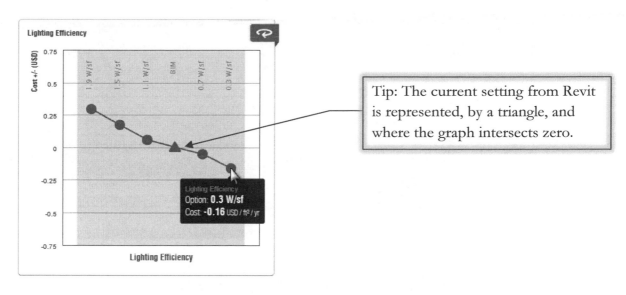

Figure 7.4-23 Cost range based on lighting efficiency

Clicking on the graph opens the cost range view. If we want to see the relative change to the **Energy Cost Mean** by designing to the upper 1/3 range, we see about a ten percent change. Similar to the building's orientation, if we get to the point where we know exactly what the lighting efficiency is, we can adjust the sliders to select a specific input. You can hover your cursor over the graph for additional information as shown in Figure 7.4-24.

As the inputs are adjusted, the **Energy Cost Mean** value continues to update. While the performance is below the ASHRAE 90.1 threshold, the circle is colored **red** as seen in all images previously. Once in the "middle" range, the color changes to **orange**. Once the Architecture 2030 benchmark has been reached, the circle turns **green** as shown in the next image (Figure 7.4-25).

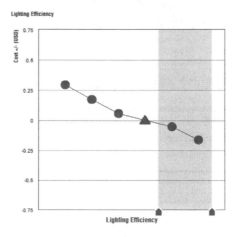

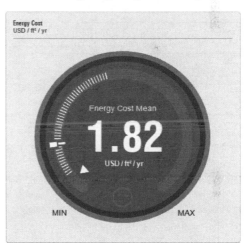

Figure 7.4-24 Cost range adjustment for lighting efficiency

Analyze

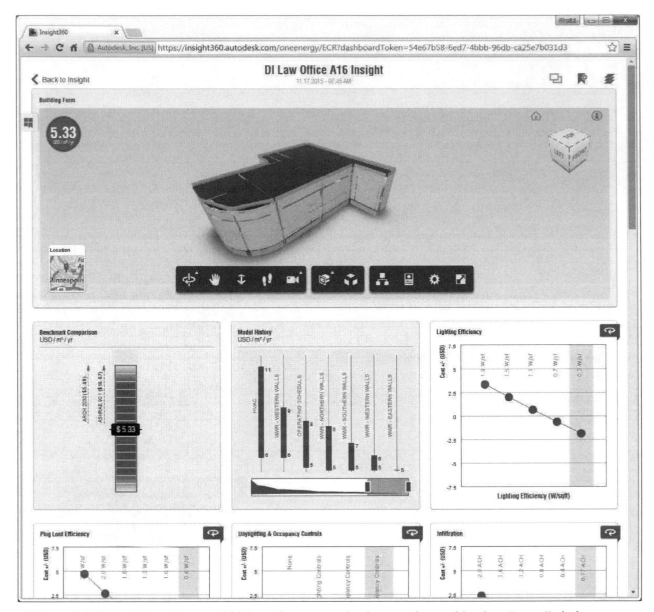

Figure 7.4-25 Insight with several high performance selections made, resulting in a "green" circle

To save scenarios, make changes to the cost range values and then click the **Add Scenario** icon in the upper right. These can then be used to compare the various effects of multiple sets of input adjustments. The example below (Figure 7.4-26) shows a comparison between a medium and high-performance scenario.

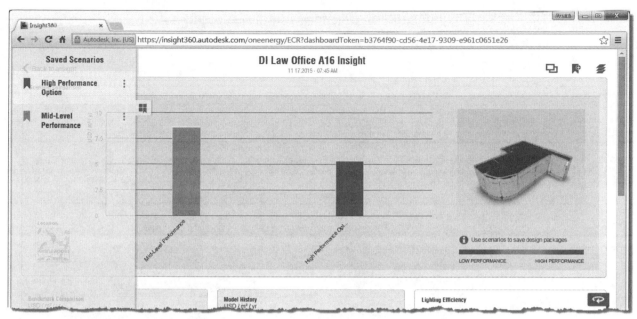

Figure 7.4-26 Comparing saved scenarios

Insight Collaboration

The Insight 360 project can be saved with other Autodesk A360 users! Not only that, but when two people are looking at the same project they both see any changes in real-time via their browser (Figure 7.4-27).

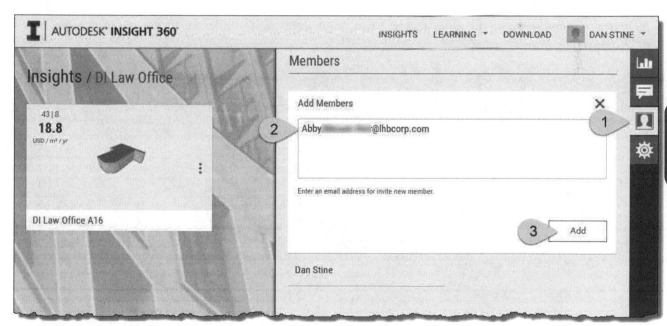

Figure 7.4-27 Sharing an Insight project with another user

The Insight home screen has an indicator to remind you that the project is shared with others. Right now there are no user rights controls. On a related note, notice the project image can be customized (Figure 7.4-28).

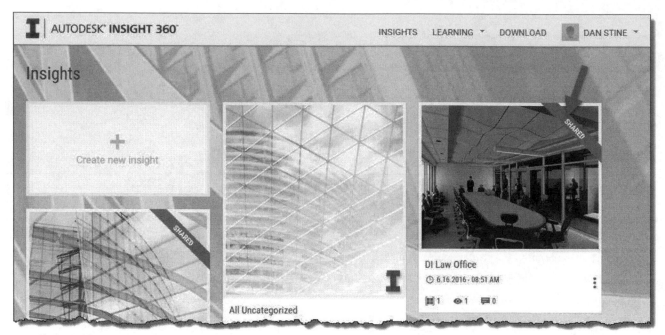

Figure 7.4-28 Indication that a project has been shared with others

Saved scenarios can be applied to all models within a project as shown in the next image (Figure 7.4-29). This allows for super-fast comparison of various massing configurations. Additionally, when scenarios are made a favorite that scenario can be accessed from any project – just click the start.

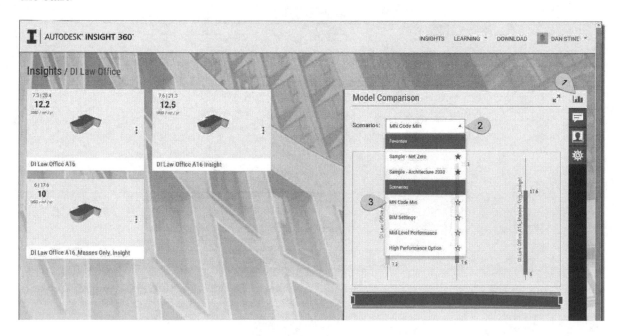

Figure 7.4-29 Saved Scenarios can be applied to multiple projects within the same Insight

Glazing Properties used in Insight

This question has come up a number of times, so I am sure this will be helpful to those who want to understand what is going on under the hood; here are the factors used relative to the glazing options in Insight.

Name	Glazing Type-Northern, Southern, Eastern, Western Walls	U-Value W/m^2K	U-value BTU/hr-ft2-F	SHGC	VLT
No Change	No change	No change		No change	No change
Sgl Clr	Single Clear 6mm	6.17	1.09	0.81	0.88
Dbl Clr	Dbl Clear 6/13 Air	2.74	0.48	0.7	0.78
Dbl LoE	Dbl Low-E (e3=0.2)Clear 3/13 Air	1.99	0.35	0.73	0.74
Trp LoE	Trpl Low-E (e2=e5=0.1) Clr 3mm/6mm Air	1.55	0.27	0.47	0.66
Quad LoE	Quadruple LoE Films (88) 3mm/8mm Krypton	0.66	0.12	0.45	0.62

Analyze

Settings

Each project has a few high-level settings as seen below (Figure 7.4-30). Sorting by "importance" means the Lighting Efficiency is going to be listed before Building Orientation using the previous examples. The **Utility Rates** can be adjusted from the default derived based on location.

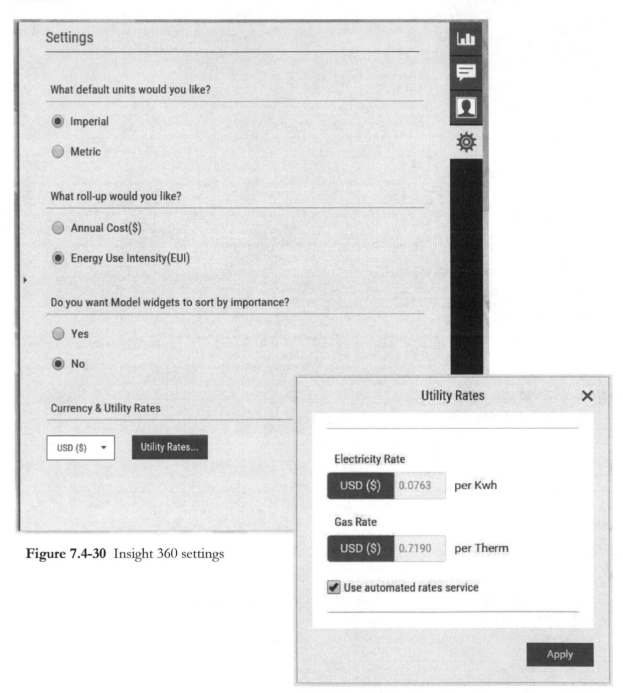

Figure 7.4-30 Insight 360 settings

Figure 7.4-31 Insight 360 utility rates

Additional EAM Validation

The next several images show how one might use the EAM 3D view to filter the analytical categories for visual validation.

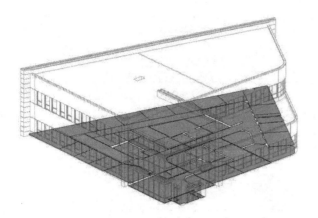

Analytical Surfaces – Slab on Grade

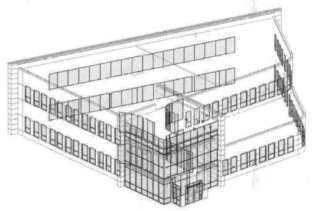

Analytical Surfaces – Exterior Walls

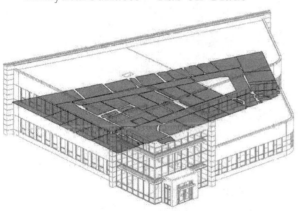

Analytical Surfaces – Interior Floors

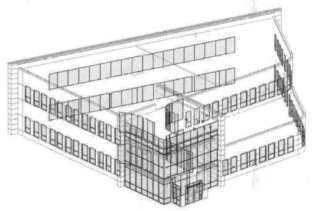

Analytical Surfaces – Windows

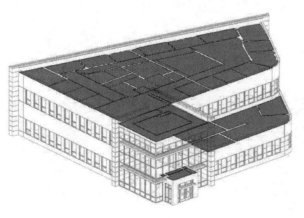

Analytical Surfaces – Roofs

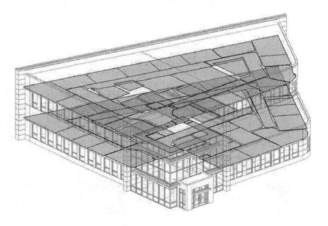

Analytical Surfaces – Ceilings

Analyze

Insight Heating and Cooling Loads - EnergyPlus

Running the Heating and Cooling loads, from within Revit, can be helpful as it provides additional information, such as **peak heating and cooling loads**, and uses EnergyPlus (rather than DOE2). The image to the right (Figure 7.4-32) shows a partial view of the results—the building summary and the first room. This tool is not meant to be used to size HVAC equipment at this time.

Running H&C with Insight 360 provides a more graphical result as shown below (Figure 7.4-33). Selecting a room in the model preview provides loads.

Load calculation summary report

Summary of heating and cooling loads for spaces

Project summary

Location and Weather	
Project	DC LAW OFFICES
Location	Minneapolis St Paul IntL Arp MN USA TMY3 WMO#=726580
Latitude	44.88
Longitude	-93.2

Building summary

Inputs	
Area (SF)	9158.89
Volume (CF)	87322.41
Calculated Results	
Peak Cooling Total Load(Btu/h)	228866.39
Peak Cooling Month and Hour	7/21 11:30:00
Peak Cooling Sensible Load(Btu/h)	220562.12
Peak Cooling Latent Load(Btu/h)	8304.18
Peak Heating Load(Btu/h)	-160135.39
Checksums	
Cooling Load Density (Btu/(h·ft²))	24.99
Heating Load Density (Btu/(h·ft²))	-17.48

Space Summary "ASSOCIATES_OFFICE_106"

Inputs	
Area (SF)	187.51
Volume (CF)	1303.10

Figure 7.4-32
Results from EnergyPlus heating and cooling analysis

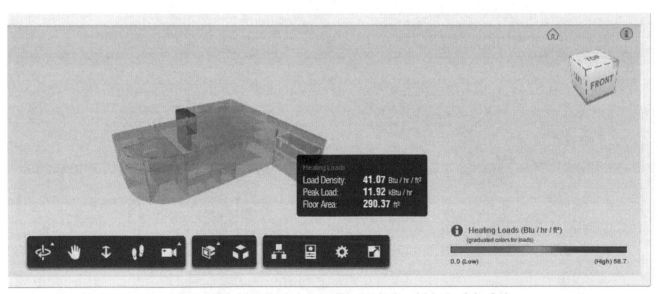

Figure 7.4-33 Results from EnergyPlus heating and cooling analysis within Insight 360

5 AIA 2030 Commitment Firms; upload to DDx

Firms participating in the AIA 2030 commitment are now able to push data from Insight directly into the AIA 2030 Design Data Exchange (DDx).

What is the AIA 2030 Commitment (from the AIA website)?

> The AIA 2030 Commitment is a growing national initiative that provides a consistent, national framework with simple metrics and a standardized reporting format to help firms evaluate the impact design decisions have on an individual project's energy performance.

For more information on this AIA initiative: http://www.aia.org/practicing/2030Commitment/

The next four images describe the process from within Insight:

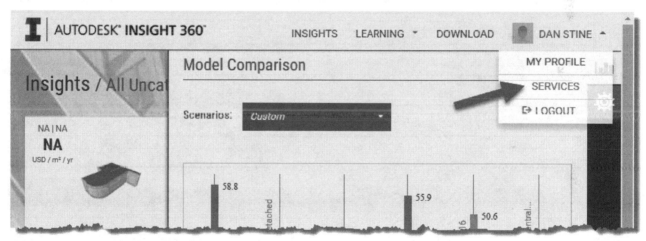

Figure 7.4-34 Verify AIA DDx service is turned on

Figure 7.4-35 Turn AIA 2030 DDx service off and then on to get login dialog (see next image)

Analyze

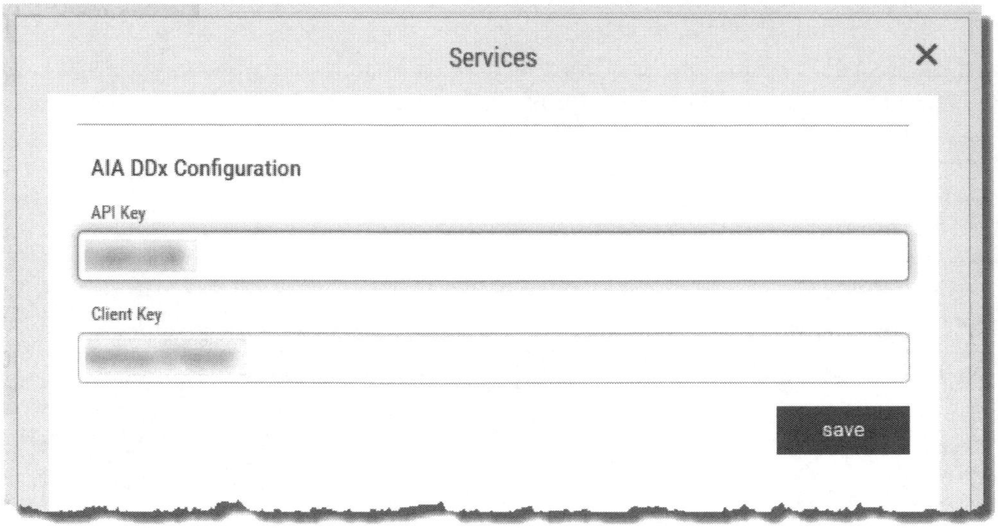

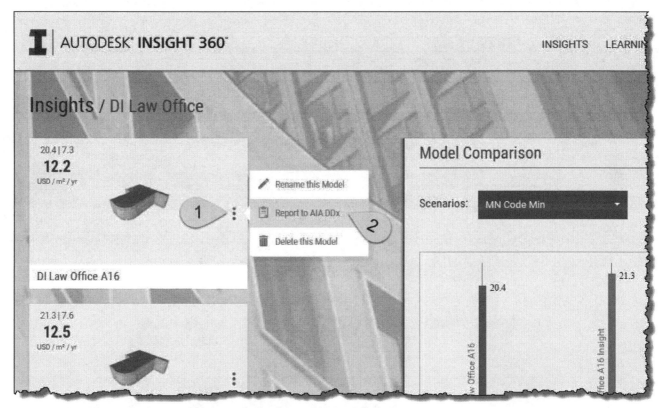

Figure 7.4-36 Insight dialog to enter AIA firm and user id numbers

Figure 7.4-37 Push current energy analysis data to AIA 2030 DDx

The following image (Figure 7.4-38) shows the exact information that will be sent to the AIA site. Notice the inputs with drop-down arrows allow information to be changed prior to clicking "send."

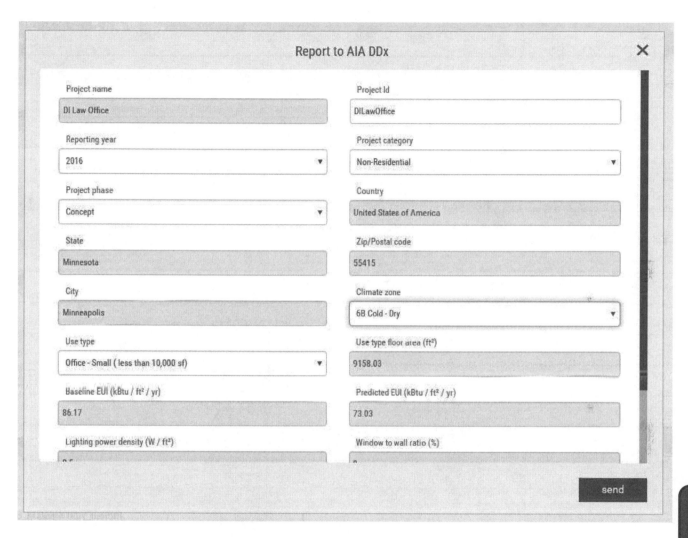

Figure 7.4-38 Review Insight data to be pushed to AIA 2030 DDX

The last image in this section (Figure 7.4-38) shows the results as seen when logged into the AIA site. Most of these inputs can be modified here as well if needed.

PROJECT VIEW

2018 - CC ▼

PROJECT SUMMARY

DI Law Office

Non-Residential

PREDICTED	BASELINE	GOAL	SAVINGS	CHALLENGE
73.03	**86.2**	**25.9**	**15%**	2030 = 100% (Carbon Neutral) 2025 = 80% 2020 = 80% 2015 = 70% 2014 = 60%
kBtu/sf/yr	kBtu/sf/yr	kBtu/sf/yr		[Architecture 2030 Challenge]
[Predicted Energy Use Intensity]	[Baseline Energy Use Intensity]	[Energy Use Intensity]		

GENERAL INPUTS	BUILDING ENVELOPE	HVAC SYSTEMS

* AIA 2030 Commitment Required Input Fields

1. Input Building Specifications   Save

Note: Basic General Inputs are required to be saved before Building Envelope and HVAC Systems screens can be accessed

Project Name *	DI Law Office	Project ID *	DILawOffice
Project Category *	Non-Residential ▼	Country *	United States of America ▼
Project Phase *	Concept ▼	State/Province *	Minnesota ▼
Year of Occupancy	2018 ▼	Zip/Postal Code *	55415
Reporting Year *	2018 ▼	City	Minneapolis
Target Certification	Select all that Apply ▼	Climate Zone	6A Cold – Humid ▼
Office Location	North Bethesda, MD, United States of . ▼		

Use Types *	Area (GSF)		Available ? [Target Finder]	BASELINE [National Avg.] kBtu/sf/yr	GOAL [2030 Challenge] kBtu/sf/yr	LPD Baseline [ASHRAE 90.1-2007] W/sf
Office – Small (less than 10,000 sf) ▼	9158.03	⊗ ⓘ	Yes	74	22.2	1.00
⊕	Total: 9.2K		WEIGHTED	74.0	22.2	1.00

2. Energy Analysis

Status of Energy Model *	HAS BEEN Modeled ▼	Responsible Party for Energy model	Please select ▼
Design Energy Code *	ASHRAE 90.1-2010 ▼	Energy Modeling Tool	Autodesk Insight 360 ▼
Energy Use Data will be collected* ☐		Time Spent On Energy Modeling	Please select ▼
		☐ Design Energy and Emissions Inputs	

3. Baseline & Target Energy Use Intensity

Define Baseline *	BASELINE	GOAL *	TARGET *
○ ENERGY STAR Target Finder™ ○ National Average ◉ Other [Source]	**86.2** kBtu/sf/yr	**25.9** kBtu/sf/yr [2030 Challenge]	**73.03** kBtu/sf/yr [TARGET EUI]

4. Additional Inputs

Lighting Power Density	0.5	Watts/sf	Occupancy Sensor Included?	☐

Figure 7.4-39 Review data on the AIA 2030 DDx website portal

Optional Settings and Workflows

The remainder of this chapter section will cover optional settings and workflows related to energy simulation in Revit.

Advanced Energy Settings

All other settings, in the Energy Settings dialog (Figure 7.4-40), not mentioned yet do not have to be set prior to creating the EAM or running a simulation. These inputs are all automatically varied over the possible ranges and can be adjusted in **Insight 360** to see instant feedback on performance impact.

Rooms or Spaces are not required, but when they are present some additional information is used (more on this later).

Revit 2021 changes the way Thermal Properties data is prioritized and used in the energy analysis. The order they are listed here relates to priority: **Conceptual Types** (lowest) to **Detailed Elements** (highest). If Detailed Elements is checked, the thermal properties saved in the building elements (e.g. walls, floors, etc.) will be used. If no thermal properties exist then any Schematic Type overrides are used, and if no overrides, the Conceptual Types settings are used.

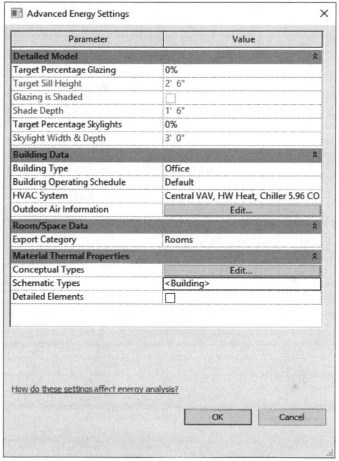

Figure 7.4-40 Advanced Energy Settings dialog

> For combined mass and building elements where the building elements define all glazing (like the combined example shown previously in Figure 7.4-40), set **the Target Percentage Glazing** to 0%.

The **Conceptual Types** dialog shown below (Figure 7.4-41) defines the construction defaults, assuming vertical surfaces are walls, top horizontal surfaces are roofs, etc. Again, it is no longer important to adjust these prior to running a simulation.

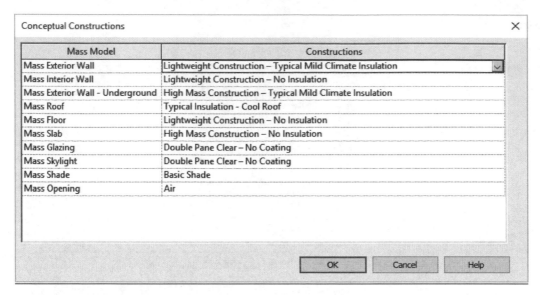

Figure 7.4-41 Define generic constrictions

The **Schematic Types** dialog shown below (Figure 7.4-42) has more detailed options from which to choose. When a category is checked, the defaults selected in the Conceptual Constructions dialog are overridden. Again, it is no longer important to adjust these prior to running a simulation.

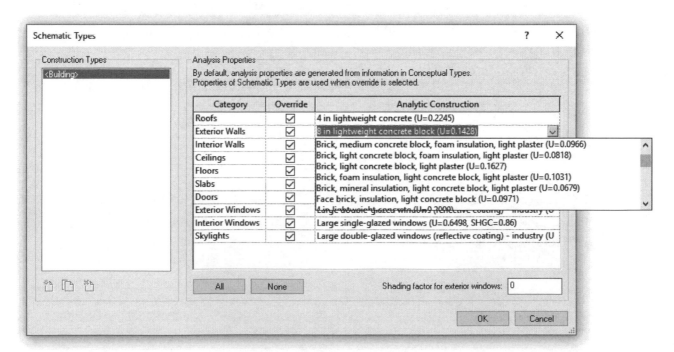

Figure 7.4-42 Use generic overrides based on element category

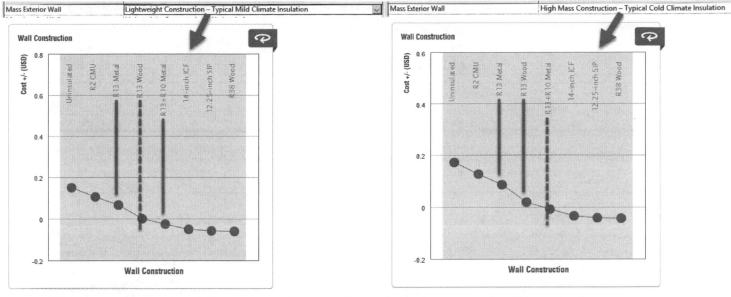

The image above, for a Minnesota project site, shows how changes to the Conceptual Constructions affect the results with Insight; remember the graph intersects "0" at the BIM setting. The image below shows we can override just the walls, in this example, with a super insulated wall.

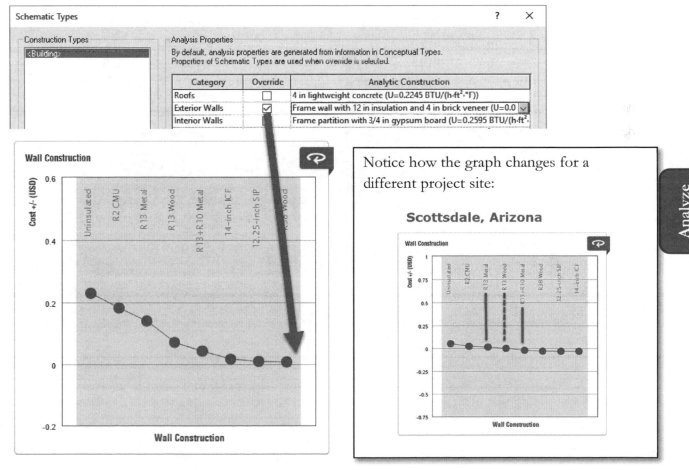

> Edit the Conceptual Constructions in your firm's <u>Revit template</u> to align with your state's minimum energy code requirements.

One final comment about building elements is that their **Thermal Properties**, assigned via materials, can be used in the energy simulation. For **layered** system families, such as walls, floors and roofs, the thermal properties are calculated for all layers (e.g., brick, air space, insulation, etc.). Notice highlighted **Resistance (R)** and **Thermal Mass** properties for the selected wall assembly in the image below (Figure 7.4-44). However, applying and/or using building element thermal properties is not required. Generic assembly overrides can be applied, as just discussed, which is great as this would be putting the "cart before the horse." We are using this process, partly, to determine what the thermal properties should be!

As the model develops to the point where building elements in Revit have the correct thermal properties, check the Detailed Elements in the Energy Settings dialog (Figure 7.4-43).

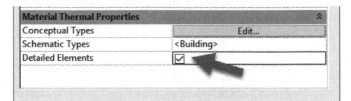

Figure 7.4-43 Use detailed elements checked

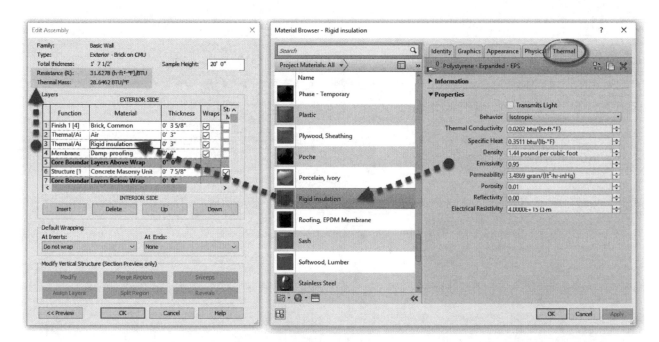

Figure 7.4-44 Thermal properties associated with building elements can be used in the energy simulation

Irrelevant Settings

Some settings have no impact on the Energy Simulation. Revit 2017 removed these irrelevant settings from the Energy Settings dialog. The image to the right, Figure 7.4-45, shows the Revit 2016 dialog with all the omitted parameters highlighted. Helpful information if you are still using Revit 2016.

These omitted settings are specific to the old room/space based gbXML export and heating/cooling loads and they are still present in those dialogs.

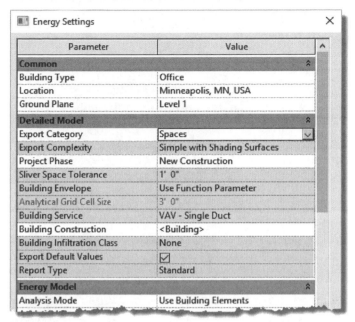

Figure 7.4-45 Parameters not used in Energy Simulation

Design Options

Revit's **Design Options** feature **partially** works for energy modeling in the preliminary design phase. The EAM is defined by elements in the **Main Model** and the **Primary** design option. To study another option, use the **Make Primary** command in the **Design Options** dialog, and recreate the EAM.

> If using design options for construction documents, e.g., a deduct alternate, it is not possible to change the primary options as annotations will get messed up.

Don't make a copy of your project and work in a separate file—it is not very BIM-like. There are always exceptions, but the energy analysis workflow is designed to work within the context of an active project.

Rooms and Spaces Not Required

It is helpful to understand that **Rooms** or **Spaces** are not required when using the **Use Building Elements** analysis mode. If they exist, some information is used. However, the EAM is generated from **Room Bounding** elements. Additionally, **Area and Volume Computations** does not have to be set to calculate volumes.

Just in case these two terms are not clear, understand that **Rooms** are typically placed by Architects and Interior Designers, while **Spaces** are placed by MEP designers. These two elements are placed and look the same in the model, but Rooms contain parameters like **Department** and **Wall Finish** while Spaces have engineering data such as **Electric Loads** and **Heating/Cooling Loads** (Figure 7.4-46). Also, when a Space exists within the same enclosed area as a Room (even when the Room is in a linked model), it has the ability to read the Room Name and Number.

When Rooms/Spaces exist, these are the parameters used:

- **Rooms**
    1. Room Name and Number
- **Spaces**
    1. Space Name and Number
    2. Occupancy; number of people
    3. Lighting and Equipment Loads
    4. Plus
        1. Building Construction (via Energy Settings dialog)
        2. Zones (i.e. collections of Spaces w/ set points)

> Use the **Space Naming Tool** add-in to make the Space names match the Room names (subscription benefit and built into Revit 2021.1).

Selecting and then using Spaces does allow for more detailed inputs, as just seen in the list above. Additionally, when **Energy Settings > Export Categories** is set to **Spaces**, the **Building Construction** option becomes available (see next section for more on this). Just remember, Rooms or Spaces are not required to start getting useful information on the performance potential of a design.

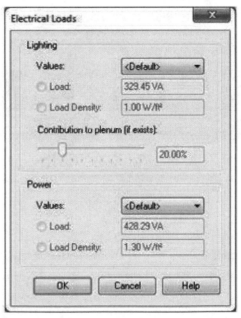

Figure 7.4-46 Additional properties embodied in EAM when Spaces are used

Spaces also have the ability to be grouped into **Zones**. These allow things like **Outdoor Air Information** and **Heating / Cooling** set-points to be entered as shown below (Figure 7.4-47).

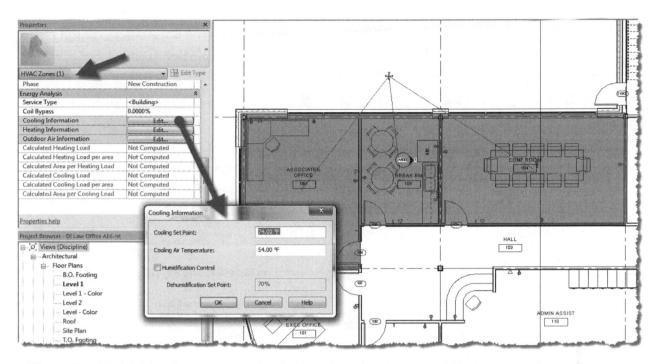

Figure 7.4-47 Additional properties embodied in EAM when Spaces and Zones are used

Going Further

Green Building Studio (GBS) is the web service that runs DOE 2.2 and EnergyPlus simulations used by Revit's **Energy Simulation** tool and by **Insight 360**. Use GBS to define custom settings for the analysis, such as currency, unit costs for electricity and natural gas, and the utility bill history with historical weather data. Once a Revit Energy Simulation has been run, it can be opened in GBS as seen below (Figure 7.4-48).

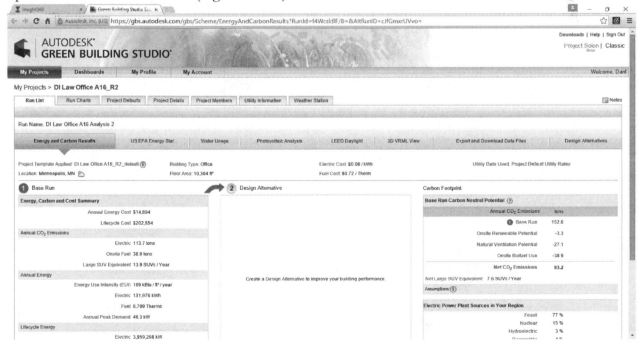

Figure 7.4-48 An example of the GBS cloud service

Analyze

7

Reference Material

When it comes to energy modeling, it is important to understand the required inputs and desired outputs, or results. The following information may be helpful to those just getting started in the "art" of energy modeling.

Results

Autodesk Insight 360 provides the primary results in either EUI or Cost. Having a feel for the actual EUI of existing buildings can be helpful. The graph below is based on research EPA conducted on more than 100,000 buildings. We obviously want to do better than these numbers…

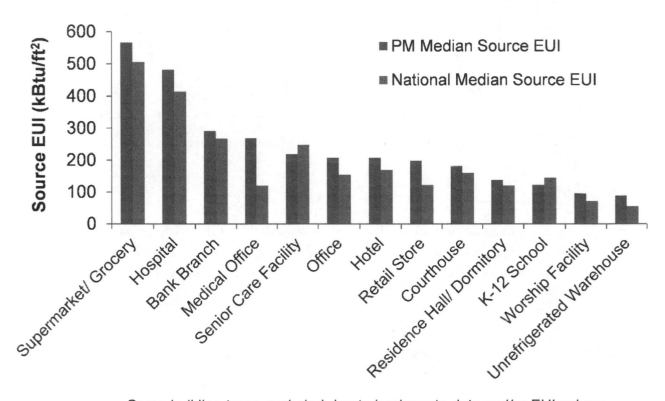

Some building types excluded due to inadequate data and/or EUI values beyond this range

Source: https://www.energystar.gov/buildings/facility-owners-and-managers/existing-buildings/use-portfolio-manager/understand-metrics/what-energy

Firm Reference Material

The following lists all the inputs available in Insight 360 and some comments (italicized text) added to help guide staff in a design firm. The comments are tips and specific requirements based on that firm's local energy code. These comments are not meant to be used directly, but as a guide to setting up one's own standards in a design firm.

Insight 360 Inputs and Comments

Firm specific comments for staff reference (not official)

Building Orientation	Rotates a building clockwise from 0 degrees, e.g., 90 degrees rotates the North side of the building to face East.
45 degree increments	*0 degrees equals True North position in Revit*

Wall Window Ratio (for N, E, S, W)	Window-Wall-Ratio (glazing area / gross wall area) interacts with window properties to impact daylighting, heating & cooling.
0%	*In the Energy Settings dialog, 'Target percentage glazing' is set to 0% as Insight 360 will vary for all options. Once*
15%	*windows or curtainwall are added, the Revit setting will*
30%	*be the point where the graph crosses 0 in Insight for Window-Wall-Ratio for each direction.*
40%	
50%	
65%	
80%	
95%	

Window Shades	Shades can reduce HVAC energy use. The impacts depend on other factors, such as window size and solar heat gain properties.
No Change	*Can also model sunshades in Revit. Use floor element in secondary Design Option. Set to Primary prior to*
1/6 Win Height	*generating Energy Analysis Model (EAM) in Revit. Switch*
1/4 Win Height	*back to secondary right after EAM creation. FYI: Revit ignores the Generic Model category.*
1/3 Win Height	
1/2 Win Height	
2/3 Win Height	

Analyze

Window Glass	Glass properties control the amount of daylight, heat transfer & solar heat gain into the building, along with other factors.
Sgl Clr	*Zone 6 (non-residential): U .55, curtainwall U .45*
* Dbl Clr	*Zone 7 (non-residential): U .45, curtainwall U .40*
Trp LoE	
Dbl LoE	
Quad LoE	

Wall Construction	Represents the overall ability of wall constructions to resist heat losses and gains.
Uninsulated	*http://www.dli.mn.gov/CCLD/codes15.asp*
R2 CMU	*Reference: 2015 MN Energy Code --> ASHRAE 90.1 --> 5.*
R13 Metal	*Building Envelope --> Tables 5.5-1 thru 5.5-8 (depending on climate zone)*
R13 Wood	
R13 + R10 Metal	
14-inch ICF	*Zone 6 (non-residential): R13*
12.25-inch SIP *(approx. R40)*	*Zone 7 (non-residential): R15.2*
R38 Wood	

Roof Construction	Represents the overall ability of roof constructions to resist heat losses and gains.
Uninsulated	*http://www.dli.mn.gov/CCLD/codes15.asp*
R10	*Reference: 2015 MN Energy Code --> ASHRAE 90.1 --> 5.*
R15	*Building Envelope --> Tables 5.5-1 thru 5.5-8 (depending on climate zone)*
* R19	
10.25-inch SIP	*Zone 6 & 7 (non-residential): R20*
R38	
R60	

Infiltration	The unintentional leaking of air into or out of conditioned spaces; often due to gaps in the building envelope.

2.0 ACH	*If building pressurized (values from Trane Trace defaults):*
1.6 ACH	*0.3 Average construction (LHB typical)*
1.2 ACH	*0.5 Poor construction*
0.8 ACH	*If not pressurized (neutral):*
0.4 ACH	*0.6 Average*
0.17 ACH	*1.0 Poor*
	2.5 Loose

Lighting Efficiency	Represents the average internal heat gain and power consumption of electric lighting per unit floor area.
1.9 W/sf	*ASHRAE 90.1 (2010), Table 9.5.1*
1.5 W/sf *Retail*	*Office 0.9 W/sf*
1.1 W/sf *Mfr Facility*	
0.7 W/sf *Dormitory*	
0.3 W/sf *Parking Garage*	

Daylighting & Occupancy Controls	Represents typical daylight dimming and occupancy sensor systems.
None	
Daylighting Controls	
Occupancy Controls	
* Daylighting & Occ Ctrls	

Plug Load Efficiency	The power used by equipment, i.e., computers and small appliances; excludes lighting or heating and cooling equipment.
2.6 W/sf	*Varies widely based on building type and use.*
2.0 W/sf	
1.6 W/sf	
1.3 W/sf	
1.0 W/sf	

Analyze

0.6 W/sf

HVAC	Represents a range of HVAC system efficiency which will vary based on location and building size.
ASHRAE Package Terminal Heat Pump	*= Typical options for education projects*
ASHRAE VAV	
ASHRAE Heat Pump	
High Eff. Heat Pump *	
High Eff. VAV *	
ASHRAE Package System *	
High Eff. Package System	
High Eff. Package Terminal AC	

Operating Schedule	The typical hours of use by building occupants.
24/7	
12/7	
12/6	
12/5	

PV - Panel Efficiency	The percentage of the sun's energy that will be converted to AC energy. Higher efficiency panels cost more but produce more energy for the same surface area.
16%	*We typically spec Suniva OPT250 which has 15.4 to 16.33 eff*
18.60%	
20.40%	

PV - Payback Limit	Use the payback period to define which surfaces will be used for the PV system. Surfaces with shading or poor solar orientation may be excluded.
10 yr	
20 yr	
* 30 yr	

PV - Surface Coverage	Defines how much roof area can be used for PV panels, assuming area for maintenance access, rooftop equipment and system infrastructure.

0%

60%

75%

90%

Conclusion

It should be clear that Autodesk is investing significant resources in the advancement of its performance design tools. It is definitely worth spending some time learning to use these tools. Hopefully this document will serve as a reference for those who are ready to jump in and get started!

Analyze

Notes:

Chapter 8
Massing & Site Tab

The Massing & Site tab contains two distinct sets of tools: **Massing** for building form conceptualization, and **Site** tools for representing the ground and related features.

The image below highlights the use of both tool sets found on this tab. The Mass tool was used to quickly define the proposed building and adjacent existing buildings. The Site tools were used to show the existing ground surface. Related commands, such as **Mass Floor** and [Site] **Subregion** were also used.

The big difference between the Mass and Site tools is their intended longevity in the life of a Revit project. Mass elements are mainly intended to conceptualize major shapes and forms of a building. Eventually these elements are replaced by model elements like walls, doors and windows. The topography is used throughout the project to define both the existing and proposed ground surfaces.

The Mass modeling tools should never be used to represent things like walls, floors, furniture or casework. Revit provides other tools for that: **Model In-Place** and the **Family Editor**. One exception is when a Mass element is needed to define and host a sloped wall—there is no other way to create a sloped wall in Revit (which is based on the wall types in the project).

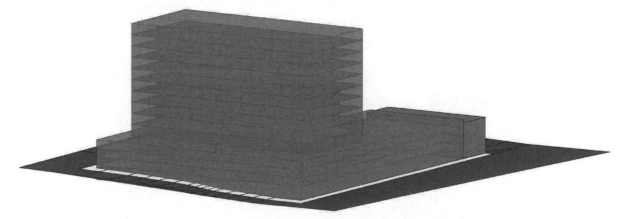

Mass elements, in the context of a site element, developed to conceptualize building form

8.1 Conceptual Mass panel

The tools found on this panel aid in creating Mass elements and controlling their visibility. The Massing tools are intended to be used to conceptualize overall building forms in the early stages of a design.

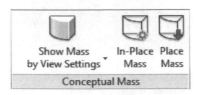

Show Mass by View Settings

When this option is selected, which is the default selection in a new project, the visibility setting in each view's **Visibility/Graphic Overrides** dialog controls the visibility of Mass elements in the current project. By default, the **Mass** category is individually turned off in every view. Thus, between these two defaults settings, Mass elements are never seen.

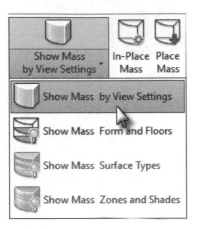

With this option selected, to see Mass elements in a particular view, open the **Visibility/Graphic Overrides** (Type VV) and on the **Model Categories** tag, turn on **Mass**.

Show Mass Form and Floors

When working with Mass elements, it is helpful to turn on this option: **Show Mass Form and Floors**. Doing so will make mass elements visible in every view, in the entire project, even when the Mass category is turned off.

Clicking In-Place Mass or Place Mass will result in the prompt shown to the right (Figure 8.1-1). Thus, just using a Massing command will enable this setting. To turn it off, simply switch back to **Show Mass by View Settings** in this same list.

As the prompt indicates, Masses will not print unless the category is turned on—which is a setting for each view. When Show Mass is active, a related prompt is offered when selecting the **Print** command (Figure 8.1-2).

> Use these two toggles to quickly see the Mass elements used to define and drive geometry, such as walls, roofs and floors (via the 'by face' commands covered later in this chapter).

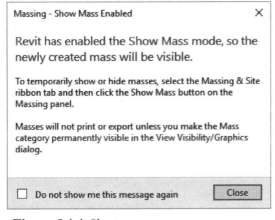

Figure 8.1-1 Show masses prompt

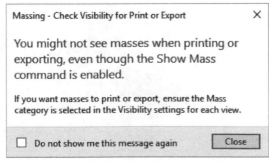

Figure 8.1-2 Printing mass prompt

Show Mass Surface Types

This command relates to older energy analysis functionality and as such is usually grayed out (or disabled). Older modeled setups with those features will likely enable these tools. In general, avoid this command as it is outdated.

Show Mass Zones and Shades

This command relates to older energy analysis functionality and as such is usually grayed out (or disabled). Older modeled setups with those features will likely enable these tools. In general, avoid this command as it is outdated.

In-Place Mass

This command is used to create a mass element in the project. This may contain one or many mass forms.

For readers familiar with SketchUp, the Massing tools in Revit offer an alternative which has a smoother transition from schematic design (SD) to design development (DD) using the **Model by Face** commands covered later in this chapter.

Masses are used to conceptualize the major forms of a building. They can also be used in **energy analysis**. Masses with Mass Floors applied are included in the energy analyses while Masses without Mass Floors become shade elements (e.g. adjacent existing buildings).

In-Place Mass vs. Place Mass:
For simple Masses, e.g. a box-shape, it would be better to use the **Place Mass** command rather than use this **In-Place Mass** option. The Place Mass is fast and can be copied around. Making changes to one changes all instances. When the In-Place Mass is copied, a new item is created in the Project Browser.

---

quick steps

In-Place Mass

1. Massing – Show Mass Enabled dialog
   - Appears first time using mass tools, if not already set
   - Ensures masses created will be visible
2. Name
   - Provide name for mass family about to be created
3. Ribbon (Create/Modify tabs)
   - Several tools to create a mass
   - Options change based on tool selected
4. Options Bar/Type Selector/Properties
   - Options change based on tool selected
5. In-Canvas
   - Must click Finish Mass or Cancel Mass to exit the command

Massing & Site

Do not use the massing tools to create items within the building, such as walls, floors, reception desks, casework, etc. There are other tools for those items, even when they are meant to be conceptual placeholders. The mass elements are more likely to be hidden and not print, and they are used in the energy analysis workflow.

In-Place Mass: **Show Mass dialog**

Clicking In-Place Mass or Place Mass will result in the prompt shown to the right (Figure 8.1-3). Thus, just using a Massing command will enable this setting – which will make masses visible in every view in the project regardless of the Mass category setting in **Visibility/Graphic Overrides**.

Click **Close** to proceed.

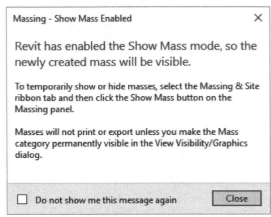

Figure 8.1-3 Show masses prompt

In-Place Mass: **Name**

Every In-Place Mass element needs a unique name (Figure 8.1-4). Enter a name and click **OK** to proceed.

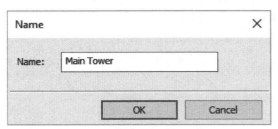

Figure 8.1-4 Mass name

In-Place Mass: **Ribbon**

When the In-Place Mass command is active, the Ribbon changes significantly; and the discipline tabs (Architecture/Structure/Systems) are all temporarily replaced by the **Create** tab.

Create tab:

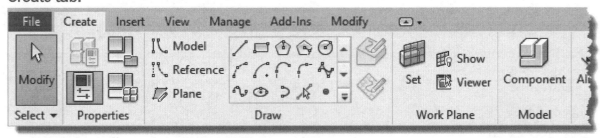

Modify tab:

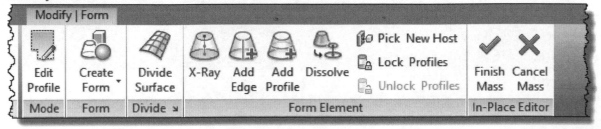

Create tab highlights:
The only way to end this command is to click
Finish Mass or **Cancel Mass**.

Notice the Draw panel has three toggles/sub-
commands on the left: **Model**, **Reference** and
Plane. The Model option is used to sketch line
work to be used in geometry creation. The
Reference Line and **Reference** Plane options are
used to create a framework to control geometry
creation and location with dimensions. These
references are not visible in the project when the
In-Place Mass is finished—but they are still
available to the Align and Dimension tools.

The **Component** command is used to place
another Family, loaded in the current project, as a
nested element within the current Conceptual Mass
family.

Modify tab highlights:
When a series of lines or an enclosed 2D outline is
selected, select the **Create Form** command to
create a solid or surface. In the example to the
right, Figure 8.1-5, five reference planes were
created in a plan view, and then the work plane Set
tool was used to make each one current to sketch
the five "contours" or framework of the desired
surface. All lines were selected and the Create Form
command was used to create the organic surface
shown below.

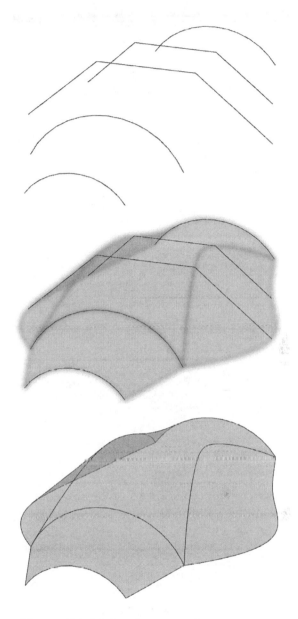

Figure 8.1-5 Mass form creation

With a Mass element selected, clicking the
Divide Surface applies a gridded system on the
surface as shown in Figure 8.1-6. Now, when
one of these surfaces is selected the **Properties
Palette** offers several grid types and related
parameters to control spacing, justification and
more (Figure 8.1-7).

Selecting the mass and then clicking **X-Ray**
reveals the original sketch lines and points.

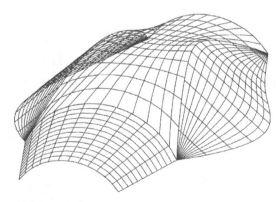

Figure 8.1-6 Divided surface applied

Massing & Site

These lines and points can be selected and used to manipulate the mass (Figure 8.1-8).

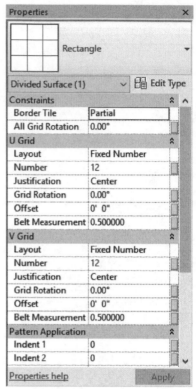

Figure 8.1-7 Divided surface properties

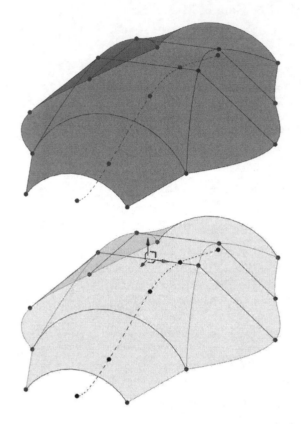

Figure 8.1-8 X-Ray mode

The **Add Edge** and **Add Profile** commands are used to add more control lines and points to the mass to facilitate further manipulation of the mass element. Finally, **Dissolve** reverts the mass back to just the original lines.

In-Place Mass: Options Bar

The Options Bar changes significantly throughout the creation of conceptual masses. Be sure to keep an eye on it to take advantage of the various options provided.

In-Place Mass: Type Selector

The main function of the Type Selector, while in the In-Place Mass command, is to select a divided surface pattern when one is applied (Figure 8.1-7).

In-Place Mass: Properties

There are many layers of properties within the context of an In-Place Mass. Here is a big picture overview:

Mass Family properties: (Figure 8.1-9)

- Elevation
  Offset value from level
- Moves With Nearby Elements (i.e., walls)
  Moves relative to nearest wall during placement
- Mass Floors
  Assign project levels to track conceptual floor area and indicate the selected mass will be used in energy analysis when performed
- Image
  Attach an image file (can appear in a schedule)
- Comments
- Mark
- Phase Created
- Phase Demolished

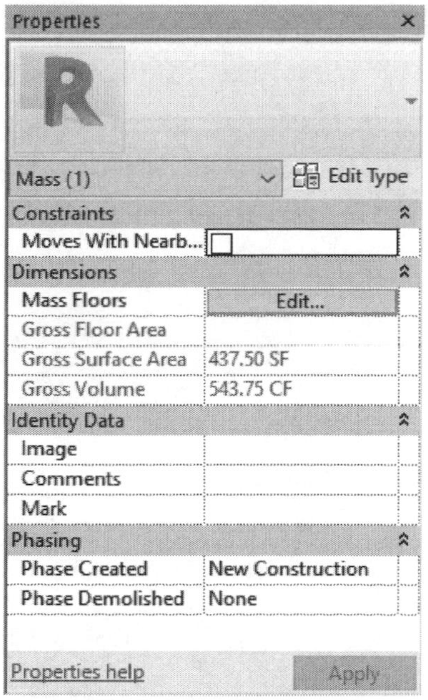

Figure 8.1-9 Mass Properties

Lines, Points and Faces properties:

- Visible
- Material
  Assign a Revit material. If none applied, the defaults assigned in Manage → Object Styles are used.
- Subcategory
  Subcategories allow more granular material and visibility control in the Visibility and Graphic Overrides dialog.
- Solid or Void toggle

Divided Surfaces properties:

- Several detailed settings: See Figure 8.1-7

Reference Planes properties:

- Wall Closure
  This setting is not relevant in a Mass element.
- Name
  Naming a reference plane is helpful as it appears in the Set work plane dialog, making it easier to select than the default "pick a plane" option.
- Is Reference
  Is this element accessible to the Align and Dimension tools in the project.

Reference Lines properties:

- Is Reference Line
  Toggle between Model and Reference Line.
- Reference
  Is this element accessible to the Align and Dimension tools in the project.

Massing & Site

In the Properties Palette, the far-right column allows many parameters to be associated with family parameters. In the image below, Figure 8.1-10, the selected Mass element's visibility will now be controlled by a type parameter called Parking Structure Visible. Several mass elements within the same In-Place Mass family can be controlled by this same parameter. In the project environment, select the mass element, and click Edit Type in the Properties Palette to access this parameter/toggle.

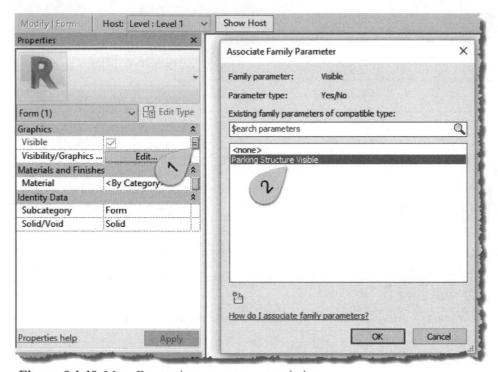

Figure 8.1-10 Mass Properties parameter association

In-Place Mass: In-Canvas

Basic workflow for In-Place Mass command:

1. Sketch 2D profile or several "contours" of the desired shape using the **Draw** options on the **Create** tab.
2. Select the 2D profile or "contours" and then click **Create Form** on the **Modify** tab.
3. For the 2D profile example, switch to a 3D view and adjust the height using the widget or temporary dimension (Figure 8.1-11).
4. Hover cursor and tap **Tab** to select an edge, a face or the entire mass.
5. Adjust edges and faces with the widget controls and temporary dimensions.

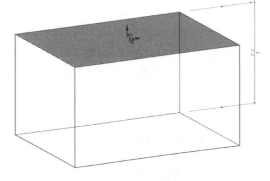

Figure 8.1-11 Control widget

6. Dimensions (Optional):
    a. Add dimensions and **Lock** to constrain
    b. Add dimensions and add **Parameter** to adjust
    c. See Figure 8.1-12

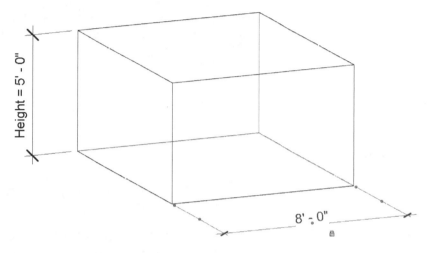

Figure 8.1-12 Dimensions added; one locked set to a parameter

7. Divide Surface (Optional):
    a. Select surface or entire mass
    b. Click Divide Surface
    c. Select pattern type from Type Selector
    d. Adjust parameters in the Properties Palette
    e. See Figure 8.1-6
8. Mass Floors (Optional):
    a. Finish the mass (i.e., while not in edit mode)
    b. Select the mass family
    c. Select Mass Floors (Properties or Ribbon)
    d. Check desired floor levels (levels must already exist)

Figure 8.1-13 Mass Floors dialog

Create a **Schedule** based on **Mass Floors** to quickly calculate the total building area based only on levels, masses and applied mass floors.

Place Mass

The **Place Mass** command is designed to place prebuilt mass families. These families are not directly editable within the project environment—only in the Family Editor.

Revit offers several mass families which represent the most basic shapes—see image below. These parametric families have several instance parameters to control the size.

The provided families can be modified or duplicated. Additionally, new custom families can be created as needed.

Using these mass families is fast and easy when basic shapes are all that are needed. They work with the mass visibility toggles, the **Mass Floors** and the energy modeling feature.

quick steps

Place Mass

1. Ribbon
   - Load Family
   - Model In-Place
   - Place on Face
   - Place on Work Plane
2. Options Bar
   - Rotate after placement
   - Placement Plane
3. Type Selector
   - Select loaded mass family to place
4. Properties
   - Offset
   - Material
   - Dimensions
5. In-Canvas
   - Spacebar key to rotate
   - Press Esc key or Modify button to exit command

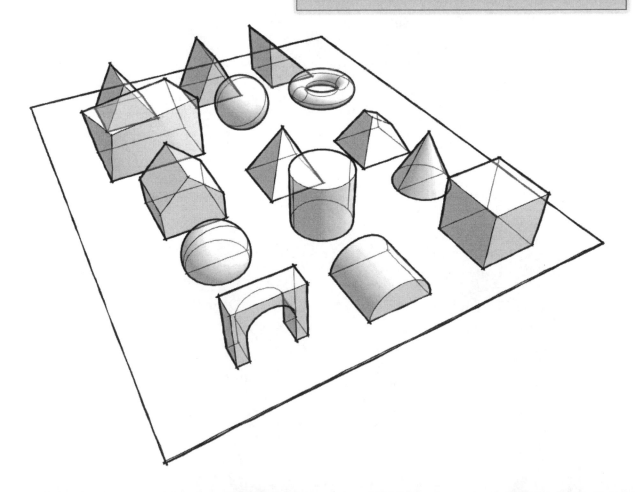

Place Mass: **Ribbon**

The Ribbon contains a **Load Family** button if the desired Mass family is not loaded in the current project. The **Model In-Place** is a quick way to leave the Place Mass command in favor of a custom Mass unique to the project.

The Ribbon also has two Placement options. The default is **Place on Work Plane**. This option will place the mass family on the current Work Plane—which is adjustable with the **Set** command (in the Work Plane panel on the Architecture tab). The **Place on Face** option allows a Mass Family to be placed on any surface below the cursor. In the example below, Figure 8.1-14, the mass "Box" family was placed on the mass "Gable" family's side using the Place on Face feature to represent windows (notice the modified instance parameters to control the size of the box). This could have been used on the roof as well—to suggest a skylight.

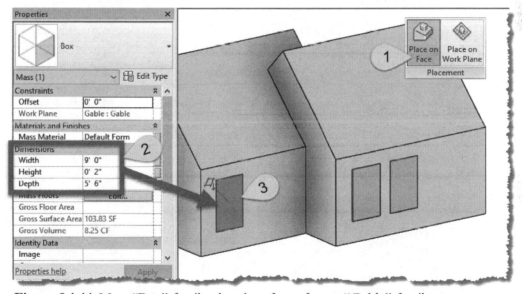

Figure 8.1-14 Mass "Box" family placed on face of mass "Gable" family

Place Mass: **Options Bar**

The Options Bar allows the element to be rotated immediately after placement. Also, when **Place on Work Place** is selected on the Ribbon, Levels, Grids and Named Reference Planes are listed in the **Placement Plane** drop-down list.

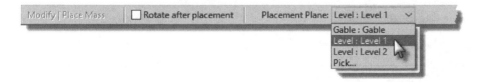

Place Mass: Type Selector

Pick from the list of Mass families loaded in the current project.

Place Mass: Properties

The **Offset** value controls the position relative to the work plane. In the "window" example in Figure 8.1-14, this would move the box away from the wall, not up from the ground.

The **Material** can be changed to any material in the current project.

Depending on the family and geometry, the **Dimension** options will vary. However, because they are instance parameters, they can be changed without the need to duplicate and create new types.

Place Mass: In-Canvas

In a model view, click to place the mass family.

Notice in the image below, Figure 8.1-16, that the **Join** command, located on the Modify tab, can be used to make two mass families appear monolithic.

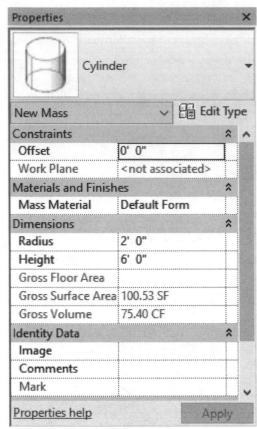

Figure 8.1-15 Mass family properties

Similarly, the **Cut** command works against mass families and will affect the listed **Gross Volume** results.

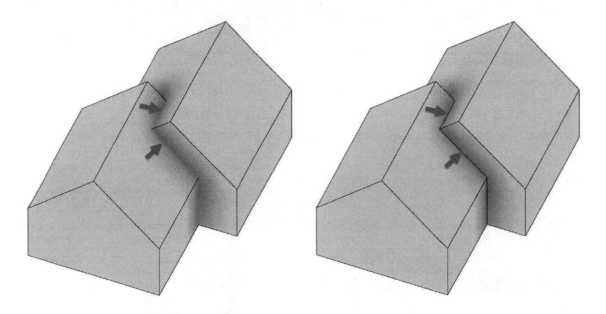

Figure 8.1-16 Join command (Modify tab) used on two masses adds lines at intersections

8.2 Model by Face panel

The Model by Face panel provides commands to aid in the transition from conceptual design, based on masses, to a more formal model using Roofs, Walls and Floors—i.e., Revit model elements. Also, due to limitations of the curtain wall "wall type" on irregular, or organic, mass surfaces, and to support the notion of a curtain grid system, the Curtain System command is provided.

Curtain System

This command is more formally called **Place Curtain System by Face**. A mass element must be created before using this command. Use this tool when a curtain wall system is needed on an irregular surface defined by a mass.

> For simple mass element surfaces, use the curtain wall types found in the **Wall by Face** command. This offers more flexibility once the mass element is no longer needed.

Place Mass: Ribbon

The Ribbon has three options when this command is active: **Select Multiple**, **Clear Selection** and **Create System**.

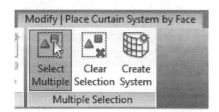

The default option is **Select Multiple**, which allows multiple surfaces to be selected before the curtain system is applied. All selected surfaces act as a single element. The number of faces selected can be changed in the future.

Clear Selection will remove all selected surfaces for the current instance of the command.

Once the desired surface(s) is selected, click **Create System** to finish the command.

---

quick steps

Curtain System

1. Ribbon
   - Select Multiple
   - Clear Selection
   - Create System
2. Type Selector
   - Select curtain system type
3. Properties
   - Room bounding
   - Grids 1 *and* 2
     i. Justification
     ii. Offset
4. Type Properties
   - Grids 1 *and* 2
     i. Set curtain grid spacing
5. In-Canvas
   - Select the face of one or more mass elements
   - Click Create System on the Ribbon to complete task

Massing & Site

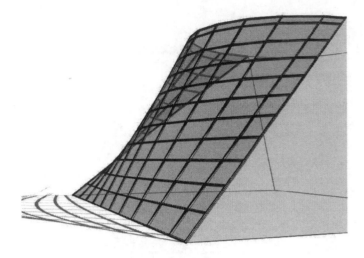

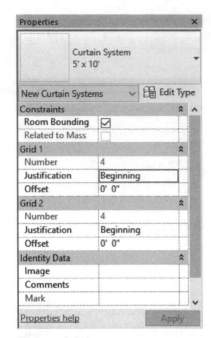

Figure 8.2-1
Curtain Grid Properties

Place Mass: Type Selector

The Type selected, in **the Type Selector**, determines the number of curtain grids applied. Click Edit Type and then Duplicate if additional options are needed.

Place Mass: Properties

Use Properties to adjust the grid justification and if the new element should be **Room Bounding** (Figure 8.2-1).

Place Mass: Type Properties

The Type Properties has several settings related to the Layout and Spacing of the grids. The Curtain Panel, Join Conditions and mullions can also be specified.

Place Mass: In-Canvas

Click on the surface of any Mass element in conjunction with the options on the Ribbon.

> Curtain Mullions can be added after the Curtain System is created. If the Mass element changes, select the Mass and click Update to Face on the Ribbon.

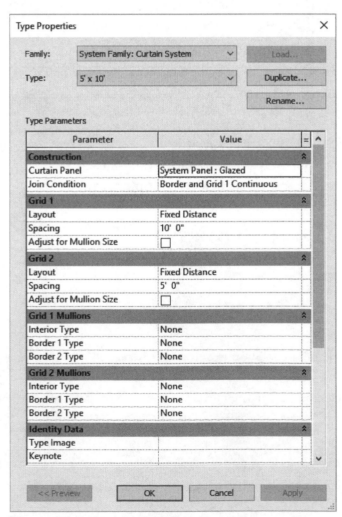

Figure 8.2-2 Curtain Grid Type Properties

Roof by Face

The **Roof by Face** command is used to quickly create a regular Revit roof element based on the face of a Mass element.

This command is the only way to create a roof, using the Roof types in the project, based on an irregular/organic surface.

The created element has a lasting tie to the Mass until the mass element is deleted. If the mass is changed, the roof does not automatically update, but selecting the element, and then clicking **Update to Face** on the Ribbon will realign the roof with the original mass face.

Roof by Face: Ribbon

The Ribbon has three options when this command is active: **Select Multiple**, **Clear Selection** and **Create Roof**.

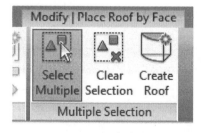

The default option is **Select Multiple**, which allows multiple surfaces to be selected before the roof is created. All selected surfaces act as a single element. The number of faces selected can be changed in the future.

Clear Selection will remove all selected surfaces for the current instance of the command.

Once the desired surface(s) is selected, click **Create Roof** to finish the command.

Roof by Face: Options Bar

The Level option is likely just for scheduling as the created element has no direct relationship with the level but, rather, the face selected (even after the mass element is deleted).

---

quick steps

> **Roof by Face**
>
> 1. Ribbon
>   - Select Multiple
>   - Clear Selection
>   - Create System
> 2. Options Bar
>   - Level
>   - Offset
> 3. Type Selector
>   - Select roof type
> 4. Properties
>   - Picked Faces Location
>   - Room Bounding
>   - Rafter Cut
> 5. In-Canvas
>   - Select the face of one or more mass elements
>   - Click Create System on the Ribbon to complete task

Massing & Site

Similarly, adjusting the **Offset** option on the Options Bar has no apparent effect on the position of the element. However, adjusting the Offset value, in Properties, after placement does cause the element to be repositioned.

> There is no overhand option for Roof by face. However, once the roof element is created, it can be selected and the grids used to extend the roof on each side.

Roof by Face: Type Selector

Select the desired Type in the **Type Selector**. These are the same options seen while using the Roof command on the Architecture tab.

Roof by Face: Properties

Use properties to set **Room Bounding** (Figure 8.2-3). Also, the **Picked Faces Location** determines which side of the Roof element is aligned with the picked face; the options are **Faces at Top of the Roof** and **Faces at the Bottom of the Roof**.

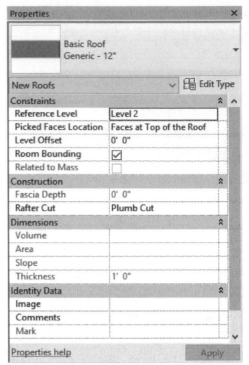

Figure 8.2-3 Roof by Face Properties

When the roof is sloped, and only when sloped, the **Rafter Cut** option can be changed. Like a normal roof, this setting only affects the lower side of the roof element.

Roof by Face: In-Canvas

Click to select a face of a mass element in the model.

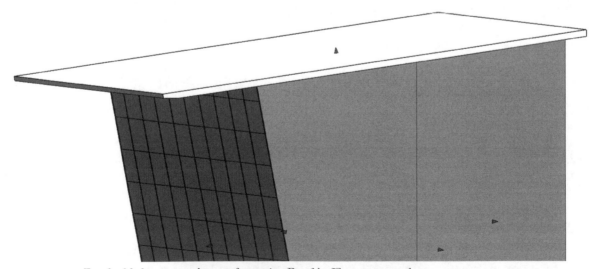

Roof added to mass element face using Roof by Face command.

Wall by Face

The **Wall by Face** command is used to quickly create a regular Revit wall element on the face of a Mass element.

The interesting thing about this "Model by Face" command is that it is 100% just starting the regular Revit Wall command found on the Architecture tab. In the Draw panel, while the Wall command is active, the **Pick Face** option is selected.

Using this command, or Draw option, is the only way to create a wall, using the regular wall types in the project, that conforms to an irregular/organic surface as seen in the image below (Figure 8.2-4).

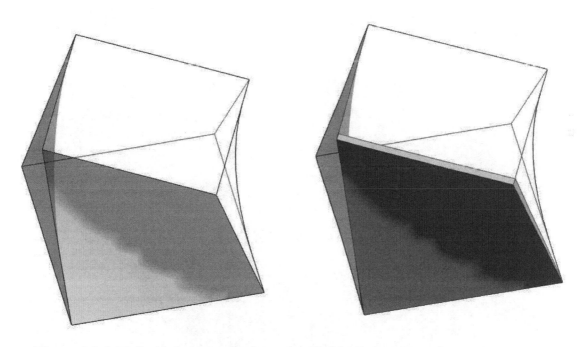

Figure 8.2-4 Wall added to irregular face using Wall by Face command

See the regular wall command, in the Architecture tab chapter of this book, for more on the wall command. Also, review the other Model by Face commands in this section for additional tips.

Massing & Site

Floor by Face

The Floor by Face command creates a regular Revit floor element on the face of a Mass element. In addition to needing a mass to use this command, the mass must also have **Mass Floors** defined as shown in Figure 8.2-5.

Floor by Face: Ribbon

The Ribbon has three options when this command is active: **Select Multiple**, **Clear Selection** and **Create Roof**.

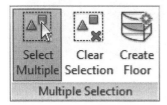

The default option is **Select Multiple**, which allows multiple mass floors to be selected before floors are created.

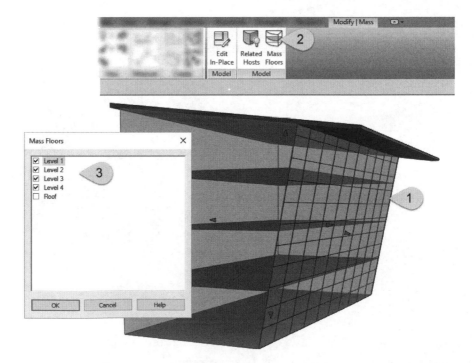

Figure 8.2-5 Mass floors must be defined prior to using the Floor by Face command

Clear Selection will remove all selected mass floors for the current instance of the command.

Once the desired surface(s) is selected, click **Create Floor** to finish the command. Separate floors are created and each is associated with the level the mass floor was associated with.

> If the original mass element is deleted, the floor becomes totally independent of any association to a mass… unlike the Roof by Face which will still offer an inactive Update to Face button on the Ribbon after the mass has been deleted.

Floor by Face: **Options Bar**

This is an unusual **Offset** option as it is from the edge, not a vertical offset. A positive number will offset the floor edge inward and a negative number outward.

Floor by Face: **Type Selector**

Select the desired Type in the **Type Selector**. These are the same options seen while using the Floor command on the Architecture tab.

Floor by Face: **In-Canvas**

With the Floor by Face command active, select one or more mass floors. When selecting multiple mass floors, they do not all have to be associated with the same mass element.

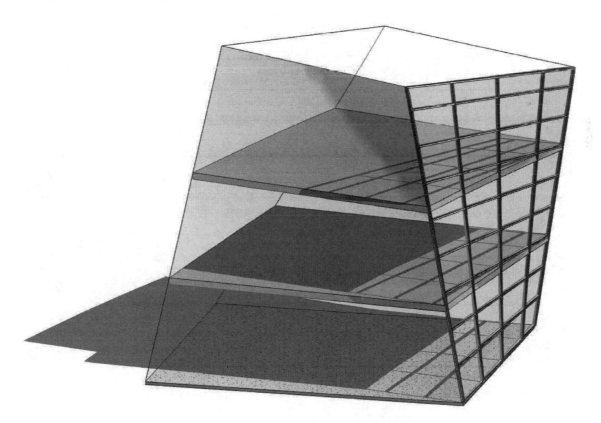

Floor added to 'mass floors' using Floor by Face command.

Massing & Site

8.3 Model Site panel

This section, and the next one, will give the reader an overview of the site tools available in Revit. The site tools are not intended to be an advanced site development package. Autodesk has another program much more capable at developing complex sites; it is **AutoCAD Civil 3D 2021**. This program is used by professional Civil Engineers and Surveyors. The contours generated from these advanced civil CAD programs can be used to generate a topography object in Revit.

Once a topography object (the topography object, or element, is a 3D element that represents part *or* all of the site) is created, the grade line will automatically show up in building and wall sections, exterior elevations and site plans. The sections even have the earth pattern filled in below the grade line.

As with other Revit elements, you can select the object after it is created and set various properties for it, such as *Surface Material*, *Phase*, etc. One can also return to *Sketch* mode to refine or correct the surface. This is done in the same way most other sketched objects are edited: by selecting the item and clicking *Edit* on the *Options Bar*.

The image below shows a Revit site with various materials and plantings placed.

Image from Residential Design using Autodesk Revit by Daniel John Stine (SDC Publications)

Toposurface

Creates a 3D surface by picking points (specifying the elevation of each point picked) or by using linework within a linked AutoCAD drawing, that were created at the proper levels.

> Once a Revit surface has been created from a linked CAD file, there is no connection between the two elements. The Revit surface will not automatically update if the CAD file linked is modified.

The ideal workflow: receive CAD-based contours, or triangulated surface – which is even better, in an AutoCAD DWG file. Link this file into Revit, and select it while using the **Create from Import → Select Import Instance**. Next, you must select, from a list of Layers, which ones have contours (or triangulated surface). Revit will automatically create a site based on the endpoints of all elements on the selected Layers. The Revit site does not have any connection back to the linked DWG file. If the site model changes, the Revit toposurface needs to be recreated.

---

quick

Toposurface

1. Ribbon
    - Place Points (default)
    - Create from Import
        i. Select Import Instance
        ii. Specify Points File
    - Simplify Surface
    - Work Plane options
2. Options Bar (for Place Point tool)
    - Elevation
    - Relative to…
        i. Absolute Elevation
        ii. ???
3. In-Canvas
    - Follow prompts

Toposurface: Ribbon

Use **Place Points**, in conjunction with the current Elevation setting on the Ribbon, to manually define, or refine, a ground surface. This option is usually used in a site plan view.

Create from Import has two options: **Select Import Instance** and **Specify Points File**.

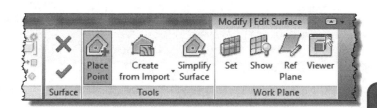

The first option is used against a linked in CAD file as discussed above. The second option is used if a "points file" is provided from a civil engineer or surveyor. This might come directly from a surveyor's data collector or exported from another application. However, the points should be reviewed and corrected, if needed, by a surveyor or civil designer before being used in Revit.

Massing & Site

Simplify Surface will reduce the number of points while trying to maintain the integrity of the original surface. When a surface is created from a CAD link, the number of points in Revit can be quite large for large sites or highly detailed triangulated surfaces. This can have an impact on performance, so reducing the number of points can help. However, the surface needs to be closely inspected to verify its integrity.

The **Work Plane** panel has the usual **Set** and **Show** options. The **Ref Plane** command is offered for convenience and the **Viewer** command opens a temporary, throw-away, 3D view.

Toposurface: Options Bar

The **Elevation** controls the vertical position of any new, or selected, points. This option can be used when creating a surface for early phase conceptual surfaces. Another use is to manually follow the contours of a scanned drawing, which has been imported into Revit.

The drop-down controls what the elevation is relative to. The default option is **Absolute Elevation**, which corresponds to the **Project Base Point**.

Toposurface: In-Canvas

Working in a dedicated site plan view can be helpful. Looking at the View Range settings you can see the site is being viewed from 200′ above the first floor level, so your building/roof would have to be taller than that before it would be "cut" like a floor plan. The View Depth could be a problem here: on a steep site, the entire site will be seen if part of it passes through the specified View Range. However, items completely below the View Range will not be visible. In this case, the **Bottom** and **View Depth** can both be set to Unlimited to avoid any visibility issues.

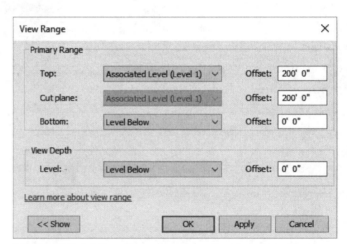

Figure 8.3-1 View Range settings for site plan view

A surface must have at least three points. Click the **Modify** button (not the tab) and select previously placed points to change their **Elevation** on the Options Bar.

Click the green checkmark to finish and the red "x" to cancel the current toposurface.

Various elements in the toposurface can be graphically controlled in **Object Styles** (project wide settings) and **Visibility/Graphic Override Settings** (view specific settings).

Site Component

Items like benches, dumpsters, etc., that are placed directly on the toposurface, at the correct elevation at the point picked.

The image below (Fig. 8.3-2) shows several trees, and two park benches, placed in a plan view, using **the Site Component** command. Each tree was automatically positioned vertically to align with the ground surface.

> After initial placement, if a Site Component is moved, it will reposition vertically to stay in alignment with the ground surface.

Site Component: **Ribbon**

Use the **Load Family** command to load additional site components from the Site or Planting folders.

The **Model In-Place** command is offered as a quick way to switch gears

Site Component

1. Ribbon
   - Load Family
   - Model In-Place
2. Options Bar
   - Rotate after placement
3. Type Selector
   - Select type to place
4. Properties
   - Level*
   - Host (read-only)
   - Offset
   - Moves With Nearby Elements
5. In-Canvas
   - *Click over a site element*; element will be hosted by, and align with, the site
   - *Items placed away from the site* will be hosted by the level listed

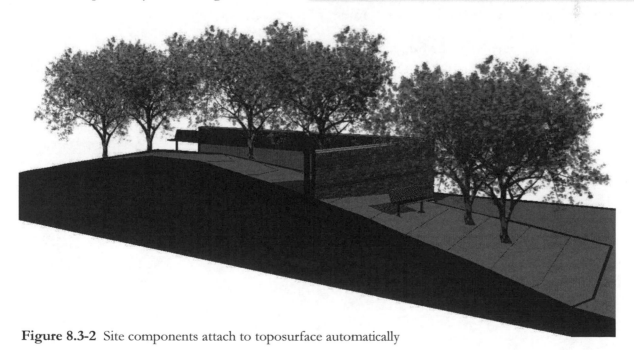

Figure 8.3-2 Site components attach to toposurface automatically

Massing & Site

and create a conceptual mass of a building in the context of a site.

Site Component: **Options Bar**

In a plan view, the singular option is **Rotate after placement**. Check this option to immediately be prompted to rotate the element after picking a point to place it.

In a 3D view, the **Level** option, on the Options Bar, is mainly for schedule purposes as the elements automatically align with the surface (vertically). The exception is when the element is moved away from, or not created, over a site… then it is positioned relative to the selected level.

Site Component: **Type Selector**

Select the desired family and type to place in the model.

Site Component: **Properties**

The **Level** option is only relevant if the element is not being placed over a toposurface. The **Host** parameter, which is read-only and thus cannot be modified, is helpful to know what the host is – a good troubleshooting clue.

> When a Site Component is selected, use the **Pick New Host** option on the Ribbon if the element loses its ability to vertically align with a toposurface.

Site Component: **In-Canvas**

For the Site Components to automatically align vertically with a toposurface, the toposurface must be visible in the current view.

Click to place elements in the current view. Working in 3D views can be tricky as the point picked on the screen may be interpreted as being beyond the surface and thus not properly positioned on the site. If this happens, consider working in a plan view.

Various elements related to the **Site** and **Plantings** categories can be graphically controlled in **Object Styles** (project wide settings) and **Visibility/Graphic Override Settings** (view specific settings).

☑ Site
☑ Hidden Lines
☑ Landscape
☑ Pads
☑ Project Base Point
☑ Property Lines
☑ Site Accessories
☑ Stripe
☑ Survey Point
☑ Utilities

Parking Component

These are parking stall layouts that can be copied around to quickly lay out parking lots. Several types can be loaded which specify both size and angle. These families also conform to the surface of the topography.

These items can be scheduled and counted by Revit, making checking parking requirements much easier than counting parking spaces individually.

Parking Component: Ribbon

Use the **Load Family** command to load additional site components from the Site or Planting folders.

The **Model In-Place** command is offered as a quick way to switch gears and create a conceptual mass of a building in the context of a site.

quick steps

Parking Component

1. Ribbon
   - Load Family
   - Model In-Place
2. Options Bar
   - Rotate after placement\Level
3. Type Selector
   - Select type to place
4. Properties
   - Level*
   - Offset
5. Properties
   - Offset
   - Moves with Nearby Elements
6. In-Canvas
   - Click to place

Parking Component: Options Bar

In a plan view, the singular option is **Rotate after placement**. Check this option to immediately be prompted to rotate the element after picking a point to place it.

Figure 8.3-3 Parking component placed on toposurface; two type options shown

Massing & Site

Parking Component: **Type Selector**

Select the desired family and type to place in the model. See two common options shown in Figure 8.3-3.

Parking Component: **Properties**

The **Level** option is only relevant if the element is not being placed over a toposurface. The **Host** parameter, which is read-only and thus cannot be modified, is helpful to know what the host is – a good troubleshooting clue.

Use **Offset** to move the element vertically. This is helpful if parts of the linework are hidden due to a warped surface (the parking family does not warp, it just slopes).

Parking Component: **In-Canvas**

Click to place the first parking element in a site plan view and then finish the command. The snaps don't work too well for subsequent locations. Instead, use the Copy command with snaps to position additional parking stalls adjacent to the first one.

The image below (Figure 8.3-4) is a site developed in Revit and then exported to a rendering and animation program called **Lumion** (https://lumion3d.com/). Animations feature moving vehicles and people when using their proprietary content in conjunction with the imported Revit model.

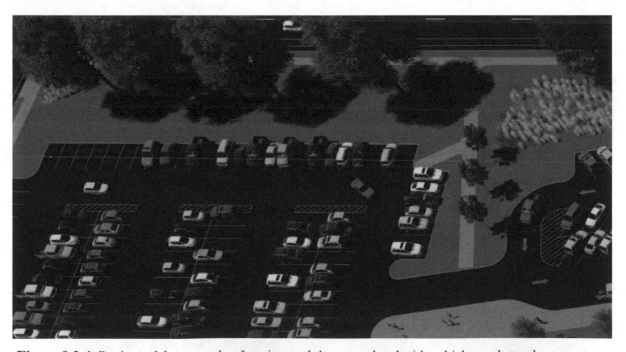

Figure 8.3-4 Revit model exported to Lumion and then populated with vehicles and people

Building Pad

Used to define a portion of a site to be subtracted when created below grade (and added when created above grade). An example of how this might be used is to create a pad that coincides with a basement floor slab which would remove the ground in section above the basement floor slab (otherwise the basement would be filled with the earth fill pattern). Several pads can be imposed on the same toposurface element–but they cannot overlap.

Building Pad: Ribbon

Use the **Boundary Line** draw tool and sketch the perimeter of the floor in a plan or 3D view. The default method is using **Pick Walls**, which automatically constrains the sketch lines to the selected walls (FYI: see Options Bar section below). Use any of the line tools to manually define the edge.

Use the **Slope Arrow** feature to define a slope for the entire floor element. When the arrow is selected, use the Properties Palette to specify the desired slope relative to the current level (and offset).

The **Set** tool, in the Work Plane panel, can be used to specify a different level for the building pad element about to be created. But this is usually not changed so the building pad will automatically be tied to the level associated with the current plan view.

Building Pad: Options Bar

When the Pick Walls draw option is selected, the Options Bar offers **Offset** and **Extend into wall (to Core)**. The "core" is the structural portion of the wall as defined in the Edit Structure part of the walls' properties. The options vary slightly based on the selected draw tool.

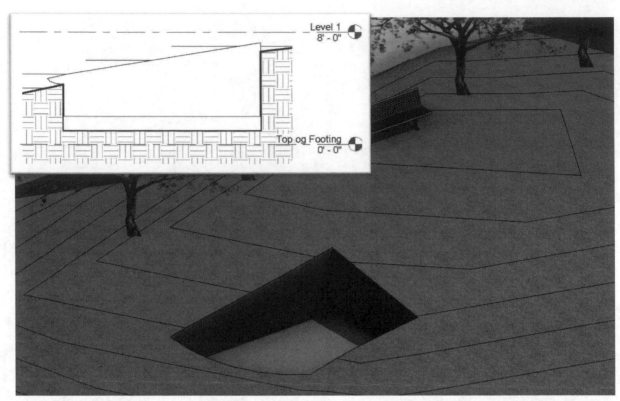

Figure 8.3-5 Building Pad positioned below grade (cuts toposurface)

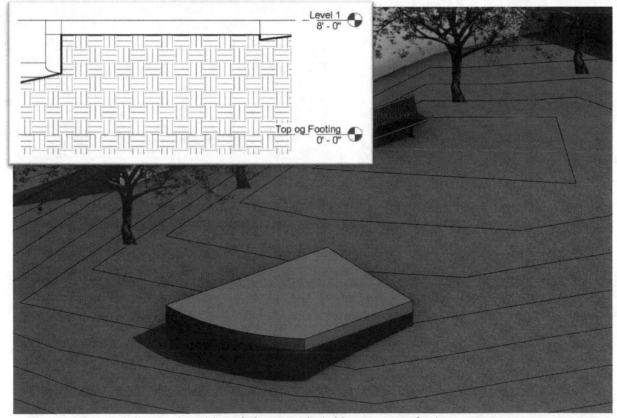

Figure 8.3-6 Building Pad positioned above grade (adds to toposurface)

Building Pad: **Type Selector**

Select the building pad type from the Type Selector. One option might be a build pad that is thin (1/8" thick) and non-room bounding. In this case, consider setting it to non-room bounding as well. If needed, create new Types via Edit Type → Duplicate (Figure 8.3-8).

Building Pad: **Properties**

In properties (Figure 8.3-7), set the Level, Offset, Room Bounding, Phase Created and Phase Demolished.

Building Pad: **In-Canvas**

In the appropriate plan view, use the **Draw** tools to sketch the perimeter of the building slab similar to how a floor element is created. Adjust the **Offset** to position vertically.

Figure 8.3-7 Building Pad properties

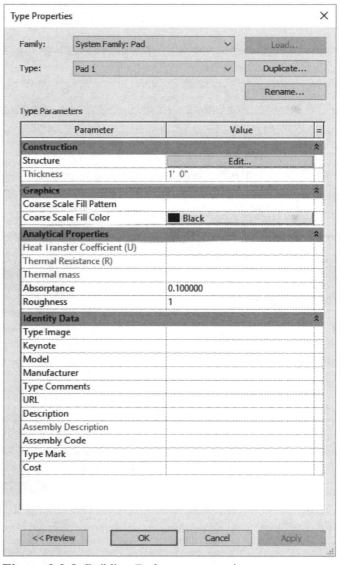

Figure 8.3-8 Building Pad type properties

Massing & Site

Site Settings dialog

The **Site Settings** dialog (Figure 8.3-9) controls a few key project-wide settings related to the site; below is a brief description of these settings.

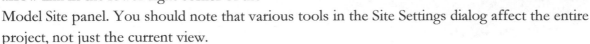

The Site Settings dialog is accessed from the arrow link in the lower-right corner of the Model Site panel. You should note that various tools in the Site Settings dialog affect the entire project, not just the current view.

Contour Line Display: This controls if the contours are displayed when the toposurface is visible (via the check box) and at what interval. If the **Intervals** is set to 1′-0″ you will see contour lines that follow the ground's surface and each line represents a vertical change of 1′-0″ from the adjacent contour line (the contour lines alone do not tell you what direction the surface slopes in a plan view; this is where contour labels are important). The **Passing Through Elevation** setting allows control over where the contour intervals start from. This is useful because architects usually base the first floor of the building on

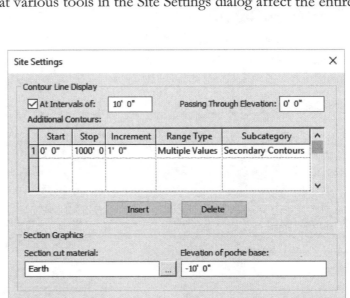

Figure 8.3-9 Site Settings dialog

elevation 0′-0″ (or 100′-0″) and the surveyors and civil engineers will use the distance above sea level (e.g., 1009.2′), so this feature allows the contours to be reconciled between the two systems. The **Additional Contours** section allows for more "contour" detail to be added within a particular area (vertically); on a very large site you might want 1′-0″ contours only at the building and 10′-0″ contours everywhere else. However, this only works if the building is in a distinctive set of vertical elevations. If the site is relatively level or an adjacent area shares the same set of contours, you will have undesirable results.

Section Graphics: This area controls how the earth appears when it is shown in section (e.g., exterior elevations, building sections, wall sections, etc.). Here is where the pattern is selected that appears in section; the pattern is selected from the project "materials" similar to the process for selecting the pattern to be displayed within a wall when viewed in section. The **Elevation of poche base** controls the depth of the pattern in section views relative to the lowest level.

Property Data: This section controls how angles and lengths are displayed or information describing property lines.

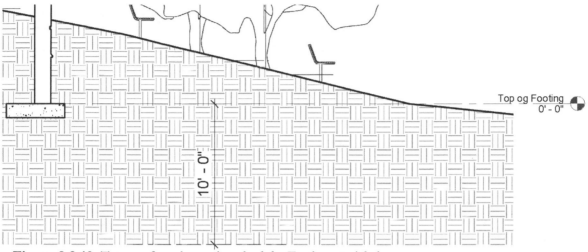

Figure 8.3-10 Toposurface shown in schedule; Earth material shown

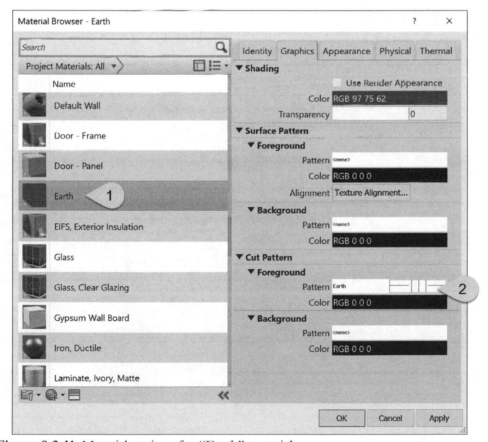

Figure 8.3-11 Material settings for "Earth" material

8.4 Modify Site panel

Tools on the Modify Site panel are used to manipulate and annotate toposurfaces.

Split Surface

This tool is similar to the Subregion tool in every way except that the result is separate surfaces that can intentionally or accidentally move apart from each other. If a split surface is selected and deleted, it results in a subtraction or a void relative to the original toposurface.

> Generally, avoid using Split Surface to remove the surface at the building location. Surveyors may be included to do this to remove all surface data in this area; however, there will be no earth shown in the section below the building.

Split Surface: **Ribbon**
Use the **Draw** tools to sketch an enclosed area.

Split Surface: **Options Bar**
Use **Chain** to quickly sketch multiple line segments. **Offset** displaces picked points while **Radius** rounds off corners as they are created.

Split Surface: **Properties**
Set the material and phase as needed.

Split Surface: **In-Canvas**
Usually work in a [site] plan view to easily position and lay out the enclosed perimeter, using trim as needed to clean up corners.

quick steps

Split Surface

1. Select a surface to edit
2. Ribbon (in sketch mode)
   - Line segment
   - *Or* Pick Lines
3. Options Bar
   - Chain
   - Offset
   - Radius
4. Properties
   - Material*
   - See warning
5. In-Canvas
   - Sketch lines to form an enclosed area.

The next four images (Figure 8.4-1 thru 4) show the process of creating a slip surface and what happens when one is deleted.

The Split Surface tool can be used to "clean up" the edges of a toposurface. For example, make the surface bigger than it needs to be and then cut back the edges so they are square to the building using this tool.

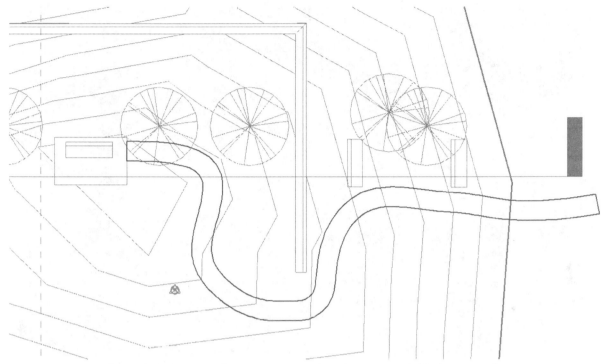

Figure 8.4-1 Split Surface command – sketching enclosed area on selected toposurface

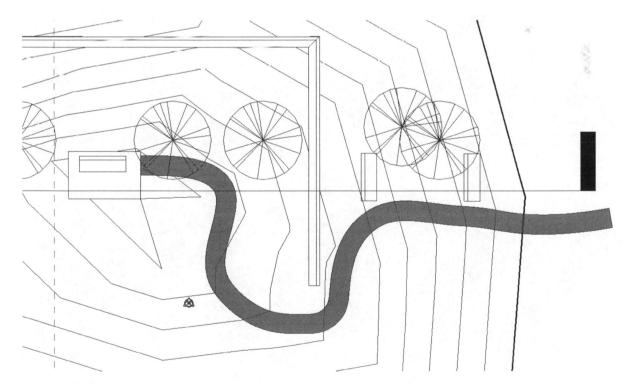

Figure 8.4-2 Split Surface command – separate new surface created and selected (plan view)

In some workflows where a site is really big you can split a surface to make editing one portion have better performance on your PC and then use merge surface to rejoin the pieces when editing is done.

Figure 8.4-3 Split Surface command – separate new surface created and selected (3D view)

Figure 8.4-4 Split Surface command – hole in surface after deleting walk surface

Merge Surfaces

After a surface has been split into one or more separate surfaces, you can merge them back together. Only two surfaces can be merged together at a time. The two surfaces to be merged must share a common edge or overlap.

If Split Surface is used to separate a surface for a parking area or walk, and one of the surfaces is edited, it becomes nearly impossible to merge the two surfaces back together. Therefore, it is often preferred to use the **Subregion** command rather than **Split Surface**.

Subregion

Allows an area to be defined within a previously drawn toposurface. The result is an area within the toposurface that can have a different material than the toposurface itself. The subregion is still part of the toposurface and will move with it when relocated. If a subregion is selected and deleted, the original surface/properties for that area are revealed.

The steps to create a Subregion are the same as creating a Split Region.

Subregion: **Ribbon**
Use the **Draw** tools to sketch an enclosed area.

Subregion: **Options Bar**
Use **Chain** to quickly sketch multiple line segments. **Offset** displaces picked points while **Radius** rounds off corners as they are created.

Subregion: **Properties**
Set the material and phase as needed.

Subregion: **In-Canvas**
Usually work in a [site] plan view to easily position and layout the enclosed perimeter, using trim as needed to clean up corners.

quick steps

Subregion

1. Ribbon (in sketch mode)
    - Line segment
    - *Or* Pick Lines
2. Options Bar
    - Chain
    - Offset
    - Radius
3. Properties
    - Material*
4. In-Canvas
    - Sketch lines to form an enclosed area.

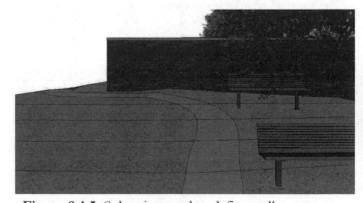

Figure 8.4-5 Subregion used to define walk

Massing & Site

Property Line

Creates property lines (in plan views only). Use this command to define property lines using simple sketch tools or by entering distance and bearings from a legal description of the property.

These lines only appear in plan views. The property line has no connection with the toposurface; a surface is not even required.

Property Line: **Ribbon**

<u>For the sketch option:</u>
Use the **Draw** tools to sketch lines. The lines do not have to form an enclosed area but cannot cross/intersect each other.

Property Line: **Options Bar**

Offset displaces picked points while **Radius** rounds off corners as they are created.

Property Line: **In-Canvas**

Select the desired input method (Figure 8.4-6) and then, based on that selection, either sketch the perimeter or enter distance and bearing in the dialog (Figure 8.4-7). FYI: An enclosed property line calculates area.

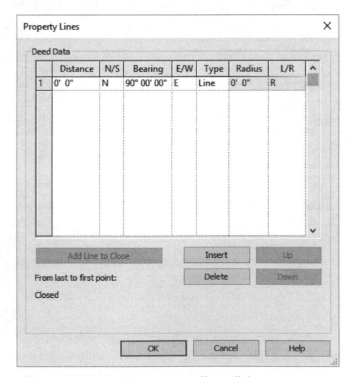

quick steps

Property Line

1. Initial creation options
   - Create by entering distances and bearings
   - Create by sketching
2. Ribbon (in sketch mode)
   - Line segment
   - *Or* Pick Lines
3. Options Bar
   - Chain
   - Offset
   - Radius
4. In-Canvas
   - Sketch lines to define property lines
   - Does not have to be enclosed
   - Cannot have intersecting lines

Property Lines

Deed Data

	Distance	N/S	Bearing	E/W	Type	Radius	L/R
1	0' 0"	N	90° 00' 00"	E	Line	0' 0"	R

Add Line to Close   Insert   Up

From last to first point: Delete   Down

Closed

OK   Cancel   Help

Figure 8.4-7 Define property lines dialog

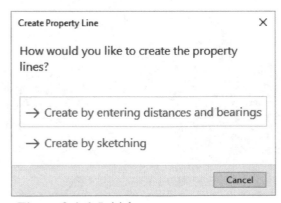

Create Property Line

How would you like to create the property lines?

→ Create by entering distances and bearings

→ Create by sketching

Cancel

Figure 8.4-6 Initial prompt

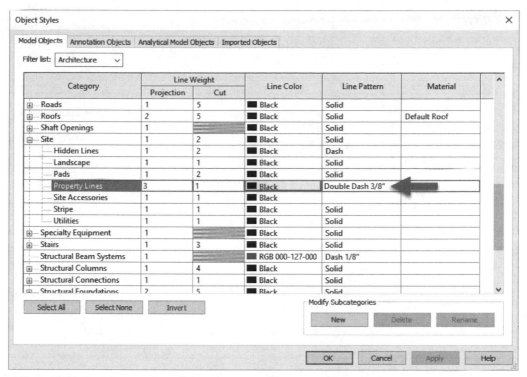

Figure 8.4-8 Object Styles dialog; Property Lines settings

Line Pattern for the Property Line is set in the Object Styles dialog - from the Manage tab on the Ribbon (Figure 8.4-8).

The image to the right shows the result of the Property Line command. Each plat of land would need to be sketched separately. Given the lines overlap at the edges, if the lines are not the same length the dashes in the two overlapping lines will not align resulting in a messy looking line. There is really no way around this.

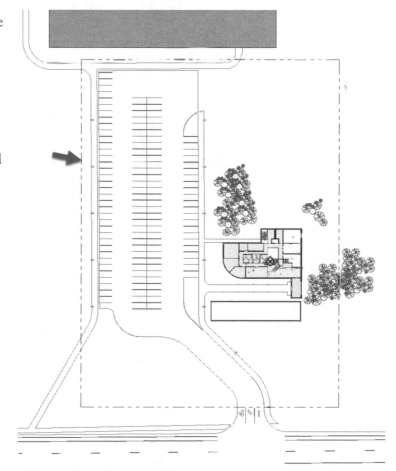

Figure 8.4-9 Property line shown in site plan

Graded Region

This tool is used to edit the grade of a toposurface that represents the existing site conditions and the designer wants to use Revit to design the new site conditions. This tool is generally only meant to be used once; when used it will copy the existing site conditions to a new Phase and set the existing site to be demolished in the New Construction Phase. The newly copied site object can then be modified for the new site conditions.

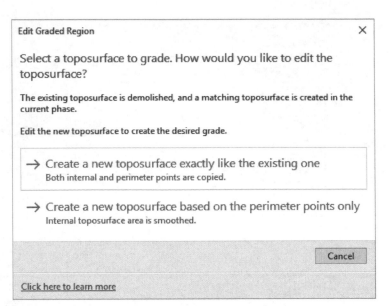

Figure 8.4-10 Graded Region - Initial prompt

When using this command, due to the relationship between the existing and new surface, Revit is able to calculate the amount of cut and fill as seen in Figure 8.4-11.

Given Revit is not fully intended to be a site design tool, this command has limited application—residential work, for example. On commercial projects, a civil engineer is typically working on a more advanced site design program like Autodesk Civil 3D.

Phasing	
Phase Created	New Construction
Phase Demolished	None
Other	
Net cut/fill	1827.39 CF
Fill	3992.31 CF
Cut	2164.93 CF

Figure 8.4-11 Cut/Fill reported

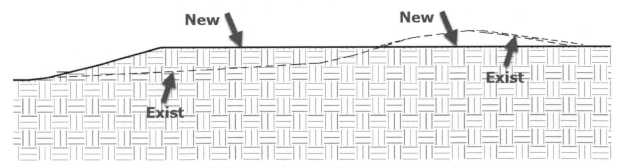

Figure 8.4-12 Section view; result is exist and new surface – two separate elements

Label Contours

This command adds an elevation label to the selected contours in a plan view (only).

Multiple elements are added for each toposurface below the defined sketch, including existing and new surfaces.

Property Line: **Options Bar**

Use **Chain** to quickly sketch multiple line segments.

Property Line: **Type Selector**

Select the desired type.

Property Line: **In-Canvas**

Pick two points to define a line perpendicular to the contours – labels are added where this line crosses each contour. The text is always oriented to be read in the uphill direction. This cannot be changed.

quick steps

Label Contours

1. Options Bar
   - Chain
2. Type Selector
   - Select type to define label style
3. In-Canvas
   - Pick two points to define an imaginary line, perpendicular to contours; the contour labels are added, one at each contour, along the imaginary line

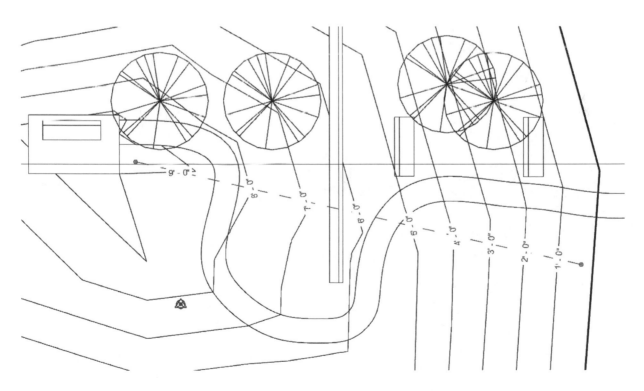

Figure 8.4-13 Label Contours added in site plan view

Massing & Site

The style of the contour label can be changed by editing the Type Properties (Figure 8.4-14). The contours are typically associated with the sea level elevation so the **Survey Base Point** needs to be set properly and the **Elevation Base** (shown below) is set to **Survey Point**. Additionally, the value is usually shown in **Decimal Feet** rather than feet and inches.

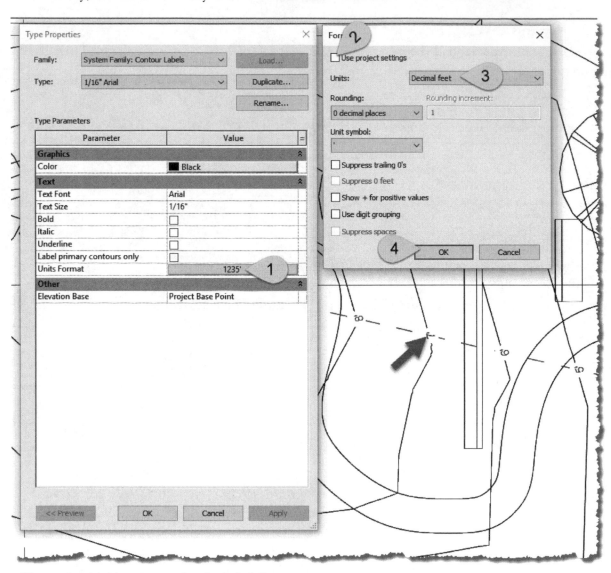

Figure 8.4-14 Type Properties for Label Contours

Chapter 9
Collaborate Tab

The Collaborate tab contains tools to facilitate working with other users in the same model. There are also special tools related to linked Revit models on this tab.

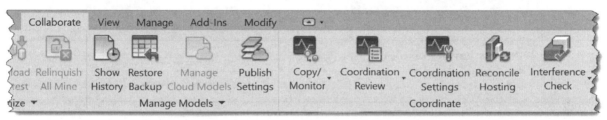

The term **Worksharing** is used to describe multiple people working in the same Revit model and is a fundamental concept of many of the commands on this tab. Thus, the following high-level overview is offered now…

Worksharing Concept

The basic concept of Worksharing can be described with the help of the graphic below. When a regular Revit project file has Worksharing enabled, it becomes a **Central File**. The Central File is stored on the server where all staff have access (normal Read/Write access). **Once a Central File is created, it is typically never opened directly again.** Individual users work in what is called a **Local File**, which is a copy of the Central File, usually saved on the local C drive of the computer they are working on. When modifications are made to the Local File, the Central File is NOT automatically updated. However, the modified elements ARE checked out in that user's name—other users cannot modify those elements until they are checked back in. When a Local File is **Synchronized with Central**, the delta changes are saved; that is, only changes the user made are 'pushed' to the Central File, and then only changes found in the Central File, since the last Sync, and 'pulled' down (thus, the two-way arrow). All elements checked out are typically relinquished during a Sync with Central. There is no technical limit to the number of users, though when there are more than 10 users the project can become sluggish.

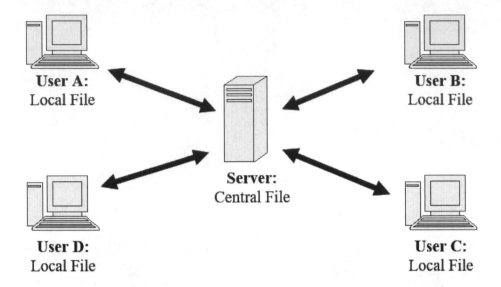

The author's blog, **BIM Chapters**, has a number of popular posts on Worksharing:

- **Opening a Revit Project Detached From Central; What You Need To Know**
  https://bimchapters.blogspot.com/2017/10/opening-revit-project-detached-from.html
- **Create a New Local File Daily**
  https://bimchapters.blogspot.com/2017/10/create-new-local-file-daily.html
- **Avoid Using Worksets to Control Visibility**
  https://bimchapters.blogspot.com/2017/10/avoid-using-worksets-to-control.html

9.1 Communicate panel

The Communicate panel has tools to facilitate working with other design team members who are using Revit. The first tool works with team members who might work for a different company and can be anywhere in the world. The second is designed to work with "internal" team members; these are usually staff working for the same company but may still be located in different offices.

Editing
Requests
Communicate

Editing Requests

The Editing Requests command is used to show any requests to edit elements that are checked out by someone other than the person trying to edit an element in a worksharing project.

Each request can be expanded to see exactly which elements are part of the request.

> Sometimes it is easier to call the person and ask them to Synchronize with Central so you know they get the request. The automatic prompt only appears for a short time and then disappears. So, it is easy to miss if not currently looking at the computer screen.

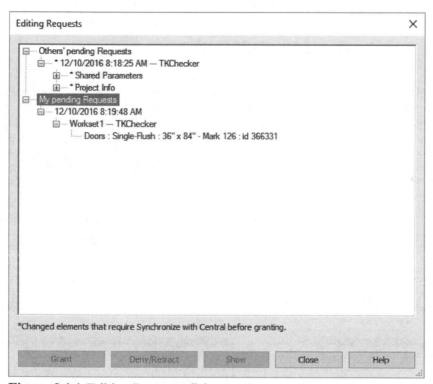

Figure 9.1-1 Editing Requests dialog

Grant

Requests can be selected and then granted. The request is then removed from this list.

Deny/Retract

A request can be denied. For example, someone wants rights to adjust a toilet room layout, but someone else is already working on it. The person currently doing the work would click **Deny**. If you placed a request that has not been granted and is no longer needed, click **Retract**.

Show

Select an individual element in the list and click **Show** to open a view and zoom in on it.

The following is not directly part of the Editing Requests dialog but is offered as a way to better understand the series of events that led up to requests being listed there.

If user TKChecker tries to move an element, but it has already been modified in some way by user DJStine, then the 'can't edit element' prompt appears (Figure 9.1-2). If TKChecker really needs to edit this element, she will click **Place Request**. A prompt indicating the request has been placed and some additional information is presented (Figure 9.1-3).

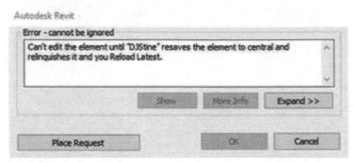

Figure 9.1-2 Worksharing prompt – can't edit

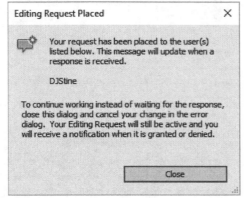

Figure 9.1-3 Request placed

If user DJStine is at the computer, he may see the temporary alert shown in Figure 9.1-4. In this case, the only way to actually 'grant' the request is to **Synchronize with Central** to update the central file and relinquish rights; thus, the Grant button is grayed out. Once the request is granted, TKChecker will receive a 'request granted' prompt (Figure 9.1-5). Now she can proceed to edit the elements under consideration.

Figure 9.1-4 Request received

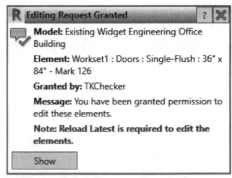

Figure 9.1-5 Request granted

9.2 Manage Collaboration panel

This panel contains commands to convert a regular, one-person project to a worksharing-based project (aka Central file) which allows multiple people to work on the same project at the same time. Tools to manage worksets are also found here; worksets are what Revit uses to segregate elements into separately editable containers.

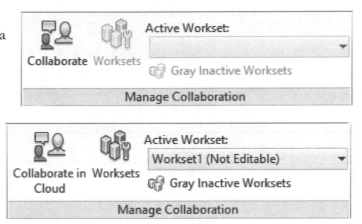

Notice, in the two images to the right, that this panel actually changes once a project is converted to a worksharing project. The **Collaborate** button is changed to **Collaborate in Cloud**.

Collaborate

Use this command to convert a project to a Central file. The initial prompt (Figure 9.2-1) offers two paths to take:

- **Collaborate within your network**
- **Collaborate using the cloud**

The first option is the most common option as the Central file is saved on a server in your office or school.

The second option works with **Collaboration for Revit** (C4R) which is a separate Cloud-based service offered by Autodesk. There is a per user charge for this service. This option does not work with any other cloud-based options.

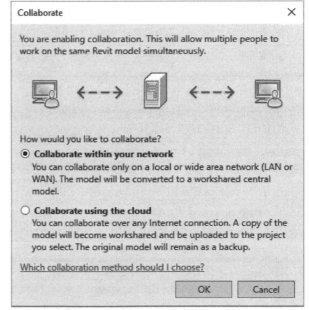

Figure 9.2-1 Convert to worksharing

Before using this command, make sure the file is in the **correct location** on a network where everyone who will be working in the file has access (i.e., rights). Everyone should have the same mapping as well, meaning the same **drive letter** and folder structure.

Collaborate

Once a Central file is created, on a local network, the first time the file is saved will result in this prompt (Figure 9.2-2). Once the file is saved, it should be closed and, in general, never opened directly again. Rather, a 'local file' is used going forward via the normal Revit **Open** command in the Application Menu.

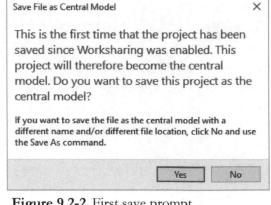

Figure 9.2-2 First save prompt

Once the Revit project file has been converted to a Central file, two folders will appear in the same location as the RVT project file.

Never move, rename or delete these folders. In fact, they should never be opened. They contain information about the various users and what they have checked out for editing.

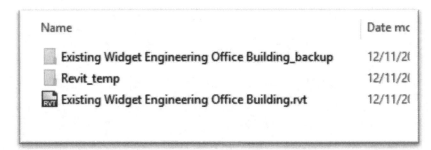

Figure 9.2-3 Folders created next to central file

It is NOT possible to save everyone's work back to a single file if this information is deleted or corrupted in any way.

> **Never copy a Central file**; it will become a Local file and connect back to the Central file location you just copied the file from. Use Detach from Central!

Collaborate in the Cloud

Access to this feature is either from the initial Collaborate option (Figure 9.1-2) or from the Collaborate in the Cloud option once a Central file has been created—as seen in the two panel images on the previous page.

> Using the collaboration in the cloud (C4R) is only possible with a separate per person service through Autodesk. This feature does not work with any other cloud-based services.

As a separate external service, this command will not be covered any further.

Figure 9.2-4 First C4R prompt

Worksets

Worksets are used to segregate the Revit database in order to control how things are loaded into memory and who can edit views and elements. They also have the ability to control element visibility, but generally should not be used just for this purpose.

Once a project has worksharing enabled, aka has a Central file, then worksets exist. The initial result is seen in Figure 9.2-5 below: two worksets—**Shared Levels and Grids** and **Workset1**. All Levels and Grids in the project were automatically placed on the Shared Levels and Grid workset, but future Levels and Grids need to be manually placed here.

> Many projects can be entirely created on Workset1. Revit checks elements out on a per element and view basis. Large or complex projects should have multiple worksets which can be turned off while opening a project—thus limiting system RAM being used.

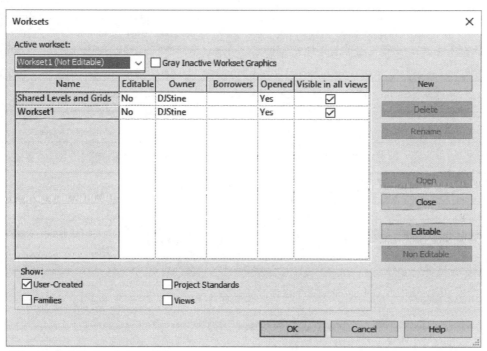

Figure 9.2-5 Worksets dialog

Active workset: see the next command covered in this section.

Use the **New**, **Delete** and **Rename** buttons to manage worksets.

The **Open** and **Close** buttons control the visibility but also if elements are loaded into memory. Only loading the worksets needed on a large project can have a substantial positive impact on system performance. For each workset, **Yes** means opened and **No** means closed.

Collaborate

The **Editable** and **Non Editable** buttons determine whether the entire workset is checked out (editable) by one user or not. This can be a little confusing, but in general the Editable column should always say **No**. This means several users can Borrow individual elements within a workset. If an entire workset is Editable for a single user, then no one else can edit anything in that workset.

The **Show** section controls which workset categories are listed. The default is to just show the User-Created worksets. The other categories are largely automatic in how they are created and managed.

Visible in all views is a project-wide toggle to control the visibility of all elements within a project.

> The visibility of elements, based on workset controls, has no impact on schedules and totals.

In addition to the project-wide or global control of visibility in the Worksets dialog, each view's **Visibility/Graphic Overrides** dialog also has the ability to control workset visibility (Figure 9.2-6). The **Use Global Setting** is tied to the setting in the Worksets dialog.

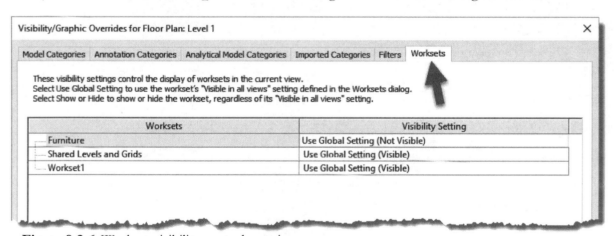

Figure 9.2-6 Workset visibility control per view

In the above example, the furniture is not part of the contract. To hide all furniture in the project, so it does not appear on the construction documents, the custom 'Furniture' workset is set to not be visible in all views in the Worksets dialog. In renderings and 'for reference only' furniture plans, this setting can be overridden in a given view.

The problem with controlling visibility by workset is that elements must be created, or placed, on the correct workset for things to work as intended. This manual effort often leads to elements showing up, or not showing up, in various views. Often, using well developed **View Templates** which control visibility by predetermined categories and sub-categories is more predictable and less prone to user error.

Active Workset

The main purpose of the Active Workset list is to know what the active, or current, workset is and change it when needed. The active workset is the workset all new elements are created on. This includes elements that are simply copied or copy/pasted; that is, even if the original item is on another workset, when copied, the new element ends up on the active workset.

Depending on a company's standards and template, and/or depending on the size of the project the active workset may never need to be changed. However, on larger projects this may change often, and needs to be watched closely to ensure elements are being created on the correct workset or they may not appear properly in the required views based on view filters, view templates and the 'select worksets' option in the Open dialog.

> Many BIM Managers encourage their designers to create a new Local file each day. In this case, the first thing done should be to check the active workset. Otherwise you may be modeling for a few hours and realize you are on the wrong workset and will need to go back and select elements and change their workset in the Properties palette (Fig. 9.2-7).

Figure 9.2-7 Workset setting

The **Active Workset** dialog lists the active/current workset (Figure 9.2-9). Click the down-arrow to select a different workset from the available list. Because this is an important setting in worksharing models, this command is also located more prominently on the status bar (Figure 9.2-8). Each workset typically has **(Not Editable)** next to it, which means the entire workset is not checked out. This is confusing, as it actually means the elements in the workset <u>are editable</u> individually.

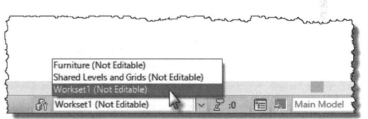

Figure 9.2-8 Active worksets on status bar

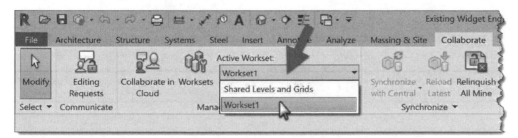

Figure 9.2-9 Active worksets on Collaborate tab on Ribbon

Collaborate

Gray Inactive Worksets

This command is a toggle to control a graphic override option for **Inactive Worksets**. When this command is selected, and is toggled on, all worksets except the **Active Workset** are grayed out in the canvas.

Clicking this command again will take the filter off.

Using this feature helps highlight a problem when something is copied and modeled and is placed on the wrong workset.

Additional Worksharing Display Settings

There are additional display override options located in the **Display Control Bar** at the bottom left of each view (Figure 9.2-10). Each of these options "highlight" elements based on specific conditions related to worksets. For example, **Owners** highlights elements in the model based on who currently owns, i.e., has them checked out, in the model. The highlight colors can be controlled in the **Worksharing Display Settings** dialog (Figure 9.2-11).

Figure 9.2-10 Worksharing display controls

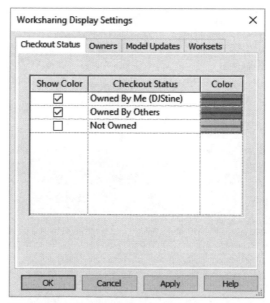

Figure 9.2-11 Worksharing display settings

9.3 Synchronize panel

Use the commands on the Synchronize panel to keep a local file up to date with the Central file. The extended panel area has a command that provides access to **Revit Server** settings.

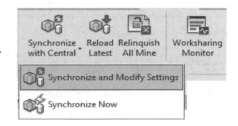

Synchronize with Central

This important command is used to upload changes made in your local file to the central file—and, at the same time, download any new changes found (made by other users). For any number of people working on the same project, only one can use this command at a time. This author recommends the **Synchronize with Central** command is used rather than the **Synchronize Now** command (which is covered next).

This command offers several options in addition to just saving to central. Most notably is the option to relinquish elements checked out (Figure 9.3-1). When elements are checked out no one else can edit them—even if the central file is up to date with your changes. Thus, using **Synchronize with Central** and making sure **all checkable boxes are checked** (in the 'relinquish' area) will save to central AND relinquish control of those new/modified elements.

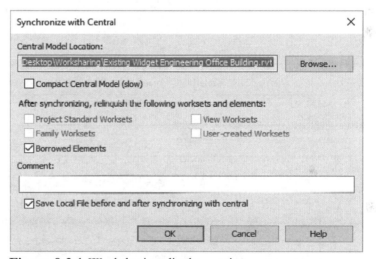

Figure 9.3-1 Worksharing display settings

On large projects, or projects nearing completion, one person should be assigned the task of checking the **Compact Central Model (slow)** option once a week (do this for each active Revit model on the project). This process cleans up the database, thus improving performance and stability.

> Another reason for each user to create a new local file each day is this is the only way the local file benefits from the compacted central file enhancements.

The **Save Local File before and after synchronizing with central** is checked by default. In some rare cases, Revit can crash during a synchronize with central, so having saved the local file first avoids losing work—this happens more often on 32bit computers as they tend to run out of

Collaborate

available RAM sooner. Also, saving the local file again, after the sync with central, will ensure the local file fully matches the current version of the central file.

The **Central Model Location** shows where Revit is looking for the central file.

Autodesk provides an add-in tool, for subscription customers, called **Worksharing Monitor**. When Worksharing Monitor is running, it lists the users working on the same Central file. Additionally, it also indicates when someone is synchronizing with central. You should never initiate a sync with central while someone else is doing it as this will slow down the process for both people trying to save.

Synchronize Now

This command is intended to be a faster, streamlined version of the **Synchronize with Central** command. When this command is selected, it uses the current settings in the **Synchronize with Central** dialog, but no dialog is shown. Thus, a single click without options or a final OK button uploads your changes and downloads anything new in the central file.

The one drawback to using the Synchronize Now command is that some categories in the "to be relinquished" section may not be checked. This means the file can be fully up-to-date and closed but still have items checked out. This can then be an obstacle for other users who need to modify the model and you are not available to open the model and use the regular, or full, **Synchronize with Central** command.

> Elements are checked out using the username specified in **Application Menu →**
> **Options** dialog. If you log into **Autodesk A360** to access Cloud services, the username
> is automatically changed to match your A360 username. If you are currently in a
> worksharing model you cannot log into A360 because that would cause problems with
> the elements already checked out. The model needs to be closed, then you log into A360,
> and then create a new local file.

Reload Latest

This command downloads any new changes to the central model, from other users, to your local file. None of your changes are pushed to the central file. Use this occasionally to quickly keep your model contextually more accurate when there are several other designers working on the same model as this command is faster than a full **Synchronize with Central**.

> Neither this command, nor the full **Synchronize with Central**, will update a linked file.
> These files, RVT, DWG, etc., are only updated when the Revit project is initially opened
> or when **Reload** is selected from within the **Manage Links** dialog.

Relinquish All Mine

In the case of a workset being manually checked out in its entirety, or elements being left checked out due to the **Synchronize Now** command being used, use this command to quickly relinquish ownership of unchanged elements. This scenario is somewhat rare, so users often just do a full **Synchronize with Central** to check things back in.

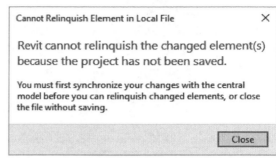

Figure 9.3-2 Relinquish All Mine prompt

This command cannot be used if there are un-synchronized changes in the local files (Figure 9.3-2).

Worksharing Monitor

The worksharing monitor tool was previously an add-in tool that now appears in the ribbon. The worksharing monitor runs in a separate process from Revit and automates or helps with some of the activities you need to keep track of while working in a workshared file.

The monitor is broken into 3 sections:

- **Central File Access** – See who is working on the project on your team and what they are currently doing in the project. You can see when team members are synchronizing or reloading latest updates from the central model, or if they have a local or central copy of the model open for editing.
- **Editing Requests** – See all of your editing requests, granted, denied, and retracted, in this pane.
- **Notifications** – Any notifications from the model that impact worksharing in Revit is listed here.

Use the options settings for the worksharing monitor to control the information displayed in each of the 3 sections of the monitor.

Manage Connection to a Revit Server Accelerator

This command is used to configure Revit to use an external Autodesk product called **Revit Server**—which is a tool used within the same company to connect users/designers in multiple offices, i.e., geographical locations.

> For users/designers from different companies, **Collaboration for Revit** (C4R) is the service used, for a fee, to allow everyone to work in the same model or with live links.

The Revit Server feature requires several things be set up and configured outside of Revit. Once that has been done, each user opens this dialog and enters the provided server name and IP address. Once connected, the accelerator name appears, and users are able to work on projects loaded into Revit Server.

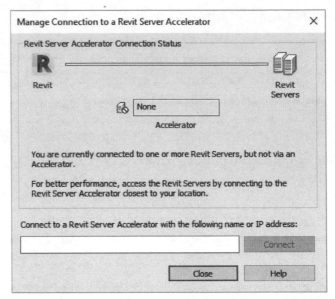

Figure 9.3-3 Revit server connection dialog

9.4 Manage Models panel

This panel contains commands used to manage the central file and settings related to publishing sheets.

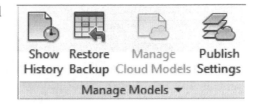

Show History

Use the show history command to browse to and select a central file to see who has worked in the model recently and how often people are synchronizing with central. This command does not work directly on the current file; you always have to browse to a central file, select it and then click Open. The Comments column contains any information entered in the comments field while using the **Synchronize with Central** command as seen in the image below (Figure 9.4-1). Entering a brief description of the change contained in the current sync can help with troubleshooting later; see **Restore Backup** command next.

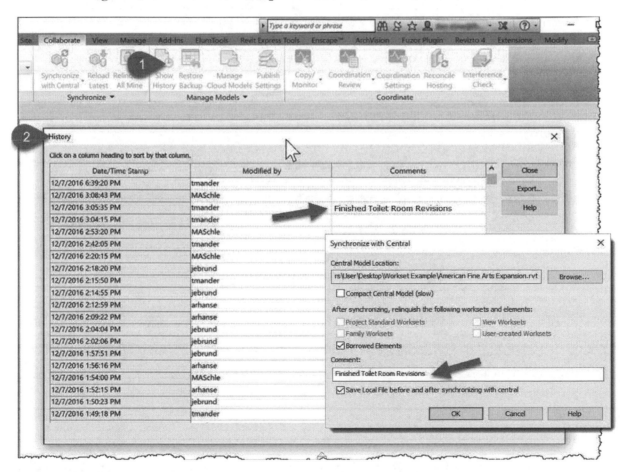

Figure 9.4-1 Show history dialog and Synchronize with Central dialog

Restore Backup

When a problem arises or some changes need to be undone in a central file, use the **Restore Backup** command. Using this command, the current central file can have several edits "rolled back" relative to when each user synchronizes with central.

When this command is started, browse to the **backup folder** with the same name as the central file and open it (you will not see any files in the 'browse for folder' dialog). Be sure to browse to the central file folder on the server and not the local files backup folder (which is the default location this command starts with).

Each user's saves are listed as seen in Figure 9.4-2. In addition to the date, comments entered in the **Synchronize with Central** dialog are provided for reference.

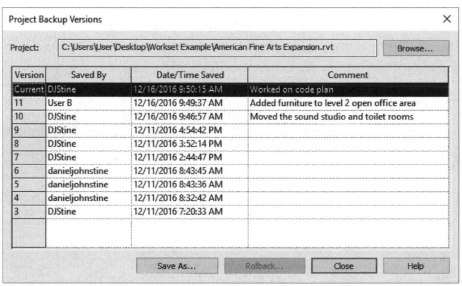

Figure 9.4-2 Project Backup Versions dialog

Save As

Use this option to save a separate file that represents a specific point in time per the selected Version (as listed in the first column). This is preferred over the **Rollback** option because the model can be opened separately and reviewed before committing to changing the central file.

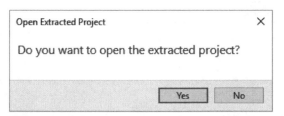

Figure 9.4-3 Open Extracted Project prompt

After the file has been saved, a prompt to open the file is given (Figure 9.4-3). Typically, click **No** and then use the **Open** command so the file can be opened **Detached from Central**. Clicking **Yes** to open the extracted project puts you in a local file which is connected to the old central file.

If the extracted project is good, after opening detached and reviewing it, simply save it over the main central model. At this point, close the central file and have everyone working on the project make a new local file. Old local files cannot, in any way, be reconnected to a new central file.

> Before replacing a central file, be sure everyone has closed their local file. Once the central file has been replaced, everyone must create a new local file.

Rollback...

The Rollback option changes the selected file whether it is a local or central file. If rolling back a local file it will become incompatible with the central file. It is therefore better to select the central file directly and then make a new local file.

> If the central file becomes corrupt, another option is to find the user who has the most up-to-date local file that is not corrupted and turn their file into the new central file. Simply open their central file detached from central and save it over the corrupt central file. At this point everyone working on the project needs to create a new local file. The other disciplines using this model as a link do not need to create new local files.

In addition to these Autodesk opportunities it is extremely important to have a backup strategy in place. Servers often run tape backups nightly, and data can also be synced in the Cloud.

Manage Cloud Models

Use this command when working with Collaboration for Revit (C4R) to manage specific settings related to Revit projects saved in the cloud.

> The term Cloud here is only in reference to Autodesk's service. Do not try to use a worksharing Revit project file in cloud storage such as Dropbox or Box as it will not work. There are options other than Autodesk, like CSG BIMCloud service, which is based on Riverbed's Steelhead accelerator hardware; however, firms and their users do not need any special hardware. http://www.cs-grp.net/about-csg/csg-bimcloud/

Publish Settings

This command is used to save a group of selected views and/or sheets, which can be used later to streamline printing and exporting – especially when using BIM 360 Design.

The next image shows the **Publish Settings** dialog with the Preview area expanded (Figure 9.4-4).

Collaborate

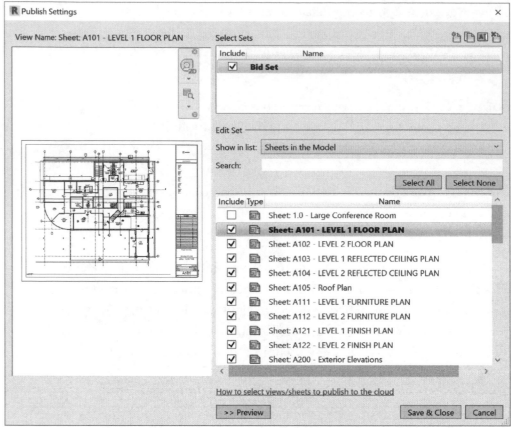

Figure 9.4-4 Publish Settings dialog

The **Set** list shows saved groupings. Add, rename, duplicate or delete items in this list using the icons below.

The **Show** in list is an important setting as it controls what views and sheets appear in the list view below. To see everything, select **All views and sheets in the Model** (Figure 9.4-5). To select only sheets, first click the **Select None** button and then select **Sheets in the Model**.

> If a view is selected before switching the filter to just show sheets, the view will still be selected and part of the group even though it is not visible when the group was created.

Figure 9.4-5 Publish Settings dialog; show in list

The **Search** option filters the list of visible Views or Sheets. As in the example shown in Figure 9.4-6 the word "finish" was entered and the current visible list was filtered to show two sheets containing that text.

The **checkboxes** in the list of views/sheets are used to determine which ones are in the selected set. Click the column header/title to sort by **Name**, view **Type**, or **Include** (in current set).

Click the **Preview** button to see a preview of the selected view/sheet in the list.

Figure 9.4-6 Publish Settings dialog; search

This saved list (Set) appears in the **Print** dialog (Figure 9.4-7).

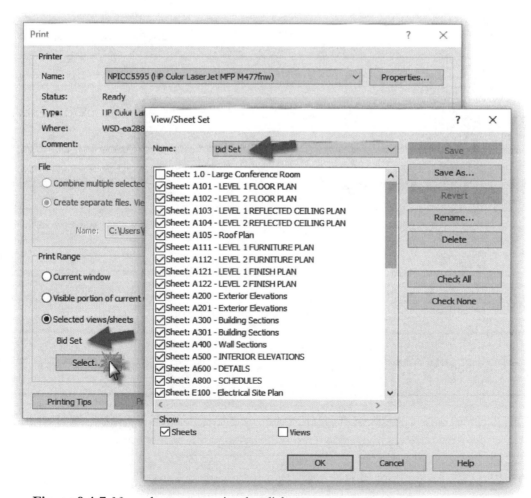

Figure 9.4-7 Named sets appear in plot dialog

Collaborate

The named Set also appears in the **DWG export** dialog (Figure 9.4-8). In fact, this dialog looks very similar to the Publish Settings dialog (Figure 9.4-4).

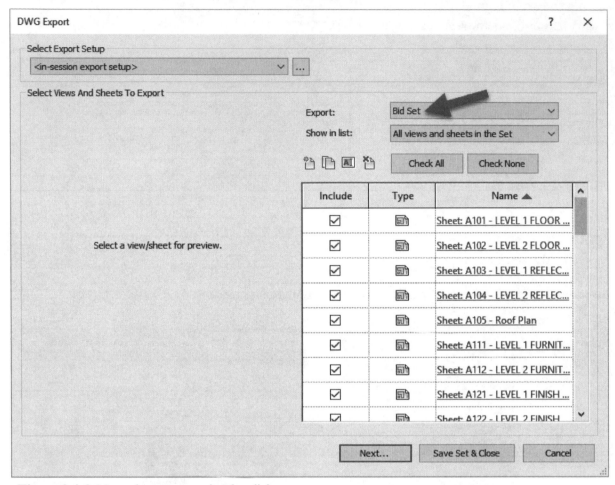

Figure 9.4-8 Named sets appear in plot dialog

Repair Central Model

This command, added in Revit 2021.1, is used to automate the process of replacing a good working local file with a problematic central file. Revit will provide a prompt indicating the central file has corruption issues and that this Repair Central Model command should be used.

As with any other steps involving the replacement of a central file, everyone working on the project needs to close their local file. When this process is done everyone then needs to make a new local file.

Once the central file has been replaced, the backup folder no longer contains any backup history prior to when the central file was replaced.

9.5 Coordinate panel

All the commands on this panel provide tools to collaborate with linked in Revit models. Additionally, the Interference Check command is used to check

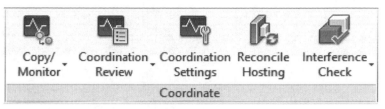

for clashes between the current model and a selected link. However, it can also do clash detection between categories in the same model (no link required).

Copy/Monitor → Use Current Project

Use this command to make sure related elements stay in a relative position to each other within the same project. For example, in Figure 9.5-1 below, the finished floor level (13'-4") and the structural bearing level (12'-8") are 8" apart. Assuming this has been determined to be the specific spacing due to the materials used or required fire rating, if either of the levels are moved a warning will appear indicating a change to two monitored elements.

The steps to use this command are the same as the next command: Copy/Monitor – Select Link. See that command for more information on Copy/Monitor.

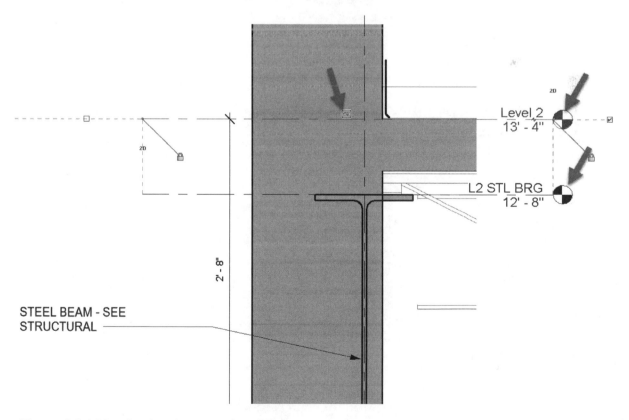

Figure 9.5-1 Two levels using Copy/Monitor to 'monitor' relative proximity of two levels

Collaborate

Copy/Monitor → Select Link

Use the Copy/Monitor tool to maintain the relative position of a pair of elements, one in the current Revit model and another in a linked Revit model. A common example is Structural, Mechanical, Electrical Copy/Monitor the Architectural **Levels**; meaning the architects control the elevation, for floor-to-floor height, of the levels. Another example is Architectural, Mechanical and Electrical Copy/Monitor Structural's **Grids**; meaning structural has the primary control over the location of grids for the project.

The **Copy** part of the tool will make a copy of a selected element, from a link, in the current model. For example, a grid or level in a linked model can be copied into the current model in the same exact position. The **Monitor** part of the tool is used to make sure two elements do not move apart from each other.

It is possible to just show the grids or levels directly from the link. However, for levels the current model must have its own levels to create views and elements. Also, when a Grid is Copy/Monitored, they cannot be laterally, but the length of the grid line can be adjusted. This cannot be done if the grid is just coming from the link.

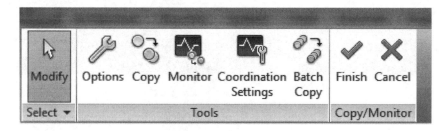

Copy Grids: basic steps…

- Open a floor **plan view** with linked grids visible
- Start the **Copy/Monitor** command
- Select the **linked model**
- Select **Copy** from the Copy/Monitor contextual tab (shown above)
- Select an element
    - o   Use the **Multiple** option on the Options Bar if needed
- Click **Finish**

Monitor Levels: basic steps…

- In an elevation or Section view, manually **create levels** (if needed)
- **Align** levels with levels from linked model (do not 'lock' alignment)
- Start the **Copy/Monitor** command
- Select the **linked model**
- Select **Monitor** from the Copy/Monitor contextual tab (shown above)
- Select a Level in the current Revit model

- Select the corresponding Level in the linked model
- Click **Finish**

For the Levels example just outlined, it is easier to use the levels already in your template as they are associated with views and may have specific settings applied such as a **View Template**. Using the Copy part of Copy/Monitor will create a new level in the current project, but it will not create views for these levels in the Project Browser. Also, using Monitor will help ensure there are never two levels at the same elevation. This is bad practice as elements are associated with levels and if a level is deleted all those elements are automatically deleted without any warning message from Revit.

Options

The Options button opens the Copy/Monitor Options dialog (Figure 9.5-2). The five tags correspond to the five categories supported by the basic Copy/Monitor command:

- Levels
- Grids
- Columns
- Walls
- Floors

This dialog controls exactly what happens when elements are copied. Notice, for grids, a prefix can be added to the grid number/letter to help distinguish it from grids in another link. Plus, grids cannot have the same designation, so if the link has a Grid A and the local file has a Grid A, adding a prefix will avoid errors.

Coordination Settings

Coordination Settings works in conjunction with the **Batch Copy** command (covered next) to perform automated Copy/Monitor functions on MEP elements like lighting fixtures and air terminals.

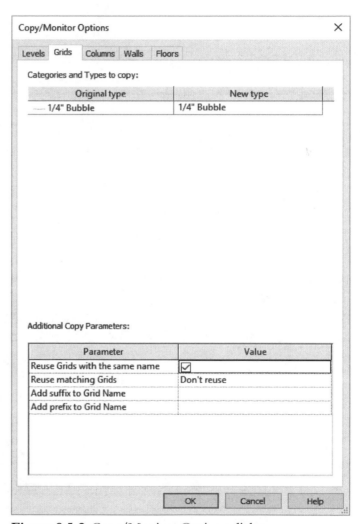

Figure 9.5-2 Copy/Monitor Options dialog

This dialog (Figure 9.5-3) controls how elements in the various supported categories are automatically copied into the current project when the **Batch Copy** option is selected.

Collaborate

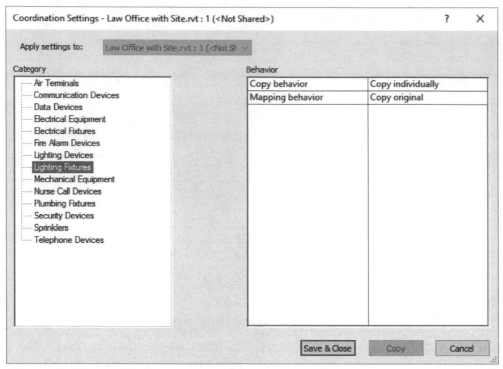

Figure 9.5-3 Copy/Monitor – Coordination Settings dialog

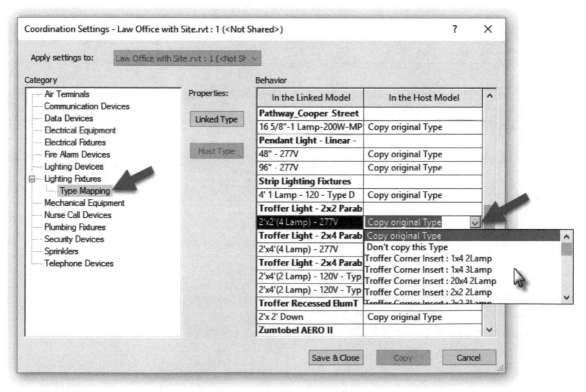

Figure 9.5-4 Copy/Monitor – Coordination Settings dialog

Figure 9.5-4 shows how a lighting fixture from a linked model can be replaced with a family in the current Revit model.

> Many of the options here to copy model elements are not typically necessary as the discipline responsible for the element created and controls it in their model. Other disciplines can often just use the linked version in their model. FYI: The light source from a linked light fixture will be used in a rendering in the host model – the only problem is linked families cannot cut anything (like ceilings) so if the light source is above the ceiling that will be a problem.

Be sure to set any categories to "ignore" if those elements are not needed directly in the current model as this will make the file size larger and potentially have an impact on performance.

Batch Copy

When the **Batch Copy** command is selected this prompt appears (Figure 9.5-5). The first option opens the Coordination Settings dialog (just covered on the previous pages). The second option uses the current settings and starts the process.

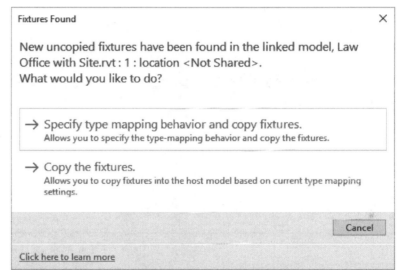

Figure 9.5-5 Copy/Monitor - Batch Copy prompt

When the copy behavior is set to **Allow batch copy** (Figure 9.5-6) elements are copied when the **Batch Copy** command is selected and whenever the linked model is reloaded—this includes anytime the host model is opened as linked Revit models are not saved in host model. This feature allows the host model to stay up to date with minimal user intervention. However, when new family types are added, it may be necessary to adjust the mapping in the **Coordination Settings** dialog.

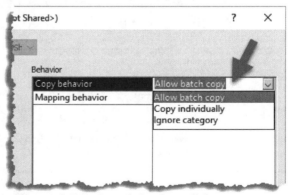

Figure 9.5-6 Coord. Settings dialog

Stop Monitoring

When an element no longer needs to be associated with another element in a linked model, select it and then click **Stop Monitoring** on the Ribbon.

Additional Information; Hiding Linked Elements

Once an element has been Copy/Monitored, it is desirable to hide the linked version directly below it. In a model view, type **VV** to open the Visibility/Graphic Overrides dialog (Figure 9.5-7). Select the **Revit Links** tab (which only exists if the current Revit project has a link) and click the **By Host View** button next to the link containing the elements you wish to hide. On the **Basics** tab (not shown in the image below) select **Custom**. On the **Annotation Categories** tab select **Custom** from the drop-down list and then uncheck **Levels**. Close both open dialog boxes.

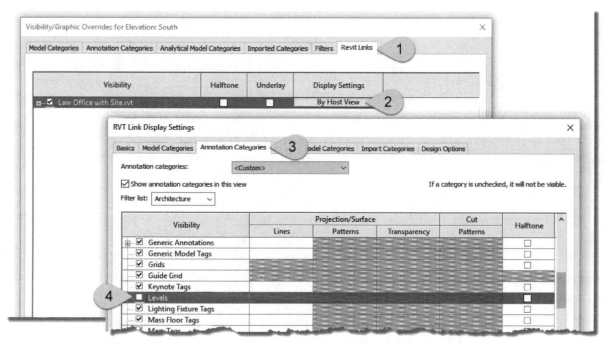

Figure 9.5-7 Hiding levels from linked model in current view

Coordination Review → Use Current Project

Use this command to identify any issues with elements which have been set up with the Copy/Monitor tool in the current project. See the next command for additional information.

Coordination Review → Select Link

When monitored elements are moved, either in the host model or the link, a new item shows up in the Coordination Review dialog.

Select the command and then select a Revit link in a model view (i.e. floor plan, ceiling plan, section, elevation, 3d view). The **Coordination Review** dialog appears (Figure 9.5-8).

There are four action options (see Revit help for additional actions possible):

- **Do nothing**: this postpones the issue to be addressed later, or by someone else
- **Reject**: indicates the change to the link is wrong; you must tell the team working on the linked model as this option will not do it. Once they make the change this line disappears.
- **Accept difference**: this resets the current relationship to be correct
- **Modify Grid "E"**: this will move, or delete, the element in the host model to match the link

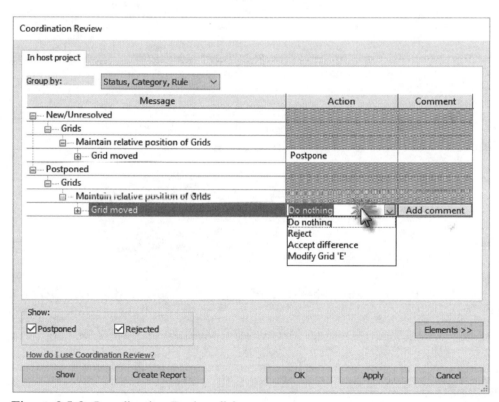

Figure 9.5-8 Coordination Review dialog

When a model is being opened, if monitored elements have changes, a prompt is offered (Figure 9.5-9). This does not provide access to the Coordination Review dialog, just notifies you to open it manually and review issues.

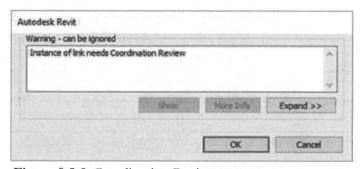

Figure 9.5-9 Coordination Review prompt at open

Collaborate

The Coordination Review dialog can be filtered to not show **Postponed** or **Rejected** items. This can be helpful on large projects when looking for new items. The list of issues to review can be exported using the **Create Report** button. This will create an HTML file which can be opened in any web browser (Figure 9.5-10).

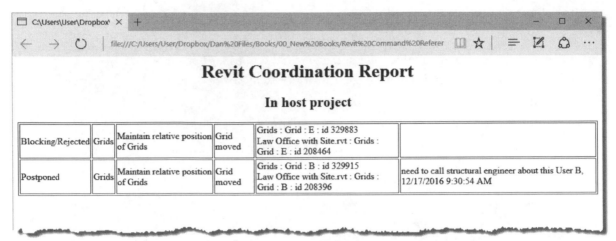

Figure 9.5-10 Coordination Review output using Generate Report command

When a linked model is selected directly (no command active) then the Ribbon will offer a **Coordination Review** button (see image below) if there are items in the list.

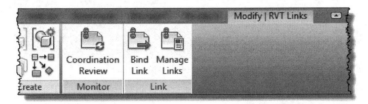

> The Copy/Monitor command has no effect on the linked file. With few exceptions, Revit cannot make changes to a linked Revit model. Thus, the comments column, for example, is just a place for "notes to self" for future reference or follow up.

Coordination Settings

For information on this command, turn back a few pages to the same command covered within the **Copy/Monitor; Select Link** command.

Reconcile Hosting

When elements from a linked model host a face-based family, or are tagged, the **Reconcile Hosting** palette can help resolve issues when elements are deleted in the linked Revit model. In the image below (Figure 9.5-11), there are two light fixtures (which are base-based) and one wall type which is orphaned—meaning the wall it was associated with has been deleted in the linked model.

Select an item in the list and then click the **Show** button to find it in the project. Once found, select the element in the project and click the **Pick New** (work plane) button on the Ribbon to select a new host. The element will then disappear from the Reconcile Hosting dialog.

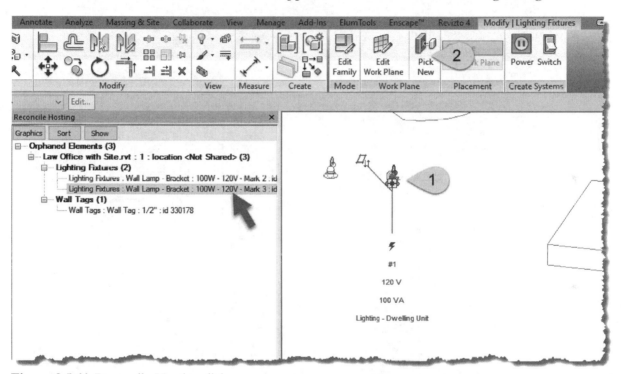

Figure 9.5-11 Reconcile Hosting dialog

Clicking the **Graphics** button provides a way to change the apply graphic overrides to the orphaned elements within the model (Figure 9.5-12). This makes them easier to find visually.

> It is a good idea to check this palette before printing a formal set of drawings to make sure there are no problems.

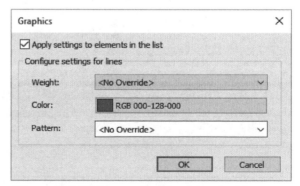

Figure 9.5-12 Reconcile Hosting graphics

Collaborate

Interference Check → Run Interference Check

This last command on the Coordinate panel does not have anything to do with tracking or hosting to elements in linked models. Rather, this command checks for interferences, aka clashes between elements based on category. This could just be elements in the current model without any links present. However, it does work with linked models as well. This command is an important part of the high value attributed to BIM, in that fewer clashes in the design mean less problems during construction.

Clicking **Run Interference Check** opens the Interference Check dialog (Figure 9.5-13).

Categories from:

This has three options:

- Current Project
- Linked Project
- Selected

If elements are selected when the command is started, the 'selected' option is the default.

There will be an item, in the list, for each Revit link in the project.

Selected categories on the left are checked against the categories selected on the right.

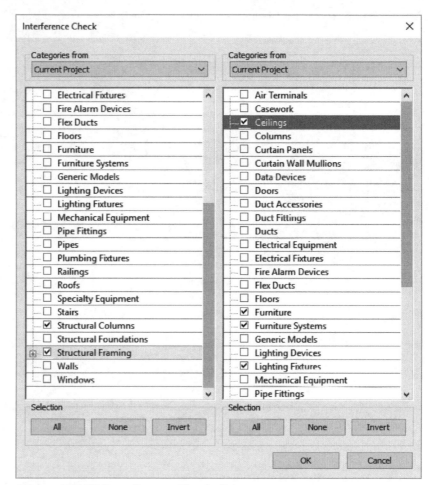

Figure 9.5-13 Interference Check dialog

There are a few combinations for **Categories From**… the left can be Current Project and the right set to a linked model. Or both sides can be Current Project. However, both sides cannot be set to a linked model—to perform that check a different model would need to be opened.

Categories

Check categories on the left to check against categories on the right.

It is best to break this task down into more manageable chunks; for example, furniture and structural columns and framing. Another example is ducts and ceilings. However, checking walls against ducts is not usually done as ducts often go through walls and no hole is added, so there would be many errors listed.

> This command is a good "minimum" tool used to check for clashes in lieu of having, or using, a more sophisticated application. For example, **Autodesk Navisworks** - *Manage* is a separate, powerful program used to find clashes. Note: **Navisworks – *Simulate*** does not do clash detection.

Consider creating a series of 3D views with just the categories to be checked. When a row is selected in the Interference Report dialog, that item highlights in the project.

Interference Check → Show Last Report

The initial check for interferences can take some time on a large project. Use the Show Last Report command to quickly see the list again (Figure 9.5-14). Use the Export button to save the report as an HTML file. This can be viewed in a web browser (Figure 9.5-15).

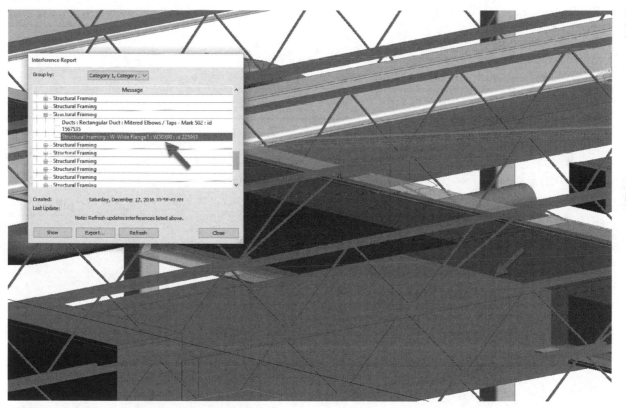

Figure 9.5-14 Interference report dialog

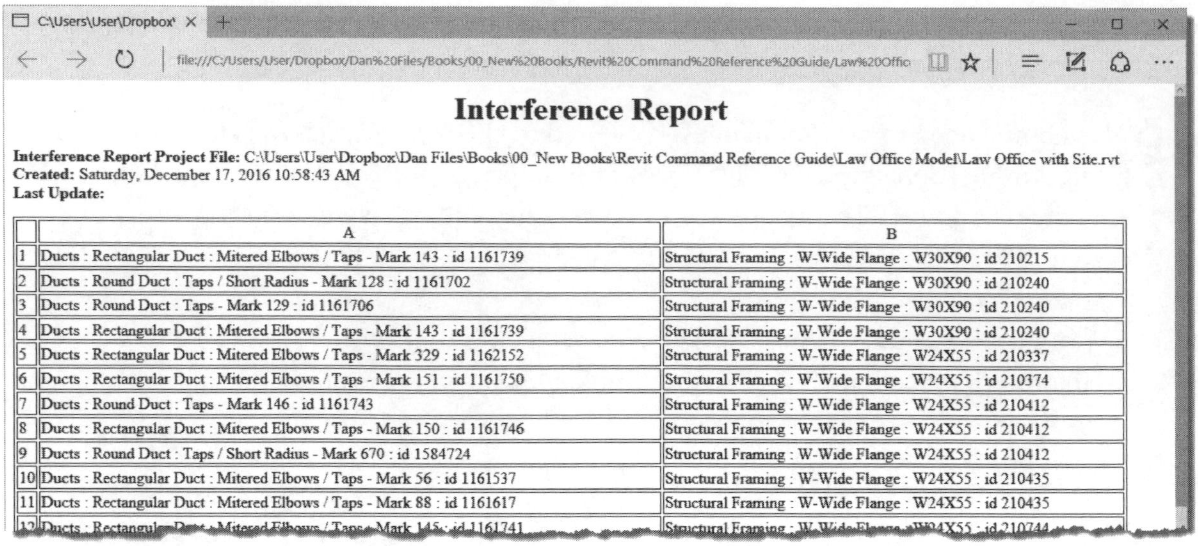
C:\Users\User\Dropbox\

file:///C:/Users/User/Dropbox/Dan%20Files/Books/00_New%20Books/Revit%20Command%20Reference%20Guide/Law%20Offic

Interference Report

Interference Report Project File: C:\Users\User\Dropbox\Dan Files\Books\00_New Books\Revit Command Reference Guide\Law Office Model\Law Office with Site.rvt
Created: Saturday, December 17, 2016 10:58:43 AM
Last Update:

	A	B
1	Ducts : Rectangular Duct : Mitered Elbows / Taps - Mark 143 : id 1161739	Structural Framing : W-Wide Flange : W30X90 : id 210215
2	Ducts : Round Duct : Taps / Short Radius - Mark 128 : id 1161702	Structural Framing : W-Wide Flange : W30X90 : id 210240
3	Ducts : Round Duct : Taps - Mark 129 : id 1161706	Structural Framing : W-Wide Flange : W30X90 : id 210240
4	Ducts : Rectangular Duct : Mitered Elbows / Taps - Mark 143 : id 1161739	Structural Framing : W-Wide Flange : W30X90 : id 210240
5	Ducts : Rectangular Duct : Mitered Elbows / Taps - Mark 329 : id 1162152	Structural Framing : W-Wide Flange : W24X55 : id 210337
6	Ducts : Rectangular Duct : Mitered Elbows / Taps - Mark 151 : id 1161750	Structural Framing : W-Wide Flange : W24X55 : id 210374
7	Ducts : Round Duct : Taps - Mark 146 : id 1161743	Structural Framing : W-Wide Flange : W24X55 : id 210412
8	Ducts : Rectangular Duct : Mitered Elbows / Taps - Mark 150 : id 1161746	Structural Framing : W-Wide Flange : W24X55 : id 210412
9	Ducts : Round Duct : Taps / Short Radius - Mark 670 : id 1584724	Structural Framing : W-Wide Flange : W24X55 : id 210412
10	Ducts : Rectangular Duct : Mitered Elbows / Taps - Mark 56 : id 1161537	Structural Framing : W-Wide Flange : W24X55 : id 210435
11	Ducts : Rectangular Duct : Mitered Elbows / Taps - Mark 88 : id 1161617	Structural Framing : W-Wide Flange : W24X55 : id 210435
12	Ducts : Rectangular Duct : Mitered Elbows / Taps - Mark 145 : id 1161741	Structural Framing : W-Wide Flange : W24X55 : id 210744

Figure 9.5-15 Exported Interference Report

Additional relevant **BIM Chapters** blog posts:

- **Copy Monitor; Levels**
  https://bimchapters.blogspot.com/2017/11/copy-monitor-levels.html
- **Copy Monitor; Grids**
  https://bimchapters.blogspot.com/2017/11/copy-monitor-grids.html
- **Copy Monitor: Light Fixtures**
  https://bimchapters.blogspot.com/2017/11/copy-monitor-light-fixtures.html
- **Copy Monitor: Light Fixtures, Part 2**
  https://bimchapters.blogspot.com/2017/12/copy-monitor-light-fixtures-part-2.html
- **Copy Monitor: Light Fixtures, Part 3**
  https://bimchapters.blogspot.com/2017/12/copy-monitor-light-fixtures-part-3.html
- **MEP Project Setup 02**
  https://bimchapters.blogspot.com/2017/08/mep-project-setup-02_8.html

Chapter 10
View Tab

The View tab contains tools to create and manage views.

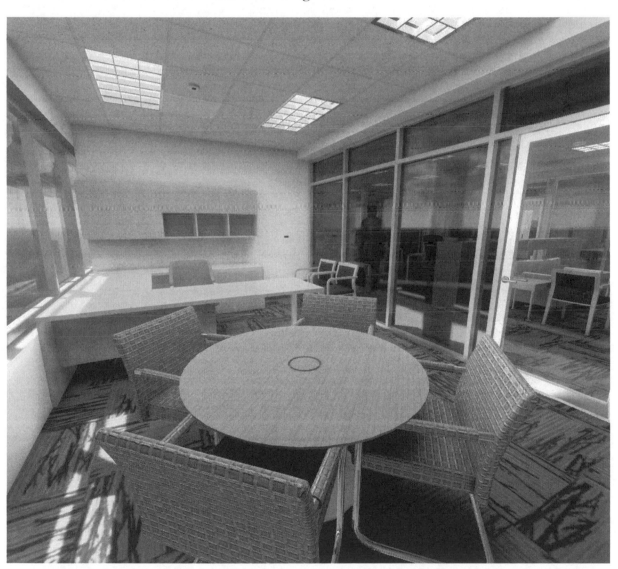

10.1 Graphics panel

The Graphics panel has a wide range of view related tools as seen in the image below.

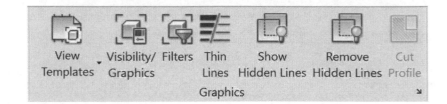

View Templates → Apply Template Properties Current to Current View

There are many settings available to control what is seen in a view and how elements look; for example, View Scale, Detail Level, Phase, Shadows, etc. A **View Template** saves all, or a portion, of those settings, making it easy to consistently control common views.

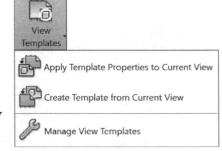

Use the **Apply View Template Properties to Current View** command to do just what it says—apply the various settings in a View Template to the current view on screen. When the command is selected, the **Apply View Template** dialog appears (Figure 10.1-1).

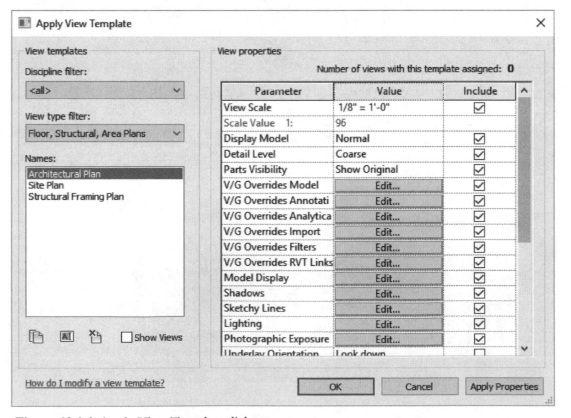

Figure 10.1-1 Apply View Template dialog

Use the **Discipline** and **View type** filters to narrow down the options. Select a named View Template from the list on the left. Use the icons to **Duplicate**, **Rename** and **Delete** a selected View Template. The **Show Views** option allows the settings for a view, selected in this dialog, to be Duplicated into a View Template. The settings of the view itself cannot be used to apply settings to the current view – which is a little confusing.

> Right-clicking on a View(s) in the Project Browser also offers the same command: **Apply Template Properties**. Similarly, right-clicking on a Sheet in the Project Browser offers the option to apply a View Template to all views currently on the same sheet: **Apply View Template to all Views**.

Using the Apply View Template to Current View command is a onetime push of settings from the View Template to the View. Future changes to the View Template will have no effect on the view.

To force a View Template to continuously control the settings of a View, the View Template needs to be assigned in the View's **Properties** (Figure 10.1-2).

> When a view has a View Template assigned, as in Figure 10.1-2, the normal settings for that view are greyed out.

Figure 10.1-2
View Template assigned to view

Be sure to take note of the **Include** setting for each row in the View Template dialog. When not checked, that setting is not applied to the view. Rather, the view's current setting is retained. This can be confusing. For example, the View Scale may have Include unchecked so the *Architectural Plan* view template can be applied to views at different scales (e.g. 1/8" = 1'-0" and ¼" = 1'-0").

View Templates → Create Template from Current View

When a view has been manually set up so the scale, detail level, visible categories, etc., are all as needed, a View Template can be created based on those current settings---thus, facilitating the ability to quickly apply those settings to other views.

Figure 10.1-3
Name New View Template

To use this command:

- Open the desired view so it is the current view
- Select **View → View Templates → Create View Template from Current View**
    - Alternate option: right-click on the view name in the project browser; select **Create View Template from view...**
- Enter a name (Figure 10.1-3)
    - To overwrite an existing View Template, enter the same name
    - If the name entered already exists, click **Yes** to replace, **No** to enter a different name or Cancel to end the current command (Figure 10.1-3)
- Click **OK**

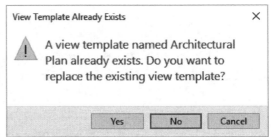

Figure 10.1-4
View Template name already exists

A new, or updated, View Template now exists and can be applied to other views as needed.

View Templates → Manage View Templates

Use the Manage View Templates command to access the View Templates dialog directly. Here, the templates can be modified, duplicated, renamed or deleted.

If the selected View Templates is assigned to any views, set in the view's properties, this dialog will indicate how many views have it assigned to them. This helps to know if any views will be immediately affected by any changes made to the selected View Template.

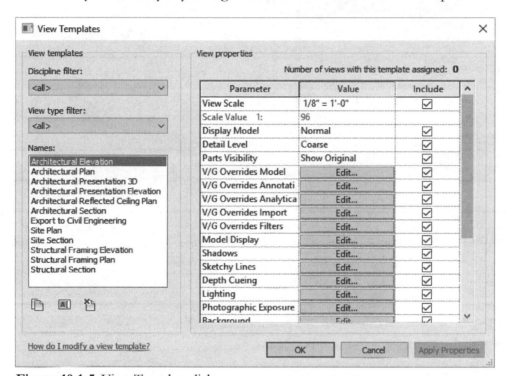

Figure 10.1-5 View Template dialog

The properties on the right correspond to the selected View Template listed on the left.

The **Include** option allows for certain settings to not be applied to a view. When not checked, that setting is not applied to the view. Rather the view's current setting is retained. This can be confusing. For example, the View Scale may have Include unchecked so the *Architectural Plan* view template can be applied to views at different scales (e.g. 1/8" = 1'-0" and ¼" = 1'-0").

Default View Template for New View

When creating a new view, it is possible to automatically apply a View Template. This can save a lot of time. Open a view, for a specific view type, such as a section (Building Section is listed in the Project Browser for the default Revit templates) and then click **Edit Type**. The type properties for the current view are now listed. Notice, an option **View Template applied to new view** (Figure 10.1-6). Setting a View Template here only affects sections created in the future; i.e., this type is selected in the Type Selector while creating a new section.

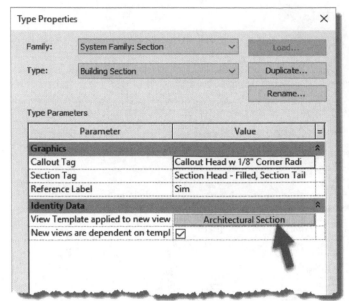

Figure 10.1-6 Set default view template   **Figure 10.1-7** VT set for a view

Assigning a View Template to a View

In addition to applying the properties of a View Template, we can also assign, or hard wire, a View Template to a view. In this case, any changes made to a View Template will automatically affect any view with that View Template assigned to it. To assign a View Template to a view, set the view current, and in the Properties Palette, set the View Template option. To clear a View Template, so it no longer controls the view, set it to **<None>.**

Temporary View Template Settings per View

When View Templates are applied directly to a view, it is often helpful to use the Temporary View Properties options on the View Control bar to troubleshoot or coordinate with items currently hidden. **Enable Temporary View Properties** allows another View Template to be applied to the current view. **Enable Temporary Properties** allows settings to be changed as if no View Template was applied to the view. In both cases, when the project is closed and re-opened the View Template will be reset.

Figure 10.1-8 Temporary View Template options per view

Visibility/ Graphics

The Visibility/Graphics command opens the **Visibility/Graphics Overrides** dialog—which controls many related settings for a single view (Figure 10.1-9). This dialog is like 'Grand Central Station' when it comes to managing what and how things are seen in a given view.

The quicker way to get to this dialog is to type **VV** or **VG** on the keyboard.

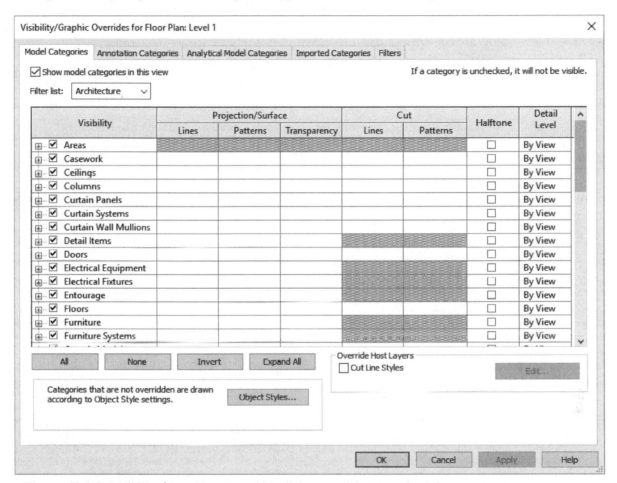

Figure 10.1-9 Visibility/Graphics Overrides dialog; Model Categories tab

Model Categories tab

This tab contains the settings to control model elements in the current view—that is, things you can put your hands on when the building is actually built. The first option is **Show model categories in this view** checkbox. This is rarely unchecked as it turns everything below completely off, including lines and room elements.

> Tags automatically disappear when its corresponding model category is turned off.

This **Filter List** option narrows the rows below to just the ones for the selected discipline(s). Keep in mind that there are some overlapping categories like Generic Model which appear for all disciplines. This means turning it off, or leaving it on, may have undesired results which can be tricky to deal with when multiple disciplines are working in the same model. This is not a problem when the disciplines are separated by a linked model.

Each row represents a model category. This list is predetermined and cannot be modified. Expanding a category, by clicking the plus symbol to the left, will reveal sub-categories. This list can vary as these can be manually created in the project or in the family editor. **Unchecking** a category or sub-category will hide any related elements in the current view.

Each row has several **override opportunities** defined by each column. If the project-wide settings in **Manage → Object Styles** are not ideal for the current view, an override can be applied. The first three columns, under **Projection/Surface**, are for elements which are visible in the view but not being cut. The next two columns, under **Cut**, are for elements visible in the view and being cut. For example, in a floor plan, all the walls that extend from the floor up through the specified cut plane (this view) are being "cut." Similarly, in a wall section, the main wall is being cut, but other walls might be seen beyond (perpendicular to the view) and are considered "Projection/Surface."

Many categories can have their line or pattern overridden or be set to transparent (Figure 10.1-10). Any grayed-out cells cannot have an override applied. For example, Revit does not allow furniture to be cut. This would look odd in plan views for sure. Anytime an element is touched by a cut plane, in plan, elevation or section, the entire element appears.

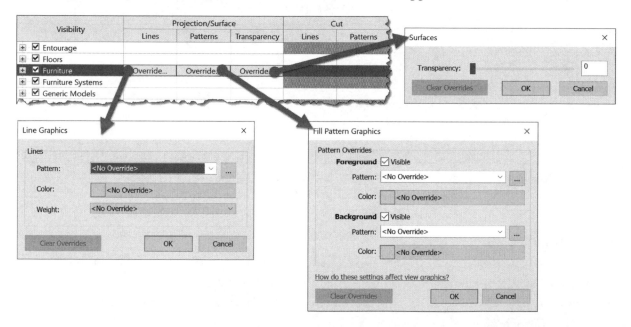

Figure 10.1-10 Visibility/Graphics Overrides dialog; Model Categories tab; Overrides

Use the **All**, **None** and **Invert** buttons to control which rows are selected. The **Expand All** button quickly exposes all sub-categories. The **Object Styles** button provides quick access to the related project-wide settings.

The **Override Host Layers** section controls the lines around each material layer within a wall based on its **Function** setting. Use this to make the structural portion of the wall stand out more. However, this does not affect the fill patterns. Note that the higher lineweight determines which properties are used when two edges overlap.

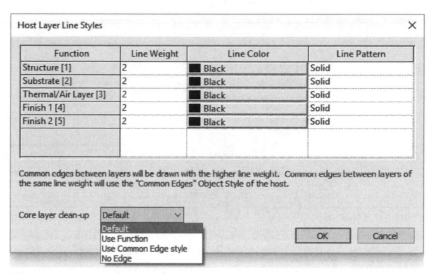

Figure 10.1-11 Host Layer Line Styles

Annotate Categories tab

This tab contains visibility control for elements used to annotate the model, typically for printed documents (Figure 10.1-12).

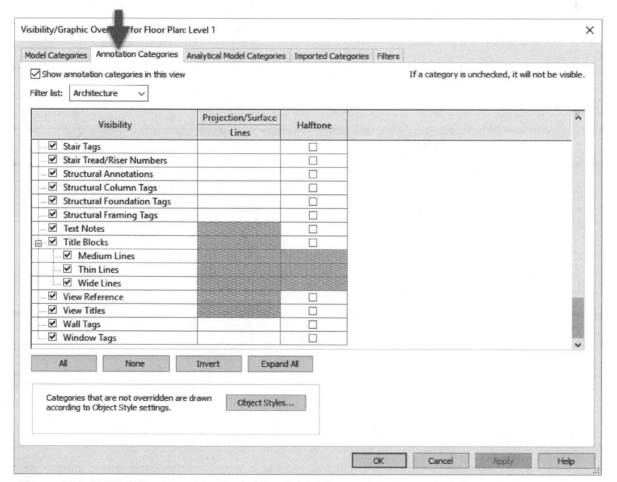

Figure 10.1-12 Visibility/Graphics Overrides dialog; Annotation Categories tab

The first option is **Show annotation categories in this view** checkbox. This can be turned off to ensure all annotation is hidden. However, it may be more appropriate to just delete all the annotation—the exception being **dependent views** where the same annotation needs to be visible in one view but hidden in another.

This **Filter List** option narrows the rows below to just the ones for the selected discipline(s).

Each row represents an annotation category. This list is predetermined and cannot be modified. Expanding a category, by clicking the plus symbol to the left, will reveal sub-categories. This list can vary as these can be manually created in the project or in the family editor. **Unchecking** a category or sub-category will hide any related elements in the current view.

> Using a View Template applied to multiple views, for example—all interior elevations, is the best way to ensure consistency in a project.

Each row has two **override opportunities** defined by each column. If the project-wide settings in **Manage → Object Styles** is not ideal for the current view, an override can be applied. The first column, under **Projection/Surface**, provides a way to control the lines within annotation. The next column is used to set the entire category as **Halftone**.

> Individual tags can be set as Halftone by right-clicking on them and selecting **Override Graphics in View → Element**. Then check halftone (Figure 10.1-13).

The project wide setting for Halftone is found at **Manage** tab **→ Additional Settings → Halftone / Underlay** (Figure 10.1-14). Note that Underlay and Halftone elements will always print grayscale even when "Black Lines" is selected in the Print Setup dialog.

Use the **All**, **None** and **Invert** buttons to control which rows are selected. The **Expand All** button quickly exposes all sub-categories. The **Object Styles** button provides quick access to the related project-wide settings.

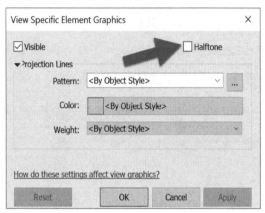

Figure 10.1-13
Halftone option per annotation element

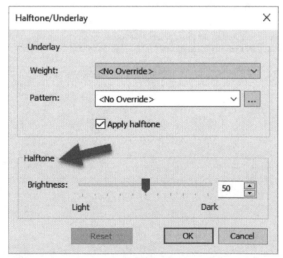

Figure 10.1-14
Project-wide Halftone control

View

Analytical Model Categories tab

This tab contains visibility control for analytical elements related to structural and energy modeling design (Figure 10.1-15).

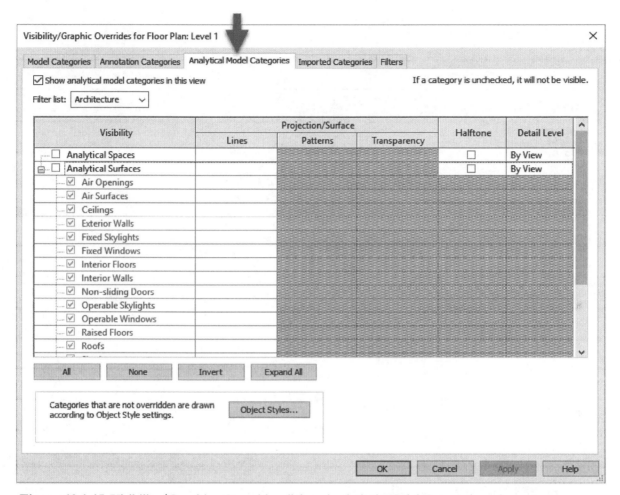

Figure 10.1-15 Visibility/Graphics Overrides dialog; Analytical Model Categories tab

Most of these categories are turned off by default. The energy model categories are static elements created and deleted by the **Generate** command on the **Analyze** tab. The analytical elements are built into the modeled structural elements and update with the model. These elements are mainly used to export information to more specialized analysis software.

The Show analytical model categories in this view checkbox is also controlled by a toggle on the view control bar in the lower left of the canvas area (Figure 10.1-16).

Figure 10.1-16 Toggle analytical elements visibility for current view

Imported Categories tab

This tab contains visibility control for analytical elements related to structural and energy modeling design (Figure 10.1-17).

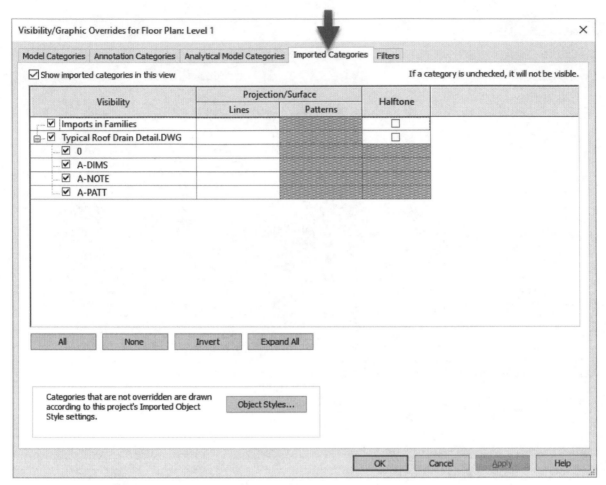

Figure 10.1-17 Visibility/Graphics Overrides dialog; Imported Categories tab

The first option is **Show imported categories in this view** checkbox. This can be turned off to ensure all imported elements, e.g. DWG files, are hidden.

The visibility of each imported, or linked, element is controlled via the **checkbox** for each row. Expanding a row reveals the AutoCAD Layers (or Microstation Levels). These sub items can have their visibility and line style controlled separately.

> If an item appears here but not in the **Manage Links** dialog, that means it has been imported rather than linked.

Use the **All, None** and **Invert** buttons to control which rows are selected. The **Expand All** button quickly exposes all sub-categories. The **Object Styles** button provides quick access to the related project-wide settings.

Filters tab

This tab is used to associate **Filters** with a view and specify what happens when the criteria in the Filter is met (Figure 10.1-18). This provides a very powerful tool to control element graphics and visibility at a very granular level. For example, a filter can make all walls with a fire rating a different fill pattern, or all furniture without a Cost entered a different color, or hide all elements with "sign" entered in the comments field.

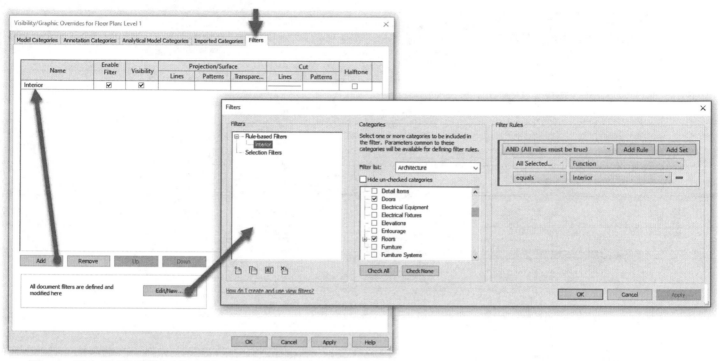

Figure 10.1-18 Visibility/Graphics Overrides dialog; Filters tab

The **Edit/New** button opens the **Filters** dialog as shown in the previous image. This is a separate command covered next in this section on the Graphics panel.

Clicking **Add** button opens the **Add Filters** dialog (Figure 10.1-19). Select a Filter and click **OK** to add a new row to the Filters tab.

Adjust what happens when a specific filter's criteria are met. The first option, **Enable**, allows you to control if the filter will be applied to a view. Use this to have a filter available in a view but turn the effects of the filter on and off. The second option, **Visibility**, will hide elements when unchecked. The remaining options are self-explanatory or covered in more detail on the previous pages for this command.

Figure 10.1-19 Add Filters dialog

Revit Links tab

The Revit Links tab (Figure 10.1-20) only appears when another Revit project file has been linked into the current project.

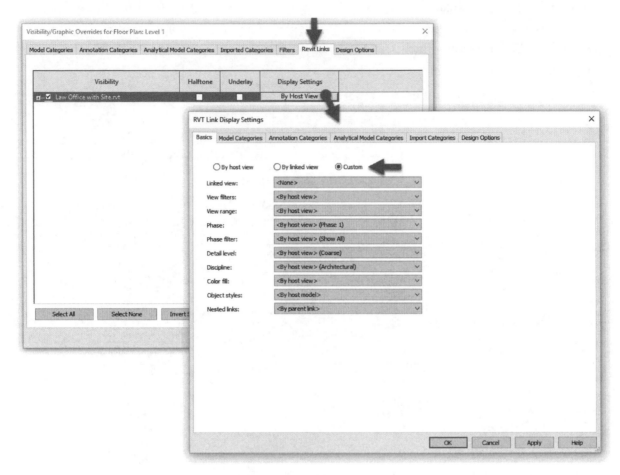

Figure 10.1-20 Visibility/Graphics Overrides dialog; Filters tab

The **Visibility** of each link can be controlled by the first checkbox. The next two control **Halftone** and **Underlay**, respectively. The settings for Halftone and Underlay are controlled from the Manage tab (Figure 10.1-14).

If there is more than one instance of the same link in the current project, expanding the row will show an indented item for each instance, thus providing a way to control the visibility settings for each instance. When only one instance exists, there is no need to expand the row… just edit the main row and this will automatically apply to the single instance.

The **Display Settings** dialog says **By Host View** by default. This means that all the settings on the previous tabs, in the Visibility/Graphics dialog, will also apply to the link in this view. Clicking on "By Host View" opens the **RVT Link Display** Settings dialog. This dialog is similar to the Visibility/Graphics dialog but controls the selected linked project.

The other two Display Settings options are **By Linked View** and **Custom**. The **By Linked View** option has the unique opportunity to see annotation and detail lines associated with a specific view (Figure 10.1-21) in the linked file. Normally, only model elements travel with a linked Revit model. The last option, **Custom**, activates the other tabs to allow for maximum adjustments. Each tab must be set to "Custom" as well, before changes can be made.

> In the **RVT Links Display Settings** dialog, the **Design Options** tab only appears when a Design Option exists in the linked model. It may be necessary to create a same-named Design Option(s) in the current project to work properly in the context of the linked model's design option(s).

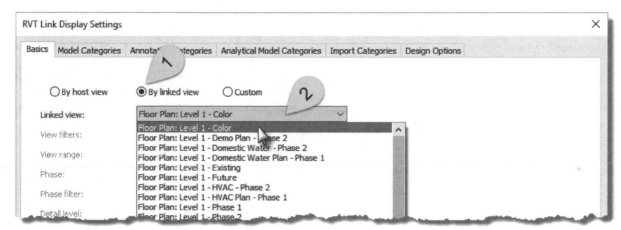

Figure 10.1-21 RVT Links

Design Options tab

The Design Option tab only appears when the current project has Design Options (Figure 10.1-22).

For each Design Option Set, there are at least two options in the drop-down list:

<Automatic>

This means whatever the **Primary** design option currently is, is what should be displayed. In the Design Options dialog, the Primary option can be changed using the **Make Primary** command. This is the default option for all views, meaning secondary design options will never appear anywhere in the project unless this setting is changed in a view.

Option Name (primary)

This is a specific design option that happens to be the Primary design option in the set. If the primary option changes, this view will not be affected.

Secondary Option Name

This is a specific design option that does not happen to be the Primary design option in the set. If the primary option changes, this view will not be affected.

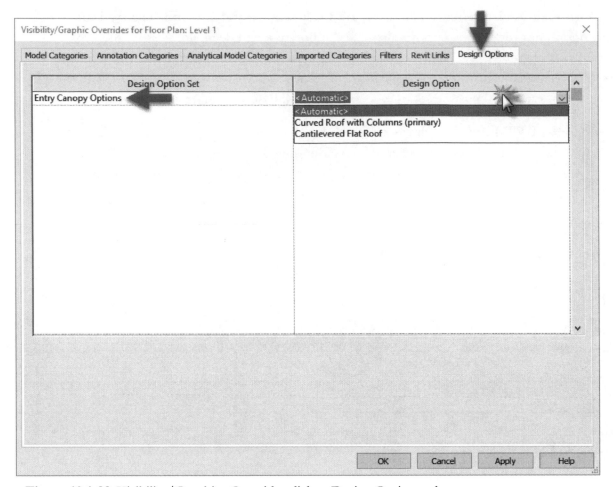

Figure 10.1-22 Visibility/Graphics Overrides dialog; Design Options tab

Design Options are used in preliminary design to maintain different designs during the same timeframe—whereas Phases are used to control different designs at different times. The two should not be confused as several problems will arise.

Design Options can be used to manage the visibility of things like entourage (RPC and plantings for renderings). Creating a Primary design option called "Empty" and keeping it empty will ensure nothing ever shows up in the construction documents views. Then, creating a secondary option called "Turn on in rendering views" and then selecting this only in the camera views to be rendered will do the trick.

Filters

The Filters command opens the Filters dialog (Figure 10.1-23).

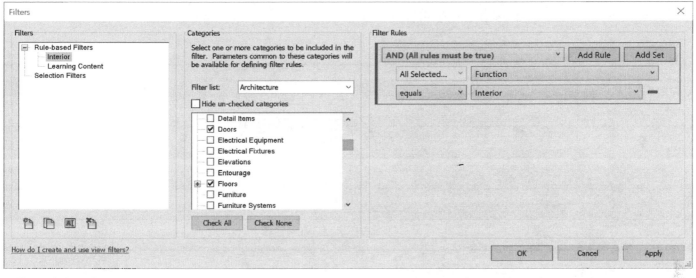

Figure 10.1-23a Filters dialog

Two types of filters can be defined in the Filters dialog;

Rule-based Filters

Creating a **Rule-Based** filter is the main use of this dialog. This is a filter which can be applied to views. When the criteria (i.e. rules) of the filter are met, the view-specific settings in the Visibility/Graphics dialog are applied to that view.

Here is how a Filter is created:

- Use the "**New Filter**" icon in the lower left to create a new Filter
- Provide a **Filter Name**
- Select which **Categories** this Filter should apply to
- Define the Filter Rules
    - Select **And** or **Or**, which has obvious implications
    - Select parameter: lists parameters common to all selected categories
    - Select the **qualifier**: equals, not equal, contains, etc.
    - Pick or type the desired **value**
    - Use the **Add Rule** button - this relates to the 'And' or 'Or' selection
    - Filter sets can be **nested** into each other to make more complex rules.

In the next two images, Figures 10.1-23b & c, we notice how a filter can be created to check for potential problems with a door. For this filter to activate, any of the items in the "Or" Set AND the one item in the "And" set must be true. For example, if the Cost is $10001.00 and the Type Comments is *blank* then this filter applied where used.

It is worth noting in this example the **Or** set cannot come before the **And** set. The filter will not work properly if those two sets are reversed.

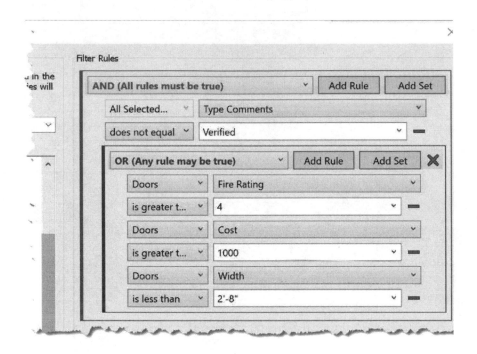

Figure 10.1-23b Filters dialog

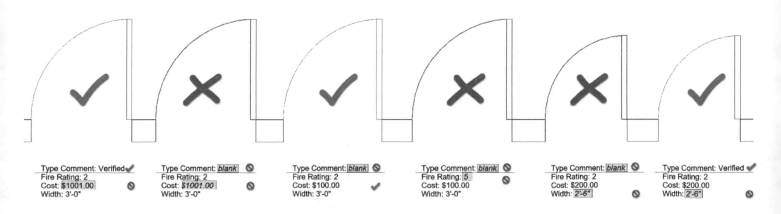

Figure 10.1-23c Results in floor plan view

Selection Filters

This dialog offers a way to manage saved **Selection Filters**: rename and delete. To create a Selected Filter, elements are selected in canvas and then the Save command is selected on the Modify tab → Selection panel (see image to right). These tools are also accessible from the Manage tab if nothing is selected in the model.

Click the **Load** button to reselect elements based on their unique code. This can save a lot of time when several elements need to be selected and a simple selection window or selection filter will not work due to the complexity of the project.

Thin Lines

This command is a toggle between showing lines at their specified thickness and showing everything as fine lines. Each mode has its benefits. Seeing line thicknesses helps to know how the drawing is going to print. Seeing all fine lines helps see additional detail, where two thicker lines close together might otherwise look like a very large single line because they merge together. The images below compare the two modes. This command is also, more conveniently, accessed from the **Quick Access Toolbar**.

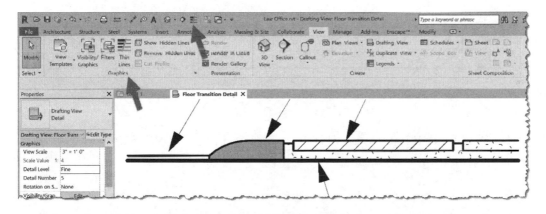

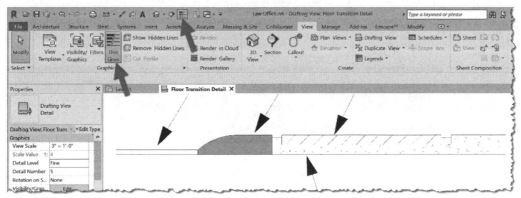

Figure 10.1-24 Thin Lines **off** (top example) and **on** (bottom example)

Show Hidden Lines

The **Show Hidden Lines** command by itself simply shows one object, which is being obscured by another, as hidden lines. Select the command, pick the element in the foreground and then pick the element being obscured. The portion of the element previously not visible is now shown with hidden lines. However, controlling hidden lines in a view is a bit more complicated. Thus, the following broader overview is offered.

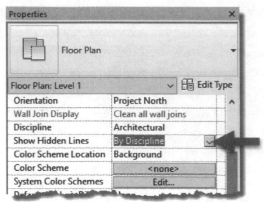

Figure 10.1-25 Show Hidden Line parameter for plan view

In addition to the command with the same name, every model view (except perspective) has an instance parameter called **Show Hidden Lines** (Figure 10.1-25). This parameter has up to three possible options available via a drop-down list: **None**, **By Discipline** and **All**. This is a high-level way in which to control how hidden lines function in a given view. *By Discipline* is the default.

> This view parameter can also be controlled by a View Template.

By Discipline

Each view has a Discipline setting as seen in Figure 10.1-25—currently set to **Architectural**. The **Mechanical** and **Electrical** disciplines each have specific hidden line options in their respective settings dialog on the Manage tab. These discipline specific settings can be overridden per the other two options, as we will discuss next.

None

This setting will turn off all hidden lines controlled by the MEP settings *AND* those controlled by the *Show Hidden Lines* tool on the *View* tab.

> The **Show Hidden Lines** tool has been around for a long time. This tool is used to reveal the outline of an item which is being obscured by another element. This appears to only work when the view's discipline is set to Architectural or Coordination.

There are a few things *None* does not affect. This setting does not affect the *Beyond* line controlled by elements falling between the **Bottom** and **View Depth** settings in the *View Range* dialog. This does not affect the hidden lines set using the Linework tool or line work assigned to the *Hidden Lines* sub-category within a family. It is interesting, however, that the *Show Hidden Lines* tool is also dependent on the *Hidden Lines* sub-category and IS affected by the *None* setting.

Figure 10.1-26 shows a simple reception desk family in plan view; note these three points:

- **Item #1** was set to use *Hidden Lines* via the **Linework** tool
  - Selected line style controls color and pattern
- **Item #2** was set to *Hidden Lines* via the line's properties within the family
  - Hidden Line color and pattern controlled by *Furniture* sub-category
- **Item #3** has *Hidden Lines* showing using the **Show Hidden Lines** feature
  - This is a low wall extending under the worksurface
  - Hidden Line color and pattern controlled by *Wall* sub-category

When **Show Hidden Lines** is set to *None*, only the lines related to Item #3 are hidden. Everything else stays the same.

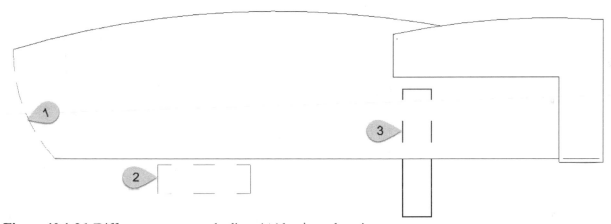

Figure 10.1-26 Different ways to make lines hidden in a plan view

All

This option is interesting in that all obscured edges appear as *Hidden Lines* in the entire view. The images below, Figures 10.1-27 & 28, show the same reception desk in plan and 3D view. Notice all the 3D geometry is shown through the work and transaction surfaces. Also note that the color and line style is being controlled by the elements sub-category (in this case, furniture). Also, the view scale controls the line pattern size.

Individual hidden lines can be removed using the **Linework** tool with the line style set to **<Invisible Lines>**.

This functionality is not available for perspective views nor does it work on linked models. The "All" option is also not an option for MEP plan views.

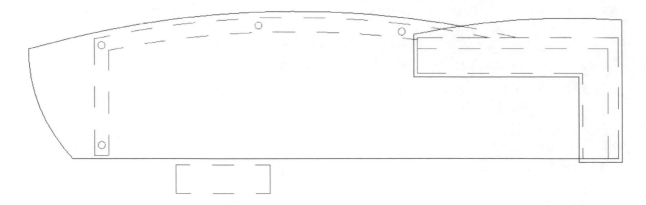

Figure 10.1-27 Show Hidden Line parameter set to All (plan view)

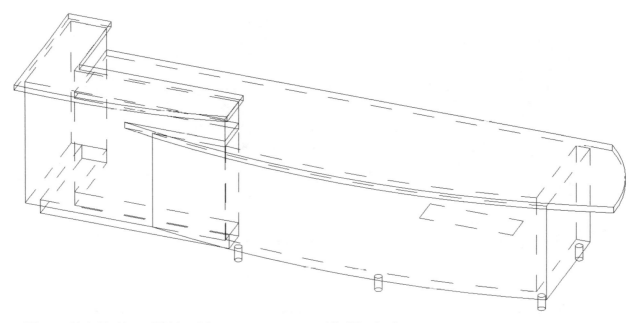

Figure 10.1-28 Show Hidden Line parameter set to All (3D view)

Remove Hidden Lines

Use the **Remove Hidden Lines** command to undo any elements modified with the **Show Hidden Lines** command. Select the command, pick the object obscuring the elements and then pick the element that has been overridden with the Show Hidden Lines command—the hidden lines disappear.

Cut Profile

This is one of those commands that should rarely be used as it defies the ideals of BIM—that is, to have changes update everywhere and work with analytical tools and clash detection.

This tool is meant to fake how a material in a wall looks in plan view. More specifically, it might be used to convey some existing condition that might otherwise be difficult to model three-dimensionally. But however realistic the plan view might look, these changes do not appear anywhere else in the model. Take note that this command is on the View tab.

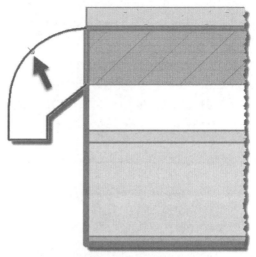

Figure 10.1-29 Cut Profile sketch

To use this command follow these steps:

- Open a plan view
- Select **Cut Profile**
- Select a **layer** (i.e., material) in a wall
  - The layer will highlight before selecting it
- Sketch lines connecting the start and end of the chain of lines to the edges of the layer
  - This is not a closed loop
- Verify the arrow direction
  - Pointing in: add material
  - Pointing out: cut material

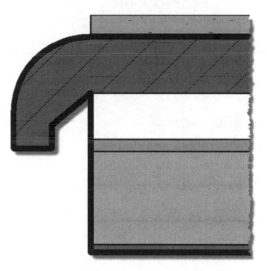

Looking at this wall in a 3D view, or even another plan view at the same level, will not show any of these edits because they are all 2D and view specific.

Figure 10.1-30 Cut Profile result

Graphic Display Options

The small icon in the lower right corner of the Graphics panel opens the **Graphic Display Options** dialog (Figure 10.1-31); this image shows all options fully expanded. This dialog can also be accessed via the **Properties Palette** and the **Visual Style** list on the **Display Control Bar** as shown below.

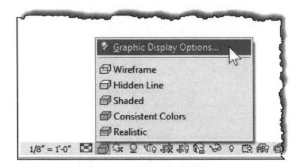

The settings in this dialog only apply to the current view, similar to the **Visibility and Graphic Overrides** dialog.

Model Display

These are the **Style** options:

- Wireframe
- Hidden Line (default)
- Shaded
- Consistent Colors
- Realistic

Use **Wireframe** to temporarily see through all solid elements in the view—but keep in mind some commands will not work in this mode, like face-based families and spot elements. This mode will make model navigation super-fast.

Hidden Line is the default, showing surfaces without color and obscuring elements behind solid surfaces.

Shaded fills each surface with a color based on the **Shading** color specified for each Material.

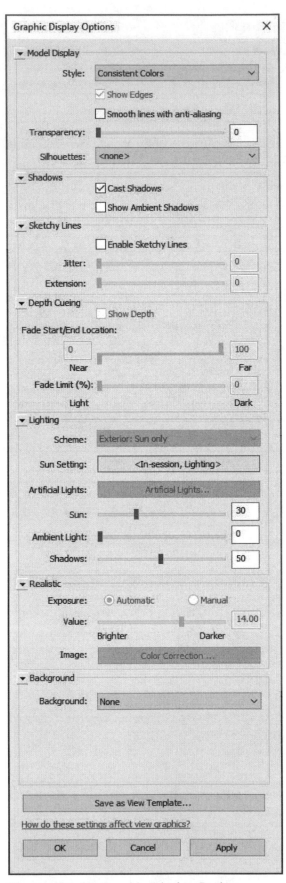

Figure 10.1-31 Graphic Display Options

In the absence of a material, a middle gray color is used. This setting adds shading to better define shapes and suggest a sense of depth.

Consistent Color is closer to the actual Shading color defined in the material and lacks the shading. Elements tend to look flat and performance is better than shaded.

Realistic mode uses the render appearance asset defined in each material. For more detailed models, this setting can significantly slow down model navigation. This is similar to a rendering, but without the shadows and reflections; shadows can be turned on, but in a Rendering shadows are not optional.

Show Edges is an option with some of the above settings. This option shows a distinct line at all edges.

Smooth lines with anti-aliasing, by default, is a view specific setting. This option smooths out angled lines on the screen. This setting is off initially for every view as it can slow down model navigation in more complex models.

In the Options dialog, there is an option to globally turn on, or off, Smooth Lines (Figure 10.1-32). Again, this can cause performance issues for complex models.

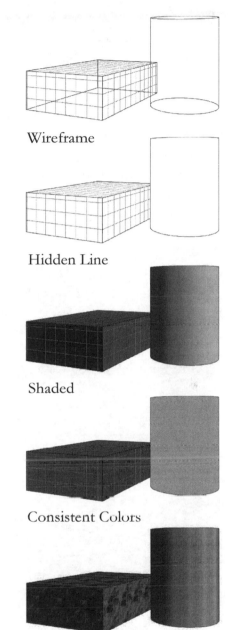

Wireframe

Hidden Line

Shaded

Consistent Colors

Realistic

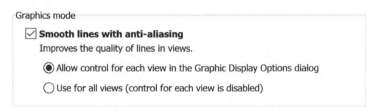

Figure 10.1-32 Options dialog settings

The **Transparency** slider has an effect similar to the Hidden Line setting. But this controls how intensely the hidden lines are displayed.

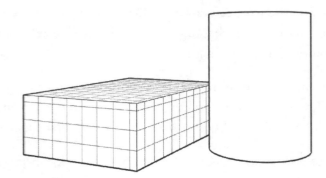

The **Silhouettes** option lists all the **Line Styles** defined in the project. When a style is selected the perimeters of objects have that line style applied. The example to the right, Fig. 10.1-33, has a heavy line applied to make each element stand out more.

Figure 10.1-33 Heavy silhouette applied

Shadows

Check **Cast Shadows** to see and print shadows in the current view. Shadows are based on the current Sun Settings for the view. For most of the **Sun Settings** options, it is also important to set **True North** for the project.

> Shadows are cast on an imaginary floor when **Ground Plane at Level** is checked in the **Sun Settings** dialog for a given view (Figure 10.1-34).

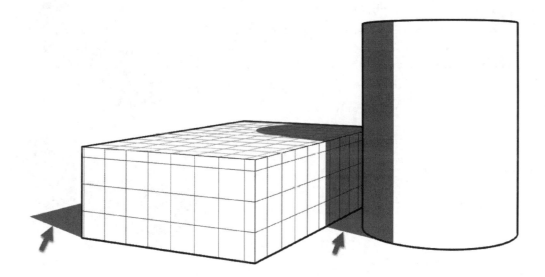

Figure 10.1-34 Shadows cast on imaginary ground plane

Check **Show Ambient Shadows** to add a subtle sense of depth and proximity (Figure 10.1-35). Use with, or without, the Cast Shadows option. This option tends to be better by itself for interior views. Like Cast Shadows, this can have a negative impact on performance.

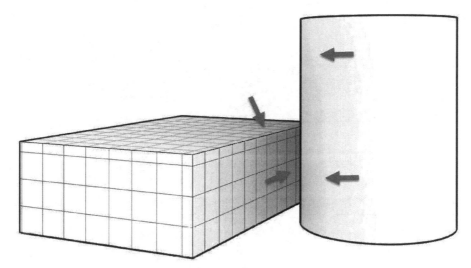

Figure 10.1-35 Ambient shadows

Sketchy Lines

Sketchy Lines is used to give a less rigid representation of a design when presenting to a client (Fig, 10.1-36). **Jitter** controls how shaky the link looks whereas **Extension** determines roughly how far a line extends past its endpoint. This feature only works when **Show Edges** is turned on.

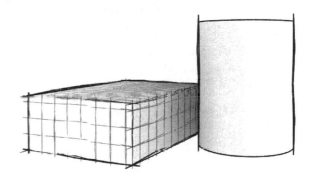

Figure 10.1-36 Sketchy Lines

Depth Cueing

This feature is able to make lines further back from the cut plane lighter. This helps to distinguish a door on the closest plane when compared to another that might be 150' back. This feature takes a little trial and error to get the settings as desired.

Lighting

Use this section to access the **Sun Settings** which affect shadows. When **Shadows** are turned on, the **Shadows** slider controls their intensity. **Sun** and **Ambient Sun** sliders only affect the view when set to Realistic.

Realistic

When the view is set to Realistic, these values can be adjusted and are similar to the Adjust Exposure (called Color Correction here) options in the rendering dialog (Figure 10-1-37).

Background

The **Background** option allows the view itself to have a constant color or image applied. Figures 10.1-38 & 39 show the various settings for Gradient and Image. Additionally, for Image, selecting Customize Image opens a dialog in which the image can be selected and sized and positioned.

Save as View Template...

This button provides a quick and convenient way to create a **View Template** based on the settings of the current view. Notice, in the **View Template** dialog that only the parameters associated with the settings in the **Graphic Display Options** dialog are checked as **Include**.

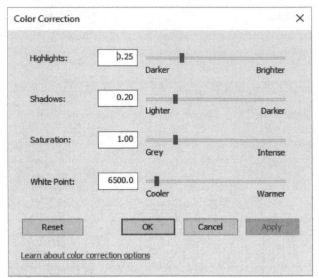

Figure 10.1-37 Color Correction

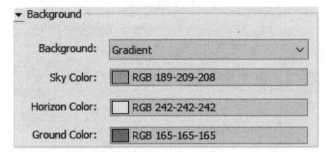

Figure 10.1-38 Background, gradient

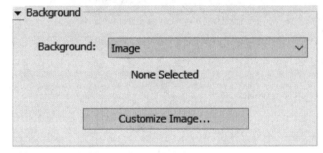

Figure 10.1-39 Background, image

10.2 Presentation panel

The Presentation panel has three tools related to creating photo-realistic renderings as seen to the right.

The image below is the Revit model based on the tutorial-style book *Design Integration Using Autodesk Revit* and was rendered using the Revit add-in **Enscape**; notice how that software represents grass and glass.

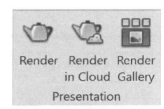

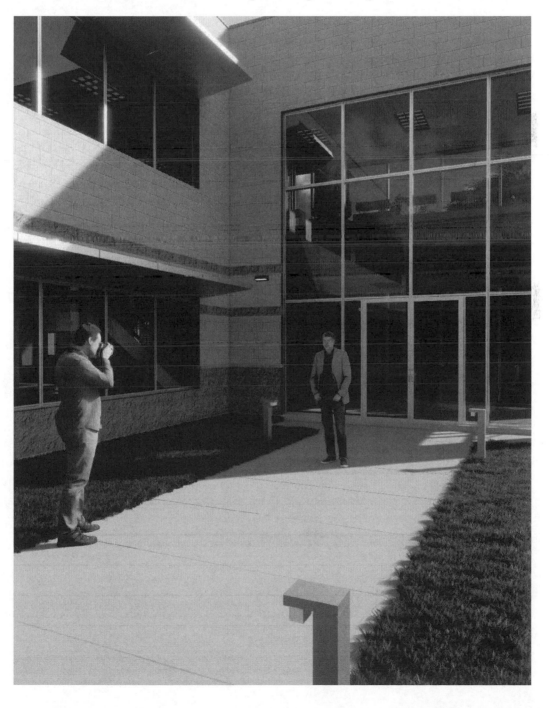

Render

Revit offers the ability to create a photo-realistic rendering from a Revit model. By adding lighting and properly applying materials, amazing results can be achieved.

Figure 10.2-1 Rendering icon

To access the rendering dialog, click the **Show Rendering Dialog** icon on the View Control Bar in the lower left (Figure 10.2-1). This opens the **Rendering** dialog (Figure 10.2-2). This dialog box allows you to control the environmental and quality settings and actually create the rendering.

> You have limited options for setting up the building's environment. If you need more control than what is provided directly in Revit, you will need to use another program such as Autodesk *3ds Max Design 2021* which is designed to work with Revit and can create extremely high-quality renderings and animations; it even has day lighting functionality that helps to validate LEED® (Leadership in Environmental and Energy Design) requirements.

A combination of Quality and Resolution determines the overall final quality of the rendered image.

Quality

Quality controls accuracy of light and materials, plus duration. Selecting Edit… from the list of Quality options opens the **Rendering Quality Settings** dialog (Figure 10.2-3). The **Custom (view specific)** option allows the settings to be modified. The other settings are predetermined, but their values are seen in this dialog.

Figure 10.2-2 Rendering dialog

Output Settings

The **Output Settings (Resolution)** determines how large the image can be viewed or printed. Increasing the DPI increases the total width and height pixels.

> To increase the image size and achieve a higher resolution, in the camera view, select the Crop Region and **click Size Crop** on the Ribbon. Select the **Scale (lock proportions)** options and then adjust the Width and Height. The width and height in the Rendering dialog will adjust proportionally.

Lighting

The **Lighting** drop-down list offers very simple choices (Figure 10.2-4): Is your rendering an interior or exterior rendering, and is the light source *Sun, Artificial* or both? You may have artificial lights, light fixtures, but still only desire a rendering solely based on the light provided by the sun. *Sun only* will be faster as artificial takes much longer to calculate.

With one of the artificial lighting schemes selected, click on the **Artificial Lighting** button to access the Artificial Lights dialog. You will now see a dialog similar to the one shown to the right (Figure 10.1-35).

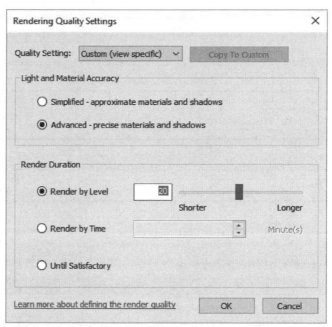

Figure 10.2-3 Render Quality Settings dialog

The light fixtures listed relate to the fixtures' families placed in the model. Here you can group lights together so you can control which ones are on (e.g., exterior and interior lights).

In the *Artificial Lights* dialog (Figure 10.2-5) the *Dimming* value can be changed for individual fixtures or for the entire group. To change the dimming value for the entire group, simply change the value to the right of the group name.

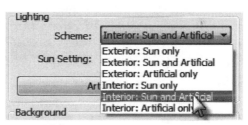

Figure 10.2-4 Lighting options

> When a Light Fixture is selected in the model, the **Options Bar** allows the Lighting Group association to be changed via a drop-down list. An **Edit** button provides a way to open Lighting Group edit mode and quickly add or remove lighting fixture families to/from the group.

Figure 10.2-5 Artificial Lights dialog

Background

Select from several **Sky** options with varying degrees of cloud cover. A solid **Color** option is also available. The **Image** option allows you to specify a photograph of the site, or one similar. It can prove difficult getting the perspective just right, but the end results can look as if you took a picture of the completed building (Figure 10.2-6).

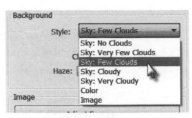

Figure 10.2-6
Background options

Image

Adjusting the ***Exposure Control*** before exporting the image can make a huge difference in how realistic the image looks. The image to the right, Figure 10.2-7, has been adjusted to increase the contrast by darkening the shadows. For interior renderings, setting the *White Point* to about 8600, or warmer, for artificial light makes the materials look more realistic.

Use the **Reset** button to return all settings in this dialog back to the defaults.

Clicking **Apply** will show the updates immediately in the rendered image—the image does not need to be rendered again.

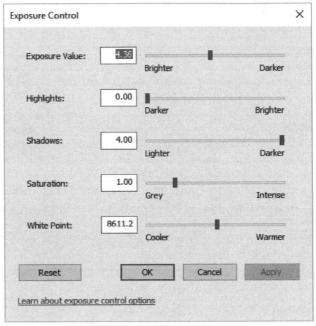

Figure 10.2-7 Exposure Control dialog

Save To Project

When a rendering is complete, a raster image can be saved in the project. Clicking the **Save To Project** button prompts for a name (Figure 10.2-8). Revit will increment the name with a numbered suffix to keep the progress images in order. This static image now appears in the project Browser (Figure 10.2-9).

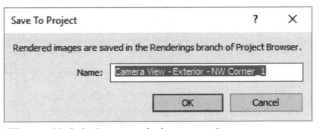

Figure 10.2-8 Save rendering to project

> Be sure to delete old renderings to keep the Revit project file clean. Consider exporting renderings and only saving final renderings needed on drawings/sheets.

Click **Export** to save the rendering as a raster image file—in BMP, JPG, PNG or TIF formats (Figure 10.2-10).

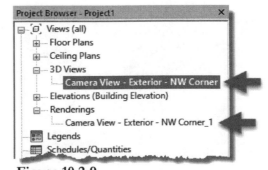

Figure 10.2-9
Renderings in project browser

The final image can be viewed with a digital projector, printed (letter or tabloid) or plotted (22"
x 34" or larger) or even printed as a high-quality photo via an online service like Shutterfly.com
or Walgreens.com (just as two examples of many options)—in addition to photo paper, they
have options like canvas prints. Exported images can also be brought back into the Revit model
with the **Insert Image command**.

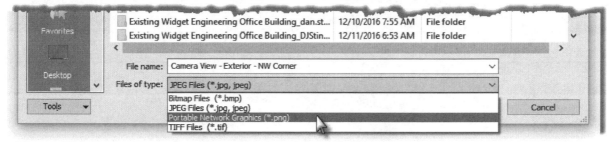

Figure 10.2-10 Export rendered image to raster image file format

Rendering Progress

To start a Rendering, click the **Render** button at
the top. You will see a progress bar while the
rendering is processing (Figure 10.2-11).

> The Region checkbox, next to the
> Render button, allows a smaller sample
> area to be rendered. This can save a little
> time when needing to just check a color.
> But the entire model still needs to be
> considered, so the time saved is usually marginal.

Figure 10.2-11 Render progress

You may see a number of warnings about missing render appearance images. These are image
files which were assigned to some of the content possibly downloaded from the internet. These
files are not always readily available when you downloaded the content. You would have to go
back and look for it, but it is more likely that you would have to request these image files or
substitute different ones.

After a few minutes, depending on the speed of your computer, you should have a rendered
image similar to the example at the beginning of this chapter. You can increase the quality of the
image by adjusting the quality and resolution settings in the Render dialog. Keep in mind these
higher settings require substantially more time to generate the rendering.

Render in Cloud

Use the **Render in Cloud** command to upload images to be rendered using Autodesk's Cloud-based rendering service. Rather than using your local computing power, the image is processed within a large data center. What might take hours on a local computer can be accomplished in a fraction of the time in the cloud.

Everything leading up to deciding which way to render an image (local or in the cloud) is the same—applying materials, placing lights, etc. Even the settings in the Render dialog (previous command covered) are used for the cloud rendering.

Rendering in the cloud consumes Cloud credits. At the time of this writing, they cost $1 (US) each and are sold in blocks of 100. Students have free access to rendering in the cloud.

> Your Revit project must be saved before rendering in the cloud.

Selecting **Render in Cloud** opens the overview dialog shown in Figure 10.2-12.

Figure 10.2-13 shows the available settings, in

Figure 10.2-12 Initial render in the cloud prompt

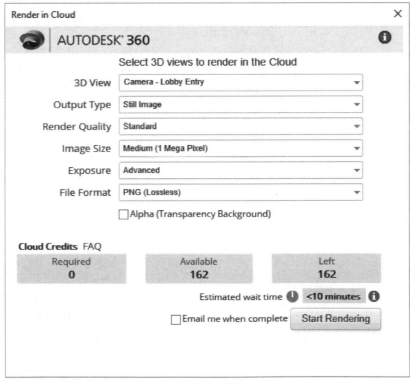

Figure 10.2-13 Render in the cloud settings

addition to those in the Rendering dialog. These specific settings relate to the final cost shown.

The default settings are low quality and free. This allows a rough version of the image to be seen before committing to the final paid version. Once in the cloud, if the free draft version looks acceptable, it can be re-rendered there… avoiding the need to go back to Revit and export again.

> You must be logged into **Autodesk A360** to use this feature. Click Login in the upper right corner of the screen. Anyone can create an account—new accounts receive a small amount of free cloud credits. To access credits associated with a firm you work for, the subscription administrator in your firm must set up your account and send you an invite.

One of the options under **Output Type** is **Stereo Panorama**. This image can be used with a **Google Cardboard Viewer** to experience a space in **Virtual Reality**. With this option selected, once the rendering is complete, there is an option in the **Render Gallery** called "View on phone" which generates a web address (URL) which can be emailed to yourself or others.

Render Gallery

To view the completed images rendered in the cloud, click **Render Gallery**. This command opens your default browser and will prompt for username and password if not already logged in. These images can be accessed anytime in the future, until you delete them. To access all the features, use a modern Browser such as Chrome. FYI: Stereo Panorama example below.

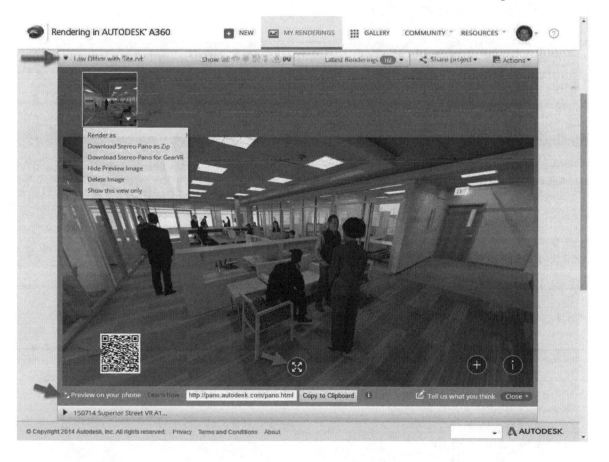

10.3 Create panel

The Create tab contains many commands to create new views, which will then appear in the Project Browser, based on the 3D model.

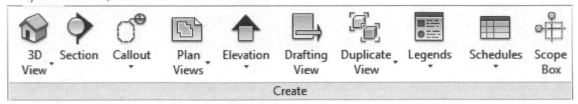

3D View → Default 3D View

This command could be more readily accessed from the Quick Access Toolbar (QAT).

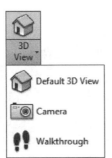

Clicking the **Default 3D View** command opens a 3D view named **{3D}** and appears under the **3D Views** heading in the **Project Browser** (Figure 10.3-1). This is a quick way to see an exterior overview of the project. If this view is ever renamed or deleted, a new one is automatically created the next time the **Default 3D View** command is selected.

For worksharing projects, the 3D view has the username, from the Options dialog, added as a suffix. This allows each user to customize the default 3D view.

The Default 3D view is the view used when the **Selection Box** command is selected. Simply turning the **Section Box** off and back on resets the results of the **Selection Box** command.

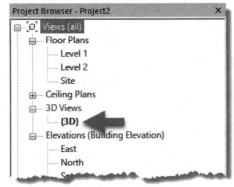

Figure 10.3-1 Default 3D view

The default 3D view can be copied, via right-click, in the Project Browser. The default 3D view can be thought of as a working view. A copy can be placed on a sheet, and even "locked" so notes, tags and dimensions can be added.

A 3D view is an isometric view.

3D View → Camera

This command could be more readily accessed from the Quick Access Toolbar (QAT).

A Camera view is a **perspective** view, which looks more realistic than the Isometric view (Default 3D View just covered).

Camera views are typically placed in a plan view. Here is the basic outline:

- Open a **floor plan** view
- Select the **Camera** command
- Uncommon change: turn off **Perspective** on the Options Bar
  - Normally this is left on
  - Idea: Turn off and create what looks like a plan view, turn on shadows and see the proper shadows based on window heads, walls and more… normally cropped in a plan due to the cut plane, or View Range.
- Notice **Offset** value on Options Bar
  - This is the "eye" height camera
- Pick a point to locate the **eye position**
- Notice **Offset** value on Options Bar again
  - This is a point at which the sightline will pass through
- Pick to define the view **target location**
- Optional: Turn off **Far Clip Active** so elements are not being omitted.

If the view needs to be adjusted:

- Use the Navigation Wheel's look and walk tool to keep the camera properly positioned off the floor.
- Depending on sun light and project design, it may be helpful to turn on Section Box in the properties to increase rendering performance.

In a plan view, the camera and section box for that camera view can be selected. Right click on the camera view, in the Project Browser, while in the plan view (Figure 10.3-2).

> The **Section Box** option only appears when the selected 3D view has the Section Box feature enabled for that view.

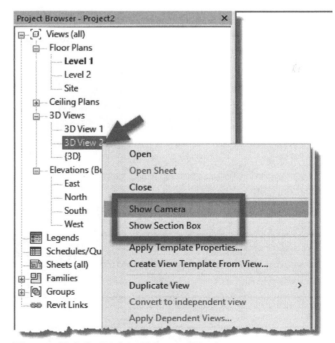

Figure 10.3-2 Right-click on 3D view

3D View → Walkthrough

Use the Walkthrough command to create a video through or around the building. The path is defined by picking points in a floor plan view – the result is saved as a view in the Project Browser (Figure 10.3-3). When this view is opened, and the Crop Region is selected, a button called **Edit Walkthrough** is available on the Ribbon.

> Use the Navigation Wheel's **Look** feature to adjust the view direction for the walkthrough. Use the **Next Frame** option and repeat if needed.

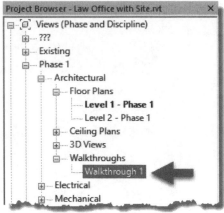

Figure 10.3-3 Project Browser

The image below, Figure 10.3-4, shows the camera path being defined in a floor plan view. While picking camera points, the eye level (i.e., Offset) can be adjusted on the Options Bar for each location. Use this technique to go up/down stairs to another level.

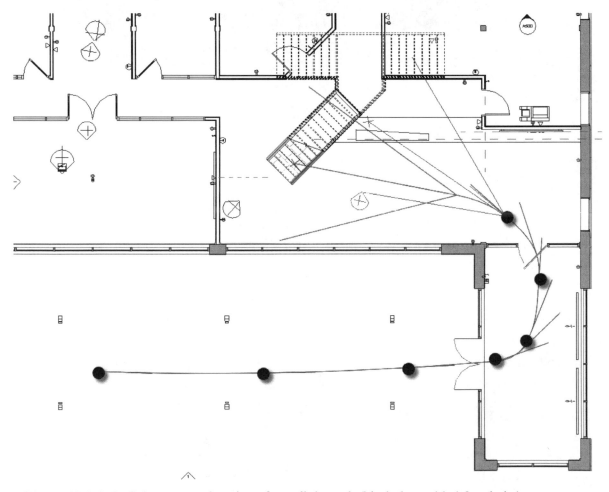

Figure 10.3-4 Defining camera locations for walkthrough (black dots added for clarity)

To edit a Walkthrough, open it from the Project Browser, and select the Crop Region and then select **Edit Walkthrough** from the Ribbon. This opens the Edit Walkthrough contextual tab (Figure 10.3-5).

Figure 10.3-5 Editing a walkthrough

Click the total number of frames button on the Options Bar. This opens the **Walkthrough Frames** dialog (Figure 10.3-6). Use this dialog to define the total number of frames and, optionally, the number of frames between each Key Frame. The total number of frames is directly proportional to how smooth the video is and how long it takes to process.

While editing the Walkthrough, click the **Play** button on the Ribbon to see a rough simulation of what the exported video will look like. Reviewing this can save a lot of time, as walkthroughs can take a long time to process. Having to do it over because of a small mistake would be unfortunate.

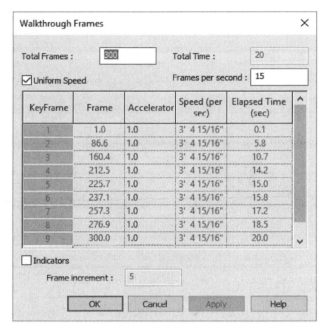

Figure 10.3-6 Editing walkthrough frames

When the walkthrough is ready, make sure the Walkthrough view is current/active—this may require clicking in the view. To export, click **File Tab → Export → Images and Animations → Walkthrough**. The first dialog verifies the **Length/Format** of the video file (Figure 10-2-7). The **Visual Style** setting has several options: Hidden Line, Shaded, Realistic. The next dialog prompts for a file name and location. This tool only supports AVI video file format which is very old and not compressed. Consider using a video editing program to convert the AVI to an MP4 file.

Figure 10.3-7 Export animation settings

View

Section

Use this command to create a section view of the Revit model—the result is a graphic in the model representing the cut plane and view direction and a new item in the Project Browser. Typically created in a floor plan view, a section view is created by simply clicking two points. The direction picked—left to right, or right to left—determines the view direction. However, the view direction can be changed later if needed.

The image below lists some terms related to the *Section Mark* graphic.

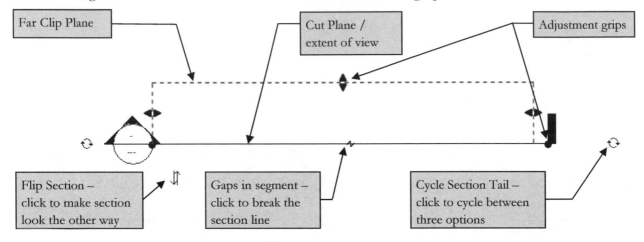

- **Cutting plane/extent of view line:** This controls how much of the 3D model is elevated from left to right (i.e., the width of the elevation). This list also acts like a section line in that nothing behind this line will show up.

- **Far clip plane:** This controls how far into the 3D model the elevation view can see. This can be turned off in the elevation's *View Properties*, not the plan's *View Properties*, making the view depth unlimited; the next page covers this more.

- **Adjustment grips:** You can drag these with the mouse to control the features mentioned above. These values can also be controlled in the elevation's *View Properties*.

- **Flip Section:** Click this to make the section look in the opposite direction.

- **Gaps in Segment:** For busy floor plans (lots of lines, notes and grids) this feature can omit the middle portion of the section line, so it only appears at each end. When using this option, the full section head, drawing/sheet numbers is toggled on for both ends.

- **Cycle Section Tail:** Toggle between full head, tail or nothing.

> When using a **Windows Selection**, anything between the cut plane and the far clip plane will get selected, even when the elements are obscured by other elements. For example, furniture in a room behind a wall will get selected.

When a section view is active, and nothing is selected, the Properties Palette lists that view's properties (Figure 10.3-8). Below are a few highlights:

- **Name:** This is the name that appears on the sheet and in the Project Browser. Consider changing the view name right after creating it as the default name is too vague.

- **Title on Sheet:** Some office naming conventions require the name that appears in the project Browser include the sheet number or some other designator which is not needed in the view name on the sheet. Use this parameter to use a different name on the sheet.

- **Crop View**: This crops the width and height of the view in elevation. Adjusting the width of the cropping window in elevation also adjusts the "extent of view" control in plan view.

- **Crop Region Visible**: This displays a rectangle in the elevation view indicating the extent of the cropping window, described above. When selected in elevation view, the rectangle can be adjusted with the adjustment grips – it will also print when set to be visible.

- **Far Clipping**: If this is turned off, Revit will draw everything visible in the 3D model, within the extent of view. Avoid Clip with line as this can be misleading – especially for sloped roofs in exterior elevations.

- **Annotation Crop**: This option hides annotations which fall outside of this rectangular area. In most cases it is better to just delete the annotation (never delete sections, elevations, grids) as these elements are view specific. However, when working in a dependent view, the notes may be used in another view and therefore should not be deleted.

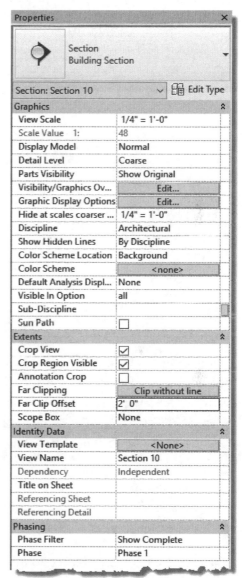

Figure 10.3-8

Section view properties

- **Edit Type** (button): It is possible to create multiple section view Types. Doing so can control the section mark graphics and how the various section views sort in the Project Browser.

Callout: Rectangle

A section smaller than a building section or wall section does not typically have a full-blown section mark in the floor plan view. Thus, the Callout command is used to reference a more detailed section (usually at a larger scale) from a building/wall section.

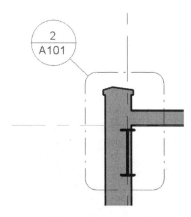

Figure 10.3-9 Callout section

quick steps

Callout

1. Ribbon
   - Reference Other View
2. Type Selector
   - Select callout type
     i. Note limitations of "detail view" option
3. In-Canvas
   - Pick two points to define rectangular area
   - Optional: Select and adjust location of reference bubble

Linear: **Ribbon**

While creating a callout, and before selecting points in the view, check **Reference Other View**, and select another view from the drop-down list (Figure 10.3-10). Rather than creating a new item in the Project Browser, the callout will simply reference another one. Use this to call out a typical condition, which occurs several times in the project, without the need to add all the same notes and dimensions for each location.

> Due to this manual selection of another view, it is possible to select the wrong view and thus create an error. As soon as the new view is created, double-click on the reference bubble to open the view and verify it is the correct reference.

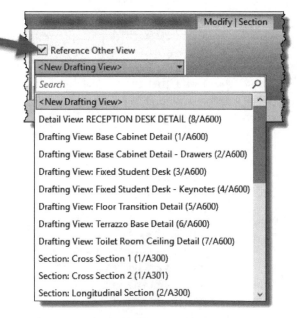

Figure 10.3-10 Reference other view

Linear: Type Selector

Select the desired **Type** from the Type Selector. Creating/using multiple types allows for different graphics in the views and refined sorting in the Project Browser.

Linear: In-Canvas

Pick two diagonal points to define a rectangular area. This is the same size as the **Crop Region** in the new view. Once created, select it and click-and-drag the **Grip** near the reference bubble to reposition the bubble.

> The Callout graphics can be adjusted in **Manage → Object Styles → Annotation Objects** tab: line weight, color and line pattern.

Callout: Sketch

Same as previous command, except you may sketch an irregular callout area – not just a rectangular area.

Plan Views → Floor Plan

Floor plan views are horizontal cuts through the building. Each plan view is associated with a level in the building. They can have different **View Range** settings, notes, dimensions, etc.

Selecting this command opens the New Floor Plan dialog (Figure 10.3-11).

The **Type** option allows several settings to be applied, including a **View Template** (Figure 10.3-12) if one is selected for a given plan Type. This selection can also be used to organize views in the Project Browser. Click **Edit Type** to manage the type properties and create new ones if needed.

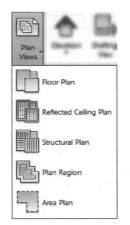

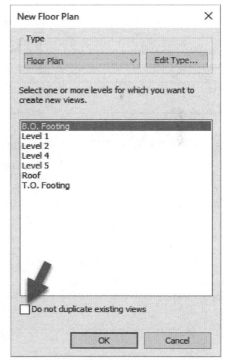

Figure 10.3-11
New floor plan view

The list in the center lists all the **Levels** in the project. However, by default, any Levels already associated with a view are hidden due to the check box below.

Do not duplicate existing views is a bit misleading. This just hides levels that are already associated with a Level in the current project. Projects often have multiple views associated with the same level; for example:

- Architectural Floor Plan
- Code Plan
- Furniture Plan
- Floor Finish Plan
- Enlarged Plans (cropped views)

Clicking **OK** created a new Floor Plan view and ends the command.

> The associated level cannot be changed. Never copy a floor plan with the intent of depicting another level. Again, the level cannot be changed and trying to adjust the View Range from one level to another will only result in several graphics issues for that view.

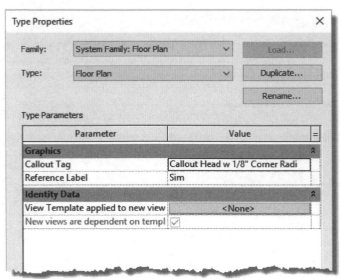

Figure 10.3-12 New floor plan type properties

Each floor plan has its own **View Range** settings (Figure 10.3-13), which are accessed via the Properties Palette (when nothing is selected in the model and the view is current).

For a plan view, architecturally, not much appears above the cut plane. The main setting here is **Cut Plane**. This determines if windows appear, where stairs are cut for example.

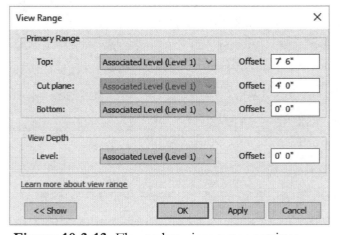

Figure 10.3-13 Floor plan view range settings

The Bottom setting is where Revit stops looking. Thus, even when a floor has not been created yet, the elements on the floor/level below will not appear.

The **View Depth** section allows some elements, which fall between the **Bottom** and **View Depth**, to be overridden with a special **Line Style** called **<beyond>**. The View Depth must be equal to, or lower than, the Bottom setting.

Plan Views → Reflected Ceiling Plan

This command is essentially the same as the previous command, Floor Plan, with the exception being the view includes a reflected representation of the elements above the cut plane, such as ceilings, bulkheads, ductwork, piping, lights and more.

For RCP (reflected ceiling plan) views, the View Range options are a little different (Figure 10.3-14). Click the "Learn more about view range" link to access more information on this topic. Also, click the **<<Show** button to see a diagram showing how this feature affects the view.

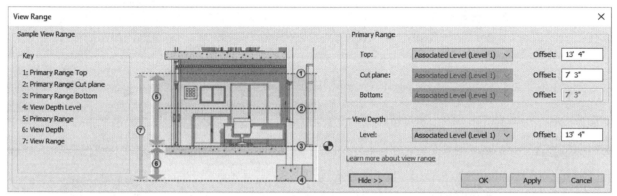

Figure 10.3-14 RCP plan view range settings with sample graphic expanded on left

Plan Views → Structural Plan

This command is similar to the Floor Plan view command just covered. The difference is the **Discipline** setting, the initial **View Range** settings (Figure 10.3-15) and an extra **Type Properties** setting called **View Direction**.

A Structural floor plan cannot be turned into a regular floor plan. However, any floor plan view template can be applied to it.

> Structural plan views are often set to Coarse so beams and joists appear as single lines rather than their actually thickness.

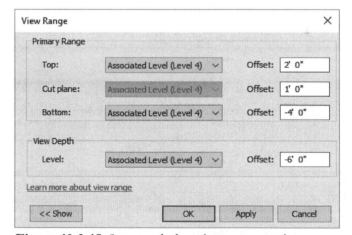

Figure 10.3-15 Structural plan view range settings

Plan Views → Plan Region

A Plan Region provides a way to override the View Range in a specific area of a plan view. Using this command, an enclosed area is sketched. During creation, and when selected later, the View Range of just this area can be controlled.

Use this to show windows, on an exterior wall, which are higher than the cut plane defined in the view. The View Range, just around the windows, can be made to appear in the floor plan view by raising the Cut Plane. In this case, raising the Cut Plane for the entire view might have a negative effect in another area – thus, a Plan Region is needed.

Plan Views → Area Plan

See the **Architecture** tab → **Room & Area** panel chapter for coverage of this command.

Elevation → Elevation

This command is used to create exterior or interior elevations.

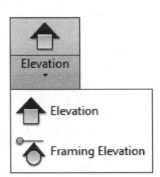

From a floor plan view, click near a wall to be elevated – before clicking, be sure the pointer on the elevation tag is properly positioned. The pointer automatically changes, during placement, to point at the nearest perpendicular wall.

Once placed in a plan view, selecting the center portion of the elevation tag causes a few graphic options to appear (Figure 10.3-16).

When the "body" of the elevation tag is selected, there is an option to "turn on" the other elevations by checking the desired boxes. Doing so will cause additional elevations to appear in the Project Browser. However, selecting and deleting the "body" part will cause all related elevations to be deleted.

Unchecking a box will also delete a view from the Project Browser.

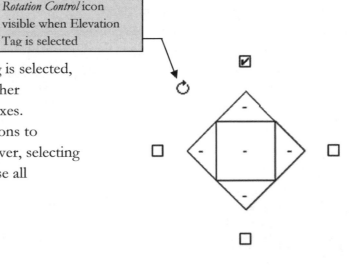

Rotation Control icon visible when Elevation Tag is selected

Figure 10.3.16 Selected elevation tag

Selecting the pointer-part of an elevation tag selected the cut plans and far clip plane as seen below.

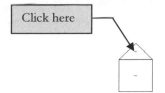

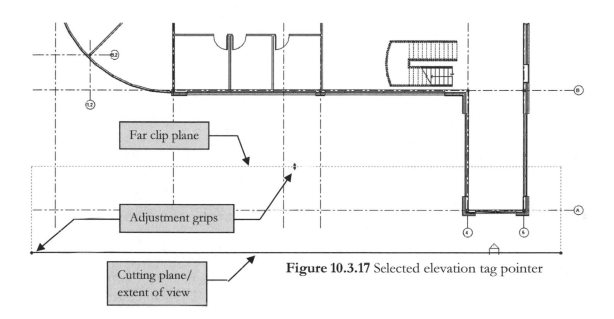

Figure 10.3.17 Selected elevation tag pointer

The elevation tag, as selected in Figure 10.3.17, has several features for controlling how the elevation looks. Here is a quick explanation:

- **Cutting plane/extent of view line:** This controls how much of the 3D model is elevated from left to right (i.e., the width of the elevation). This list also acts like a section line in that nothing behind this line will show up.

- **Far clip plane:** This controls how far into the 3D model the elevation view can see. This can be turned off in the elevation's *View Properties*, not the plan's *View Properties*, making the view depth unlimited; the next page covers this more.

- **Adjustment grips:** You can drag these with the mouse to control the features mentioned above. These values can also be controlled in the elevation's *View Properties*.

Notice how the "extent of view" line was created to match the width of the building.

You have several options in the Properties Palette (Figure 10.3-18). Under the heading *Extents,* notice the three Crop/Clipping options—these control the following:

- **Crop View**: This crops the width and height of the view in elevation. Adjusting the width of the cropping window in elevation also adjusts the "extent of view" control in plan view.

- **Crop Region Visible**: This displays a rectangle in the elevation view indicating the extent of the cropping window, described above. When selected in elevation view, the rectangle can be adjusted with the adjustment grips. See Figure 10.3-19.

- **Far Clipping**: If this is turned off, Revit will draw everything visible in the 3D model, within the extent of view.

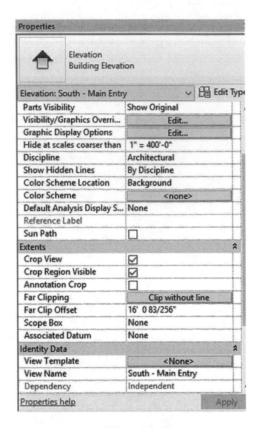

Figure 10-3.18 View properties

The next image is the elevation view (Figure 10.3-19) that resulted from creating the elevation in plan view (Figure 10.3-17).

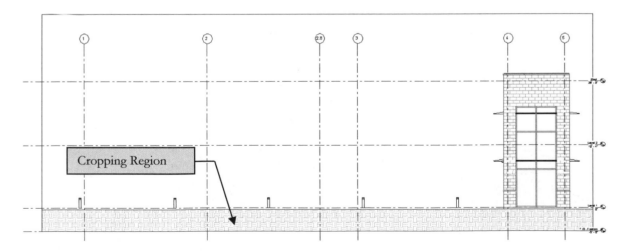

Figure 10.3-19 Elevation with far clipping plane through main entry

As you can see, the **Grid** lines appear automatically in elevations, as do the **Levels**. Take a minute to observe how the far clip plane controls the visibility of Grid line **1.2** at the curved wall; it must be intersecting the grid lines in plan for it to show up in elevation. This is good as it just clutters the drawing to see grid lines that have nothing to do with this part of the drawing.

Notice that much of the elevation is not visible. This is related to the "Far Clip Plane" set in the plan view, or in the view's properties.

Clicking the ***Hide Crop Region*** icon on the View Control Bar makes the Crop Region disappear.

Turning off the View Name on the Elevation Tag:

The interior elevation tag has the view name showing in plan view by default - when using the project templates provided with Revit. You can turn this off as it is not typically required on construction drawings.

1. Select **Manage → Settings → Additional Settings** (down-arrow)**→ Elevation Tags**.

2. Set the Type to **½" Circle** (Figure 10.3-20).

3. Select **Elevation Mark Body_Circle: Filled Arrow** from the Elevation *Mark* drop-down list (Figure 10-2.6).

4. Click **OK** to close the dialog box.

Figure 10.3-20
Adjusting the interior elevation

The view name should now be gone from the *Elevation Tags* in the floor plan views.

> If you right-click on an elevation view name in the Project Browser, you will notice you can **Duplicate** the view (via Duplicate View → Duplicate). This is similar to how a floor plan is duplicated with one major variation: when an elevation view is duplicated, Revit actually just copies another elevation tag on top of the one being duplicated in the plan view. This is because each elevation view requires its own elevation tag. If two views

could exist based on one elevation tag, then Revit would not know how to fill in the drawing number and sheet number if both views are placed on a sheet.

Elevation → Framing Elevation

Use this command to create a new elevation view based on a Grid line or Named Reference Plane. These are the required steps:

- In a plan view, select the command
- Click on a Grid line or Named Reference Plane
- An elevation tag is placed near the point clicked
- A new view is created
  - The crop region of this view matches the extents of the Grid/Ref. Plane
  - The Work Plane is set to the Grid-Ref. Plane

Use this view to place bracing and document vertical framing (columns and beams) at a specific plane in the building (Figure 10.3-21).

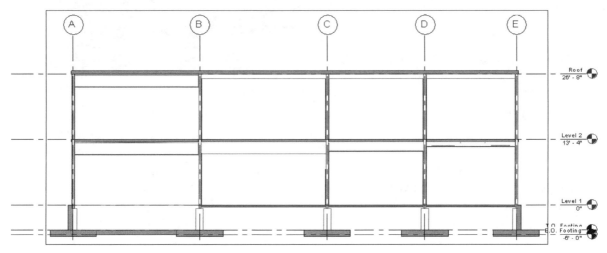

Figure 10.3-21 Framing elevation view

Drafting View

Drafting views are 2D drawing areas which have no connection to the 3D model. These views are used for creating diagrams and generic, or typical, details.

This command prompts for a view **Name** and **Scale**, both of which can be changed later if needed (Figure 10.3-22).

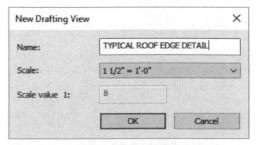

FIGURE 10.3-22 New drafting view

When creating an Elevation, Section or Callout, use the **Reference Other View** on the Ribbon, to document a reference in a view to a Drafting View. Be sure to verify the selection, as this can be a source of user error.

Given the 2D nature of a Drafting View, only commands from the Annotate tab are used to develop a drawing in one of these views. Use the provided Detail components to quickly add various elements of detail to the drawing.

Drafting Views are placed on sheets like any other view; open a sheet and then drag the view from the Project Browser onto the sheet. A Drafting View can only be placed on a single sheet. The view name is the drawing title on the sheet. Use the **Title on Sheet** parameter to specify a different name from the view name (Figure 10.3-23).

The **Visual Style** for a Drafting View is limited to **Wireframe** and **Hidden Lines**. Wireframe will make masking regions (but not solid fill regions) transparent—which can be helpful for troubleshooting.

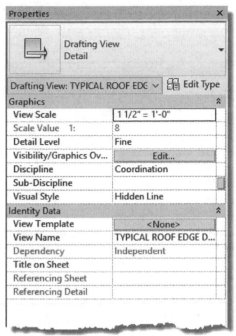

Figure 10.3-23
New drafting view properties

Duplicate View → Duplicate View

Use this command to make another copy of the current view, model or drafting view without including the annotation and detail lines in the view.

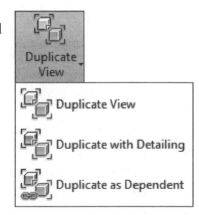

For <u>model views</u> - i.e., floor plans, sections or elevations – the new view has no tags, dimensions or text; Grids and Levels are exceptions—they will appear in the new view. For <u>drafting views</u>, the view is completely empty, as everything in the view is considered "detail" elements.

This new copy can be modified separately and placed on a different sheet.

▌ This command is also accessible by right-clicking on view name in the Project Browser.

The **Project Browser** can be sorted by **View not on Sheets**. This helps identify views no longer needed, which can be deleted. Or, the view has been missed and should be placed on a sheet.

Duplicate View → Duplicate with Detailing

Use this command to make another copy of the current view and include the annotation and detail lines. This new copy can be modified separately and placed on a different sheet. The annotation in the new view has no connection to the original.

█ This command is also accessible by right-clicking on view name in the Project Browser.

Duplicate View → Duplicate as Dependent

Use this command to create a view whose properties and detailing are tied to the original view, thus, a dependent view. This is often done to manage large buildings that need to be broken up onto multiple sheets.

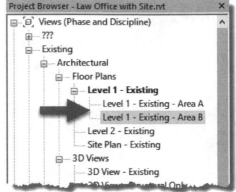

█ This command is also accessible by right-clicking on view name in the Project Browser.

Dependent views appear indented in the default Project Browser organization (Figure 10.3-24). When the original view is deleted, the dependent views are also deleted. When the Project Browser is customized, the dependent views may not appear indented or even in close proximity to the original view... thus, use caution when deleting views.

Figure 10.3-24 Dependent Views

All view settings, such as View Scale, are the same between the original view and all dependent views. The only thing that can be different are elements hidden in a dependent view using the select/right-click option to hide them.

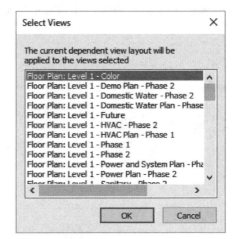

Once a floor plan is set up with one or more dependent views, this configuration can be quickly copied to other floor plan views. Simply right-click on the original view, that already has the dependent views, and select **Apply Dependent Views**. A list of other floor plan views appears (Figure 10.3-25).

Figure 10.3-25
Apply dependent Views

The master, or original, view can show its crop region and the crop region for each dependent view (Figure 10.3-25).

█ The Crop Region can be rotated to align an angled wing of a building with the sheet (and computer screen). Just select the Crop Region and use the Rotate command. This does not actually rotate the building. Also, this is different than the option, on the Options

Bar, when the view is selected on the sheet.

The Annotation Crop option is automatically turned on so text outside of the cropped area is hidden. In the dependent views the grids and sections will automatically resize.

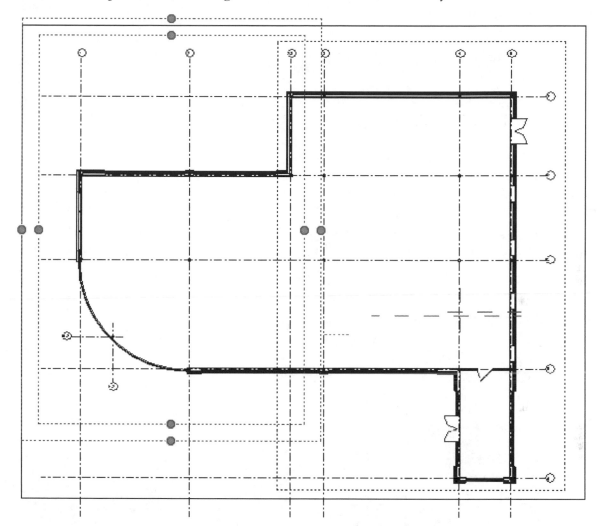

Figure 10.3-26 Crop region for two dependent views, one on left selected

Additional related commands include the **Matchline** command and the **View Reference** command. These help the person looking at the printed drawings know which sheet to flip to in order to see the continuation of that part of the floor plan.

Legends → Legend

This command is nearly identical to the Drafting View command, with the following exceptions:

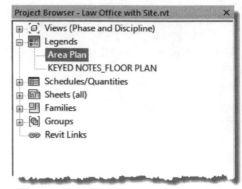

- A single Legend can be placed on multiple sheets
- No Drawing Number or Sheet Number tracking
- Legends have a separate branch in the Project Browser (Figure 10.3-27).

This command is also accessible by right-clicking the word Legend in the Project Browser.

Figure 10.3-27
Legends in Project Browser

Doors, Windows, Walls, etc., can be placed in a Legend view. Drag a type from the project Browser into a Legend view. Notice, before clicking to place, that the Options Bar gives the option to switch from plan or elevation view. This is a great way to create a live legend of modeled elements; however, tags cannot be added.

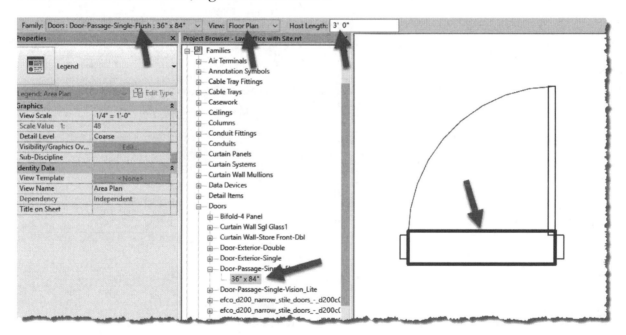

Figure 10.3-28 Adding model-based content to a legend

Legends → Keynote Legend

Create a schedule which lists all the elements which have been tagged with a keynote anywhere in the project (Figure 10.3-29). A keynote legend can be placed on multiple sheets.

> This command is also accessible by right-clicking the word Legend in the Project Browser.

Revit can only use one Keynote text file at a time. In this example, the keynote file was changed after tagging multiple elements. The Key Value was saved in the Keynote parameter for each element, but the Keynote Text data is no longer available, so the field is blank. To correct the issue, in the schedule view, click in the Key Value cell and select the small icon that appears to open the keynote mapping dialog.

Keynote Legend	
Key Value	Keynote Text
06 16 00.C3	
06 16 00.C7	
06 22 00.A2	
06 22 00.A4	
06 40 00.A2	
09 22 16.13.D1	
09 22 16.13.I4	
IE2	COUNTERTOP WITH BACKSPLASH
IE3	BASE CABINET
IE5	GYP BD SOFFIT
IE6	SOFFIT ABOVE WALL CABINETS
IE8	SUPPORT BRACKET
IE9	REF
IE10	VENDING

Figure 10.3-29 Keynote legend

A keynote legend can be filtered but the parameters are limited (Figure 10.3-30).

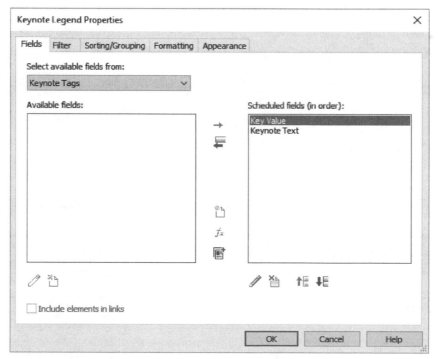

Figure 10.3-30 Keynote legend parameters

Schedules → Schedule/Quantities

Schedules are incredibly powerful in Revit. We will not be able to cover this feature in its entirety within the constraints of this book. What follows is a detailed overview, with tips and tricks, which should prove helpful for Revit users of any skill level.

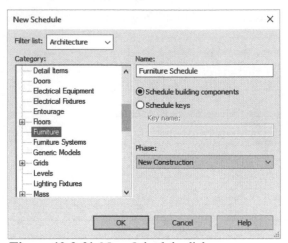

When starting the command, the first step is to select the **Category** (Figure 10.3-31). The **Name** and **Phase** can be changed later, but the Category option is a one-time deal.

Every element in Revit goes in a specific Category, which is defined when the content was initially created.

Figure 10.3-31 New Schedule dialog

The list of Categories is predefined by Revit and cannot be changed.

Most of the *Category* names are pretty straightforward; walls go in the *Walls* category, doors in the *Doors* category, and so on. Below are a few examples that might not be as easy to deduce.

- **Casework**: Base cabinets and wall cabinets, reception desk, built-in bookshelves
- **Furniture Systems**: Cubicles and other built-in office furniture
- **Specialty Equipment**: Toilet room accessories, lockers, ladders, etc.
- **Topography**: Site or ground surface

When a Category is selected for a new schedule, Revit knows to look for all the elements that have been placed in the model which are in that Category. A **<Multi-Category>** schedule has the ability to look at every element in the model and report on parameters they all have in common, such as Cost.

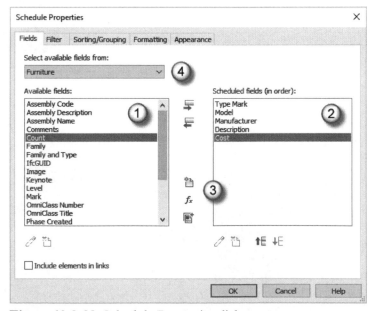

Schedule Properties

Once a schedule is created, you are immediately brought to the **Schedule Properties** dialog (Figure 10.3-32). This is the place you select

Figure 10.3-32 Schedule Properties dialog

what information you want listed in the schedule and what it should look like. We will take a few minutes to review the various options.

You will notice the Schedule Properties dialog is divided into five separate areas; each area has its own tab near the top. In a moment we will take a look at each tab's contents.

While in a schedule view, the Properties Palette shows information related to the current schedule (Figure 10.3-33). Clicking any of the **Edit** buttons will open the **Schedule Properties** dialog.

Before Revit will create the schedule, you have to select one or more parameters to be scheduled. To do this you add **Available Parameters**, list on the left, to the **Scheduled Fields (in order)** list, on the right. Simply select a parameter on the left and click the **Add →** button, or just double-click the parameter name.

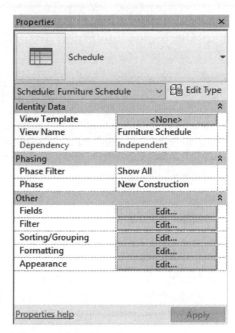

Figure 10.3-33 View Properties

If you simply selected the five parameters shown on the right (Figure 10.3-32, #2) and clicked OK the schedule would be created. A new item would appear in the Project Browser, under **Schedules/Quantities**. If you are in a new or empty project, the schedule would be empty (Figure 10.3-34). On the other hand, if the model has 30 items in the selected category, you should see 30 lines/rows, one for each instance in the model. Additionally, if those 30 elements have information entered into any of the scheduled parameters, this information will appear in the schedule automatically.

> *Tip:* When working with schedules with many elements you may find it helpful to zoom the schedule. The schedule view can be zoomed in and out using **Ctrl +** and **Ctrl –** respectively.

If you close the Schedule Properties box, this is how you get back to it: open the schedule view and then select one of the **Edit** buttons in the **Properties Palette** (Figure 10.3.33). You do not click the Schedules/ Quantities command as this is only to create a new one.

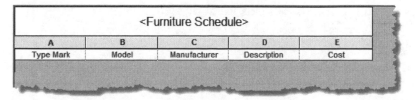

Figure 10.3-34 New schedule created in an empty model

Schedule Properties: Fields Tab

Looking at the **Fields** tab (Figure 10.3-32) we will take a look at the four areas highlighted.

1). **Available Fields:** Each element type, or *Category*, has a number of fields, or parameters, associated with them. Some of these parameters are hard-wired into the software, some have been added initially in the content, and some are created in the project as needed (see item 3 below).

   In addition to common parameters such as *Cost* and *Manufacturer*, each category has a unique list of options. Here are a few selective examples:

   Casework: Construction Type, Depth, Finish
   Lighting Fixture: Circuit Number, Ballast Loss, Wattage
   Windows: Width, Height, Sill Height

2). **Scheduled fields (in order):** This list determines what information appears in the schedule. Each parameter listed here equals a column in the schedule. The top item is the first column on the left in the schedule. Compare the order of parameters in Figure 10.3-31 with the schedule in Figure 10.3-32.

   These parameters can be reordered using the **Move Up** and **Move Down** buttons below; you cannot drag the parameters. It is also possible to **Edit** and **Delete** from parameters; however, most built-in parameters cannot be deleted nor edited.

   > Edit allows you to change the parameter Name, select a different Group in which the parameter should fall, and select which Categories this parameter should be available in.

3). **Custom and Calculated Parameters, plus Combine Parameters:** When you have information that needs to be stored in an element but there is no appropriate parameter, you can create one. You will see two examples, custom and calculated, on the next two pages. A newer option is to combine multiple parameters into a single column – for example, a table width and depth could be listed as 4'-0" x 6'-0".

4). **Select available fields from:** In the example shown, we are looking at a **Furniture Schedule**, thus we are able to select *Available Fields* from the *Furniture* category. You may also change this drop-down list to *Room*. This allows you to list which room number and room name each piece of furniture falls in.

   > Changing the drop-down from Furniture to Room also changes the fields that appear in the Available fields list above.

Creating a Parameter:

When you click the **Add Parameter...** button (Figure 10.3-32), the **Parameter Properties** dialog opens (Figure 10.3-35). Here you specify the following:

Project or Shared parameter: The main thing you need to know is if you want the information stored in this parameter to appear in a tag, it has to be a *Shared Parameter.*

Instance or Type parameter: Do you want the value entered to be unique for each instance, or the same? You might want the "Chair Fabric" to be an instance because the same model number can have different colors, but the "Recycled Content" to be the same value per model number. To state it another way: if you have four chairs in the model, two may have a red fabric and two blue, but they all have the same **.35** value for recycled content. Changing the recycled content value for one chair automatically updates the other three.

Parameter Name: Be sure to use a simple name everyone can understand. Do not use dashes as they are seen as a minus sign in formulas.

Discipline: *Common*, *Structural*, and *Electrical* are the options in Revit. This determines what is listed in the next drop-down list: *Type of Parameter*.

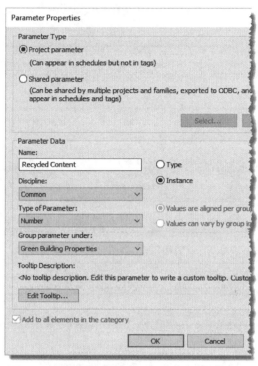

Figure 10.3-35 Parameter Properties; creating a new parameter

Type of Parameter: This option requires you to know what type of information you will be storing in the parameter. Will it be Currency, Volume, Slope, URL, Material, etc.? **This is the one thing in this dialog you cannot change once you click OK.** If it needs to change, you have to delete the parameter and recreate it. Selecting *Text* is the most flexible as it can hold words and numbers. But, selecting something specific, such as *Integer*, will prevent someone from entering bad data (e.g., a word or a decimal value).

Group Parameter Under: This simply determines which heading the parameter falls under in the *Properties* window. Looking back at Figure 10.3-33, you see three group headings: Identity Data, Phasing and Other.

Creating parameters is somewhat like "graphical" computer programming. This is part of what makes up the "I" in BIM - **B**uilding **I**nformation **M**odeling.

Creating a Calculated Value:

Whereas *Parameters* show up in the *Properties Palette* and schedules, also in tags when using shared parameters, **Calculated Values only appear in schedules.**

> Tags can now also have calculated values, but they are not the same parameters created in scheduled.

A Calculated Value can be used for many situations. Here are just a few examples:

- Calculate sales tax
- List percentage difference between actual and programmed square footage
- Determine occupant load for a room based on building code

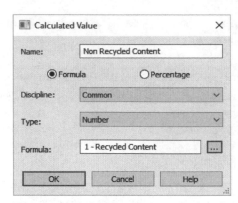

Figure 10.3-36 Creating a calculated value

Parameters used in a formula are *case sensitive*. That means they need to be typed exactly as they appear in the project. To help avoid making a mistake, you can click the "..." button to the right of *Formula* to select from a list of available parameters, which can include other *Calculated Values.*

Basic mathematical formulas may be used. Here are a few examples:

Parameter A * Parameter B
Parameter A + Parameter B
Parameter A - Parameter B
(Parameter A + Parameter B + Parameter C) * Parameter D

It is also possible to use constant values:

Parameter A * .1   *or*   Source_Distance - 500

Sometimes you need to strip the feet and inches from a parameter so it can be used in a formula. For example, you cannot multiply 5'-6" x $5 because Revit is confused by the feet and inches, but 5.5 x $5 works. Here is how that is done:

Parameter X = 10'-6" (Parameter X is a *Length* parameter type)
Parameter Y = Parameter X / 1' = 10.5 (Parameter Y is a calculated value set to *Number*)

Here is a more complicated set of *Calculated Values* used in a schedule to calculate the occupant load for each space in the building; focus on the mathematical expressions rather than the specific example at this time:

Control = if(OLF_Calc > OLF_Integer, 1, 0)
Occ_Load_Calc = if(Control = 1, OLF_Calc + 1, OLF_Integer)
OLF_Integer = (Area / SF_Per_Occupant) / 1 SF
OLF_Calc = ((Area / SF_Per_Occupant) / 1 SF)

To see an example of what we have covered thus far, notice the drawing below (Figure 10.3-37). The drawing depicts two desks and two chairs in a model. For this example, this is all the furniture in the entire model at this time.

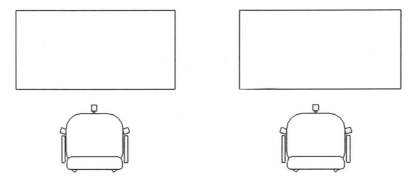

Figure 10.3-37 Four furniture elements in a Revit file

<Furniture Schedule>						
A	B	C	D	E	F	G
Type Mark	Model	Manufacturer	Description	Cost	Recycled Content	Non Recycled Con
D1	SC2401F	STEELCASE	OFFICE DESK	2550.00	0.35	0.65
D1	SC2401F	STEELCASE	OFFICE DESK	2550.00	0.35	0.65
C1	LEAPV2	STEELCASE	OFFICE CHAIR	1325.00	0.1	0.9
C1	LEAPV2	STEELCASE	OFFICE CHAIR	1325.00	0.1	0.9

Figure 10.3-38 Furniture elements automatically appear in the furniture schedule

The schedule we talked about creating, shown in Figure 10.3-38, now has four rows which relate to the elements placed in the model (Figure 10.3-37). Additionally, a new Parameter was created to track **Recycled Content**. A Calculated Value was created to determine the **Non Recycled Content**.

The width of a column may need to be adjusted to see the title (as in the example above) or the contents of the cells. Simply place your cursor between the columns and drag. Later we will see how the column heading in the schedule can be different from the actual parameter name (e.g., Manufacturer can be abbreviated to MFR.).

One last comment about the *Fields* tab: notice the **Include elements in linked files option**. When you have other Revit models linked into your model, structural, mechanical and electrical, etc., you can check the box in the lower left to be able to schedule elements located in the linked models.

Schedule Properties: Filter Tab

Schedules may be filtered by one or more parameter criteria. In the example below (Figure 10.3-39), the furniture schedule is being filtered by any **Type Mark** parameter which begins with the letter "C." In our two chairs and two desks example drawing, the desks are now hidden because their *Type Mark* begins with "D." This filter assumes the *Type Mark* for all chairs will begin with the letter "C." Obviously, someone could key in the wrong value for *Type Mark* and create a problem. See the filtered result in Figure 10.3-40.

It should be pointed out that each additional filter condition added makes the final result more restrictive. Note the "**and**" next to the 2nd, 3rd and 4th filters (not an "**or**").

Revit will only filter by a parameter that is added to the schedule. Looking back at Figure 10.3-32 we cannot filter by furniture located on Level 1 because the parameter *Level* is not in the schedule, even though it is "available" in the model.

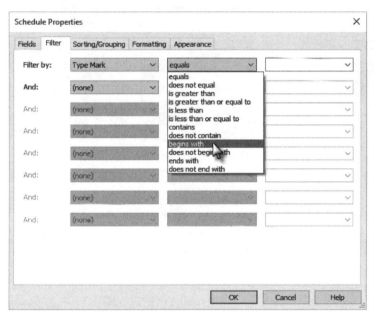

Figure 10.3-39 Schedule Properties: Filter tab

<Furniture Schedule>						
A	B	C	D	E	F	G
Type Mark	Model	Manufacturer	Description	Cost	Recycled Content	Non Recycled Con
C1	LEAPV2	STEELCASE	OFFICE CHAIR	1325.00	0.1	0.9
C1	LEAPV2	STEELCASE	OFFICE CHAIR	1325.00	0.1	0.9

Figure 10.3-40 Furniture schedule filtered for chairs

Having a schedule, such as this example, which just reports chairs, does not mean there is no way to now schedule desks. It is possible to have multiple "furniture" based schedules: one schedule for each level, one for chairs, one for desks, one for estimating, etc. Each furniture-based schedule would just be reporting a different set of parameters from those available.

Schedule Properties: Sorting/Grouping Tab

There is a lot more to the **Sorting/Grouping** tab than meets the eye. As the name implies, we can control sorting. Looking back at Figure 10.3-38, we can see the *Type Marks* are not sorted alphabetically. In Figure 10.3-41, we have told Revit to sort by *Type Mark* in ascending order. The result is seen in Figure 10.3-42; notice the *Type Marks* are now listed in ascending order.

> Note that we have turned off the Filtering applied on the previous page.

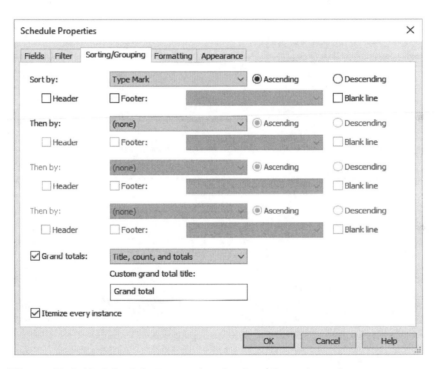

Figure 10.3-41 Schedule Properties: Sorting/Grouping tab

			<Furniture Schedule>			
A	B	C	D	E	F	G
Type Mark	Model	Manufacturer	Description	Cost	Recycled Content	Non Recycled Con
C1	LEAPV2	STEELCASE	OFFICE CHAIR	1325.00	0.1	0.9
C1	LEAPV2	STEELCASE	OFFICE CHAIR	1325.00	0.1	0.9
D1	SC2401F	STEELCASE	OFFICE DESK	2550.00	0.35	0.65
D1	SC2401F	STEELCASE	OFFICE DESK	2550.00	0.35	0.65
Grand total: 4						

Figure 10.3-42 Furniture schedule sorted by parameter "Type Mark"

In a sense, when a list is sorted it is grouped. But what Revit means by **Grouping** relates to the **Header**, **Footer** and **Blank Line** check boxes. The two images below show how we can *Group* by the floor level and then *Sort* by the *Type Mark*. The next page has more information.

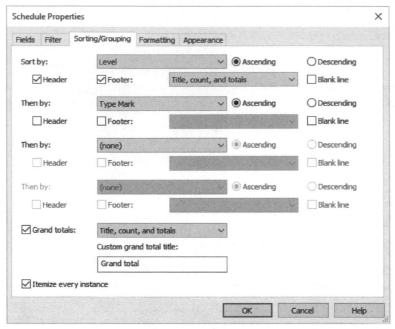

Figure 10.3-43 Schedule Properties: Sorting/Grouping tab revised

			<Furniture Schedule - Group by Level>			
A	**B**	**C**	**D**	**E**	**F**	**G**
Type Mark	Model	Manufacturer	Description	Cost	Recycled Content	Non Recycled Con
Level 1						
C1	LEAPV2	STEELCASE	OFFICE CHAIR	1325.00	0.1	0.9
C1	LEAPV2	STEELCASE	OFFICE CHAIR	1325.00	0.1	0.9
D1	SC2401F	STEELCASE	OFFICE DESK	2550.00	0.35	0.65
D1	SC2401F	STEELCASE	OFFICE DESK	2550.00	0.35	0.65
Level 1: 4						
Level 2						
C1	LEAPV2	STEELCASE	OFFICE CHAIR	1325.00	0.1	0.9
C1	LEAPV2	STEELCASE	OFFICE CHAIR	1325.00	0.1	0.9
C1	LEAPV2	STEELCASE	OFFICE CHAIR	1325.00	0.1	0.9
D1	SC2401F	STEELCASE	OFFICE DESK	2550.00	0.35	0.65
D1	SC2401F	STEELCASE	OFFICE DESK	2550.00	0.35	0.65
D1	SC2401F	STEELCASE	OFFICE DESK	2550.00	0.35	0.65
Level 2: 6						
Grand total: 10						

Figure 10.3-44 Furniture schedule grouped by parameter "level"

Of course, for this example, three more chairs and three more desks have been added to Level 2 in our Revit model. Remember, you don't need to be following along in Revit at this time. In order to sort or group by level, the *Level* parameter had to be added to the **Scheduled Fields (in order)** list on the *Fields* tab. Normally, this would create a new column in the schedule. You will see how this column was hidden when you review the *Formatting* tab.

When *Header* was checked in Figure 10.3-43, a header space was added with the word **Level 1** listed, and another above the Level 2 furniture items. The footer allows you to list totals by level (because 'level' is the current sort by parameter). The default is only the number of elements listed in the schedule per level with a grand total near the very bottom. When you get to the *Formatting* tab you will also learn how to make specific columns add up automatically. For example, what is the total for the *Cost* column?

Finally, for the *Sorting/Grouping* tab, you will look at the **Itemize Every Instance** option at the bottom of the *Sorting/Grouping* tab.

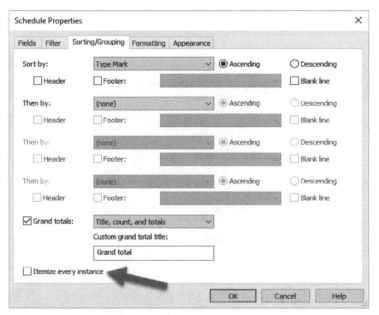

Figure 10.3-45 Schedule Properties: Sorting/Grouping tab revised

<Furniture Schedule - Do not itemize every instance>							
A	B	C	D	E	F	G	H
Type Mark	Model	Manufacturer	Description	Count	Cost	Recycled Content	Non Recycled Con
C1	LEAPV2	STEELCASE	OFFICE CHAIR	5	1325.00	0.1	0.9
D1	SC2401F	STEELCASE	OFFICE DESK	5	2550.00	0.35	0.65
Grand total: 10							

Figure 10.3-46 Furniture schedule NOT itemizing every instance

The two examples above, Figures 10.3-45 and 10.3-46, show sorting just by *Type Mark* and with *Itemize every instance* turned off. When *Itemize every instance* is turned off, the schedule only shows one row for each element in the model. That is, there is one row for the five chairs and one row for the five desks.

This example in Figure 10.3-46 has had the **Count** parameter added to the *Scheduled Fields* list on the *Fields* tab. Notice how the *Count* column automatically shows the correct number of items. This is a built-in parameter, not a *Calculated Value*. Finally, the Cost parameter still lists the per unit cost rather than a total cost for five units. This may well be the information you are looking for and nothing needs to change. However, in the *Formatting* tab discussion you will learn how to make the *Cost* column report the total for five units.

Schedule Properties: Formatting Tab

There are a number of items which can be controlled from the *Formatting* tab (Figure 10.3-47).

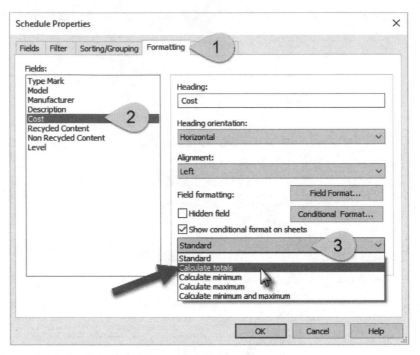

Figure 10.3-47 Schedule Properties: Formatting tab

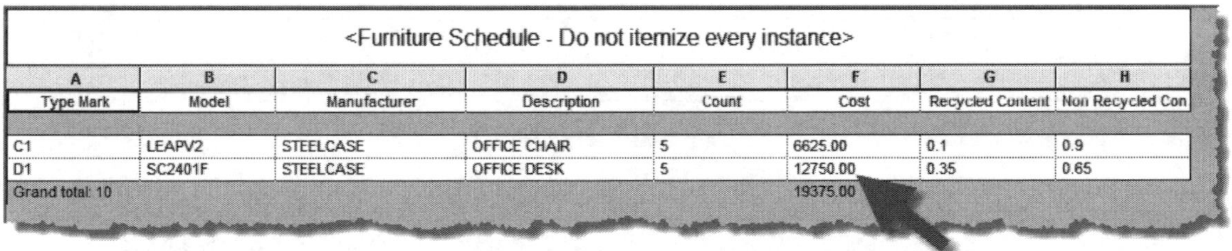

			<Furniture Schedule - Do not itemize every instance>				
A	B	C	D	E	F	G	H
Type Mark	Model	Manufacturer	Description	Count	Cost	Recycled Content	Non Recycled Con
C1	LEAPV2	STEELCASE	OFFICE CHAIR	5	6625.00	0.1	0.9
D1	SC2401F	STEELCASE	OFFICE DESK	5	12750.00	0.35	0.65
Grand total: 10					19375.00		

Figure 10.3-48 Furniture schedule with Cost parameter set to "Calculate Totals"

Following is a brief description of the various parts of the *Formatting* tab:

Fields: This is a list of the *Fields* which have been added to the schedule on the *Fields* tab. All the settings on the right relate to the parameters (or fields) selected from this list.

Heading: This is the text that is used at the top of the column in the schedule. By default it matches the parameter name. However, it is possible to change the heading and not affect the parameter name. This is often done to shorten the heading, which saves room as the column does not need to be as wide. For example, **Room Number** might be changed to **RM. #**. This allows the column to be narrower because the numbers are often 3-4 digits (e.g., 1004).

Heading orientation: The header text may be horizontal, which is the default, or vertical (Fig. 10.3-49).

> Changing the heading to vertical is only apparent when the schedule is placed on a sheet.

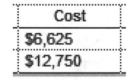

Figure 10.3-49 Type Mark heading set to be vertical

Alignment: This determines whether the information in a given column is left, right or center justified.

Field Format... (button): This option only applies to *Fields* with numbers. When the project-wide format settings (*Manage → Project Units*) do not work, they may be overridden for this one *Field*. Figure 10.3-50a shows the **Cost** parameter being changed and Figure 10.3-50b shows the result in the schedule.

Conditional Format... (button): The background for a cell can have its color changed if certain conditions exist. For example, if the number is below 20, make the background red; if the *Type Mark* is blank, make the background yellow, etc. The background color does not appear on the sheet, nor does it print.

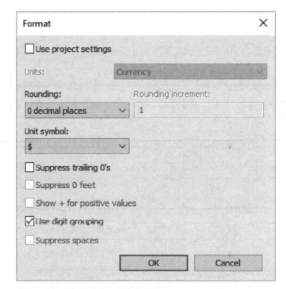

Figure 10.3-50A Field Format options

Hidden field (check box): Figure 10.3-44 requires the *Level* field be in the schedule for grouping, but we do not want to also see a column listing the schedule as this would be redundant. Thus, we select the *Level* field and set it to be hidden.

> You may also right-click in a column and see the hidden column options.

Figure 10.3-50B
Field Format results

Show conditional format on sheets (check box): This will make the filled cells, as defined via the *Conditional Format* button just discussed, actually print. If this is not checked, the filled cell will still appear in the schedule view, but it will not appear nor print on the sheet it is placed on.

Calculate totals (check box): This option does two things. If *Itemize every instance* is <u>not</u> checked, any fields set to **Calculate totals** will display the number times the count. Also, when a *Footer* with totals, or *Grand totals*, is selected on the *Sorting/Grouping* tab, a total value will appear below the column when **Calculate totals** is checked for that field, or column (see Figure 5-1.22).

Schedule Properties: Appearance Tab

This last tab in the *Schedule Properties* dialog has several settings which are very straightforward. Following are a few things to note:

- **Build schedule**, Top-down, Bottom-up is only available for revision schedules.
- Some appearance settings do not show until placed on a sheet.
- Text options are based on text styles defined in the Revit project.

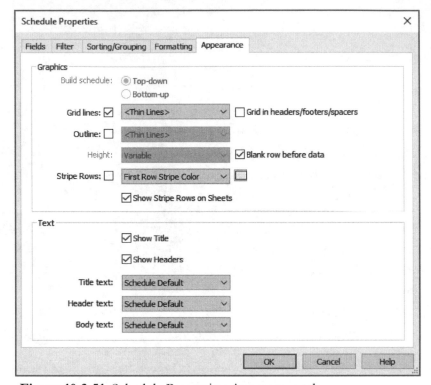

Figure 10.3-51 Schedule Properties: Appearance tab

Unfortunately, you cannot print a schedule unless it is on a sheet (see Exercise 18-1 for steps on how to place a view on a sheet). It is possible to export a schedule so it can be used, viewed or printed in a spreadsheet or database application. You must be in the schedule view, not the sheet, and then select **File Tab → Export → Reports → Schedules**.

Stripe Rows is a feature that shades every other row, which helps the user of the schedule track the correct row as they look from left to right on larger complex schedules. This shading can be set to show (and print) on sheets via the checkbox: **Show Stripe Rows on Sheets**. Clicking the icon to the far-right of the strip rows heading allows a custom color to be selected. Clicking the adjacent drop-down, an option exists to select the **Second Row Stripe Color**, where the same icon now picks a custom color for the second row.

Schedules → Graphical Column Schedule

Revit can create a Column Schedule similar to the Door and Room Schedules just covered; most elements can be scheduled like this. However, Revit can create a graphical schedule of columns and their related grids. This type of schedule is typical industry practice for structural engineers on a set of Construction Documents (CDs).

Simply select the command and you now have a Graphical Column Schedule (Figure 10.3-52) which can be placed on a sheet. Notice the grid intersections listed along the bottom. Also, each level is identified by horizontal lines within the schedule. The concrete foundation is shown, and then the steel. Some of the concrete columns look odd because they are missing the thickness for the foundation wall which passes by it. In this example, the columns were set to start 8″ below the Level 1 floor slab; this can be seen and double-checked in this view.

The Project Browser now has a new category called **Graphical Column Schedules**. It is nice that the various branches only show up when something exists in that section; otherwise the browser list would be difficult to navigate and manage.

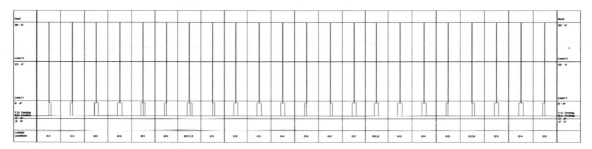

Figure 10.3-52 Graphical Column Schedule created

Next, we will look at a few ways in which you can control the graphics of the *Column Schedule*.

Click **Edit** for *Hidden Levels* (Figure 10.3-53): Revit allows you to hide levels that are not needed in the *Column Schedule*, for example, if someone on the design team added a level to manage the height of the top of masonry walls or window openings. It should be pointed out that *Level Datums* should only be used to define surfaces you walk on to avoid unintended issues. In this example, you could turn off the **B.O. Footing** level as it is not needed in this schedule.

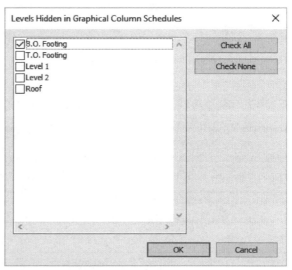

Figure 10.3-53 Hiding Levels

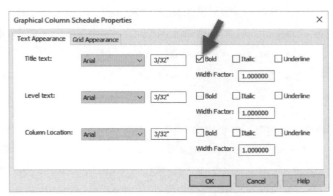

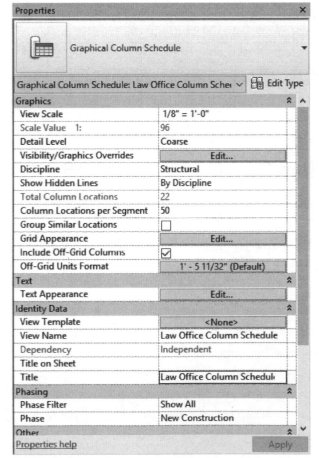

Figure 10.3-55 Text Appearance settings

Figure 10.3-54 Structural materials

Click **Edit** for *Material Types* (Figure 10.3-54): If you unchecked *Concrete* here, Revit would remove the concrete columns from the schedule.

Click **Edit** for *Text Appearance* (Figure 10.3-55). Various options are offered here to control the appearance of text in the schedule.

Notice a few more things about the Column Schedule's *Properties* (Figure 10.3-56).

The **total number of columns** considered in the schedule is listed (22 in our example).

Also, **Include Off-Grid Columns** is checked. This will include columns that do not fall on any grid lines; most columns fall on grid lines as that is the main purpose of grid lines.

The **Group Similar Locations** option will show one graphical column for each column size and then list all the grid intersections below it. This makes the schedule smaller, but if your project only has one column size, the schedule would actually be too small.

Figure 10.3-56
Graphical Column Schedule properties

Below is a final example with the edits described in this section applied.

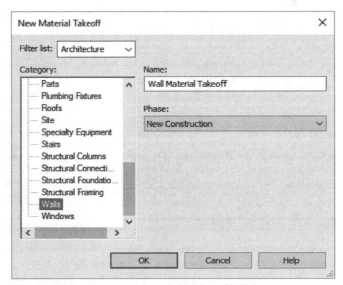

LAW OFFICE COLUMN SCHEDULE

| Column Locations | A-4 | A-5 | B-2 | B-3 | B-4 | B-5 | B-2-1.2 | C-1 | C-2 | C-3 | C-4 | C-5 | D-1 |

Roof 26' - 8"

Level 2 13' - 4"

Level 1 0' - 0"

T.O. Footing -5' - 0"

Figure 10-3.57 Graphical column schedule (partial view)

Schedules → Material Takeoff

Use this command to create a schedule which lists all the materials within in-place families, loaded families and/or system families. A regular schedule can only schedule parameters in a category, whereas the Material Takeoff schedule looks for materials used AND parameters. Thus, a Material Takeoff schedule can be made to look exactly like a regular schedule but also have applied material information.

The only way a material can be listed in a regular schedule is if a shared material parameter is added to a family, and some geometry in that family has its material mapped to that material parameter.

Clicking to start the command opens the **New Material Takeoff** dialog (Figure 10-2.58). Using the **Multi-category** options provides access to all materials in the project. Using a specific **Category** narrows the scope of the schedule.

As an example, here the wall category is selected. Click **OK** to proceed.

Figure 10.3-58 New Material Takeoff dialog

The next step, similar to regular schedules, is to select the material related information to list in the schedule. This is done in the **Material Takeoff Properties** dialog (Figure 10.3-59).

All the regular parameters for the selected category are listed, plus the material parameters—which are all prefixed with "Material:."

Figure 10.3-59 shows an example of a Material Takeoff Schedule.

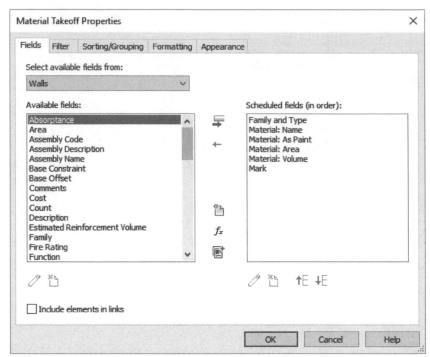

Figure 10.3-59 Material Takeoff Properties dialog

\<Wall Material Takeoff\>					
A	**B**	**C**	**D**	**E**	**F**
Family and Type	Material: Name	Material: As Paint	Material: Area	Material: Volume	Mark
Gypsum Wall Board					
Basic Wall: Interior - 4 7/8" Partition (1-hr)	Gypsum Wall Board	No	724 SF	37.73 CF	A
Basic Wall: Interior - 4 7/8" Partition (1-hr)	Gypsum Wall Board	No	512 SF	26.69 CF	B
Basic Wall: Interior - 4 7/8" Partition (1-hr)	Gypsum Wall Board	No	327 SF	17.03 CF	C
Basic Wall: Interior - 4 7/8" Partition (1-hr)	Gypsum Wall Board	No	94 SF	4.89 CF	D
Basic Wall: Interior - 4 7/8" Partition (1-hr)	Gypsum Wall Board	No	523 SF	27.25 CF	E
			2181 SF	113.58 CF	
Metal Stud Layer					
Basic Wall: Interior - 4 7/8" Partition (1-hr)	Metal Stud Layer	No	362 SF	109.42 CF	A
Basic Wall: Interior - 4 7/8" Partition (1-hr)	Metal Stud Layer	No	256 SF	77.39 CF	B
Basic Wall: Interior - 4 7/8" Partition (1-hr)	Metal Stud Layer	No	163 SF	49.37 CF	C
Basic Wall: Interior - 4 7/8" Partition (1-hr)	Metal Stud Layer	No	47 SF	14.17 CF	D
Basic Wall: Interior - 4 7/8" Partition (1-hr)	Metal Stud Layer	No	262 SF	79.02 CF	E
			1090 SF	329.38 CF	
Paint - Red					
Basic Wall: Interior - 4 7/8" Partition (1-hr)	Paint - Red	Yes	262 SF	0.00 CF	E
			262 SF	0.00 CF	
Grand total: 11			3533 SF	442.95 CF	

Figure 10.3-60 Material takeoff schedule example

The sorting and grouping in this example schedule are achieved with the following settings (Figure 10.3-61).

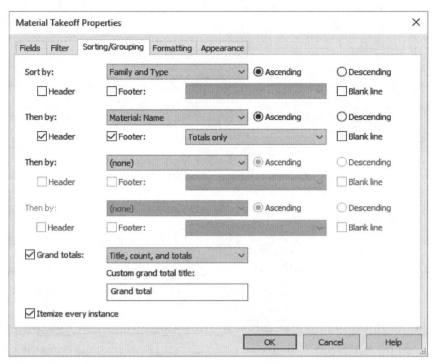

Figure 10.3-61 Material Takeoff Properties dialog; sorting

Schedules → Sheet List

This special schedule command is hardwired to only schedule sheet related information. Starting this command goes directly to the Sheet List Properties dialog (Figure 10.3-62). The options here are similar to the regular Schedule Properties dialog.

In addition to the common use of listing sheet number and name, this schedule can also list revision information as well as assembly data on a sheet.

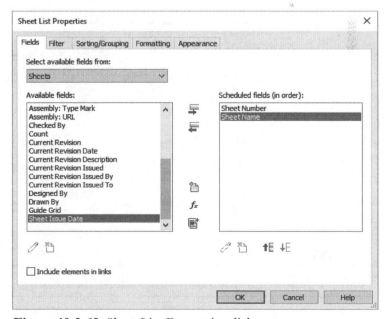

Figure 10.3-62 Sheet List Properties dialog

Schedules → Note Block

A **Note Block** schedule lists all the annotation families, which are placed using the Symbol command, in the current project. For example, in the plan view below, the diamond symbol is an annotation family calling out general construction items. In addition to listing this family, the schedule can list custom parameters such as the number and description.

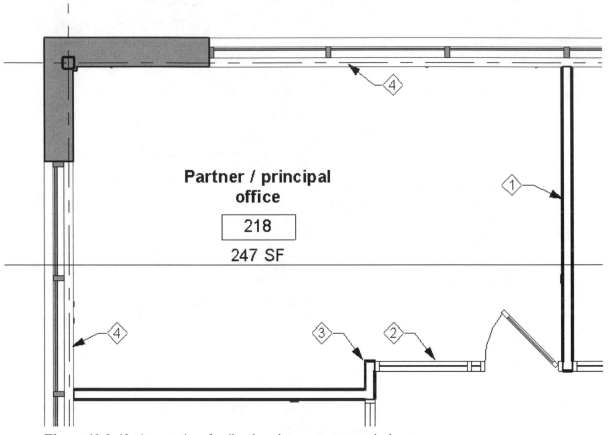

Figure 10.3-63 Annotation family placed to represent typical notes

Clicking the **Note Block** command opens the **New Note Block** dialog (Figure 10.3-64). Here, all of the annotation families loaded in the current project are listed. The schedule name is defined in the **Note block name** text box.

Clicking **OK** opens the **Note Block Properties** dialog (Figure 10.3-65). Here, the parameters in the families can be added to the schedule.

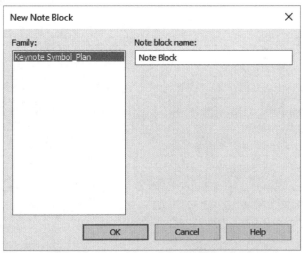

Figure 10.3-64 New Note Block dialog

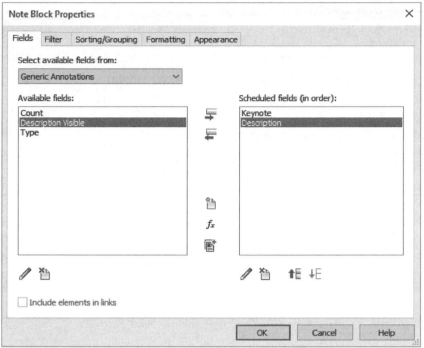

Figure 10.3-65 Note Block Properties dialog

A	B
	\<Note Block\>
Keynote	**Description**
1	COVER THIS WALL WITH WC.1 FROM FLOOR TO 32" AND PROVIDE WOOD TRANSITION TRIM PER SPECIFICATION
2	CURTAINWALL SYSTEM WITH SAGE ELECROCHROMIC GLASS SYSTEM
3	PROVIDE SS CORNER GUARD
4	WOOD WINDOW SILL
4	WOOD WINDOW SILL

Figure 10.3-66 Note Block schedule example

In the above example, keynote #4 appears twice because it is used twice in the model (Figure 10.3-63). This would not need to appear twice in the schedule. On the **Sorting/Grouping** tab, uncheck **Itemize Every Instance**—the schedule must also be set to sort by the Keynote parameter. Now there will only be one row for each keynote mark.

> Another parameter could be added, or a separate family, to keep track of keynotes by sheet, thus creating a Note Block schedule for each sheet, to limit the list to elements notes on a given sheet rather than all the keynotes used in the project.

Schedules → View List

A View List schedule lists the views in the current project and view parameters. This schedule can be used to review for naming conventions, view templates applied, and to see which views are placed on which sheets—or if they are placed on sheets at all.

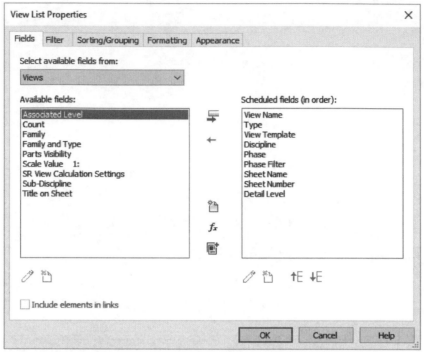

Figure 10.3-67 View List schedule

The following image, Figure 10.3-68, is an example based on the settings shown above.

Notice the following:

- The schedule is being sorted by View Name
- The Detail Level, View Template, Phase and Phase Filter settings can be changed here
- Drafting Views (i.e., Detail) do not have a Phase setting
- Type, Sheet Name and Sheet Number are read-only values

View

\<View List\>								
A	**B**	**C**	**D**	**E**	**F**	**G**	**H**	**I**
View Name	Type	View Template	Discipline	Phase	Phase Filter	Sheet Name	Sheet Number	Detail Level
3D Energy Model	3D View	None	Coordination	Phase 2	Show Complete			Medium
3D RECPTION DESK DETAIL	3D View	None	Architectural	Phase 2	Show All			Medium
3D View - Existing	3D View	None	Architectural	Existing	Show Complete			Fine
3D View - Existing - HVAC Only	3D View	None	Coordination	Phase 2	Show Complete			Fine
3D View - Existing - Structure Only	3D View	None	Coordination	Phase 2	Show Complete			Fine
3D View - Future Addition	3D View	None	Coordination	Future	Show Complete			Fine
3D View - MEP and Struct - Phase 2	3D View	None	Coordination	Phase 2	Show Complete			Fine
3D View - Overall	3D View	None	Coordination	Phase 2	Show Complete			Fine
3D View - Phase 1	3D View	None	Architectural	Phase 1	Show Complete			Fine
3D View - Phase 2	3D View	None	Coordination	Phase 2	Show Complete			Fine
3D View - Structural Only	3D View	None	Architectural	Existing	Show Complete			Fine
Base Cabinet Detail	Detail	None	Architectural			DETAILS	A600	Medium
Base Cabinet Detail - Drawers	Detail	None	Architectural			DETAILS	A600	Medium
Cafe - Cropped View	3D View	None	Architectural	Phase 2	Show Complete	INTERIOR EL	A500	Fine
Camera - Hallway	3D View	None	Architectural	Phase 2	Show Complete			Medium
Camera - Impromptu from Hallway	3D View	None	Architectural	Phase 2	Show Complete			Medium
Camera - Level 2 Open Office Looking	3D View	None	Architectural	Phase 2	Show Complete			Medium
Camera - Lobby Entry	3D View	None	Architectural	Phase 2	Show All			Fine
Camera - Lobby Spot Light	3D View	None	Architectural	Phase 2	Show Complete			Fine
Camera - Lunch Room	3D View	None	Architectural	Phase 2	Show All			Fine
Camera - Mech Room View	3D View	None	Architectural	Phase 2	Show Complete			Fine
Camera - Office -Phase 2 - New Sun S	3D View	None	Architectural	Phase 2	None			Medium
Camera - Small Conf Rm - Section Box	3D View	None	Architectural	Phase 2	Show All			Fine
Camera - Wide Impromptu Looking Sout	3D View	None	Architectural	Phase 2	Show Complete			Medium
Cross Section 1	Building Section	Building Section	Architectural	Phase 2	Show Complete	Building Secti	A300	Fine
Cross Section 2	Building Section	Building Section	Architectural	Phase 2	Show Complete	Building Secti	A301	Fine
Displaced Elements View	3D View	None	Coordination	Phase 2	Show Complete			Fine
East	Building Elevation	Exterior Elevatio	Architectural	Phase 2	Show Complete	Exterior Elev	A201	Fine
ElumTools_WorkingView_Exterior	3D View	None	Coordination	Future	Show Complete			Medium
Fixed Student Desk	Detail	None	Architectural			DETAILS	A600	Medium
Fixed Student Desk - Keynotes	Detail	None	Architectural			DETAILS	A600	Medium
Floor Transition Detail	Detail	None	Architectural			DETAILS	A600	Fine
Graphical Column Schedule 1	Graphical Column	None	Structural	Existing	Show Complete			Coarse
Level 1 - 3D View - Existing - HVAC On	3D View	None	Coordination	Phase 2	Show Complete			Fine
Level 1 - Color	Floor Plan	None	Architectural	Phase 2	Show Complete			Medium
Level 1 - Demo Plan - Phase 2	Floor Plan	None	Architectural	Phase 2	Show Previous			Medium
Level 1 - Domestic Water - Phase 2	Floor Plan	Plumbing Plans	Plumbing	Phase 2	Show Complete	Level 1 Dome	M201	Medium
Level 1 - Domestic Water Plan - Phase	Floor Plan	Plumbing Plans	Plumbing	Phase 1	Show Complete			Medium
Level 1 - Existing	Floor Plan	None	Architectural	Existing	Show Complete			Medium
Level 1 - Future	Floor Plan	None	Architectural	Future	Show Complete			Fine
Level 1 - HVAC - Phase 2	Floor Plan	Mechanical Plan	Mechanical	Phase 2	Show Complete	Level 1 Ventil	M301	Medium
Level 1 - HVAC Ceiling Plan - Phase 1	Ceiling Plan	Mechanical Ceili	Mechanical	Phase 1	Show Complete			Medium
Level 1 - HVAC Plan - Phase 1	Floor Plan	Mechanical Plan	Mechanical	Phase 1	Show Complete			Medium
Level 1 - HVAC RCP - Phase 2	Ceiling Plan	Mechanical Ceili	Mechanical	Phase 2	Show Complete			Medium

Figure 10.3-68 View List schedule example

In the example above, a fairly consistent naming convention can be seen. Views that look out of place can be quickly renamed just by clicking in the cell and typing.

Scope Box

Scope boxes can be used to create consistent cropped regions for many floor plan views.

To create a Scope Box:

- Open a plan view
- Click the Scope Box command
- Enter a name and height on the Options Bar
- Define a rectangular area

In a floor plan view, there is a **Scope Box** property in the Properties Palette. Setting this to a named Scope Box will change the Crop Region to align with the extents of the Scope Box.

Levels, **Grids** and **Sections** also have a **Scope Box** property setting…this will make its extents conform to the extents of the Scope Box. Doing so will ensure they all have the proper width and height, so they appear in the proper views. This can also be used to omit these elements from a view. For example, the large base of a high-rise building likely has several grids which do not need to appear at the higher levels.

> The ability to turn cropping on or off is disabled when the view is associated with a Scope Box.

The visibility of Scope Boxes can be controlled on the **Annotation Categories** tab in the **Visibility/Graphics Override** dialog for a given view.

10.4 Sheet Composition panel

The group of Sheet Composition commands helps to create and manage sheets and revisions in the current project.

Sheet

Use the **Sheet** command to create a new sheet or turn a **Placeholder Sheet** into a real sheet.

> Right-click on the **Sheets** branch title in the **Project Browser** for another option to create sheets.

When the Sheet command is selected, the **New Sheet** dialog appears (Figure 10.4-1).

Select titleblocks:

Any titleblock families loaded into the current project are listed here. The selected title block is placed on the sheet. **None** is an option, but the titleblock family defines the size of the sheet so this option is typically not used.

> A titleblock can be selected in a sheet view and swapped out with a different one via the Type Selector.

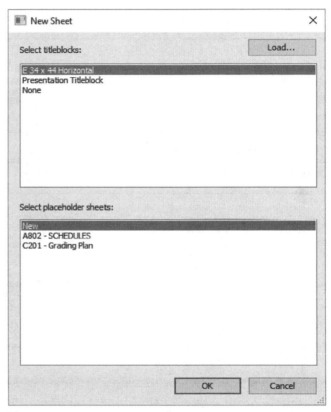

Figure 10.4-1 New Sheet dialog

Select placeholder sheets:

A placeholder sheet is a sheet created in a schedule using the **Insert Data Row** command. This adds a row to the sheet schedule but does not actually create a sheet in the project. This can be done to list sheets, which are part of the project, but by another firm and/or not being developed in Revit. For example, the civil design for a building project is typically done in AutoCAD Civil 3D. If the architects are responsible for the project title sheet, which has the master sheet list, then placeholder sheets can be used to list them but not complicate the printing process as placeholder sheets are not real sheets and thus do not appear as an option when printing.

However, if a placeholder sheet needs to become a real sheet, use the Sheet command; select the desired titleblock, the placeholder sheet and then click **OK**.

▌ Placeholder sheets can only be deleted from a sheet schedule.

View

This is the formal way to add a View to a Sheet. The alternative is to drag a view from the Project Browser onto a sheet. The benefit of using the Views command is only views which are not already placed on a sheet are listed.

Title Block: **Open a Sheet View**

This command is only available on the Ribbon when a Sheet view is active.

Title Block: **View dialog**

Starting the View command opens the **Views** dialog (Figure 10.4-2).

Title Block: **In-Canvas**

When a view is selected, click the **Add View to Sheet** button.

Move the cursor around to position an outline representation of the view on the sheet.

Click to place the view.

quick steps

Place Views on Sheet

1. Open a Sheet view
2. Views dialog
   - Select available view to be placed on current sheet
   - Only views not already placed on sheets will appear in the list; a view may only be placed on a single sheet
   - Legends and Schedules can be placed on more than one sheet
3. In-Canvas
   - Outline of draw appears as the cursor is moved about the sheet
   - Click to place view

Figure 10.4-2 Views dialog

Title Block

If a sheet is created without a titleblock or a titleblock is deleted from a sheet, use the **Title Block** command to add one to the current sheet.

Title Block: Open a Sheet View

This command is only available on the Ribbon when a Sheet view is active.

Title Block: Type Selector

Once the command is started, the **Type Selector** listed the titleblock families loaded in the current project.

Title Block: In-Canvas

Click in the sheet view to place the titleblock.

Title Block

1. Open a Sheet view
2. Type Selector
   - Select a Title Block to be placed
3. In-Canvas
   - Click to place title block

> It is possible to place more than one titleblock on a sheet. One use for this would be to have a Letter or Tabloid sized revision titleblock with a large marking area to hide the rest of the larger sheet. Once printed, this revision titleblock can be deleted or hidden in view.

The titleblock defines the sheet size.

The Starting View can be set to a sheet view. Consider doing that and using a custom titleblock family for this view that lists the Project Name, Project Number and other project details. All this information is live. When changes are made here or on a sheet, it updates everywhere. This is better than using a static drafting view with text as the starting view.

Revisions

The Revisions feature is used to document changes made to the design. This command is used in conjunction with the **Revision Cloud** and **Tag by Category** commands.

Basic Overview of Revisions:

Revision
Cloud

- Specific revisions are managed via the *Revisions* dialog (Figure 10.4-3).
    - ○ *Tool location:* View → Revisions
- Revision clouds are added and assigned to a specific revision.
    - ○ *Tool location:* Annotate → Revision Cloud
    - ○ Each cloud can be a separate instance or multiple clouds can be created in a single sketch (per sheet).
- A tag is added to a cloud to identify it.
    - ○ *Tool location:* Annotate → Tag by Category (or *QAT*)
- The revision schedule, in the titleblock family, updates automatically.
    - ○ Edit sheet properties to show revision without cloud.
- When a revision will no longer be changed, it is marked as issued.
    - ○ Issued revisions cannot have new clouds added to them.
- Older revisions can be made invisible on the sheets via *Show* options.

In the big picture of BIM, Revit revisions are superficial in that the actual model elements being changed do not know which revision they belong to.

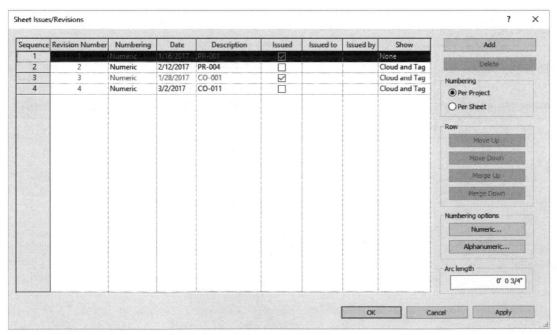

Figure 10.4-3 Sheet Issues/Revisions dialog

Revisions during Bidding

- Set up the *Revision* dialog for the anticipated change to the model.

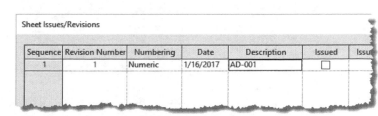

- Make changes to model; e.g., edit *Break Room* (plan and elevation)

Figure 10.4-4 Default initial revision

- Duplicate **elevation view** (with *Detailing* or as *Dependent*)
  - With <u>Detailing</u> option (Fig. 10.4-5):
    - Can move view to another category in *Project Browser*;
    - All view specific elements are now copied; i.e., text, dimensions, detail lines (harder to keep in sync);
    - Crop region can be adjusted if needed.
  - With <u>Dependent</u> option (Fig. 10.4-6):
    - View stays "close" to the original view;
    - Does not duplicate view specific items (less chance for errors later);
    - Scale cannot be changed from master view.
  - Both scenarios have the following issues:
    - New view cannot have the same name as the original.
      - Workaround: add a period at the end.
    - A new elevation tag added in the plan views.
      - To hide these (steps required for each view):
        - Select each one and right-click → Hide in view, - *or* -
        - Create a new elevation tag type and filter it out.

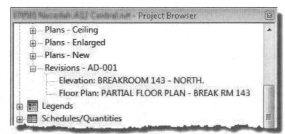

Figure 10.4-5 Revision in Project Browser

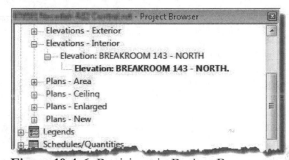

Figure 10.4-6 Revisions in Project Browser

- Duplicate **plan view** as *Dependent*
  - View can be cropped.
    - Grids follow crop region.
    - Section heads can be adjusted manually (but do not affect master view).
    - *Annotation Crop* must be turned on.
      - Some annotations extend outside of crop.
        - These can be hidden via select/right-click/*Hide element*.
        - Slight problem for inline posting of plan rack set.

        o   Annotations are not duplicated.

        o   Notice all callouts get reset (and have to be manually adjusted).

- Add revision clouds and tags to each sheet as needed.
    - Revision clouds/tags can be added directly in most views (rather than placing them on the sheet).
    - Adjust revision cloud line weight and color (if desired).
        - Manage → Object Styles → Annotation Objects
            - Could also change a given view typing VV.

        WARNING: Changing color can affect printing.

- Create a Revision sheet and place duplicated/dependent view on it.
    - Revit only allows a view to be placed on one sheet.

You should archive your project at end of bidding.

Revisions during Construction:

During construction you will most likely want to clear out all the bidding phase revisions and have the numbering start over for construction phase revisions.

- Archive project.
- Delete all revision clouds and tags from project.
    - Must un-check the "issued" option first.
- Merge all revisions in *Revision* dialog.
    - Must un-check the "issued" option first.

        FYI: Revisions can now be deleted starting in Revit 2021.

- Start documenting construction phase revisions (e.g., PR-01, ASI-01, CO-01).
- Consider getting rid of any *Design Options* used for Bid Alternates.
    - This makes revisions easier during *CA*.
    - Set selected option to *Primary* and *Accept* it (via *Design Options* dialog).
    - Delete any unneeded sheets/views related to alternates.
        - Dimensions and tags can be deleted.

        WARNING: This can get messy for large complex alternates.

    - The deleted options will be available in the archived project.

There are add-in tools available for Revit which can automatically remove all the revisions in a given project. This can save a lot of time. Some of these tools are free and others cost money.

All *Construction Administration* (*CA*) phase revisions should be made to a single model (i.e., the main model). This way any implications CO-04 might have on PR-57 will be evident to all parties.

In an IPD scenario (or design build) the best approach is to utilize a single model. This can be accomplished utilizing technology such as Remote Desktop (RDP) or Autodesk's Revit Server.

Example 1: Proposal Request

The biggest challenge related to *Proposal Requests* (PR) is: Should the main model be edited or not? This is due to the fact that the *PR* might not ultimately be accepted due to contract implications (time/cost). If it is highly likely that the *PR* will be accepted (e.g., code or errors/omissions required change), then the main model should be edited.

If the change is not likely (e.g., client is curious what another 50'-0" of building is worth), then a temporary copy of the model could be made (via *Open Detached, Save-As*). However, in the unlikely event the *PR* is accepted, then the main model needs to be updated (some copy/paste might help but will not cover everything).

When in doubt, make changes to the main model and undo them if the change is not accepted. Of course, you can do whatever you want in this case.

Example 2: Schedules

- Make a change to the *Door Schedule*.
  - o Note that revision clouds cannot be added to the view.
- Add a door which causes the clouded items to shift down in the *Schedule*.
  - o Notice clouds do not move with highlighted area.
  - o Designer must manually move the cloud when changes such as this occur (if they can remember!)
- Move the revision cloud.
  - o Revision must be unissued before the cloud can be moved.

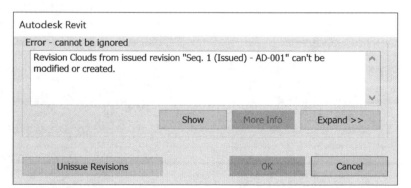

Figure 10.4-7 Error when trying to change issued revision

Guide Grid

Use the Guide Grid command to create a new **Guide Grid** type in the project. A guide Grid is like grid-paper for a Revit sheet. This can be used to manually align views between sheets.

Start the command options the **Assign Guide Grid** dialog (Figure 10.4-8). Pick from previous created types or create a new one. A new one can be selected and its name or spacing adjusted in the Properties Palette (Figure 10.4-9).

Figure 10.4-8
Assign Guide Grid dialog

> Selecting and deleting a Guide Grid will delete from the entire project, not just current sheet. Also, if moved on one sheet, it will move on all sheets.

In a sheet view, a Guide Grid can be selected in the Sheet's properties or set to None. The Guide Grid command is really only needed when a new Guide Grid type is needed.

The image below shows the Guide Grid superimposed on a sheet (Figure 10.4-10). The overall size and position can be changed.

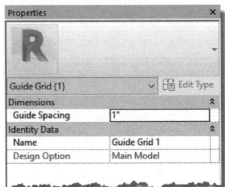

Figure 10.4-9
Guide Grid properties

Grids, Reference Planes, Levels and Crop Regions are able to snap to Guide Grids on a sheet with the **Move** command.

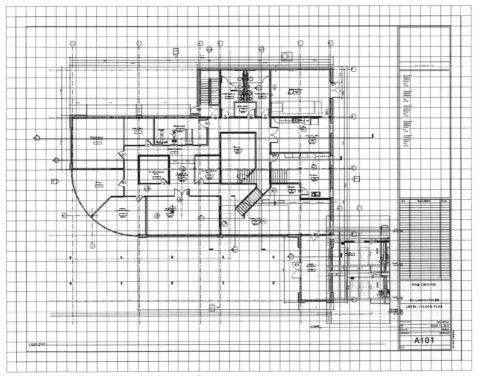

Figure 10.4-10 Guide Grid on a sheet

Matchline

When the building floor plan is too large to fit on a single sheet you need to use the **Matchline** tool and the **dependent views** feature. Each can be used exclusive of the other, but they generally are used together.

The *Matchline* tool can be found on the *View* tab. It is similar to the *Gridline* tool in that you add it in plan view, and it will show up in other plan views because it has a height. When the *Matchline* tool is selected, you enter *Sketch* mode. Here you can draw multiple lines which do not have to form a loop or even touch each other.

Matchline: **Ribbon**
Select to draw a **Line** (default selection) or **Pick Lines** to sketch a new Matchline.

Decal: **Options Bar**
Use the **Chain** option to sketch consecutive lines. The **Offset** value displaced the sketch from the point picked on the screen.

Matchline: **Properties**
A Matchline can have an upper and lower limit (or extent) defined to limit where it appears in the project. A plan view's cut plane must pass through the matchline for it to appear in a view.

Matchline: **In-Canvas**
Sketch the lines which will represent the matchline.

Regarding Dependent Views:
A *dependent view* is a view which is controlled by (i.e., dependent on) a main overall view but has its own *Crop Region*. A main view can have several *dependent views*. Any change made in *Visibility/Graphics Overrides* is made to the main view and each dependent view. The dependent view's *Crop Regions* generally coincide with the *Matchlines*. However, the *Crop Regions* can overlap.

Additionally, if a dependent view's *Crop Region* is rotated, the model actually rotates so the *Crop Window* stays square with the computer screen. However, the other dependent views and the main view are unaffected. This is great when a wing of the building is at an angle. Having that wing orthogonal with the computer screen makes it easier to work in that portion of the building and it fits better on the sheet in most cases.

---

quick steps

Matchline

1. Open a plan view
2. Ribbon (in sketch mode)
   - Sketch line segments
   - Or Pick lines
3. Options Bar
   - Chain
   - Offset
4. Properties
   - Top and Bottom constraint/offset
5. In-Canvas
   - Sketch lines
   - Sketch can be disconnected lines

To create a dependent view, you right-click on a floor plan view and select **Duplicate View > Duplicate as Dependent**. The image to the right shows the result in the *Project Browser;* the **dependent views** are indented (Figure 10.4-11).

Once the *Matchline* is added and the dependent views are set up, you can add *View References* (from the *Annotate* tab) which indicate the sheet to flip to in order to see the drawing beyond the *Matchline.*

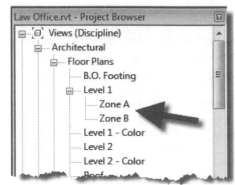

Figure 10.4-11 Dependent views

Finally, on large projects with multiple levels and floor plan views, you only have to set up the dependent views once (i.e., number and crop regions). With one plan set up, you right-click on the main view and select *Apply Dependent Views*. You will be prompted to select all the floor plan views for which you would like to have the dependent views created.

View Reference

Used for matchlines and for notes to refer to other drawings. Can also be used for reference to 3D views.

View Reference: **Ribbon**
Select the View Type and Target View (Figure 10.4-12).

View Reference: **Type Selector**
Select the desired family type.

View Reference: **In-Canvas**
Click to place the family.

quick steps

View Reference

1. Ribbon (in sketch mode)
   - View Type
   - View Target
2. Type Selector
   - Select type to place
3. In-Canvas
   - Click to place view reference

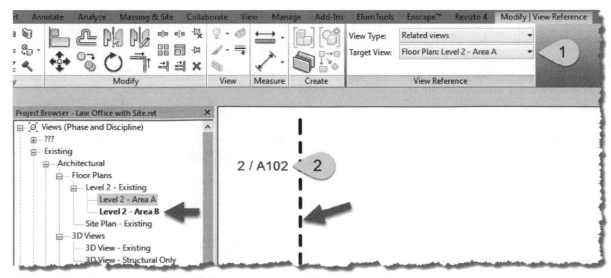

Figure 10.4-12 Placing a View Reference at a matchline in plan view

Custom View Reference for 3D view

When a 3D view is preferred over 2D for interior elevations, use a custom **View Reference** family that looks like an elevation tag. The healthcare group at my firm does this, as seen in the image below. The View Reference feature streamlines references to these views from the plans—because the elevation tag cannot point to a 3D view.

The View Reference family can be made to look like an elevation tag. When the tool is selected, pick the view type and specific view to reference on the Ribbon, as seen in the next image. You will need to predefine the angles you want as the arrow cannot be rotated with grips (Figure 10.4-14).

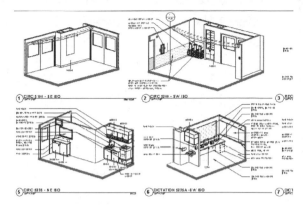

Figure 10.4-13 3D views used in place of 2D interior elevation views

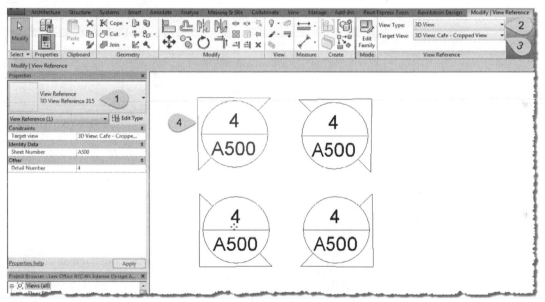

Figure 10.4-14 View reference family with multiple angles to mimic elevation tag

The View Reference feature can also be used to create more intelligent references to other drawings within your notes. The example here (Figure 10.4-15) shows a note using regular text and then a View Reference family positioned at the end.

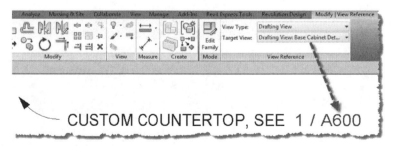

Figure 10.4-15 View reference used with note

Of course, a person could accidentally pick the wrong reference, so care should be used with this feature.

Viewports → Activate View

Activates the selected viewport. Alternate method: select viewport and right-click menu or simply double-click within viewport to activate.

When a view is active, it is the same as if the view were opened directly. The elements in the view can be changed. The view's properties can also be changed.

Viewports → Deactivate View

When a view is active, the sheet properties cannot be accessed. Also, views cannot be repositioned on the sheet. Thus, to deactivate the view, use this command or right-click and select the same named command on the contextual menu which appears next to the cursor.

10.5 Windows panel

The grouping of commands on the Windows panel are used to manage the open views, sheets, legends and schedules.

Switch Windows

This drop-down list provides a way of switching between views—this command is also on the Quick Access Toolbar. Alternatively, pressing **CTRL + Tab** cycles through the open views. **CTRL + SHIFT + Tab** cycles through views in the opposite direction. Think of one as 'forward' and the other as 'backwards'. However, with the advent of the **View Tabs**, these other techniques may not be used as much as in previous versions of Revit.

Close Inactive

This command will close all views except the current view. This command does not close project or family files… so one view for each open file will also remain open.

> A project can become sluggish or even prone to crash when too many views are open as this uses more system resources such as RAM.

Tab Views

Using this command will, for lack of a better description, un-tile views. The result is View Tabs aligned across of the top of the drawing windows. However, views dragged out of the Revit application window manually, will not get pulled back in line – that would need to be done manually.

Tile

This command will rearrange all the views on the screen, so they are all visible (Figure 10.5-1). Typically, you would close any irrelevant views before using this command. This is a quick way to see more detail of a specific area in which you are working. Views manually positioned outside of the application will not be pulled into the tiles views within the application – that has to be done manually first.

> The view tabs can be pulled out of the Revit application, to another screen, which allows for better on-screen coordination between a schedule and floor plan, for example. Views can also be "stacked" on top of each other resulting in tabs. See the user interface section towards the front of this book.

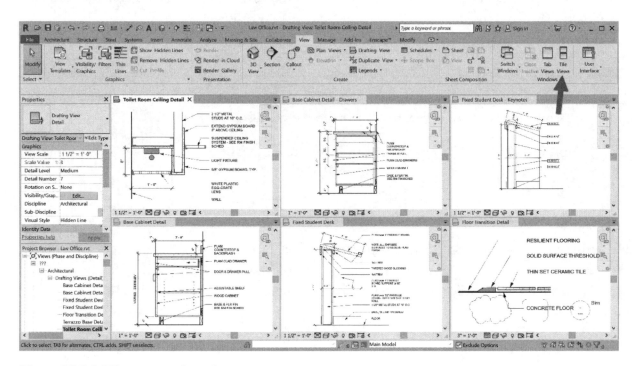

Figure 10.5-1 Tile command result

User Interface → Toggles

The **User Interface** toggles the listed palettes. Each of these items can remain open while working on the design. Some of these can be turned on via additional methods. For example, right-clicking on an element offers a Properties command which opens the Properties Palette. However, the Project Browser can only be opened here if it is closed (intentionally or accidentally).

User Interface → Browser Organization

The **Project Browser** can become very cumbersome to navigate on large projects using the default organization. There are some limitations on how much things can be adjusted, but making a few changes in a firm, or student, template can help a lot.

Figure 10.5-2 UI menu

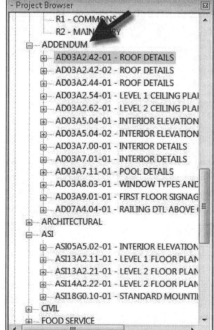

Figure 10.5-3 Project Browser

For example, sorting sheets can be done by creating a *Project Parameter* such as *Sheet Group*. In the image to the left, all sheets listed under ***Addendum***, in the *Project Browser*, have their *Sheet Group* parameter set to the word "Addendum" (Figure 10.5-3).

Here is how to create a project parameter associated with sheets:

- Click **Manage → Project Parameters**.

- **Add** a parameter (Figure 10.5-4):
  a. *Name*: **Sheet Group**
  b. *Type of Parameter*: **Text**
  c. *Instance*: **checked**
  d. *Categories*: **Sheets**

- Close the open dialog boxes.

Now all sheets will have a parameter called "Sheet Group."

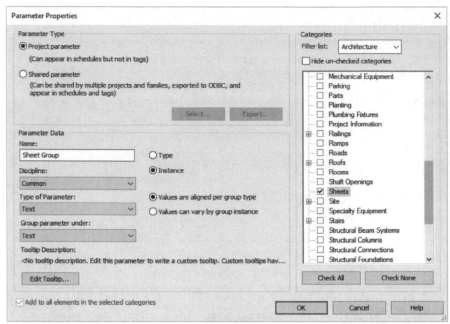

Figure 10.5-4 Creating a project parameter

The following steps show how to sort the sheets by the *Sheet Group* parameter:

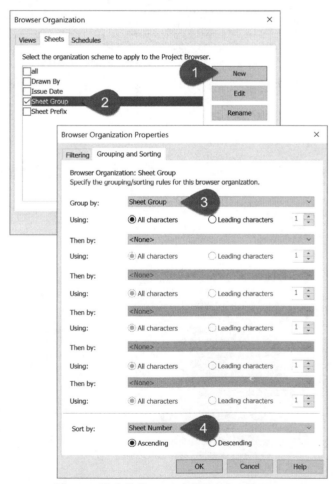

- Right-click on the word **Sheets (All)** in the *Project Browser*.

- Select *Browser Organization...* from the pop-up menu.

- On the *Sheets* tab, click **New**.

- Enter **Sheet Group** for the name. (The name is case sensitive.)

- Edit the *Grouping and Sorting* tab (Figure 10.5-5):
  a. *Group By*: **Sheet Group**
  b. *Sort By*: **Sheet Number**

Similar steps can be used to sort views and schedules as well. Views can also be sorted by View Type rather than creating a custom parameter.

Figure 10.5-5 Browser Organization

User Interface → Keyboard Shortcuts

Keyboard shortcuts can save a significant amount of time as it takes less time than selecting commands on the Ribbon.

The Keyboard Shortcuts command opens the **Keyboard Shortcuts** dialog (Figure 10.5-6). Selecting a command, a keyboard shortcut can be assigned, or removed, via the buttons and textbox below.

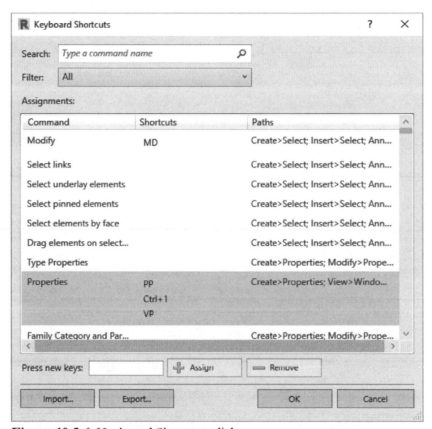

Figure 10.5-6 Keyboard Shortcuts dialog

Notes:

Chapter 11
Manage Tab

The Manage tab contains tools to help you set up and maintain project wide settings for your projects. Visiting the Manage tab early in a project can help you establish standards for a project. In most cases the template file you start a project with will establish most project settings, but when something needs to be altered you can make changes using tools on the Manage tab. With strong settings and standards, you will find your work to be faster, more consistent, and of a higher quality as you develop your projects.

11.1 Settings panel

The Settings panel is where project wide settings are controlled. Establishing settings early in a project or by using template files helps you work faster and more efficiently as well as helps maintain drawing and office standards as you develop your projects.

Materials

The Materials button opens the **Material Browser** (Figure 11.1-1) where you can find the material's definitions currently loaded into your project as well as a library of materials you can import for use in your project. The Material Browser dialog also is used to make changes to the assets that make up a material definition.

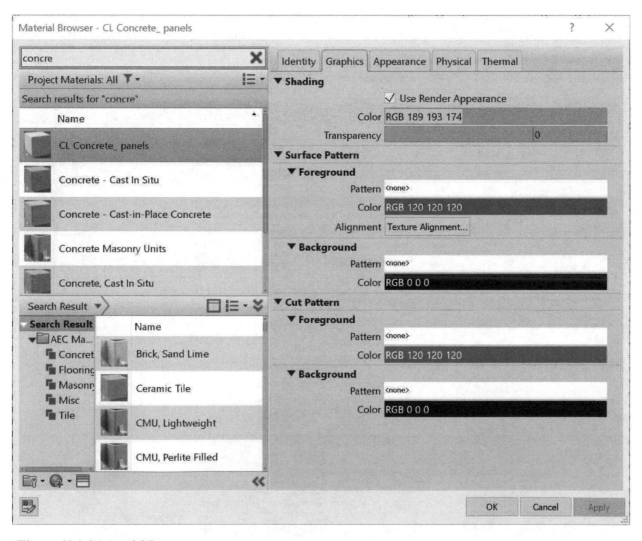

Figure 11.1-1 Material Browser

The Material Browser is broken into 2 major parts. On the left side of the browser is the library of materials. The right side is composed of tabs defining different assets that make up the material definition. The assets of the material selected in the left portion of the browser can be altered on the right side of the browser.

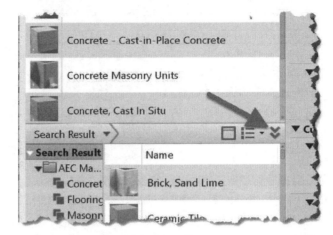

On the left side, the top portion of the materials library lists the materials currently loaded into your project. The lower section of the library lists materials found in your external materials library. These materials are not currently in your project and cannot be used on elements, but you can load these materials from the library into your project

Figure 11.1-2 Show/Hide Library

so you can use them on elements. If the lower section of the library (external libraries) is not visible, click the show/hide button near the bottom of the project materials area to expose the **external libraries** (Figure 11.1-2).

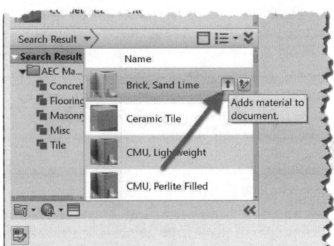

The **Search** tool located above the libraries is the easiest way to locate a material for your project. When you search with this tool both the project materials and the external libraries are searched. Both lists are filtered to show only those materials matching the search criteria. If the material you want to use is in an external library, it must be loaded into the project so it can be used. **To load the material** from the external library into your project, select the material you want to load and then click the small arrow appearing on the material definition (Figure 11.1-3). This will load

Figure 11.1-3 Load material from library into project

the material into your project library where it can be used and assigned to elements in the project.

The 3 buttons below the library panes are used to create, load and manage external material libraries, add and duplicate materials in your project, and open the asset browser to help you create new materials or edit existing materials in your project library.

The default external libraries are read only but you can create your own custom external libraries for different project types or libraries can be shared (Figure 11.1-4). **Create a new library** and select a location on your network to store the library file. Then click and drag materials from your project library into the custom library to populate it. This new custom library can now be shared with those who need access to a custom palette of materials. Use the other tools in the drop down to manage which external libraries are loaded and searched. Custom libraries can also have categories created within them to help manage the library organization.

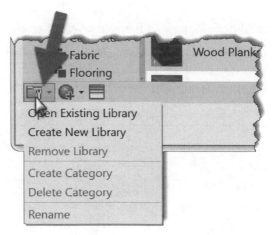

Figure 11.1-4 External library management

Use the **new/duplicate buttons** to create or duplicate materials in your project (Figure 11.1-5). When you duplicate, a new material is created using the exact same assets and settings as the selected material. Use this to make subtle changes to a material or a material that looks exactly the same but has different thermal or physical assets. Create a new material to start a material definition from scratch. You then define all of the assets for the material and give it a graphic appearance.

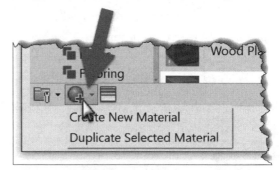

Figure 11.1-5 Create/Duplicate material

The **Asset Browser** button (Figure 11.1-6) opens and closes the Asset Browser. The Asset Browser is helpful when making edits to an existing material in your project. A material definition in Revit is made up of a number of "assets."

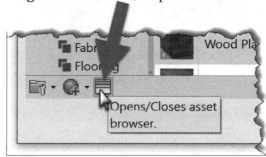

Figure 11.1-6 Open/Close Asset Browser

- **Identity** – Parameters associated with the material.
- **Graphics** – How the material looks in the canvas when the view style is not set to Realistic or Raytrace.
- **Appearance** – How the material looks when rendered or when the view style is set to Realistic or Raytrace.
- **Physical** – Parameters associated with the structural behavior of the material. Used during structural analysis.
- **Thermal** – Parameters associated with the energy performance of the material. Used during energy analysis.

The **Asset Browser** allows you to search the asset library for different assets to make up and define a material in Revit.

On the right side of the Material Browser is a set of tabs (Figure 11.1-7). The different tabs contain information about the different kinds of assets making up the material selected on the left side of the Material Browser. A single Revit material is defined by all of the assets. A single material may share an asset with another material. Defining a material in this way makes them more flexible for multiple uses. For example, you may have 3 different concrete materials in your project to represent different stained concrete or different finishes. Each of these concrete materials will have different *Appearance* assets, but probably share *Graphic*, *Physical*, and *Thermal* assets.

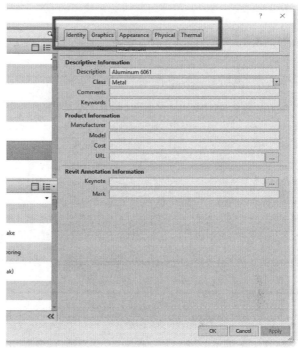

Figure 11.1-7 Asset Tabs

Identity

On the Identity tab the parameters of the material are listed. The parameters here are able to be tagged and scheduled in your model. Information added to the Identity of the material helps you find it when searching in the Material Browser and helps when creating a material takeoff for the project.

Graphics

On the Graphics tab you find the properties that define what the material will look like in your model when the view style is set to hidden, shaded, or consistent colors.

- **Shading** – Shading defines the color to be used for elements using the material in a Shaded or Consistent Colors view style. Set the color directly or check the "**Use Render Appearance**" to use an average color of the image file used in the Appearance asset of the material.
- **Surface Pattern** – Define a pattern applied to the surface of an element. If an element is projected (not cut) in the view, the surface pattern will be applied to the element. The surface pattern can be aligned with the texture of the image used for the Appearance asset of the material. You can specify a foreground and background pattern to create a more complex surface appearance.
- **Cut Pattern** – The pattern assigned to the material that will be shown when the element is cut in the the view. An element is "cut" in a view when the projection plane of the view (elevations and sections) intersects the element. In a plan view, the element is "cut"

when the cut plane of the plan intersects the element. You can specify a foreground and background pattern to create a more complex cut appearance.

Appearance

The Appearance tab defines how the material will appear when the visual style is set to realistic or raytrace. The appearance is also applied when the view is rendered. Use the Appearance tab to assign different images to the material and apply transforms to the images in order to make the rendered appearance appear more realistic. Some appearance assets may contain different transforms that can be used to define the material. Assign an asset from the **Asset Browser** that closely matches your material in order to make a duplicate with the required transforms.

Physical

Physical assets define the structural strength and other physical properties of the material. Change the properties based on the real world physical properties of the material being defined. Physical properties of a material are used when a structural analysis is run on the model.

Thermal

Thermal assets define the energy related aspects of the material. Change the properties based on the real world thermal properties of the material being defined. The values of the thermal assets of materials will be used when performing an energy analysis on the model.

In the upper right corner of the assets portion of the Material Browser you will find controls to **Replace**, **Duplicate**, and **Delete** the current asset (Figure 11.1-8). Use these controls to manage the material assets in your model. Keep in mind a single asset can be used for multiple material definitions.

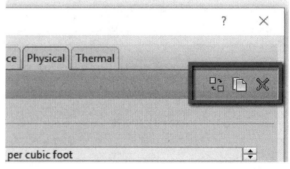

Figure 11.1-8 Material asset management tools

Object Styles

The Objects Styles dialog is where you define default display settings for elements in the model (Figure 11.1-9).

For each element category, define the **Line Weight** used for element edges. A column is provided for both **Projection** and **Cut** line weights. If a category has a gray box for the cut line weight this indicates the category is not cuttable. Annotation and Imported Objects only list projection line weights because they are not able to be cut in any views.

One **Line Color** and one **Line Pattern** is defined for each category. Both projected and cut edges of elements will use the defined line color and line pattern.

The material defined in the **Material** column will be assigned to an element if the material parameters of the element are set to <By Category>. If the element has specific material assigned in the parameters, those will be used instead of the material assigned to the material column in the Object Styles dialog.

Use controls in the dialog to help you sort and find the categories you want to modify in the dialog.

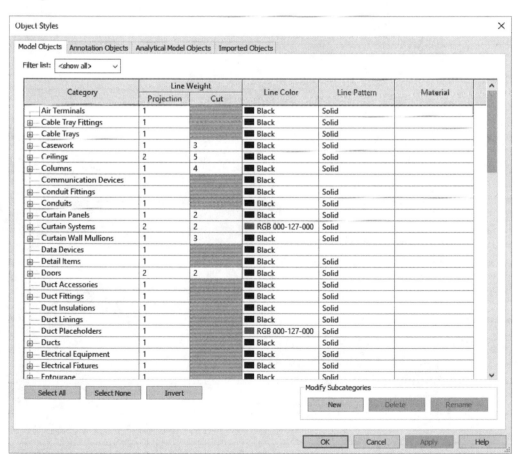

Figure 11.1-9 Object Styles dialog

Snaps

Snaps help you draw more accurately. Use the Snaps dialog to control the snap behavior (Figure 11.1-10).

All snaps can be turned off by checking the "Snaps Off" box or while working by using the keyboard shortcut "SO."

The dimension snaps for length and angles are sensitive to the zoom level of the view. As you zoom in the snap increment of distances and angles will get smaller. The lengths and angles list in the dialog are the increments that will be used. Add values here by including them in the sequence and separating the values with a semicolon.

Object snaps can be turned on and off in the snaps dialog, or the keyboard shortcut listed here can be used to toggle a snap without entering the dialog. Keyboard shortcuts to temporarily override snaps in your model are listed in the lower section of the dialog for convenience.

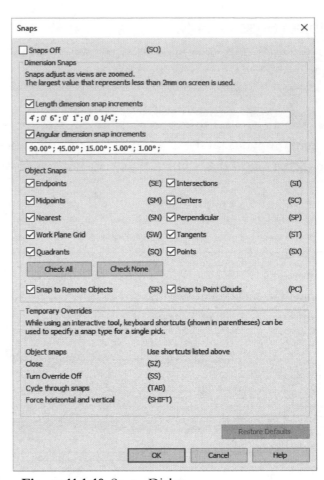

Figure 11.1-10 Snaps Dialog

Project Information

Information about the project itself (Figure 11.1-11). The parameters here and their values can be used in titleblock families. Use project information labels in titleblocks to keep information up to date and consistent.

The energy analysis section of the dialog includes default settings for the project when an energy analysis is run. Click edit to open the energy setting dialog and change the default energy values for the project.

In the **Energy Setting** dialog set the basic defaults for energy analysis. The mode setting controls of building elements, mass elements, or both will be

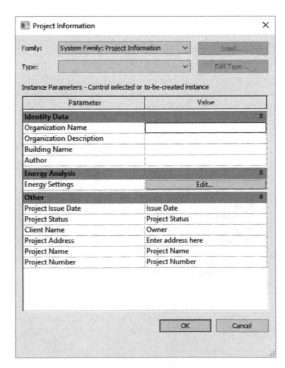

Figure 11.1-11 Project Information

used for analysis. When using actual building elements, the element constructions (walls, floors, roofs) will be used as well as the openings placed in the model. When masses are used an assumed glazing percentage and conceptual constructions will be used during the energy analysis. The other basic options in the Energy Settings dialog control the size of the analytical surfaces and volumes in the energy model. Smaller values will result in a more detailed model but will increase analysis and energy model creation time. You must consider those factors when making changes to the sizes of the analytical surfaces and volumes. If there is not a specific reason to change the values, the default values will give a good balance of performance versus accuracy.

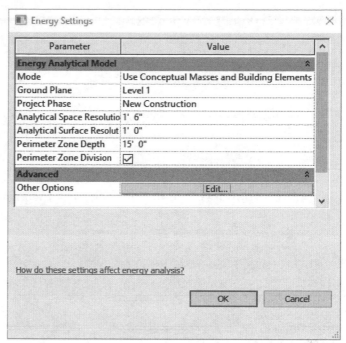

Figure 11.1-12 Energy Settings dialog

In the **Other Options** dialog additional changes can be made to the energy model creation but only need to be changed when there are specific requirements.

> See the **Analyze** tab chapter for more information on Energy Settings.

Project Parameters

Project parameters offer you a way to add information onto elements in your model that were not included in the original family definition (Figure 11.1-13).

When you add a Project Parameter you are asked to define the parameter in the **Parameter Properties** dialog (Figure 11.1-14). First you decide if you are going to define a project parameter or if you will pull the definition of the parameter from a shared parameter file. A project parameter is information added to elements in the model, but the information can only appear in a schedule in Revit. Use project parameters when the added information does not need to be visible in a tag. A **shared parameter** can

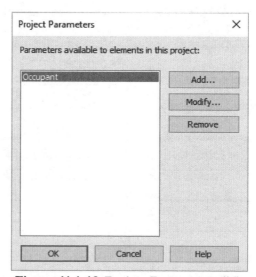

Figure 11.1-13 Project Parameters dialog

be added instead of a project parameter. The shared parameter is a predefined list of parameters that can be used across multiple projects and families. Shared parameters can be used for both tags and schedules. Because a shared parameter is defined ahead of time, the definition of the parameter can be predicted and controlled.

If you are adding a project parameter, the parameter data needs to be defined. Give the parameter a **Name** and use the drop down lists to define the **Discipline**, the type of parameter, and the group. The discipline selected will define the list of parameters that can be chosen under **Type of Parameter**. The **Group** setting will define where the parameter is displayed in the properties palette (instance parameters) or in the type properties dialog (type parameters).

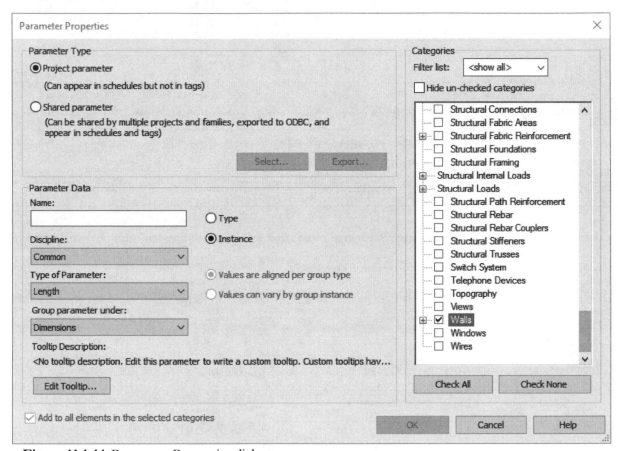

Figure 11.1-14 Parameter Properties dialog

Use the radio button selection to define if the parameter will be applied as an **Instance** parameter or a **Type** parameter. Instance based parameters can vary the value on every instance of the element in the model. A type parameter value will be applied to all instances of a specific type in the model. When choosing if a parameter will be instance or type decide how you want to have control over it in the model. Type parameters are easy to change a large number of elements at one time in the model. Instance parameters will need to be changed one by one on each element, but offer more flexibility. When you choose an instance parameter you can also choose if the instance parameter can be set independently on an element included in a **Group** or if the instance parameter value is constant in every group instance. If you are not using groups in your model, this setting can be ignored as it will not have an impact on your model.

As part of the parameter definition you need to select the category, or categories, that the parameter will appear on. Once set, the parameter is applied to all of the elements in the model. If an element of the checked category is selected in your model the new parameter will be visible in the properties palette (instance parameters) or in the type properties dialog (type parameters).

Shared Parameters

Use the Shared Parameters dialog to add parameters to a shared parameter file (Figure 11.1-15). A shared parameter file is a text file used to hold parameter definitions so they can be used in multiple projects and families. Because the definition of the parameter comes from the same source file the parameter can be both scheduled and tagged in a project.

In the top portion of the dialog **Browse** to a location where a shared parameter file exists on your network. If you do not have an existing shared parameter file, you can also create a new one from scratch and place it on a network location. You may consider having different shared parameter files for different project types or different

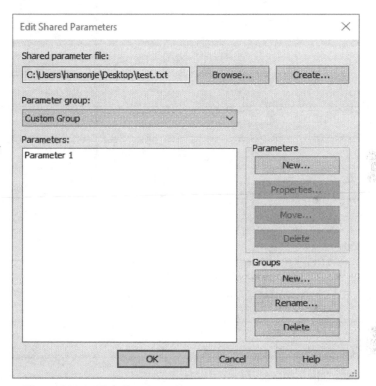

Figure 11.1-15 Edit Shared Parameters

clients depending on your needs, or you may choose to have one shared parameter file with standard parameter definitions. If you came into this dialog while creating a project parameter from a shared parameter, you change the parameter file location in this dialog.

Use the controls on the right side of the dialog to create different group definitions to organize the parameters. Once a group or a number of groups are created in the file, use the drop-down in the dialog to select a group. The parameters in the group will be displayed in the **Parameters** section of the dialog box.

Create new parameter definitions in each group for use in your projects and families.

Global Parameters

Global parameters allow you to create a parameter and a value that can be used throughout the model to help you control positioning and values related to multiple elements in the model (Figure 11.1-16). For example, a global parameter could be used to hold a value that could be assigned for a door offset. This is then assigned to all doors in the project to keep the offset consistent. If it needed to change, the value for the global parameter is changed and then all elements assigned to the global parameter are changed.

The top of the dialog allows you to **Search** the parameters in the model. Use the controls at the bottom of the dialog to add, modify and sort the parameter list.

Figure 11.1-16 Global Parameter dialog

To assign a global parameter to a dimension string in the model, select the dimension and use the label drop-down on the contextual tab of the ribbon. Then select the global parameter.

Transfer Project Standards

Use Transfer Project Standards to move settings and family types from one model file to another model file (Figure 11.1-17). The source model file will retain the settings and family types transferred.

Open the target model and the source project in your Revit session. With a view from the target project open and active start the Transfer Project Standards tool. In the dialog use the top drop-down control to select the source project (if you have more than one open). Now **check** the settings and/or family types to be transferred to the target model and click

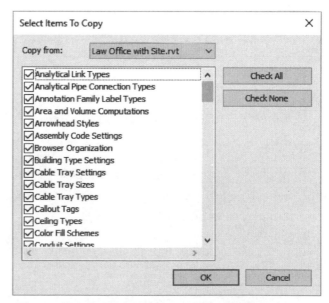

Figure 11.1-17 Transfer Project Standards dialog

OK. If the type already exists in the target model file, you will be given the option to overwrite or transfer in only new types to the target model (Figure 11.1-18). In the overwrite dialog you cannot specify to overwrite a subset of type. You will have to overwrite all. If you do not want to do this, you can cancel the action and start the transfer process again selecting only the categories you wish to overwrite and doing the transfer again.

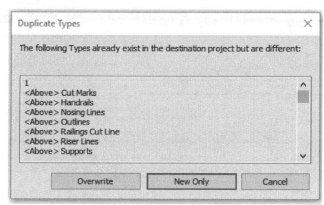

Figure 11.1-18 Duplicate type dialog

Using this tool when family types are transferred, all of the types in a category will be transferred. It is not possible to select individual types for transfer.

Purge Unused

Periodically clean your model file from extra family types and styles to make working in the model easier (Figure 11.1-19). Having a list of types in a model that are unused makes finding the right types difficult and can increase file size.

Purge unused finds all families, type, and styles in the model which are not used by anything and allows you to remove them from the model. Start the Purge Unused command and then in the dialog select the items to remove. By default, all unused items are checked. You must uncheck an item in the list in order to keep it in the project.

Each unused item is listed by a family in the dialog. Click the small "plus" icon next to each family to expand the selection and see the individual family types. You can choose to purge some individual types of a family while

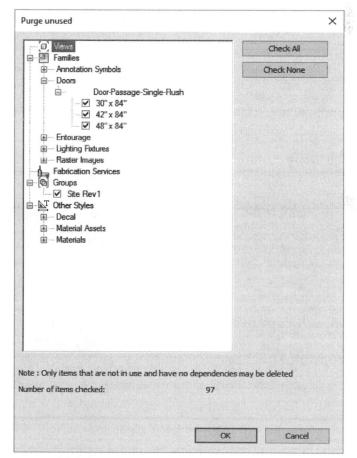

Figure 11.1-19 Purge dialog

keeping others. This may occur when you loaded a family with multiple types and only plan on using one or two of them. The total number of items that will be purged is listed at the bottom of the dialog.

In some cases, you may want to consider running the command once or twice in succession. Because of nested families and relationships between a family and some styles, the first purge may remove a parent element or style leaving behind more unused elements. Purge a second time to remove these "extra" families and styles.

Project Units

Use the Project Units dialog to control how units for different values and parameters will be displayed throughout the model (Figure 11.1-20). Use the dropdown list at the top of the dialog to select the discipline of units you want to make changes to.

Click the Format value on the right side of the dialog to change the Units for the selected type. In the Format dialog the units and rounding values are set along with other aspects of the way the units will be displayed. The other aspects of the way project units can be formatted are as follows:

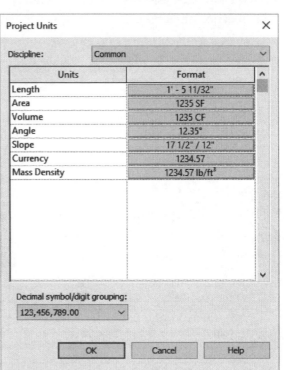

Figure 11.1-20 Project Units dialog

- **Suppress trailing 0's** – If a value ends with 0's in non-significant positions they will be removed. For example, 10.150000 would be changed to 10.15.
- **Suppress 0 feet** – If the value is in feet and inches, any values under 1'-0" will not show the feet unit. For example, 0'-8" is shown as 8". This only affects length and slope units.
- **Show + for positive values** – Any positives are marked with a "+" indicator.
- **Use digit grouping** – When selected decimal, values will be grouped as indicated in the Project Units dialog.
- **Suppress spaces** – Removes spaces between feet and inches display. For example, 8' – 6" is tightened to 8'-6". This only affects length and slope values.

Figure 11.1-21 Units format dialog

Each unit type can be formatted to use different units if desired. Change all units to match your project requirements.

Structural Settings

Due to the architectural scope of this book, this command is not covered in this book.

MEP Settings

Due to the architectural scope of this book, this command is not covered in this book.

Panel Schedule Templates

Due to the architectural scope of this book, this command is not covered in this book.

Manage

Additional Settings: Fill Patterns

Fill patterns are used to define surface and cut patterns of materials and filled region types. All fill patterns available for use are found in the Fill Patterns dialog (Figure 11.1-22). From this dialog use the controls on the right side of the dialog to create new patterns or manage existing patterns. At the bottom of the dialog use the radio buttons to control which pattern types are displayed in the dialog, Drafting patterns or Model patterns.

Drafting patterns will maintain the spacing of elements in the pattern no matter what the scale of the view is where the pattern is visible. For example, if you had a cross hatch pattern with a spacing of 1/8" between the lines in each direction, the spacing of the lines will be 1/8" in a view assigned a scale of 1"=20'-0" and in a view assigned a scale of ½"=1'-0". The spacing on the drafting pattern will remain at 1/8th in between each line.

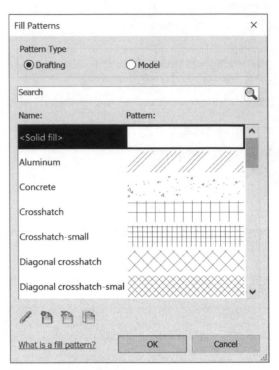

Figure 11.1-22 Fill pattern dialog

Model patterns are intended to represent a "real" world size. The pattern will be scaled as the scale of the view is changed. For example, a cross hatch pattern representing 6" tile is created with lines 6" apart. In a view assigned a scale of ¼"=1'-0" the lines will be shown 1/8" apart. The same pattern shown in a view assigned a scale of 1"=1'-0" will have the lines shown ½" apart.

When creating a new pattern or editing an existing pattern the pattern editor is displayed (Figure 11.1-23a). The top portion of the editor shows a preview of the pattern at a predefined scale. Select the radio button to define the pattern as basic or custom.

Basic patterns are defined by giving the pattern a name and then setting the line angle and line spacing. A second spacing value is required if the crosshatch radio button is selected. Remember, if defining a drafting pattern, the spacing values are the distance the lines will be spaced no matter the view scale. If defining a model pattern, the line spacing is the "real"

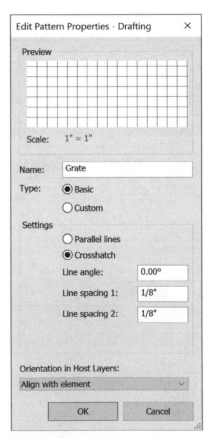

Figure 11.1-23a Pattern editor creating a simple pattern.

world spacing of the lines and will be scaled per the view setting.

If you choose to create a **custom** pattern, the dialog will change to offer a Browse button and scale input. Click the button to navigate to a PAT file on your network to define the pattern. A PAT file is a special text file that defines the lines of the pattern. PAT files that can be used for model patterns must be formatted so Revit can use them as a model pattern. Once the pattern is imported, use the import scale settings to get the pattern scaled as needed. Use the preview window to see your changes. Revit provided a sample PAT file with the installed files here: C:\Program Files\Autodesk\Revit 2021\Data, while many others may be downloaded from the internet.

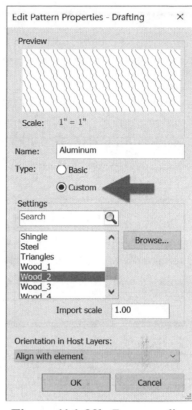

Figure 11.1-23b Pattern editor creating a custom pattern.

When making a drafting pattern there is an additional control located under the preview section. The dropdown **Orientation in Host Layers** sets how the pattern will be displayed based on the element in the view. There are 3 possible settings:

- **Orient to View** – The pattern direction will be determined by the rotation of the view. "Top" on the view will always be aligned to the "top" of the pattern.

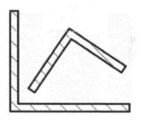

- **Keep Readable** – The pattern will be rotated as the view or the host element is rotated so it can always be "read" in the view. The pattern will remain aligned to the top of the screen.

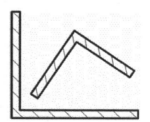

- **Align with Element** – The pattern will rotate along with the element it is used on to keep aligned within the element.

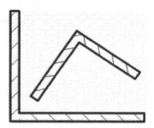

Additional Settings: Material Assets

Click to open the stand alone Asset Editor. The asset editor allows you to import material appearance assets into your model and make changes to them without assigning them to a material in the model. The assets can later be assigned to a material's appearance asset.

When you first open the asset editor nothing appears in the dialog. In the lower left corner of the editor window click to open the Asset Browser (Figure 11.1-24). In the Asset Browser window, select a library of assets and then click the arrow control shown on the right side of the assets to load it into the editor (Figure 11.1-25). Loading an asset from an external library into the editor will load the asset into your current model document. Any edits made to the asset will only affect the document asset.

Make any changes required to the asset in the editor. The aspects available in the editor may be slightly different depending on the asset selected. Use the arrow control on the left side of each aspect to expose the possible changes to each aspect. Click **OK** to finish editing assets, and close the editor.

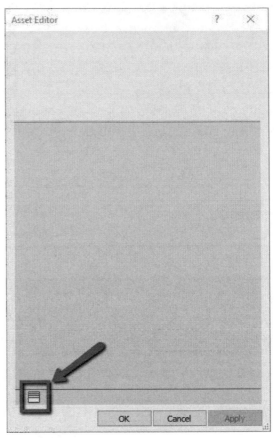

Figure 11.1-24 Open the asset browser with the control on the editor.

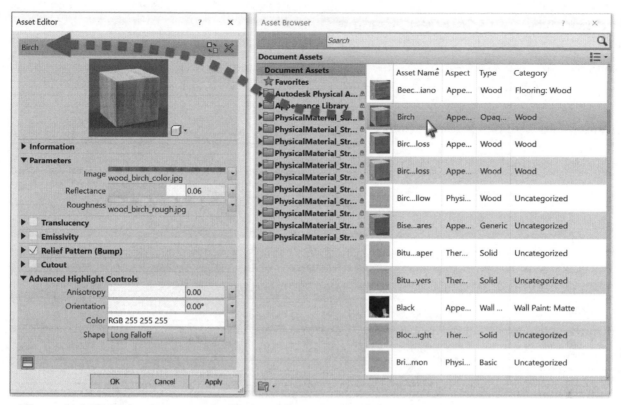

Figure 11.1-25 Make changes to the loaded asset in the editor dialog

The assets with the orange/gold triangle in the lower left corner are the original material types. Ones without the mark are the new advanced asset types, which are higher quality and will look much more realistic when rendered.

Additional Settings: Analysis Display Styles

An analysis display style is accessed by an add-in you may have installed to display the results of the analysis in a view of your model. In most cases the add-in will create the analysis display style/styles it will use. You can edit these default display styles to better suit your presentation needs. For example, in this view the 'Lighting Analysis Default' display style installed with **Autodesk Insight 360** is altered to include grid lines as part of the display style (Figure 11.1-26).

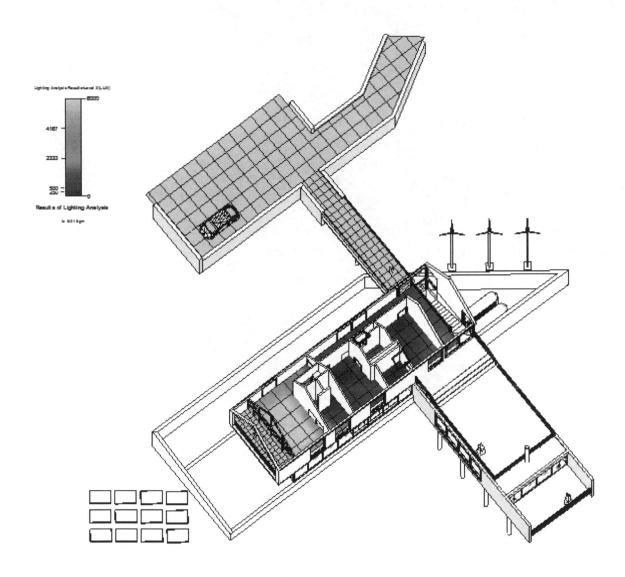

Figure 11.1-26 3d view showing results of a lighting analysis

In the Analysis Display Styles dialog, the left side of the dialog lists the display styles available for use in the model. Use tools at the bottom of the list to manage analysis display styles. Click the corresponding button to create a new style (Figure 11.1-28), duplicate an existing style, rename a style, or delete a style. Use the controls below the list on the right to control the way the list is displayed, as a list or a gallery of preview images showing the appearance of the display style (Fig. 11.1-27). Click Properties to reveal or hide the settings to change the appearance of a style.

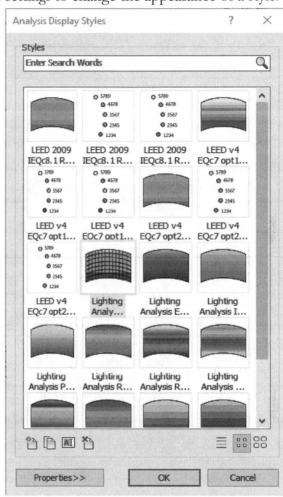

Figure 11.1-27 Analysis display styles shown as list on the left and shown as a gallery on the right

With a style selected, reveal the properties on the right side of the dialog. The 3 tabs of properties define the analysis display style (Figure 11.1-29).

The first tab is the **Settings** tab. This defines how the analytic element will be displayed. The settings are different for each kind of analytic element.

- **Colored Surface** – Set gridlines and contour display for the colored surface.

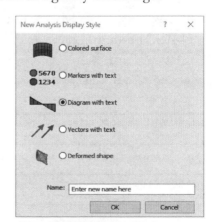

Figure 11.1-28 Analysis display style types

- **Markers with text** – The marker shape and size is defined and the text font, size, and values are defined.
- **Diagram with text** – Diagram edges and line weight is defined and the text font, size and values are defined. The fill transparency of the diagram is also set here.
- **Vectors with text** – Text size, font and values are defined. The arrow line weight and arrowhead size are set. The direction of the vector arrow is set to point at the data point or originate from the data point.
- **Deformed shape** – Set the gridlines and contour lines for the shape. Control the text size, font and values. The transparency of the fill on the deformed shape is also controlled here.

On the **Color** tab, the color for the analysis element is set. The color can be mapped to the element as a gradient or as a range. When using the gradient option, colors are blended together into a continuous gradient. Assign values for data along the gradient to set the colors for data point values. Data points will be mapped to a color along the gradient.

When the color is set by ranges, colors are set and values given to each color. All data points falling into the range will be given the same color.

Use a gradient color when you want to illustrate subtle changes in values. Use the range when it is more important to see absolutes or when a value reaches a specific important threshold.

Define how the analysis display style will show a **Legend** on the third tab in the dialog. Use the check boxes to toggle labels in the legend. Units, Name, and Description can all be toggled on and off. You may also toggle the display of the

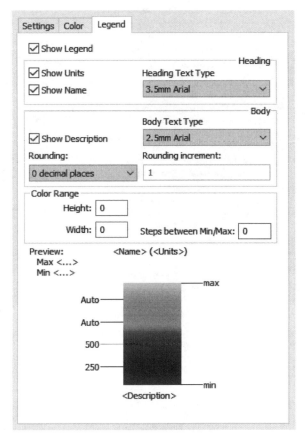

Figure 11.1-29 Legend display settings

legend entirely if you do not want it included in the display style. Each label that is toggled on can have the font controlled from the pulldown list. Use the preview at the bottom to understand what the legend will appear like in the view.

In many cases the add-in using the analysis display style will create the style and define all of the settings. You only need to adjust these settings if you need to alter the default way the analysis is displayed by the add-in.

Additional Settings: **Sheet Issues/Revisions**

See the View tab chapter for coverage of this tool; look for **Revision**.

Additional Settings: **Line Styles**

Line styles are used to control how the edges of many elements in a model are displayed (Figure 11.1-30). For example, detail lines, edges of filled regions, the line work tool, and sketch lines all use line styles to define how they will be displayed.

Each line style is defined as a subcategory of lines in the project. Each line style has a line weight, a line color, and a line pattern defined. Whenever this line style is used, these values will be assigned to the line. Click into each respective column cell in the dialog and select from the drop down or the color picker to make a change to the value assigned to a line style. Click **New** under Modify Subcategories in the dialog to create a new line style for use in your project.

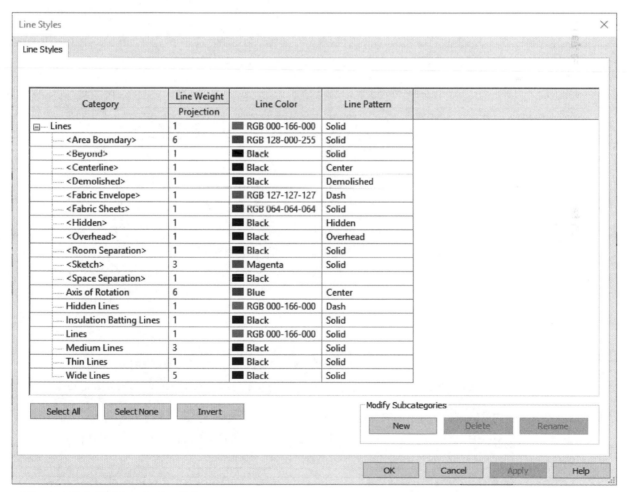

Figure 11.1-30 Line Styles dialog

Additional Settings: Line Weights

Line weight for elements is defined in the object style for the element's category. When a line weight is assigned, it is given a numerical value from 1 to 16. The **Line Weights** dialog defines the thickness of each line when printed (Figure 11.1-31).

The line weight dialog has 3 tabs to define line weights.

- Line Weights dialog shows line weights assigned to model elements in a view. **Model Line Weights** - Line weights for model elements will be based on the scale of the view. For example, in the table, the lines of an element using a 5 line weight (walls cut in plan typically are defined as 5) in a view at 1/16"=1'-0" scale will be printed 0.0090". The lines of the same element in a view at ½"=1'-0" scale will be printed 0.0250". Scales not specifically listed in the table will use the smaller line width of the 2 scales it falls between. For example, a 5 line weight in a view scaled to 3/8"=1'-0" will be printed 0.0220" (the size applied to views of ¼"=1'-0").

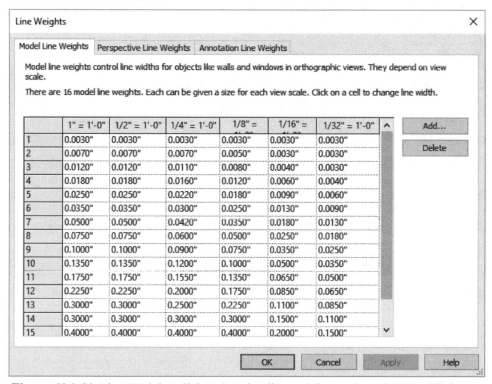

Figure 11.1-31 Line Weights dialog showing line weights assigned to model elements in a view

- **Perspective Line Weights** – Line weights 1 thru 16 are still used in perspective views, but because by the nature of a perspective view it has no "scale," the line weights table just assigns values of each line weight 1 thru 16.
- **Annotation Line Weights** – Like perspective views, in Revit annotation elements are scaled with the view proportionally so there is no scale needed for line weights 1 thru 16. Annotation elements will always be printed the same thickness no matter the scale of the view based on the line weight of the element.

To change the thickness assigned to a line weight, click in the corresponding cell in the table and change the thickness. When making a change to the thickness of a line weight, you may need to change other line weights proportionally as well. It is also possible to have multiple line weights assigned the same thickness. This effectively changes the maximum number of line weights you are able to use in your model. When making modifications to the model line weights table, use the **Add** and **Delete** controls on the right side of the dialog to add and delete scale values to the table. This can be done in addition to changing the thickness values directly in the table cells.

When viewed on your screen, depending on the zoom level of the view, it may be difficult to see the difference in line weights because of your screen resolution. When printed the line weight is applied (Figure 11.1-32).

> Some elements have a "hard coded" line weight assigned at the system level. For example, all fill patterns are drawn using line weight 1. You cannot change this assignment. You can change the thickness of line weight 1, but it may affect other elements in the model.

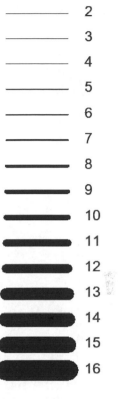

Figure 11.1-32 Example of line weights 1 thru 16

Additional Settings: **Line Patterns**

Line patterns for elements are defined in the object style for the element's category. Each line pattern consists of a series of dashes, dots, and spaces (Figure 11.1-33). The length of each dash and each space is defined. The size of the dot is defined at the system level. Use the controls on the right side of the dialog to manage the line styles available in the project.

> Line patterns <u>cannot</u> contain text characters like "GAS" defined as a repeating element.

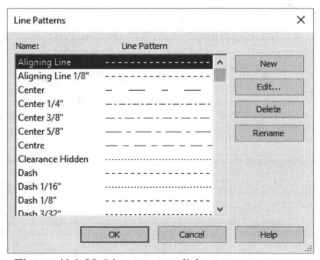

Figure 11.1-33 Line pattern dialog

Click **New** or **Edit** to open the **Line Pattern Properties** dialog (Figure 11.1-34). In this dialog, define the pattern for the line. Use the pulldown to define the type of segment in the pattern and then use the value column to define the length of each segment. A **dash** or **dot** type must be followed by a **space** type in the pattern definition. Define up to 20 segments in the pattern. The line type will repeat the pattern over the length of the line when used in the model.

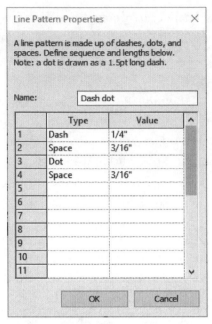

Figure 11.1-34 Example of a line pattern definition

Additional Settings: Halftone/Underlay

Elements and categories can be overridden to display in halftone, and when working in plan views you can underlay another level plan view. The Halftone/Underlay settings control how these will be displayed in a view when selected (Figure 11.1-35).

The underlay settings allow you to set a line weight and pattern override for elements of the underlay as well as apply the half tone setting to the elements of the underlay. The settings being overridden are the object style of the elements in the view. Typically underlay elements will at a minimum be halftoned so they can be distinguished visually from the non-underlay

Figure 11.1-35 Underlay/Halftone dialog

elements. Additional weight and pattern overrides provide additional visual distinction for underlay elements if needed.

The halftone value is set in the lower section of the dialog. The halftone value represents a percentage applied to halftoned elements. For example, a halftone brightness value of 100 on a black element in a view will have no effect but a halftone brightness of 0 applied to the same element would make the element completely white. Between 0 and 100 would make the element appear in a shade of gray.

Halftone is applied as a visibility and graphics override to the element or category in a view. There is only one global half tone setting. All elements with the halftone override applied will be half toned using the same brightness value.

▌ Halftoned elements still print gray even when Black & White is selected for printing.

Additional Settings: Sun Settings

The Sun Settings dialog controls how shadows will be cast on the model when shadows are turned on in a view or when creating a solar study (Figure 11.1-36). The sun settings will have a slight effect on the display of the surfaces of elements when using the shaded visual style, but to more easily see the effects of change in sun settings, turn on shadows for the view.

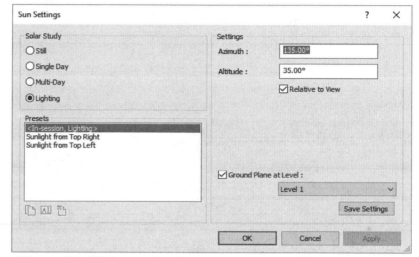

Figure 11.1-36 Sun setting dialog

In the dialog, define one of 4 kinds of solar study types for the sun setting.

- **Still** – In a still solar study you set a real world location, date, and time in the settings portion of the dialog. If your model does not have a toposurface or an element representing the ground, use the control to cast shadows at a specified level in the project. Click save settings to add a preset to the list in the dialog.

- **Single Day** – This will allow you to create an animated solar study of the view (Figure 11.1-37). The single day setting is similar to the Still setting, in that you set a location and date for the study, but the time setting has a **beginning** and **end** point. This sets where the sun position will be at the start and end of the exported animation. Either set the values explicitly, or check sunrise to sunset to calculate the time between sunrise and sunset based on location, date, and time. The number of frames of animation to be

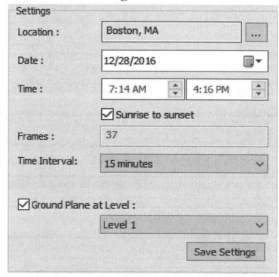

Figure 11.1-37 Single day solar study settings

generated by the animation is a read only value and dependent on the time interval setting. Smaller time intervals will result in an animation with more frames.

- **Multi-day** - A multi-day solar study is created the same as the Single Day study except the date includes a range as well as the time range.
- **Lighting** – With lighting settings, the azimuth and altitude of the light source is set directly. This might be used to create shadows for a plan view to provide depth to the view.

It is a good practice to make views specifically for solar studies and apply presets from the sun settings dialog to them for easy exporting. The Sun Settings dialog is also accessible from the View Control Bar.

Additional Settings: Callout Tags

Opens the type properties dialog for the System Family **Callout Tags**. Callout Tag families have two parameters that control how the callout appears when placed in a view. Create different view types to use different callout tags as required for your project standards.

- **Callout Head** – Annotation added to the callout to report view information about the callout view. The dropdown allows you to select different callout head annotation families loaded into your model. You can also set it to **<none>** to not display a callout head.

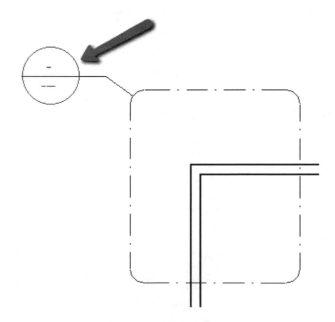

- **Corner Radius** – The callout graphic typically has rounded corners. The value here is the printed radius size used for corners on the callout. A callout is an annotation so the radius is always printed at the set radius based on the view scale.

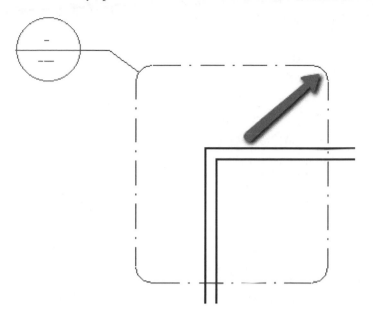

Manage

Additional Settings: Elevation Tags

Opens the Type Properties dialog for the System Family **Elevation Tags**. Elevation Tag families have a parameter that controls how the elevation tag appears when placed in a view. The drop down allows you to select different elevation mark families loaded into your model.

Create different view types to use different elevation tags as required for your project standards. For example, you may create an elevation type for exterior elevations and a different elevation type for interior elevations. The exterior elevation tag could be a square tag with a directional arrow while the interior elevation tag would be a circle that could have multiple directional arrows.

Additional Settings: Section Tags

Opens the Type Properties dialog for the System Family **Section Tags**. Section Tag families have 3 parameters that control how the callout appears when placed in a view. Create different view types to use different section tags as required for your project standards. For example, a section tag for a building section may need to appear differently than a section tag for a wall section.

- **Section Head** – Section head family used at the first click when creating a section. Use the dropdown to select from different section head families loaded into your model.
- **Section Tail** - Section head family used at the second click when creating a section. Use the dropdown to select from different section head families loaded into your model.

> The Section Head and Section Tails defined to the section tag family can be swapped on the section tag in the canvas using the small control located near the section end points.

- **Broken Section Display Style** – Set this to Gapped or Continuous. When a section mark is broken in the canvas using the small "lightning bolt" control in canvas this setting is used (Figure 11.1-38). Gapped removes the center of the Section Tag line and adds controls to make adjustments to the gap. Continuous will change the section line to use the line pattern set for the **Broken Section Line** subcategory of Section line in object styles of the model.

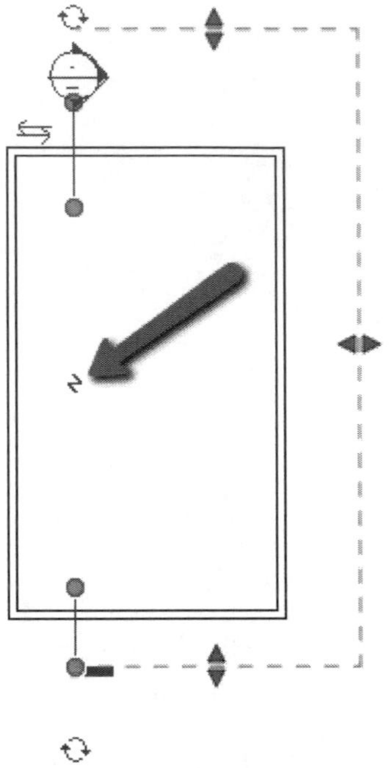

Additional Settings: **Arrowheads**

Opens the Type Properties dialog for the System Family **Arrowheads**. Arrowheads are applied as parameters to a number of objects such as leaders and spot elevations. Make arrowhead types matching your graphic standards. Arrowhead types have 6 parameters that control how the arrowhead appears. Depending on the **Arrow Style** used, not all of the parameters will be available.

The **Arrow Style** is selected from the drop down list. The 8 Arrow Styles available are set at the system level (Figure 11.1-39). Any further customization of the selected arrow style is done with the additional parameters for arrowhead types.

Figure 11.1-38 Broken section with gap. Lightning bolt break control highlighted

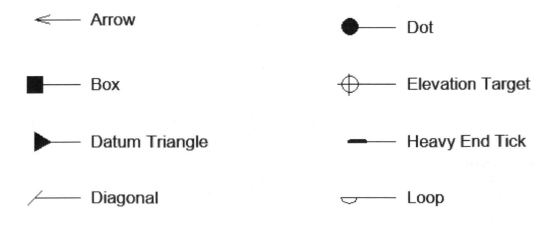

Figure 11.1-39 Arrow Styles available for use in Revit

Additional Settings: **Temporary Dimensions**

When elements are selected in the canvas, or while adding elements to the model, temporary dimensions are displayed to help you position elements as they relate to each other. To help control what temporary dimensions reference, set the temporary dimension preferences (Figure 11.1-40). Only preferences for walls, doors, and windows can be set. Temporary dimensions to other elements are controlled by the family geometry and reference plane definitions in the family.

Manage

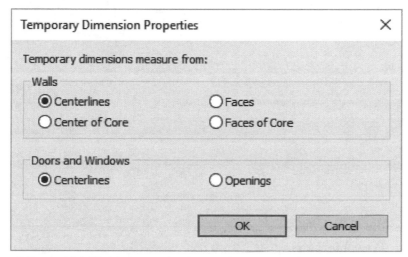

Figure 11.1-40 Temporary Dimension Properties dialog

Temporary dimensions to walls can be set to 1 of 4 logical locations of the wall.

- **Centerlines** – The temporary dimension will be placed at the geometric center (inside to outside faces) of the wall assembly.
- **Center of Core** – The temporary dimension will be placed at the geometric center of the core layers as defined in the wall structure.
- **Faces** – The temporary dimension will be positioned at the wall faces, either interior or exterior, depending on the context.
- **Faces of Core** – Temporary dimensions will be positioned at the faces of the wall core as defined in the wall structure.

Temporary dimensions for doors and windows can be set to the centerline of the opening or to the edge of the opening defined in the family.

Set the temporary dimensions to measure to the walls and openings as it makes sense to your design workflow. Setting this before you begin laying out your project can save considerable time, minimizing the need for you to reposition temporary dimensions while working.

Additional Settings: **Detail Level**

When callout views are created Revit automatically increases the scale of the new callout by a factor of 2. For example, if you place a callout on a view scaled at ¾"=1'-0", the new callout view is automatically set to use the scale of 1 ½"=1'-0". The Detail Level of the new view may also be changed as well depending on where the scale lands on the detail level table.

This table defines the scales at which the Detail Level will be changed when a callout view is placed (Figure 11.1-41). As in the example above, where the new view scale is set to 1 ½"=1'-0", because this new scale appears in the "Fine" column, the detail level of the new view will be set to fine. Select a scale value and use the arrows between the columns to move a scale value from one column to another column.

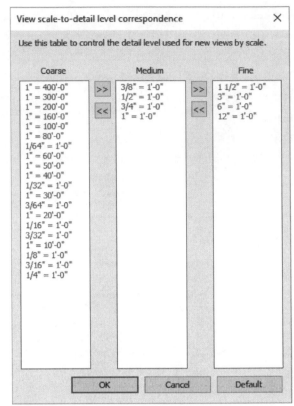

Figure 11.1-41 View scale/detail level dialog

Additional Settings: **Assembly Code**

Use this dialog to manage the assembly code file used for elements in your model (Figure 11.1-42). The assembly code is a type parameter on elements which is defined by a text file with assembly code values. This text file, and the assembly codes used, may vary based on your project specifications. **Browse** to an assembly code txt file or **Reload** the file from this dialog.

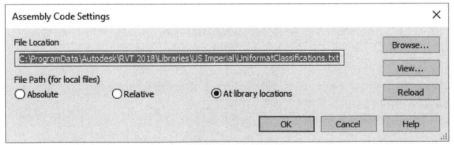

Figure 11.1-42 Assembly Code Settings dialog

11.2 Project Location panel

The Project Location panel provides tools to locate the project in the real world location on the earth as well as tools to adjust the true north and project north of the project. There are also tools to coordinate the position of linked models and files to make sure they are in the right positions related to each other.

Location

In the location dialog you set the real world position of your project (Figure 11.2-1). Use the interactive map portion of the dialog, pan and zoom the map and locate the project by clicking and dragging the red "pin" icon to the exact project position or type the address directly to instantly move the map and the pin icon to the project address.

If you are planning on doing energy analysis on the project, select a weather station near the project location. The panel on the left of the dialog lists weather stations near the project location and approximates distances away. Select a weather station from the list to highlight it on the map with a pin. In some cases, you may not want to choose the closest weather station to a project because of micro climate conditions of a specific weather station.

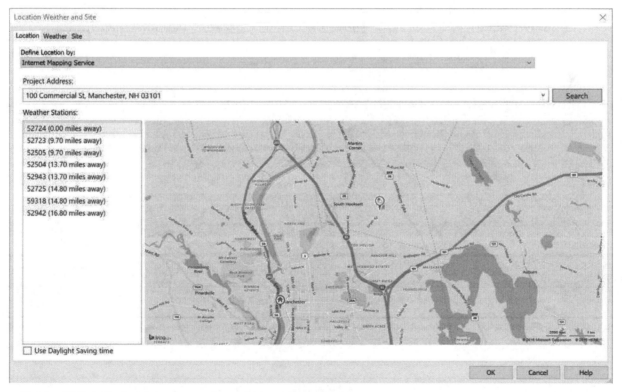

Figure 11.2-1 Project location and weather station selection

Manage

The information provided by the weather station is found on the **Weather** tab of the dialog. The information here will be used for HVAC design (if Revit is used for the MEP design of the project). This information will also be used during energy analysis. Un-check the Use HVAC design data from weather station to make the values in the table editable. Make edits to the weather station information if you have more accurate information about your project site.

The **Site** tab in the dialog provides a way for you to create different named positions in a model file. This may be used when you are referencing in a number of buildings into one common file, or when you have multiple instances of a model placed in different positions on the site.

Coordinates → Acquire Coordinates

Use Acquire Coordinates when you link another Revit model into your model or link a DWG file into your model. When you Acquire Coordinates, you select the linked file to acquire from. The shared coordinate origin of the host model will be moved to the origin of the coordinates in the linked model.

Once coordinates have been acquired the files can be linked using the "By Shared Coordinates" option. For example, you may link in a DWG survey file and position your model in relation to the survey and then acquire the coordinates of the DWG. The same process is done for the structure model of the project. Now when the structure model is linked to the Architecture model, the "By Shared Coordinates" option is used and the models will align with one another.

Use this option when everyone working on the project will be linking the same file where coordinates will be acquired from.

Coordinates → Publish Coordinates

Use Publish Coordinates when your project has another linked Revit model or a linked DWG file and you want to change the origin of the linked models coordinate system to match the origin of the host model's coordinate system.

Select Publish Coordinates and then select the linked model to publish to. Once complete, the two models share a coordinate system and can be linked together using "By Shared Coordinates."

Use this option to share coordinates when you have access to the files linked together and are the person controlling where the shared location for the files will be.

Manage

Coordinates → Specify Coordinates at Point

Use specify coordinates at a point as a method to share coordinates between two files (Figure 11.2-2). An agreed point in both files is selected and a coordinate value can be specified. Once this is done, models can be linked and aligned together.

Click a point in your model and provide northing and easting coordinate values. The angle to True North is set in the lower section of the dialog.

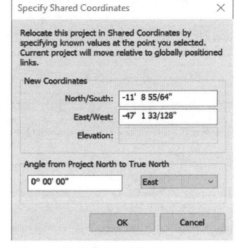

Figure 11.2-2 Specify shared coordinates dialog

Coordinates → Report Shared Coordinates

This is an inquiry tool used to help you reconcile and understand positioning of your model. Select Report Shared Coordinates tool and then select a point or edge in the model. The northing and easting position of the selection is reported in the options bar. When coordinating a linked model, use the Report Shared Coordinates to help you resolve positioning issues.

Position → Relocate Project

Relocate Project will move all elements of the project together to a new location in the coordinate system. Use this option when everything needs to be repositioned. All constraints will be maintained.

Select Relocate Project and then in the canvas select a base point for the move and then select a new position for the point. The entire project will be moved to this point. For example, you have a project located at elevation 0'-0" and you need it to be located at 10'-6". Relocate Project can be used in a section or elevation view to raise the entire project up to 10'-6". If this was done using the move tool you would encounter many constraint errors. Relocating the project will move the elements and the constraints to the new position.

Position → Rotate True North

By default, the Project North and True North are aligned in a project and set to the top of the application window. To get accurate shadows and accurate energy analysis you may need to rotate the True North position away from the Project North position.

In order to rotate the True North of the model, the view you are working in needs to be set to the true north orientation. With a plan view open, set the **Orientation** parameter to True North. Now select the Rotate True North tool. If necessary, click and drag the center of rotation to a logical position in your model. It is a good idea to select a position for the center of rotation where you can easily use a linear element in the model that is oriented to True North.

Once the center of rotation is set, click once to define an angle from North (top of the application window), then click another point when the guide is pointing North (top of the application window). Now the view is oriented to True North as up in the application window.

To change the view back to Project North, change the Orientation setting in the properties palette back to Project North. Different views can be set to use Project North and True North as required.

Position → Mirror Project

Mirror Project takes the entire project and reverses it along the chosen axis. All elements in the model are flipped along the axis. All views of the project will be mirrored as well as the annotation elements in any of the views. The axis for the mirror runs east west or north south as it relates to the application frame, north being toward the top of the screen. The axis line will be placed at the internal origin of the model file.

When you mirror a project the dialog offers 4 options (Figure 11.2-3):

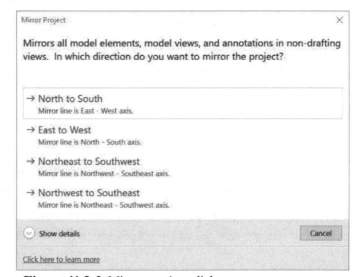

- **North to South** – Project is flipped along a mirror axis line running east west.
- **East to West** – Project is flipped along a mirror axis line running north south.
- **Northeast to Southwest** – Project is flipped along an axis running east west and then along an axis running north south. The project will end up below and to the left of the original position.

Figure 11.2-3 Mirror project dialog

- **Northwest to Southeast** - Project is flipped along an axis running east west and then along an axis running north south. The project will end up below and to the right of the original position.

Position → Rotate Project North

By default, the Project North and True North are aligned in a project and set to the top of the application window. In order to make working in the project easier you may need to rotate the Project North so the major axis of the project is aligned to the application frame.

In order to rotate the Project North of the model, the view you are working needs to be set to the Project North orientation. With a plan view open, set the **Orientation** parameter to Project North. By default, views in the project are created with a Project North orientation.

Now select the Rotate Project North tool. If necessary, click and drag the center of rotation to a logical position in your model. It is a good idea to select a position for the center of rotation where you can easily use a linear element in the model that is oriented to Project North.

Once the center of rotation is set, click once to define an angle from North (top of the application window), then click another point when the guide is pointing North (top of the application window). Now the project is oriented so north is up in the application window.

To change the view back to True North, if needed, change the orientation setting in the properties palette back to True North. Different views can be set to use Project North and True North as required.

11.3 Design Options panel

When working on your model you may find the need to explore design alternates. Revit provides design options tools so you can create and manage different options within one model. Elements are authored into a design option and then visibility and graphics controls are used to display different options in different views. Only the Design Options button is available on the panel until design options have been created in your model.

Design Options

The Design Options button opens the Design Options dialog where you manage the design options in your model (Figure 11.3-1). The left side of the dialog lists the option sets and the options in your model. The right side of the dialog gives controls to manage the design options.

First you must make an option set for different options to populate. You cannot create any options until an option set is created. An option set really serves as a "container" for your options. To create an option set, click **New** in the dialog. An option set is created, and a single option is also created. To rename the option set to something more relevant to your project, select it in the dialog and click **Rename** under the option set controls.

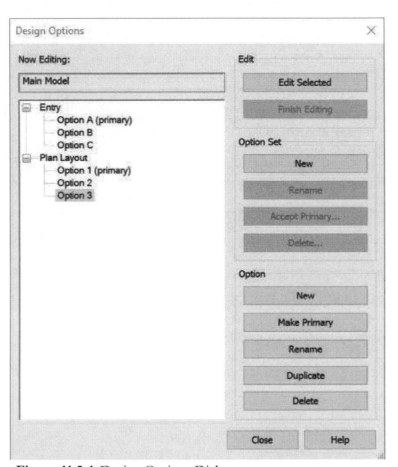

Figure 11.3-1 Design Options Dialog

The concept of option sets is important because it allows you to mix and match different options together. For example, your design has 3 entry options as well as 3 different internal layout options. You may want to mix and match these together. Entry option A with internal layout 2, or a different option with entry option A and internal layout 3. The option sets allow you to do this. Remember this functionality when setting up the option sets for your project. Add and rename as many option sets as needed for your project.

Once an option set is created you can populate and create different options. Select an option set from the list on the left and then click new under the Option controls. Use the rename tool with the option selected to make the option name more relevant to your model. Use the **Duplicate** control to make a copy of the option and copy all of the elements authored into the option. Add and rename options as needed for your project.

When you have multiple options in an option set, one of the options is marked in the dialog as (primary); this is the option you most favor for the final design. The primary option will be automatically displayed in your project views. Other options can be displayed in a view but they have to be manually applied. In the design options dialog you can use the **Make Primary** control to change the primary design option. This will change all of the views of the model to use the new primary design option.

The controls for the option set also have a control related to the primary option. Select the **Accept Primary** control under the options set area to select the primary option from an option set, delete all other options from the model, and delete the option set from the model. If any views are specifically assigned an option that is not the primary option, they will be deleted as well. Accepting the primary option of an option set will merge all of the elements in the option into the main model.

To author elements in an option, you must set the option for editing. The top of the design options dialog lists a **Now Editing** window to notify you of the option you are currently editing (Figure 11.3-2). The name of the option set followed by the name of the option will be listed here when editing an option. If not editing an option specifically, this area will report Main Model. To select an option for editing, select it from the list on the left and click

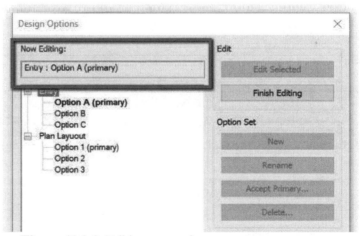

Figure 11.3-2 Editing control

Edit Selected in the dialog. Close the dialog to begin editing the option. By default, the element in the main model will be shown in halftone and the option will be shown at 100%.

Add to Set

When working with options you may find the need to move an element from the main model into a number of options in order to allow the options to work. A good example of this is when you are adding an opening, such as a door into an option but the host wall is part of the main model. The door is the option, but the wall hosting the door must also exist in the option. Because the wall needs to be different, one without a door, and one with a door, the wall also plays a role and has 2 "options." In a case such as this you need to move the wall into the different options involving the door.

Use **Add to Set** to move elements from the main model into options (Figure 11.3-3). Select the elements that need to be moved and click Add to Set. You may find it more convenient to access the Add to Set from the controls at the bottom of the application frame. The ribbon button is grayed until an element is selected and then you will need to click back to the Manage tab to access the Add to Set button from the ribbon.

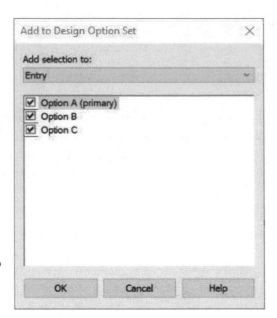

Figure 11.3-3 Add to Design Option Set dialog

In the Add to Set dialog, place a check mark on the options where the element will be added and click ok. The element(s) will be moved from the main model and added to each of the selected design options.

Pick to Edit

Pick to Edit offers an alternative way to change the active design option you are working on. Click **Pick to Edit** and then select an element that is part of a design option. The design option assigned to the clicked element will now become the active option you are editing. You can choose to edit options in this way, but you are not able to change editing an option back to the main model using this tool.

Active Design Option

The pull-down menu shows the active option being edited (Figure 11.3-4). Use the drop down to select from the list of options in the model and switch to editing the selected option. Use this control as an alternative to setting the option for editing in the Design Option Dialog.

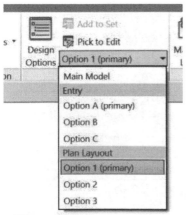

Figure 11.3-4 Active design option

11.4 Generative Design panel

Generative Design tools allow you to run Dynamo graphs that are created with rules that allow you to explore many outcomes for a design based on rules you define. The generative design tools are broken into 2 dialogs, Create Study and Explore Outcomes. In the first dialog you select a study you would like to use a generative design graph to create options. The explore outcomes dialog allows you to visualize the different solutions generated by the generative design graphs. Once you are happy with an outcome you can apply the geometry/solution to your Revit model. Revit ships with 3 example studies you can use to explore how generative tools work, but you can use Dynamo to create additional study types to use on your designs.

Note: To use this service, you need to have a subscription to the AEC collection. If you do not, you will have to use Dynamo directly to create similar kinds of studies. The main difference between using Dynamo and the Generative design tools is Generative Design allows the Dynamo graph to run multiple times with different values for the variable automating the process of running the graph many times to create many potential solutions. Using Dynamo alone you will need to run the graph manually changing the setting for each run.

Create Study

 Opens the Create Study window (Figure 11.4-1). This window is a separate application running outside of Revit. The window opens on top of Revit and stays there, but you can still use Revit behind the application window.

Select a study type from the dialog to define the variables for the study. The variable that you will set once a study is selected will be controlled by the inputs needs for the Dynamo graph being used by the study. See Revit and Dynamo documentation for additional information on how to create Dynamo graphs for use in creating a study type for Generative Design.

Figure 11.4-1 Create Study Panel

Explore Outcomes

 The Explore Outcomes panel (Figure 11.4-2) lets you see the options that have been generated by a design study once it is complete. Select a completed study on the left side of the panel then based on the study type you will see results displayed on the right side of the panel. Each study result can be selected to explore the details of the particular result. The lower line graph can be filtered to display study results that meet specific criteria.

With a study selected and the Revit project open that the study applies to, click Create Revit Elements in the lower right of the panel to create Revit elements shown in the selected study in your model.

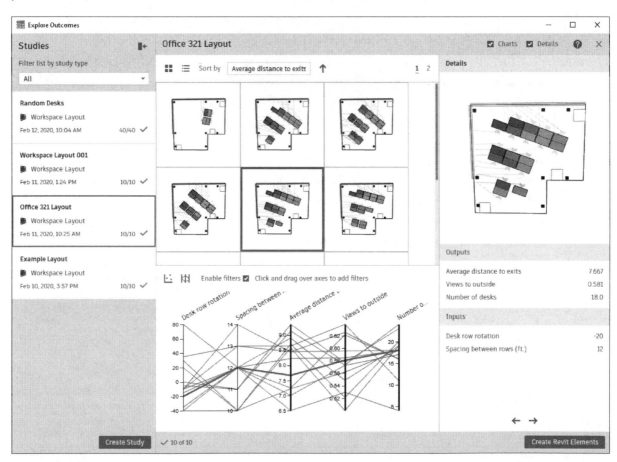

Figure 11.4-2 Explore Outcomes Panel

11.5 Manage Project panel

The Manage Project panel contains tools for managing external files linked into your current model file. Use the tools to manage RVT and CAD links as well as image files used in parameter definitions and decal types.

Manage Links

The Manage Links dialog (Figure 11.5-1) is where you control all files linked to the host model. Across the top of the dialog there are a number of tabs. Each tab corresponds to a different file type that can be linked to the model. The tabs organize the file type for you so the dialog is easier to work with.

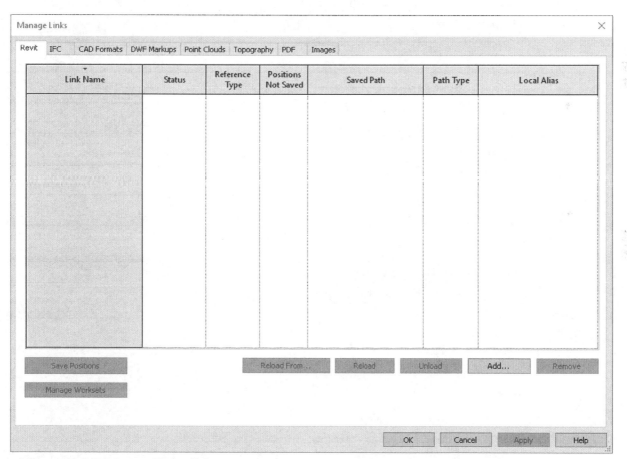

Figure 11.5-1 Manage Links dialog

Below the tabs for the file types, the links are all listed—with properties associated with each link—in columns beside the link name. Some columns are not used for some file types. If a column is not shown, that property is not relevant for that file type.

- **Status** – The status can be Loaded or Unloaded. When a link is loaded it will be visible in views of the host model. A link can be unloaded so it is not visible. Links can be left attached but unloaded to help with performance but easily loaded again to check relationships in the host model. If the path to the link is not available or the file for the link is not available, the status will be "Not Found."

- **Reference Type** – Overlay or attached. This setting affects links when the host file is linked to another file. When linked, any files set to overlay will be unloaded. Any files that are attached will remain loaded and visible when the host is linked to another file.

- **Positions Not Saved** – Indicates if the location of the link has been saved to the host's shared coordinate system. Used when you have multiple instances of the same link positioned in the host model file.

- **Saved Path** – Network location of the linked file. This path must be accessible to the PC where the host file is opened in order to load and display the link. If the path is not accessible the status of the link will read "Not Found."

- **Path Type** – The path can be set to relative or absolute. An absolute path is a path to the linked file and must match exactly in order for the link to be found. When a relative path is specified the folder location of the host and the link needs to be the same but the drive letter and pathing could be different than when the link was created.

- **Local Alias** – When worksharing is enabled on the host file, copies of the links will be created and stored on the local PC. The local alias is the location of this local copy of the link.

The controls below the list of links (Figure 11.5-2) allow you to manage the status of a link and add/remove links from the list. The controls for the different file types are the same with the exceptions listed below.

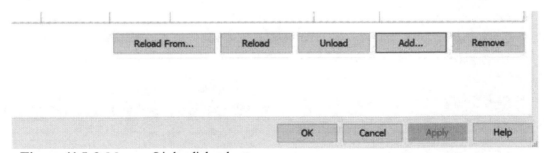

Figure 11.5-2 Manage Links dialog buttons

- **Reload From** – Use this control to reload the link from a different file path than that listed for the link. For example, this would be used if you have received an updated file from a consultant and it has a new file name. The reload from will allow you to select the new file name and path to replace the current link.

- **Reload** – Use this control if the linked file has had changes made since you last opened the host file. The linked file is reloaded from the saved path and the changes will be visible. Use this when the linked file has changed but the name and file path have not been altered. Links are always reloaded when the host file is opened in a session.

- **Unload** – Use this to unload a linked file. Unloading the link only makes the link not visible in the host file. The link is still part of the model, but the geometry has been unloaded. Typically, this would be used to increase performance, or temporarily make a link not visible so it is easier to do your work, but you will need to keep the link for coordination with the host model.

- **Import (CAD links only)** – The import control changes the CAD link into an import. The geometry will be imported into the host file and the link will be removed. Use this option if you need to manipulate CAD geometry in the model file. It is recommended to keep the CAD file as a link if possible and make changes there and reload rather than importing the CAD geometry into the model. CAD geometry may contain many elements impacting performance, or contain geometry that will make it difficult to work in Revit.

- **Add (Not available for DWF Markups)** – Add a linked file directly from the manage links dialog. Select the control and then choose the file to link.

- **Remove** – Remove the link and the geometry from the host model. The file will need to be relinked and positioned if needed in the future for coordination.

Decal Types

See the Insert Tab section for more information on Decal Types.

Starting View

When a model is first opened, by default the view that was active when the model was last closed is the view that will be opened the next time the model is opened. In some cases, this may be the desired behavior, but in others it is not. For example, when multiple people are working in a file or when worksharing is enabled, it may be better to open to a specific view each time the model is opened.

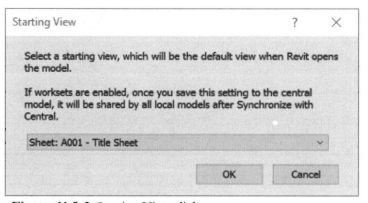

Figure 11.5-3 Starting View dialog

The **Starting View** function allows you to select any view or sheet in the project to be the view that is opened when the model file is opened (Figure 11.5-3). Select the button and then select the view to open from the drop down list. When setting a starting view, you may want to consider making a view in the project act specifically as the starting view and use this to add project related notes or information you want to communicate to different members of the team. A sheet with a custom title block, with the project name and number, can be used as the start up view; thus, the project name and number are tied to the regular sheets and project info.

11.6 Phasing panel

Phasing tools in Revit allow you to create a time line for projects that may not be completed all at one time, or for renovation projects where there are existing conditions and demolished work as part of the project. Phasing controls help you manage the visibility and graphics of elements as they exist across phases of the project on the time line.

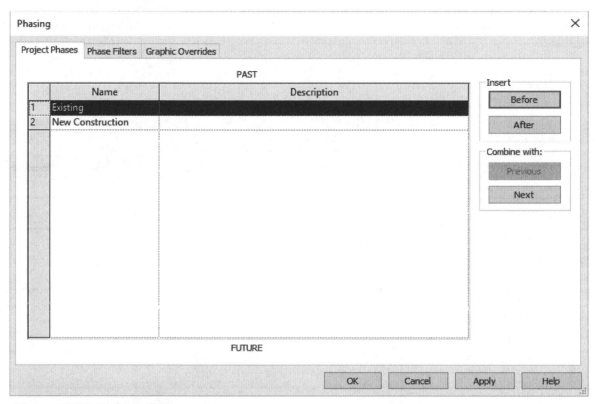

Figure 11.6-1 Phasing dialog

Phasing

Most existing/remodel projects have two phases (Figure 11.5-1):

- Existing
- New Construction

Additional phases can be created, such as Phase 2 and Phase 3 (in this case, New Construction might be renamed to Phase 1). There should never be a phase called 'Demolition', as Revit handles this automatically (more on this later).

Revit manages Phases with two simple sets of parameters:

- **Elements:**
  - o Phase Created
  - o Phase Demolished
- **Views**
  - o Phase
  - o Phase Filter

Among these four parameters, Revit is able to manage elements over time and control when to display them.

Element Phase Properties

As just mentioned, every model element in Revit has two phase-related parameters as shown for a selected wall in Figure 11.6-2.

The **Phase Created** parameter is automatically assigned when an element is created—the setting matches the phase setting of the view the element is created in. Put another way, in a floor plan view with the Phase set to *New Construction*, all model elements created in that view will have their Phase Created parameter set to *New Construction*.

The **Phase Demolished** parameter is never set automatically. Revit has no way of knowing if something should be demolished.

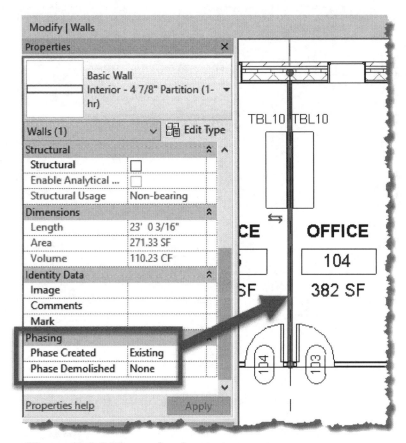

Figure 11.6-2 Phase related parameters

View Phase Properties

Every view has two phase-related parameters, as shown for the floor plan view in Figure 11.6-3. Any elements added in this view would be designated as existing.

The **Phase** parameter can be set to any Phase that exists in the project. This setting can be changed at any time; however, it is best to have one floor plan view for each phase—thus, the Phase setting is not typically changed.

This setting represents the point in time the model is being viewed.

> Create existing and demolition views in a company/personal template.

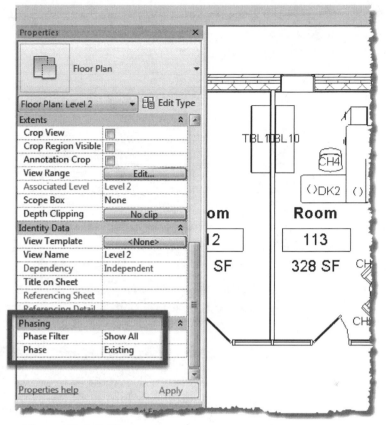

Figure 11.6-3 Phase properties

Phase Filters:

The **Phase Filter** setting controls which model elements appear in the view based on their phase settings.

To understand how the Phase Filter setting affects the view, look at **Manage → Phasing → Phase Filters** (See Figure 11.6-4).

Elements will appear in one of three ways:

- **By Category**   - Displayed as normal, no changes
- **Overridden**   - Modified based on Graphic Overrides tab settings
- **Not Displayed**   - These elements are hidden

Be careful not to confuse the *Phase Filter* column headings, New, Existing, Demolished and Temporary, as literal phases—they are not. Rather, they are a 'current condition' (aka **Phase Status**) based on a view's Phase setting and the element's Phase Created & Phase Demolished settings.

For example, the 'current condition' of an *Existing* wall in an *Existing* view is considered <u>New</u> because the phases are the same. Think of it as if you are standing in the year 1980, looking at a wall built in 1980—it is a new wall. Similarly, a *New Construction* wall, in a *New Construction* view is also considered <u>New</u> in terms of how the Phase Filters work (now all *Existing* walls are considered <u>Existing</u>). You are now standing in the year 2021 looking at a wall built in 1980.

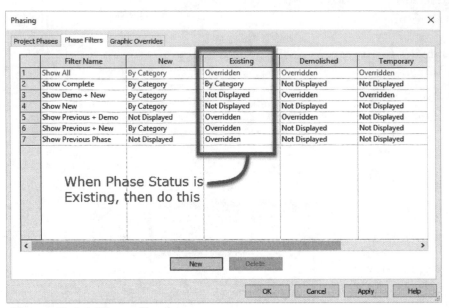

Figure 11.6-4 Phase filters

The three images below are of the same model as seen in three different views—each with a different combination of Phase and Phase Filter settings. Each condition will be explained in depth on the next pages.

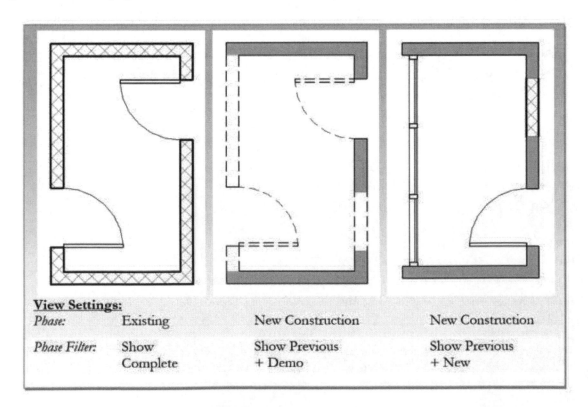

Existing Conditions:

When modeling an existing building, create a **Level 1 – Existing** plan view; do this for each level in the building. This view will have the following phase-related settings:

- Phase: **Existing**
- Phase Filter: **Show Complete**

Any model element created in this view will automatically have Phase Created set to Existing.

Everything in the drawing to the right has the Phase Created set to Existing.

> One wall and both doors have their Phase Demolished set to New Construction—however, we cannot visually see that here.

The Phase Status is actually considered New in this view. The phase of the element matches the phase of the view. Thus, because the Show Complete Phase Filter has New set to By Category, there are no overrides applied to this view. We see the normal lineweights and fill patterns (Figure 11.6-5).

Demolition Conditions:

When demolition is required, create a **Level 1 – Demo** plan view; do this for each level in the building. This view will have the following phase-related settings:

- Phase: **New Construction**
- Phase Filter: **Show Previous + Demo**

Often, new Revit users think it strange that the demo view needs to be set to New Construction—given no new elements appear in this view. However, looking at the existing view just covered—all existing elements (even ones set to be demolished) are considered "new." Therefore, the

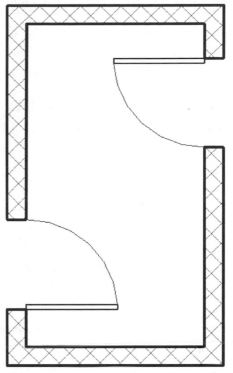

Figure 11.6-5 Existing conditions

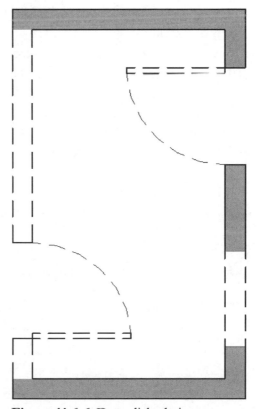

Figure 11.6-6 Demolished view

"time slider," if you will, needs to be moved past Existing to invoke the Phase Demolished setting. If this is still confusing, it should make more sense in the tutorial.

In Figure 11.6-6, the wall and two doors which appear dashed have their Phase Demolished set to New Construction. Recall that the view's phase is set to New Construction. In the lower right, a new door (i.e. Phase Created = New Construction) automatically demolishes the existing wall.

Notice the Phase Filter **Show Previous + Demo** has the <u>Demolished</u> *Phase Status* set to **Overridden**. So, any element with the Phase Created setting set to a phase which occurs prior to the Phase setting of the current view, **and** the same element has a Phase Demolished setting that matches the current view's Phase setting, then it will show demolished.

> If this project had a Phase named Phase 2, no existing elements set to be demolished in Phase 1 (i.e., New Construction in our example) could be shown in a view that's phase is set to Phase 2. Those elements simply do not exist anymore, and there is no reason to ever show them in this future context.

One final note on this demolition view: because Revit inherently understands the need to demolish elements, **it is never necessary to have a Demolition phase**.

New Conditions:

All Revit templates come with the default plan views set to New Construction. For projects with existing and demolition conditions, it might be helpful to rename these views to Something like **Level 1 – New** (similar for each level). This view will have the following phase-related settings:

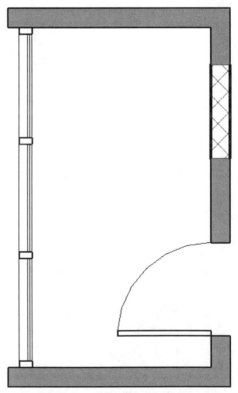

- Phase: **New Construction**
- Phase Filter: **Show Previous + New**

The door and curtain wall shown in the Figure 11.6-7 have their Phase Created set to New Construction. The existing walls are shaded due to the Phase Filter having an override applied to existing elements (more on this later).

Demolished openings present a unique situation in Revit. When an opening, e.g., door or window, is demolished, Revit automatically infills the opening with a wall. By default, this wall is the same type as the host wall (this can be selected and changed to another type). This special wall

Figure 11.6-7 New Construction view

does not have any phase settings and it cannot be deleted. If an opening in the wall is required, then the wall needs to be hidden or an opening family added. If the demolished door is deleted this special wall will also be deleted.

The Phase Filter used here has the Demolished items set to Not Displayed. Also, the Existing items are set to Overridden, which is why the existing walls are filled with a gray fill pattern. The overrides will be covered next.

> Smaller projects might show the demo in the new construction plans, in which case the Show All Phase Filter might work better.

Phase Related Graphic Overrides

When the Phase Filter has a **Phase Status** set to **Overridden**, we need to look at the Graphic Overrides tab in the Phasing dialog to see what that means (Figure 11.6-8).

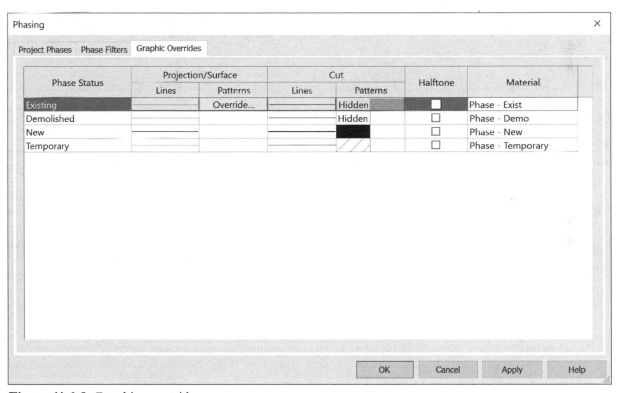

Figure 11.6-8 Graphics overrides

In the context of a given view, if elements are considered Existing, and the selected Phase Filter has existing set to Overridden—then the settings for the Existing row shown above are applied.

All the walls being cut by the View Range in plan will have **black** lines with a lineweight of **3** (Figure 11-5.9). Simply click in the box below Cut/Lines to see these settings.

Existing walls, in this example, will also have a solid fill pattern set to gray (Figure 11.6-10). This convention is helpful to clearly delineate existing walls from new walls in printed bidding and construction documents.

> Phase state overrides will override nearly all other graphic settings in the view, View Filter being an exception.

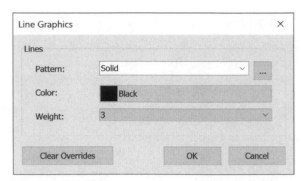

Figure 11.6-9 Line graphic overrides

In Figure 11.6-8, Graphics Overrides tab, the **Hidden** setting for the Demolished row (Phase Status) means that the Visibility option has been unchecked (compare Figure 11.6-10). Any blank boxes have no overrides and Revit will still use the By Category equivalent.

Also in Figure 11.6-8, Graphics Overrides tab, notice there are several overrides applied to the New phase status. However, notice in Figure 11.6-4, Phase Filters tab, nothing in the New column is set to Overridden. Thus, the overrides for New do not apply to anything in the entire project currently.

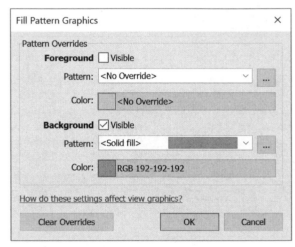

Figure 11.6-10 Fill pattern overrides

Phasing and Rooms

The Room element has only one phase parameter: **Phase**. This parameter is set based on the Phase setting of the view it is placed in. However, this parameter is read-only and cannot be changed.

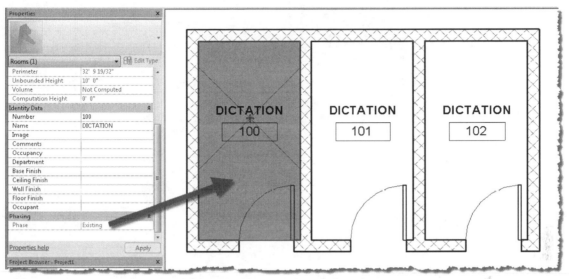

Figure 11.6-11 Phase setting for Room elements in existing view

In the image above, 11.6-11, the view phase is set to Existing. If the view's phase is changed to anything other than existing, the Rooms and Room Tags will be hidden. Rooms are not really model elements that are built in the real world, so they are handled a little differently. As we will see in a moment, when walls are set to be demolished, Revit would not be able to maintain to existing Room elements in the new, larger area in the building. Thus, Rooms only exist per phase. The unfortunate side effect to this is that existing Rooms which have no changes need to have another Room element added for each Phase in the project. The room name and number are manually entered each time and are not connected in any way between phases.

> It is possible to Copy/Paste existing Rooms with no changes into a new view. This will save time retyping room names and numbers. However, there is still no connection between the two phases.

In the image below, Figure 11.6-12, we see the same model with a wall and door demolished. The Rooms and Room Tags shown are completely separate elements from those shown in the existing view.

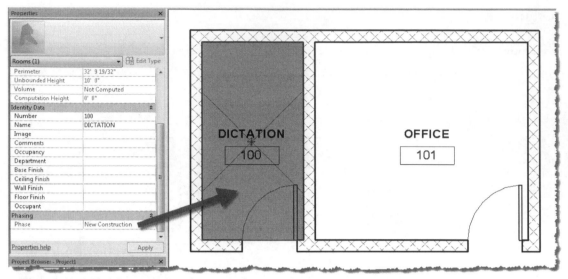

Figure 11.6-12 Phase setting for room in new view

Phasing and Annotation

Most annotation elements are not affected by phasing; they are view specific and can tag an element whether it is New Construction or Existing. However, beware of a few exceptions.

View tags such as Sections, Elevations and Callouts all have Phase and Phase Filter settings. Actually, when these elements are selected, the properties presented in the Properties Palette are the same as the view properties seen when that item's view is active. The tag and view are connected—that is why the view is deleted when the section tag is deleted.

If a View's Phase setting is changed from New Construction to Existing, all the view tags will be hidden in that view (not deleted).

One more exception is that tags will disappear if the element they are tagging disappears due to phase-related view setting adjustments.

11.7 Selection panel

Selection tools make it possible to quickly select elements in your model. A selection filter is based on selected elements that are saved as a logical group for use later in the design process. For example, you might create an element selection filter for a stair, the railing elements, and the surrounding walls so you can easily select these elements and change the location in the model to try different design solutions.

Selection: Save

Select the elements in the model that make up a logical group; you will need to quickly select again and click Save Selection. The tool is grayed out until elements are selected, then the Save selection tool is on the contextual tab or if you switch back to the Manage tab the tool will no longer be gray.

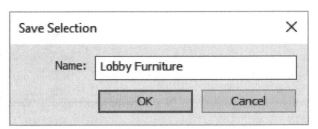

Figure 11.7-1 Save selection dialog

In the **Save Selection** dialog, Figure 11.7-1, give the selection a name that makes sense to you in the context of the project so you can load the selection at a later point in time.

Selection: Load

To recall a saved element selection filter, click Load. In the Retrieve Filters dialog select the saved selection set from the list and click **OK**. The selection filters are listed in the dialog in alphabetical order.

> Using Load will add to the current selection set if elements are already selected.

Selection: Edit

Click edit to edit the filters in the model. In the Edit Filters dialog both Rule-based Filters and Selection filters are listed and can be edited. The Rules-Based Filters are used for Visibility and graphic controls. For more information on Rules-Based Filters see the View Tab section.

Figure 11.7-2 Retrieve Filters dialog

To edit a Selection Filter, choose the filter from the list and click Edit. A selection mode will be opened in the canvas and the elements currently in the selection filter will be highlighted; other elements in the view will become grayed out. You can use the tools on the ribbon in the Edit mode to add or remove

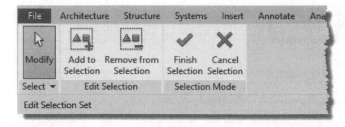

Figure 11.7-3 Edit selection mode

elements from the selection filter (Figure 11.7-3). Click the tool from the contextual ribbon and then click elements in the canvas to respectively add or remove them from the selection filter.

When editing the selection you are able to choose one element at a time. To select multiple elements using a window selection method, click on the options bar to enable multiple (11.7-4).

Figure 11.7-4 Select multiple on options bar

Once done making modifications click Finish Selection on the ribbon.

11.8 Inquiry panel

The Inquiry panel has tools for working directly with the GUID (global unique identifier) values of elements in the model. You can use these GUID values to locate elements in the model for trouble shooting warnings and diagnosing errors you may receive while working in your model.

IDs of Selection

Each element in a Revit model is assigned a GUID (global unique identifier) as it is created. The GUID value is a six-digit number that is unique to that element. No two elements will have the same GUID value.

Select an element(s) and then click back to the Manage tab so you can click the **IDs of Selection** tool. The GUID value of the selection will be displayed in the dialog (Figure 11.8-1). If more than one element is selected, the GUID values will be separated by a comma. Copy the GUID values to the clipboard from the dialog with a right click for use in troubleshooting.

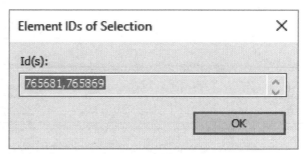

Figure 11.8-1 Element IDs of Selection dialog

Select by ID

In some cases, you may have a GUID value of an element you would like to investigate. The GUID may have come from a warning or error you were given, from a journal file, or from another source. Use the Select by ID tool to find the element in the model.

In the **Select by ID** dialog, provide the six-digit GUID (Figure 11.8-2). Multiple IDs can be selected at one time by separating the GUIDs with a semi-colon. Click show in the dialog to highlight the element in the model without closing the dialog. Click OK to close the dialog and keep the element selected.

Figure 11.8-2 Select Elements by ID

Review Warnings

As you work in your model different warning conditions may arise in the model. They do not prevent you from working but are conditions in the model that may require some attention. Over time warnings may build up in a model file and affect performance. It is a good practice to periodically use the Review Warnings tool to check the warnings in the model file and make sure the conditions causing the warning are intentional, or make corrections to the model to clear the warnings.

The Warnings dialog will list each warning in the model with a brief description why the element(s) are causing the warning (Figure 11.8-3). Click the "plus" symbol in front of the warning to expand the selection and reveal the elements causing the warning. The category, family, and type for each element is listed. The element listing also includes the six digit GUID value to identify the specific element in the model reporting the warning. You can select the element from the list and click the "Show" control at the bottom of the dialog to highlight the element in the model. If the element is not visible in the currently active view, Revit will

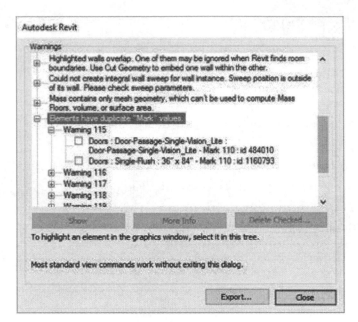

Figure 11.8-3 Revit warnings dialog

search for a view where the element can be seen and then open that view and zoom the view to the area around the element. This may help you understand the context of the warning so you can take an action to correct it. To simply delete the element causing the warning, check it in the dialog and click Delete Checked to remove the element from the model. Be careful when taking this action without understanding the context of the warning.

The model may have a large number of warnings that are more easily reviewed in another tool or in another session of Revit. You can export the warnings to an HTML file for review at a later time.

rac_basic_sample_project Error Report (1/1/2017 6:57:28 PM)

Error message	Elements
Highlighted walls overlap. One of them may be ignored when Revit finds room boundaries. Use Cut Geometry to embed one wall within the other.	Walls : Basic Wall : Wall - Timber Clad : id 427092 Walls : Basic Wall : CL_W1 : id 627729
Highlighted walls overlap. One of them may be ignored when Revit finds room boundaries. Use Cut Geometry to embed one wall within the other.	Walls : Basic Wall : SIP 202mm Wall - conc clad : id 428745 Walls : Basic Wall : Cavity wall_sliders : id 977133
Stair top end exceeds or cannot reach the top elevation of the stair. Add/remove risers at the top end by control or change the stair run's "Relative Top Height" parameter in the properties palette.	Stairs : Precast Stair : Stair : id 513254

11.9 Macros panel

Using the Revit API macro tools can be created to automate tasks in Revit. Macros can be accessed at the application level or embedded into projects at the project level. The macros panel provides access to the tools to create and run macros.

Macro Manager

The Macro manager (Figure 11.9-1) is how you launch any macros you have access to and how you can launch the editing environment for creating a macro. You must have knowledge of the Revit API and programming languages such as C# or VB.net to create macros.

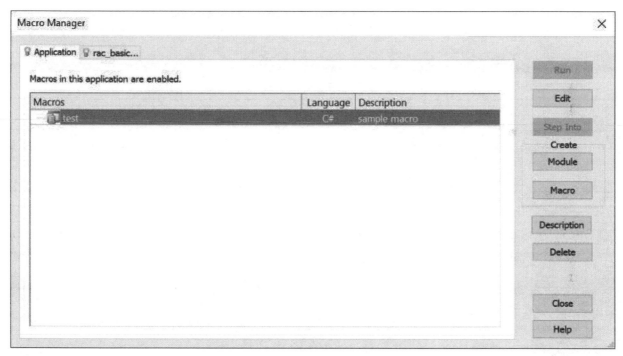

Figure 11.9-1 Macro Manager dialog

Macro Security

See the Options dialog for more information on this topic.

11.10 Visual Programming panel

Dynamo provides a way to utilize visual programming to automate tasks in Revit. Dynamo scripts are created and run in the dynamo environment and the results are then used in the Revit environment.

Dynamo

When you click the Dynamo button the Dynamo application installed with Revit is launched (Figure 11-9.1). You will then be able to create a dynamo script to run in your Revit application. Visual programming with Dynamo is beyond the scope of this book. For additional information on Dynamo visual programming see www.dynamobim.org.

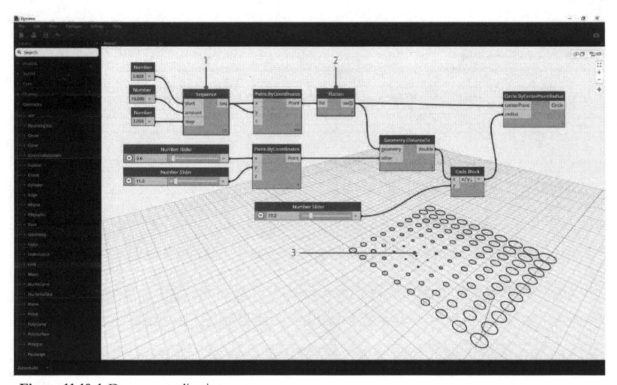

Figure 11.10-1 Dynamo application

Dynamo Player

The Dynamo Player is a tool that allows you to connect to scripts created with Dynamo and run them without opening the Dynamo application.

The Dynamo scripts are placed into a folder on your network and then the player is directed to the folder. Each script populates the player with an entry and a play button. Click the play button to execute the Dynamo script in your model.

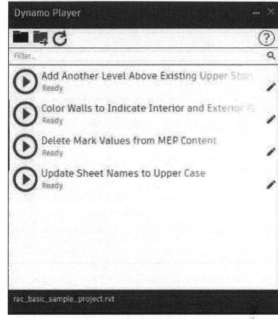

Figure 11.10-2 Dynamo Player dialog

Manage

Notes:

Chapter 12
Modify Tab

The Modify tab contains tools to make modifications to elements already placed in your model.

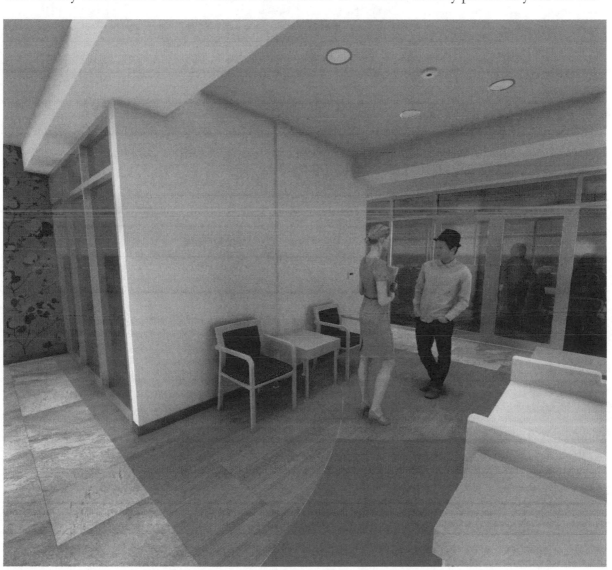

12.1 Properties Panel

The Properties panel, on the Modify tab, has two controls: the Type Properties button and the Properties button.

Type Properties

When an element is selected in the canvas, click the **Type Properties** button, on the Modify tab, to open the type properties dialog for the selected element (Figure 12.1-1). If no element is selected, the button on the ribbon is grayed out. In the type properties dialog make changes to type properties of the selected element. Changes made to the type properties will be applied to all elements of that type in your model.

The Type Properties dialog can also be accessed from the **Edit Type** button located on the Properties Palette (Figure 12.1-1).

> This dialog must be closed before work can resume on the model.

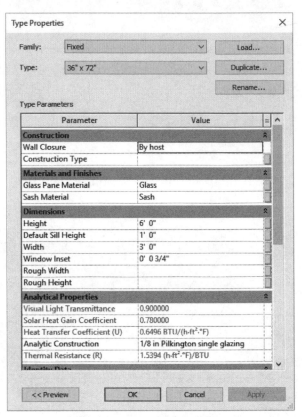

Figure 12.1-1 Type properties dialog

Properties

Click the **Properties** button, on the Modify tab, to turn the Properties Palette on and off. The Properties Palette displays instance properties of the selected element (Figure 12.1-2). A change to an instance property of an element only affects the single instance in your model. When nothing is selected, the current view's properties are displayed.

> View properties are not editable when a View Template is applied.

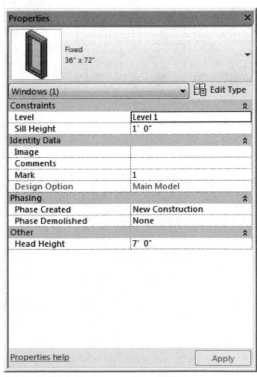

Figure 12.1-2 Properties dialog

12.2 Clipboard panel

FYI: Cannot paste between two separate sessions of Revit but can paste between two projects opened in the same session.

Paste → Paste from Clipboard

Pastes contents of Windows clipboard. Non-Revit elements (i.e., images and text) are not supported. Text copied to the clipboard can be pasted but the text command in Revit must be started first. When pasting Revit geometry, the cursor will be positioned at the lower left corner of the geometry.

Paste → Aligned to Selected Levels

Contents of the clipboard will be placed in the exact same position (x,y coordinates) but to the selected level datums in the model (Figure 12.2-1). Multiple levels can be selected at one time and the geometry is pasted to all selected levels. Use this option to quickly create geometry to many levels of the project at one time. For example, this option can be useful when creating restroom layouts that are the same from floor to floor in a building.

Figure 12.2-1 Paste: Selected Levels

Paste → Align to Selected Views

Pastes contents of the clipboard to the views you select (Figure 12.2-2). This option is only enabled when annotation elements are copied to the clipboard. Model elements can't be pasted with this option. When pasting the annotation will appear in the same x, y position it was in when copied to the clipboard. Select multiple views in the dialog to paste the elements into more than one view at a time.

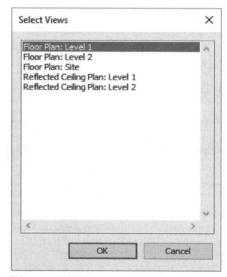

Figure 12.2-2 Paste: Selected Views

Modify

Paste → Aligned to Current View

Paste elements from the clipboard into the current view using the same x,y coordinates from the view where the elements were originally cut or copied from. This option only works when pasting into a plan view.

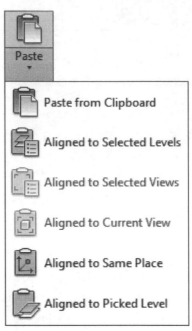

Paste → Aligned to Same Place

Paste elements to the exact same position as they were copied or cut from. Use this option to copy elements to other worksets, design options, or another model. This option can also be used to troubleshoot elements. An element cut and then pasted aligned to the same place will create a duplicate of the original item but with a new element ID.

Paste → Aligned to Picked Level

Paste elements to the same x, y coordinates to the selected level. You must be in a section or elevation view to use this option so the level can be selected in the view.

Cut to Clipboard

Removes the selection from the model and places it on the clipboard. Use one of the paste options to place the selection back into the model.

Copy to Clipboard

Places selection onto the clipboard without removing it from the model. Use one of the paste options to place a copy of the selection back into the model. Use this option when you need to copy an element but need to change views or open a different model before placing the copied geometry.

Match Type Properties

Transfers type properties from the first selected element to the second selected element. Type properties cannot be transferred across sessions of Revit or between elements in different categories. Elements must be in the same view in order to use the Match Type Properties tool.

12.3 Geometry panel

Paint

Paint will apply the selected material to the selected face of the element. Any material loaded into the model can be used as a paint (Figure 12.3-1). Paint only affects the surface appearance of the face. An element face can be split so different portions of the face can be individually painted. Material applied as paint can be scheduled using a material takeoff schedule.

> With the Paint tool active, hover the cursor over a surface and a tool tip will indicate if it has been painted.

Remove Paint

Restores a painted surface back to the originally defined material. Select the remove paint tool and then click the painted face to remove the painted material.

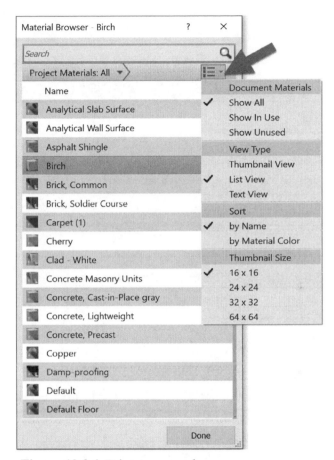

Figure 12.3-1 Paint command

Cut (Geometry)

Cut geometry can be used to remove portions of geometry when they intersect each other. Select the element to be cut first and then select the intersecting geometry to perform the cut action. The geometry will be removed from the cut element.

Uncut Geometry

Use uncut geometry to reverse the action done with the cut geometry tool. When a void creates an automatic cut condition you can use uncut geometry to define the specific elements cut by the void geometry.

Modify

Join (Geometry)

Creates a clean joint between two elements such as a floor and a wall when they overlap. The join tool will remove the edges between the two elements.

Unjoin Geometry

Removes the join condition between elements. The original boundary lines of the element are restored.

Switch Join Order

Changes which element in the join condition is primary. This will change the way the internal lines of the elements are shown.

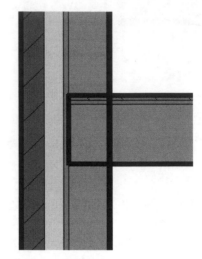

Un-Joined Wall and Floor

Apply/Remove Coping

Applies or removes a coping condition to a steel beam end. A coped steel beam end has the flanges trimmed so it can fit into an adjacent beam.

Join/Unjoin Roof

Connects and disconnects a roof element to an adjacent roof or wall condition. For example, the back edge of a dormer roof can be joined to the roof face of the main roof. This tool lets you use multiple smaller roof elements to make up a larger over all roof plan.

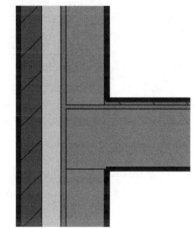

Joined Wall and Floor

Beam/Column Joins

Use this tool to control how beam ends are cut back to adjacent beams or columns. By default longer beams are cut back to shorter beams. The Beam/Column Joins tool allows you to make adjustments to the cutback conditions and override the default behavior of cutbacks.

Wall Joins

The wall joins tool allows you to make changes to the default cleanup of wall join conditions. First select the intersection of walls you want to make an adjustment to and then use the options bar to adjust the join condition. First set the configuration to use Butt (one wall is butt into the adjacent wall), Miter (the angle between the adjacent walls is calculated and mitered), or Square Off (walls at angles other than 90 degrees are squared as they connect). When there are multiple walls in the intersection there may be multiple possible configurations. Use the next and previous buttons to cycle through the join options. Once the configuration is set adjust the display as required. The display can be set to clean the join or not clean the join as an override or the view setting can be used to determine if the join is cleaned. If the walls should not join at all the disallow join option is selected.

Split Face

Use **Split Face** to create separate areas on a face (Fig. 12.3-2). Once a face is split use the paint tool to apply a different material appearance to the area. The split face tool does not affect the underlying element; it only creates a region that can accept a painted material.

> In the project environment, split face only works on systems families such as Walls, Floors and Ceilings. It does not work on loaded families. Within the family editor environment, split face works on rectilinear solid geometry.

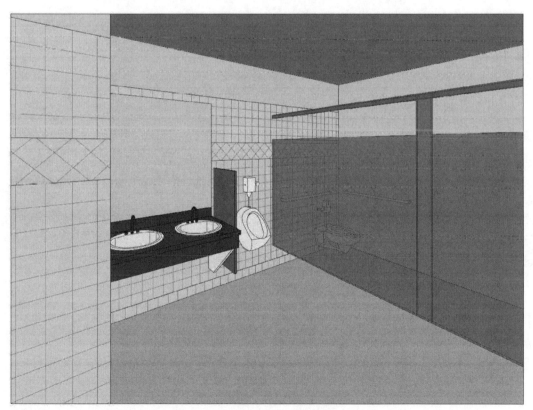

Figure 12.3-2 Split Face used to define separate tile patterns on walls

Demolish

Applies the phase of the current view to the "Phase Demolished" instance property of the selected element. Phasing settings of the model will alter the appearance of the demolished element.

12.4 Modify panel

The Modify panel, on the Modify tab – not to be confused with the Modify button – contains commands used to edit previously created geometry.

Align

Allows you to make the selected elements co-planar. Once elements are aligned the lock icon can be clicked to create a constraint between the aligned elements. If constrained when one of the elements is moved, the aligned element will move to remain co-planar with the first element. It is common to use the align tool in the Family editor to create alignment constraints between reference planes and family geometry in order to make parametric families.

Align: **Options Bar**

Select the Multiple Alignment box when you need to align multiple elements to the same position. You can select the alignment plane once and then select all of the elements to align without needing to select the alignment plane for each element.

quick steps

Align

1. Options Bar
   - Multiple Alignment
   - Prefer
        i.   Wall Centerlines
        ii. Wall Faces
        iii. Center of Core
        iv. Faces of Core
2. In-Canvas
   - First pick: Pick element to align to
        i. *OR* the element that should NOT move
   - Second pick: Pick the element to move into alignment
   - Use Tab key to cycle through "Prefer" options listed above without changing Prefer setting

The Prefer setting relates specifically to wall alignments. You choose to have specific planes within the wall to be the preferred alignment face.

- Wall centerlines – Geometric center of the wall assembly.
- Wall Faces – The interior and exterior faces of the wall will be selected first.
- Center of Core – The center line of the wall core (as defined in the structure of the wall type) will be selected first.
- Faces of Core – The faces of the layers that define the wall core (as defined in the structure of the wall type) will be selected first.

Set the Prefer setting as best suits your workflow when aligning wall elements.

Align: In-Canvas

Align element by first picking the element you want to maintain its position. The second element picked will move or extend as necessary to become co-planar with the first selection.

When making selections use the tab key to cycle through selection candidates near the cursor. For example, when near a wall use the tab key to cycle through the different Prefer options for wall alignment.

Offset

Allows you to select an existing wall or line and move or copy it a specified distance away from the current location. Used to make parallel sets of elements.

Offset: Options Bar

Select graphical or numerical for the kind of offset you will create. A graphical offset lets you pick points to define the offset distance in the canvas. When you use a numerical offset the distance of the offset is entered directly in the options bar.

Check the copy box if you want a copy of the element created. Clear the copy option if you want the selected element to move the offset distance.

Offset: In-Canvas

Select the wall or line to be offset. When using a numerical offset, as

> **quick steps**
>
> **Offset**
>
> 1. Options Bar
>   - Graphical
>   - Numerical
>   i. Offset
>   - Copy
> 2. In-Canvas
>   - For graphical:
>   i. Select element to offset
>   ii. Select a start point for the offset
>   iii. Select a 2nd point for the offset distance
>   - For Numeric
>   i. Enter Offset value on option bar
>   ii. Hover cursor over element to offset
>   iii. When dashed reference line is on correct side, click to offset element

you hover over an element a dashed line will appear indicating where the offset element will be placed. The dashed line position is sensitive to the cursor position relative to the element being offset. Move the cursor to the other side of the line or wall to offset the opposite direction. The dashed line will change position as the cursor is positioned to each side of the line or wall.

When using a graphical offset, 2 clicks are required to complete the offset. The first establishes a start position for the offset and the second click is the distance of the offset.

> ▌ Press Tab to highlight chain of walls or lines to offset them all at once.

Modify

Mirror - Pick Axis

Creates a symmetrical set of elements along a picked axis of reflection. Elements can either be copied or flipped to the other side of the selected axis. Use this mirror option when the geometry being mirrored already has an element that can be used as the axis for reflection.

Mirror - Axis: **Options Bar**

Check copy on the options bar if you want the mirror action to create a copy of the selected elements on the other side of the mirror axis. Clear the copy check box if you want to flip the selected elements to the other side of the reflection axis.

quick steps

Mirror – Pick Axis

1. Pick items to mirror first
   - The steps vary slightly if the tool is started prior to selecting elements
2. Options Bar
   - Copy (checked by default)
3. In-Canvas
   - i. Select element to use its centerline as the axis of reflection
   - ii. Using the Tab key allows for additional surfaces to be used as the axis of reflection

Mirror - Axis: **In-Canvas**

Typically, the Mirror – Pick Axis command is used after elements have been selected. In this case you are only required to pick the axis of reflection and a symmetrical set of elements is created on the other side, or the elements are flipped to the other side of the axis of reflection. When the command is selected first, the elements to be mirrored are selected and then the enter button is pressed to end the selection mode. Then select the axis of reflection.

Mirror - Draw Axis

Creates a symmetrical set of elements along a drawn axis of reflection. Elements can either be copied or flipped to the other side of the selected axis. Use this mirror option when the geometry being mirrored does not have an element that can be used as the axis for reflection. Instead you will click 2 points to draw an axis of reflection, an imaginary plane, during the command.

Mirror – Draw Axis: **Options Bar**

Check copy on the options bar if you want the mirror action to create a copy of the selected elements on the other side of the mirror axis. Clear the copy check box if you want to flip the selected elements to the other side of the reflection axis.

quick steps

Mirror – Pick Axis

1. Pick items to mirror first
   - The steps vary slightly if the tool is started prior to selecting elements
2. Options Bar
   - Copy (checked by default)
3. In-Canvas
   - Pick two points to define the axis of reflection
       i. The two points picked will define an imaginary line

Modify

Mirror – Draw Axis: **In-Canvas**

Typically, the Mirror – Draw Axis command is used after elements have been selected. In this case you are only required to draw the axis of reflection and a symmetrical set of elements is created on the other side, or the elements are flipped to the other side of the axis of reflection. When the command is selected first, the elements to be mirrored are selected and then the enter button is pressed to end the selection mode. Then draw the axis of reflection.

Move

Allows you to move the selection from one position to another position. The elements are first selected and then the move command is used. Elements can be moved with the in canvas control and dragged. The Move command has added functionality to move specific distances and constrain and disjoin controls.

Move: Options Bar

When using the move tool two check box controls are available on the Options Bar:

- Constrain
- Disjoin

Move

1. Pick items to mirror first
   - The steps vary slightly if the tool is started prior to selecting elements
2. Options Bar
   - Constrain
   - Disjoin
3. In-Canvas
   - Pick two points to move the selected elements "from" (first pick) and "to" (second pick)

A third check box for Multiple is visible but grayed out. The multiple check box cannot be used during the move command.

The Constrain option will limit the way the elements can be moved in the canvas. With the Constrain option checked the selection can only be moved in the x and y axis. Diagonal movement is not possible.

The Disjoin option has an effect when moving elements that can be joined, i.e., walls or a column and a beam. When disjoin is selected, when moving a joined element, the connection will not be maintained. If unselected the geometry of the joined elements will be altered to maintain the joined condition.

Move: In-Canvas

Pick two points to move the selected elements the distance and direction of the two click points. Alternatively, you can pick one point, move the cursor in the direction of the move, and use the keyboard to specify a distance.

> Picks do not have to snap to other geometry. If you want to move a wall 10'-0" to the East (right), simply pick your first point in empty space and then point the cursor to the right (snapped to the horizontal) and type 10 and then press Enter.

Copy

Create duplicates of the selected elements.

Copy: **Options Bar**

When using the copy tool two check box controls are available on the Options Bar:

- Constrain
- Multiple

A third check box for Disjoin is visible but grayed out. The Disjoin check box cannot be used during the Copy command.

The Constrain option will limit the way the copied elements can be moved. With the Constrain option checked the copy can only be placed in the x and y axis. Diagonal placement away from the original is not possible.

The Multiple option allows you to create multiple copies during the command. Each click will create a copy of the selection. Un-checking the second click will create a copy and the command will end.

> You can create a copy of a selection by holding the CTRL key and dragging the element in canvas. A copy will be placed at the dragged position rather than the selection being moved.

Copy: **In-Canvas**

Pick two points to move the selected elements the distance and direction of the two click points. Alternatively, you can pick one point, move the cursor in the direction of the move, and use the keyboard to specify a distance.

> You can create a copy of a selection by holding the CTRL key and dragging the element in canvas. A copy will be placed at the dragged position rather than the selection being moved.

quick steps

Copy

1. Pick items to mirror first
   - The steps vary slightly if the tool is started prior to selecting elements
2. Options Bar
   - Constrain
   - Multiple
3. In-Canvas
   - Pick two points to copy the selected elements "from" (first pick) and "to" (second pick)

Modify

Rotate

Change the angle of the selection.

Rotate: **Options Bar**

When using the Rotate tool two check box controls are available on the Options Bar:

- Disjoin
- Copy

The Disjoin option has an effect when moving elements that can be joined, i.e., walls or a column and a beam. When disjoin is selected when moving joined elements, the connection will not be maintained. If unselected the geometry of the joined elements will be altered to maintain the joined condition.

When Copy is selected the original will remain in position and a copy will be placed at the new angle.

On the options bar the angle control allows you to specify an angle and press enter to rotate the selection without using the in canvas controls.

By default the center of rotation is placed at the center of the bounding box of the selection. Use the Place control to reposition the center of rotation to a point you click in the canvas.

Rotate: **In-Canvas**

During the Rotate command you can click and drag the center of rotation grip to reposition the center of rotation from its default location. Once the center of rotation is positioned click once to establish the beginning point of rotation. The second click indicates the angle of rotation for the selection.

> The center of rotation can be placed on the cursor for placement by pressing the spacebar once the rotate command is started.

quick steps

Rotate

1. Pick items to mirror first
   - The steps vary slightly if the tool is started prior to selecting elements
2. Options Bar
   - Disjoin
   - Copy
   - Angle
   - Center of rotation
     i. Place
     ii. Default
3. In-Canvas
   - Change center of rotation if desired
   - First pick: A point away from the center of rotation
   - Second pick: Define the angle for rotation, relative to your first pick

Trim/Extend to Corner

Close two lines or walls into a corner condition. The command will either trim or extend the elements to create the corner condition as necessary.

Trim/Extend to Corner: In-Canvas

The element clicked establishes the plane for the corner condition; the second element clicked is projected to the plane of the first selection. When hovering over the second element a blue dashed line is projected showing the corner condition that will be created. Once clicked, both elements are either trimmed or extended to meet at the

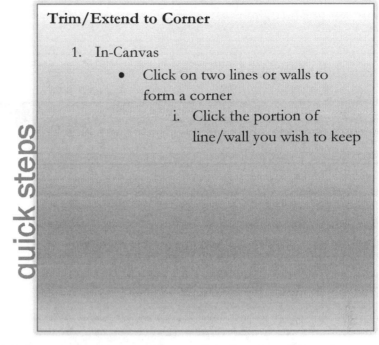

corner condition. When using this tool it is important to remember to click the element on the side of the corner condition you wish to keep. The clicked points do not need to be close to the corner.

Trim/Extend Single Element

Use the Trim/Extend Single Element command to either Extend or Trim:

- **Extend** an element to another as shown below
  Pick the element to extend to
  Pick the element to be extended

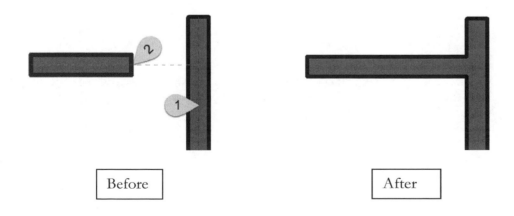

Before

After

- **Trim** an element to another as shown below

  Pick the element to trim to

  Pick the element to be trimmed, selecting the side to remain

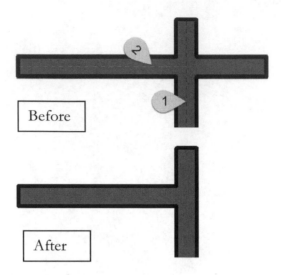

Trim/Extend Multiple Elements

Use the Trim/Extend Multiple Elements command to either Extend or Trim:

- **Extend** multiple elements to another element as shown below

  Pick the element to trimmer/extend to

  Pick the elements to be trimmed/extended

    - Drag a window
    - Pick individual elements
    - Or combination of both

With this command, trim versus extend is automatically implied. If an element crosses the first element selected, then it is trimmed. Otherwise it is extended, if it will intersect the first element selected.

The Trim/Extend commands work on Walls, lines, Ducts, Cable Tray, Pipe, Conduit and Beams.

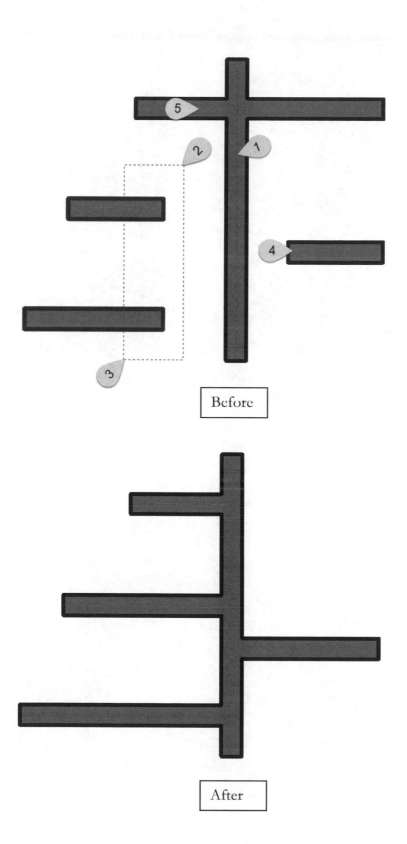

Before

After

Modify

Split Element

Divide a wall or line into multiple parts. Each click will create a break point on the element.

Split Element: **Options Bar**

Select the check box Delete Inner Segment to remove the portion of a split element between two splits. For example, this can be used to create two walls with an opening in between. One wall is split twice with delete inner segment selected. When this option is cleared the element is only split leaving an edge between 2 adjacent pieces.

Split Element: **In-Canvas**

Click the split tool on the ribbon to

> **Split Element**
>
> 1. Options Bar
>   - Delete inner segment
> 2. In-Canvas
>   - A single click splits the element into two
>   - The tool remains active, and the previous step may be repeated
>   - When "Delete inner segment" is checked, a second pick will cause a portion of the element to be deleted
>   - Press the Esc key to finish

split elements. Each click will create a split in the selected element at the point of the cursor. Use temporary dimensions to position the split accurately. The split tool remains active until Modify or Esc key is pressed.

> When splitting walls a joint between walls is created but because the wall end points are still coplanar, grip editing a wall may cause the two walls to become joined after the split. To maintain a gap between split walls consider using Split with Gap.

Split with Gap

Divide a wall into multiple parts. Each click will create a break point and a gap of a specified distance between wall ends. The distance between wall ends will be constrained to maintain the gap and dis-allow join will be applied to the split end of the walls on each side of the split. This tool can be used to create a wall made of precast panels, splitting the wall at the panel joints.

Split with Gap: Options Bar

Set the gap size created at the time of the split.

Split with Gap: In-Canvas

Click the split with gap tool on the

quick steps

Modify

> **Split with Gap**
>
> 1. Options Bar
>   - Join Gap
>   i. Enter joint size
> 2. In-Canvas
>   - A single click splits the wall, but keeps the gap between the two segments constrained

ribbon to split elements. Each click will create a split in the selected element at the point of the cursor and create a constrained gap with a size set on the options bar. Use temporary dimensions to position the split accurately. The split tool remains active until Modify or Esc key is pressed.

Unpin

The Unpin tool is used to unpin elements which have previously been pinned. Pinned elements cannot be moved or deleted, thus they need to be unpinned, permanently or temporarily, prior to editing. Some elements such as curtain wall panels, mullions, and grids or a handrail element might be pinned because they are defined by the element type. Unpin these elements to make unique changes to these elements.

When selected, pinned elements will show a "pin" icon next to each element (unless there are too many items selected, then the pin icons are omitted for clarity). Clicking the Pin icon will also unpin the element, but there is a selection mode preventing the selection of pinned elements. If you are not able to select a pinned element to unpin it, check to make sure your selection mode allows pinned elements to be selected.

Pinned elements may not be selectable depending on the **Select Pinned Elements** toggle on the status bar – see tooltip in the image to the right.

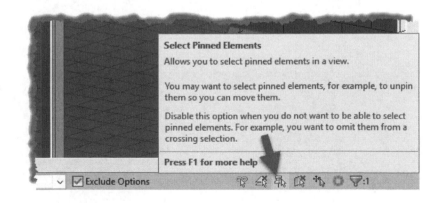

Array

Create multiple copies of a selection in one step. An array creates the copies for the selection in a linear direction or a radial pattern. Arrays are not able to be created in two directions (x and y) in one step. To create an array like this (i.e., auditorium seating) two separate arrays need to be created.

Array: **Options Bar**

After making a selection and clicking the array tool, click Activate Dimensions to show temporary dimensions for the selection. This may help you position the array as required.

Select the array type **Linear** or **Radial**. In a linear array you specify the direction and distance of the array. In a radial array you specify center of rotation and the angle of the array.

quick steps

Modify

Array

1. Pick items to array first
   - The steps vary slightly if the tool is started prior to selecting elements
2. Options Bar
   - Activate Dimensions
   - Linear Radial
   - Number
   - Move to:
     i. 2nd
     ii. Last
   - Constrain
3. In-Canvas
   - Steps vary depending on options bar selections – see section notes

The **Group and Associate** check box will create a group from the selection before the array is created. This will let you later make a change to one element of the array by doing a group edit. This change will be propagated out to the other elements of the array.

The **Number** is the number of elements created in the array. This can be changed in the canvas after the array is created if necessary.

Set the **Move To** option depending on your requirements. Use 2nd when you want to define the distance or angle (spacing) between each element of the array. Set this option to Last when you know the total distance or angle to be filled by the array and want the number of elements spaced equally.

When creating a linear array select the **Constrain** check box to limit the array direction to the x or y direction; no diagonal movement is possible.

When creating a radial array, you can set the **Angle** specifically from the options bar and press Enter to array the elements.

When creating a radial array you can click **Place** to move the center of rotation away from the default location. You can also press **Space Bar** to move the center of rotation or click and drag the center of rotation in the canvas.

Array: In-Canvas

Linear Array

Click to specify the distance and direction of the array. First click establishes a start point and the second click defines both distance and direction. You can also click once and move the cursor in a direction and use the keyboard to enter a distance directly. To make a change to the number of elements in the array use the in canvas control to specify a number. This control can be changed after the array is created by selecting a member of the array, hovering over the adjacent area and making a change to the number displayed.

Radial Array

Change the center of rotation for the array if necessary and then click once to set the start point of the array. The second click establishes the angle for the array. To make a change to the number of elements in the array use the in canvas control to specify a number. This control can be changed after the array is created by selecting a member of the array, hovering over the adjacent area and making a change to the number displayed.

Scale

The first several versions of Revit did not have the Scale tool. The thought was that scaling building elements did not make sense—which is true. A 3D wall or door would never be scaled as it would make the wall layers (e.g. Brick, Insulation, etc.) and the door thickness larger. Revit still does not allow model elements to be scaled. However, line work in details and other views may be scaled. Wall elements can also be scaled for length.

Scale: **Options Bar**

When you select the **Graphical** scaling option, in canvas controls are used to establish the scale factor. This option is good for scaling something to a visual reference in the canvas.

Numerical allows you to set a scale factor (multiplier) for the selected element. For example, if a wall 10' long was scaled numerically by .5 it would be 5' long. If scaled numerically by 2 it would be 20' long.

Scale

1. Pick items to scale first
   - The steps vary slightly if the tool is started prior to selecting elements
2. Options Bar
   - Graphical
   - Numerical
     - i. Scale
3. In-Canvas
   - For Graphical:
     - i. First Pick: Origin; scale from here
     - ii. Second Pick: The distance between this point and the next will determine the scale
     - iii. Third Pick: Sets the scale
   - Numerical
     - i. Type value; 2 makes the lines twice its original size, and 0.5 makes them half sized
     - ii. Click to define the origin

Modify

Scale: **In-Canvas**

Set the start point for the scale action. The first point clicked is the base point for the scale; element will scale from this point (Figure 12.4-1). When scaling numerically this is the only point clicked in canvas and the scale factor is applied. When scaling graphically after setting the start point, the second click establishes a reference distance. The third click defines the new distance. The selection will be scaled by setting the reference distance to the new distance.

Scale on options bar will list previous scale used (graphical or numeric).

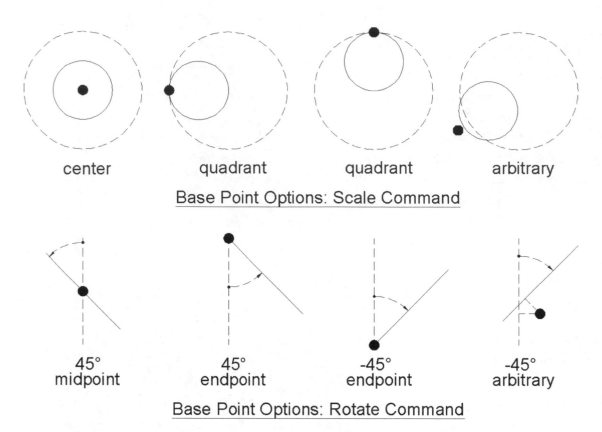

Base Point Options: Scale Command

Base Point Options: Rotate Command

Figure 12.4-1 Various base points with results for scale and rotate commands

Pin

Use the **Pin** tool to help ensure elements, including linked models, are not accidentally moved or deleted. A selection mode can be set to prevent pinned elements from being selected as an extra measure of protection against accidental changes to a pinned element.

Delete

Remove selection from the model.

12.5 View panel

Hide in View

Make a selection not visible in the current view. Elements can be hidden in three ways: by **Element**, by **Category**, and by **Filter**. When an element is hidden in the view and annotation, elements (tags and dimensions) associated with the hidden element are also hidden. The **Hide in View** tools can also be accessed from the right click menu as shown to the right. When the elements are un-hidden the annotation elements are made visible as well.

Use the **Reveal Hidden** elements mode from the view control bar to see elements hidden in the view. Hidden elements are revealed in a magenta tone. The elements can then be selected and un-hidden from the right click menu or from the Reveal Hidden tool on the contextual tab.

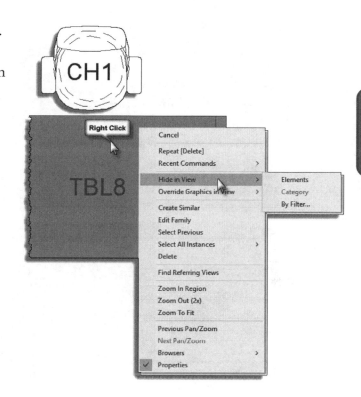

Hide Element

The single item or items in the selection set are hidden in the view.

Hide Category

The category of the selected element(s) is hidden. For example, if a wall element is selected and hidden by category all of the walls in the view are hidden. When multiple categories of elements are in the selection set all categories selected will be hidden.

Hiding elements this way is the same as using the visibility and graphics overrides dialog and unchecking the visibility of a category.

Hide by Filter

The Visibility and Graphics dialog is opened to the filters tab. If your model contains filters you can apply them from this location to hide elements. If there are no filters yet created for the model you can create one from this dialog location.

Filters are a powerful way to hide or override the visibility and graphics of an element based on parameter values of the element. For example, you can use a filter to find all of the walls with a 1hr. fire rating assigned to them and color them with a red fill pattern.

Displace Elements

In a 3d view, move elements from their original position in the model. The displacement of elements in a view is a view specific override. In other views the element will remain in the original position. Displace elements is used to create presentation views where pulling apart or exploding elements away from each other will help you present the model. Once elements have been displaced, you can use the tab key to select sub elements in a displacement set and use the displace elements tool again to make a nested displacement set.

Displace Elements: **Ribbon**

When using the displace elements tool the contextual tab of the ribbon has 3 controls.

quick steps

Displace Elements

1. Pick items to displace first
2. Ribbon
   - Edit
   - Reset
   - Path
     i. Click edge of displaced element to add dashed path reference line back to original location
3. Options Bar
   - Activate Dimensions
4. Properties
   - Type in specific X/Y/Z displacement values
5. In-Canvas
   - Use Move or the 3D widget near the selected elements to reposition in the current view

- **Edit** – The edit control allows you to add or remove elements from the displacement set. All elements in a displacement set will be moved together. Click edit and then add or remove elements from the set and click finish when complete.

- **Reset** – Reset will move any elements in the displacement set back to the original position in the model. When you reset a displacement set, any nested displacements will be reset as well.

- **Path** – Allows you to click on logical points of the displaced elements and a path is drawn back to the original position. The path can be drawn straight or in a jogged pattern if the element is displaced in 2 directions. Set straight or jogged path by selecting the displacement path and changing the style on the properties palette. The line style used for the displacement path is set in the Object Styles of the model and can be overridden with visibility and graphics controls.

Displace Elements: **Options Bar**

When a displaced element is selected you can click activate dimensions on the options bar to see dimensions and constraints related to the object.

Displace Elements: **Properties**

When a displaced element is selected, set specific offset distances for X, Y, Z displacements on the properties palette.

Displace Elements: **In-Canvas**

When a displaced element is selected a 3d movement widget is displaced near the element. Use the widget to visually drag the displacement set into a new position.

Override by Element

Override the graphic appearance of the single item or items in the selection set. Graphic overrides are applied to the current view only. This command can also be accessed from the right click menu after selecting elements.

Override by Category

Override the graphics of the selected item's category den. For example, if a wall element is selected and overridden all of the walls in the view are overridden. When multiple categories of elements are in the selection set all categories selected will be overridden with the same values.

This command can also be accessed from the right click menu after selecting elements.

Override by Filter

The Visibility and Graphics dialog is opened to the filters tab. If your model contains filters you can apply them from this location to hide elements. If there are no filters yet created for the model you can create one from this dialog location.

Filters are a powerful way to hide or override the visibility and graphics of an element based on parameter values of the element. For example, you can use a filter to find all of the walls with a 1hr. fire rating assigned to them and color them with a red fill pattern.

Linework (LW)

Use linework to override the edge lines of elements in a view. Each individual line or edge can be overridden separately.

Select the linework tool from the Ribbon and then on the contextual tab use the dropdown list to select the line style to apply. Each object edge clicked in canvas will receive the assigned line

Modify

style. While applying line work in 3d, elevation, and section views, you can use the grip controls to modify the extents of the applied linework.

To reset a linework change applied to an edge use the line work tool again and select **<By Category>**. This will change the edge back to the original linestyle as defined by the object styles for the element's category.

> When using the Linework tool on an underlay, when the underlay is turned off the linework lines will still appear and move with the original objects no longer visible in the current view. This is helpful when showing bulkheads/soffits above or a roof overhang.

12.6 Measure panel

The measure tools are used to derive the distance between, or along, elements.

Measure Between Two References

Measure distance between 2 selected points. The 2 selected points can be anywhere in the canvas and do not have to be on an element. The distance between the 2 points is reported by temporary dimensions. A linear dimension reports the length and an angular dimension reports the angle of a line projected between the 2 points from the horizontal. The dimensions will remain displayed until you start the next measurement or end the measure command.

quick steps

Measure Between Two Points

1. Options Bar
   - Total Length
   - Chain
2. In-Canvas
   - Pick two points on screen to find dimension
   - Picked points do not need to be on elements
   - Using the "chain" option allows multiple segments to be added up

Modify

Measure Between Two Points: **Options Bar**

Select chain on the Options bar to enable multiple picks using the measure command. With chain selected, the command will report a length for each pair of picks you make and keep the dimensions displayed for all picks. When chain is used the total length of all picked points is displayed on the options bar.

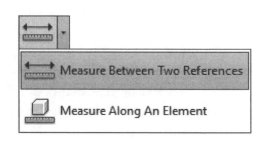

Measure Between Two Points: **In-Canvas**

Pick 2 points in the canvas area to see the dimension between the points and the angle from horizontal.

> Measure is a good tool to use to get room diagonal measurements.

Measure Along An Element

Measure the distance along a linear wall or model line or a chain of walls or model lines. The measure along element tool is limited in function because it can only measure walls or lines. All segments must be linear. When the linear element is selected, the length and the angle from horizontal (if the element is not horizontal) is reported by temporary dimensions until the next element is selected or the command is ended.

quick steps

Measure Along An Element

1. Options Bar
   - Total Length (read only)
2. In-Canvas
   - Pick linear wall or model line

Measure Along An Element: Options Bar

The options bar displays the Total length of the selection. If you use the tab key to select a chain of walls or lines, the total length of all selected elements is displayed in the options bar.

Measure Along An Element: In-Canvas

Select the element to be measured or use the tab key while hovering over an element to select the entire chain of elements. If there are non-linear elements in the selection, i.e., an arc shaped wall, it will be excluded from the measurement.

Dimension Tools

These tools are repeated here for convenience. Please see the Annotation tab chapter for detailed coverage of the dimension tools.

12.7 Create panel

Create Assembly

An assembly is created from the elements selected in the canvas. An assembly is a group of elements that are associated with one another in some way. Assemblies are typically used for portions of the model that will be constructed in a shop and then delivered to the site. For example, a wall panel with doors and windows that will be prefabricated and put in place as the building is constructed.

To create an assembly, select all of the elements that will be included and click **Create Assembly**. Some elements can't be included in an assembly:

- Annotation elements
- Other assemblies
- Complex elements (Truss, Beam Systems, Curtain Walls, Stacked Walls, Railings with continuous top rail or handrail)
- Groups
- Imports and Links
- Masses
- Rooms
- Stairs

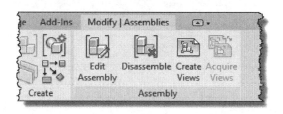

You then are prompted to give the assembly a **Name** and assign a **Naming Category**. The naming category can be any of the categories from the elements making up the assembly. The naming category is an additional way for you to organize assemblies but has no effect on the visibility and graphics of the elements in the assembly. Each assembly in the project is unique. If there are any differences between assemblies or edits made to an existing assembly, a new assembly will be created automatically.

Assemblies are created in their own branch of the **Project Browser**. If you do not have an assemblies section of the Project Browser, one will be created when the first assembly is created in the model (see image below). Place additional instances of the assembly by dragging it from the Project Browser into the canvas, or by right clicking the name of the assembly and selecting **Create Instance**.

To help create shop drawings for an assembly, you can create assembly views for an assembly. Assembly views will generate plans, sections, elevations, 3d views and schedules for the elements in the assembly. Right click the assembly name in the Project Browser and select **Create Assembly** views. Then choose the views you want to be created (see image on previous page). All views are placed under the assembly name in the Project Browser.

> An **Assembly Sheet** can be confusing as it appears in the Sheet List but not in the sheets category of the Project Browser. If the sheet should not be part of the sheet list, un-check **Appears on Sheet** in the sheet view's properties.

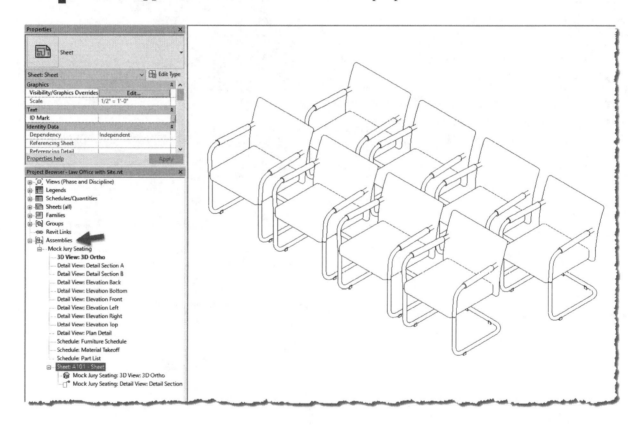

Create Parts

Parts break elements down into smaller portions that can be individually tagged, scheduled, filtered and exported. Typically, parts are used to divide larger elements into smaller pieces to help with the construction process. For example, you may have a large concrete floor that will be completed in a number of different concrete pours; parts can be used to help the model match the construction process.

Parts can be created from elements with layered structures and some other specific categories of elements:

- Standard Walls
- Floors, Foundation Slab, Building Pad
- Roofs
- Ceilings
- Slab edges
- Fascias and Gutters
- Structural Framing
- Architectural and Structural Columns

To create parts, select objects in the canvas and click create parts. Parts may be created from elements in the current model or linked models. If the element used to create the parts is changed, the parts are automatically updated to match the original element. Deleting the original will remove the parts as well. Parts can be further divided by intersecting elements if necessary.

Parts can be edited by enabling the shape handles for parts on the properties palette. Select a part and on the properties palette select Show Shape Handles. Parts that have been edited like this may not match the original element. Editing parts like this can be used to "peel back" layers of an element showing the construction of the host element.

While in **Edit Sketch** mode for the select Part, which is the slab on grade, a line is being added to divide the part into separate concrete pour areas.

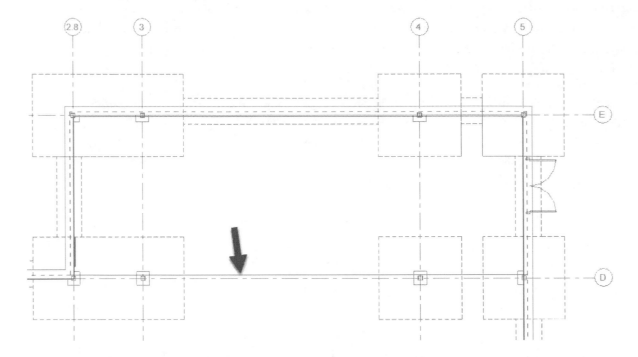

In order to see parts in a view you must set the **Parts Visibility** parameter of the view to **Show Parts** (left image below). You can also set a view to **Show Original** and **Show Both**. This setting of the view allows you to understand the relationship between the parts created from an element and the original element. Also, note that walls converted to Parts, when displayed as Parts, do not join (i.e. cleanup) like the Original wall type does (right image below).

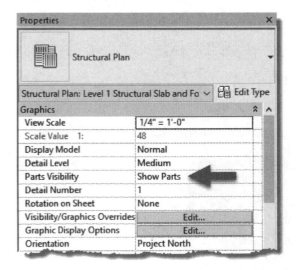

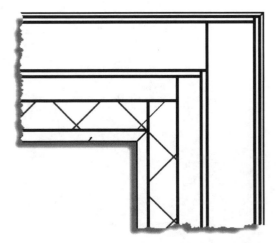

Create Group (GP)

Group a selected set of elements into a set of related elements. If the group is later edited, all instances of the group in the model are updated with the change. Use groups to create sets of elements, such as a standard office or restroom layout, that will be used multiple times in a model. Because when a group is edited all instances of the group are changed, you can make updates to many layouts at one time.

Select all elements to be included in the group and click Create Group on the ribbon. Give the group a name. An X,Y origin for the group is positioned at the geometric center of the selection. If necessary, click and drag this origin point to a position that will help you with placement of additional instances of the group. The group is created in the Groups branch of the Project Browser.

Model elements and annotation elements can't be added to the same group. You can create groups of model elements or groups of annotation elements. If your selection set includes both, an attached detail group (annotation elements) will be automatically created. When you place an instance of the model group, the attached detail group will also be placed but not visible. To make the attached detail group visible, select the group in the canvas and select Attached Detail Groups.

> See **Place Model Group** in the Architecture tab chapter for more information on Groups.

Create Similar (CS)

Create an element matching the type of the selected element. Select a single element in the canvas and then click **Create Similar**. The command to create the selected object will be started using the type of the selected element. For example, if a wall is selected the wall tool will be started with the matching wall type selected. You will still have to click points to place the wall and are able to set options during placement. If you want an exact copy of the selection, use the copy tool or copy/paste element.

> This command can also be accessed via the **right-click** menu when an element is selected.

Notes:

Chapter 13
Contextual Tab

The Contextual tab provides access to several tools used to modify previously modeled elements. This chapter aims to provide a brief overview of the various contexts one might come across while working in Revit.

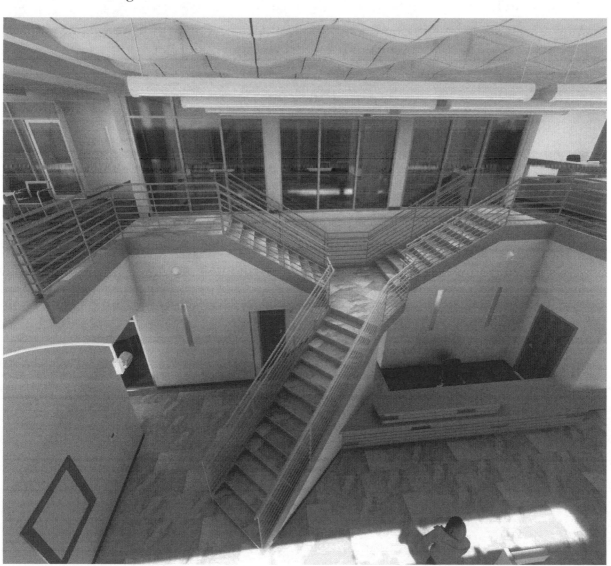

Modify | Area Boundary

- **Moves With Nearby Elements (Options Bar)** – Toggles the Moves with Nearby Elements parameter for the element. When this is set, the element will be constrained to other elements, typically walls, around it. When the element it is constrained to (wall) is moved, the element will also move. The element can be moved independently of the constraint to create a new relationship with a different nearby element. If an element is moving when walls in the model are moved, check to see if the Moves with Nearby Elements parameter is checked.

Modify | <Area Boundary>   ☐ Moves With Nearby Elements

Modify | Area Tags

- **Edit Family** – Opens the selected family in the family editor. The model where the family originated will remain open. Use Switch Windows to go back to a view in the open model.

- **Select Host** – Change the selection from the tag to the element it is tagging. This is a way you can see or modify the parameters of the tagged element.
- **Pick New Host** – Choose another element for the tag. Properties for the new element will be displayed in the tag.
- **Reconcile Hosting** – When tags are used on elements form a linked file because of changes in the linked file, a tag can become out of position or orphaned. Reconcile Hosting opens a review palette so you can locate these tags and take the appropriate action. If the reconcile hosting palette is already open this tool will be grayed out.

Modify | Ceilings

- **Edit Boundary** – Open the sketch editor mode and use the draw and modify tools to change the boundary sketch. Cancel in the sketch mode to exit without changes. Click Finish in the sketch mode to apply the new boundary.

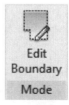

Modify | Color Fill Legend

- **Edit Scheme** – Opens the color scheme dialog. In the color scheme dialog change the colors and fill patterns for the applied color scheme. You can also change the color scheme of the view to use a different category or parameter to use as the criteria for the color scheme of the view.

Modify | "Component"

- **Edit Family** – Opens the selected family in the family editor. The model where the family originated will remain open. Use Switch Windows to go back to a view in the open model.
- **Pick New Host** – Change the host for the selected element. Depending on the element and the view you are using, valid hosts can vary. Look at the status bar to determine a valid host for the element.

Modify | Columns

- **Edit Family** – Opens the selected family in the family editor. The model where the family originated will remain open. Use Switch Windows to go back to a view in the open model.

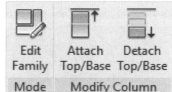

- **Attach Top/Base** – Attach the top or the bottom of the column to a valid target. The column and the target will be constrained. When the target is moved, the column attachment will be maintained. Column must be able to physically intersect with the target in order to attach. These are valid targets for column attachment:
  - Roof
  - Floor
  - Ceiling
  - Beam
  - Foundation
  - Reference Plane
  - Level

  On the **Options Bar** select the options for the attachment.

  - **Top/Base** – Choose the attachment point for the column.

  - **Attachment Style** – Use the pulldown menu to choose the cutting behavior of the column and the attachment target. The column can be cut, the target can be cut, or neither will be cut.

  - **Attachment Justification** – Use the pulldown to set how much of the column will intersect the attachment target. The Minimum intersection will stop the column attachment as soon as any part of the column geometry touches the target. Column midpoint will stop attachment at the mid-point of the column geometry. Maximum intersection will stop the attachment when all of the column geometry intersects the target.

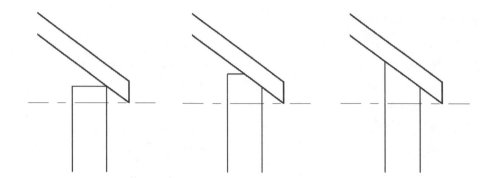

    o **Offset from Attachment** – Set a distance away from the intersection point between the column and the target where the column stops attachment.

- **Detach Top/Base** – Remove an attachment target from a column. Click **Detach Top/Base** and then select the target in the canvas. The column will no longer be constrained to the target but will keep any height adjustment that was made when it was attached. Click **Detach All**, on the Options Bar, to quickly remove all attachment references. Use the properties palette to reset any offsets.

Modify | Contour Labels

- **Select Host** – Change the selection from the tag to the element it is tagging. This is a way you can see or modify the parameters of the tagged element.

Modify | Curtain Systems

- **Edit Face Selection** – Add or remove faces of a mass element used to create the curtain system. In canvas, hover over the face of the mass to add or remove. If a curtain system is already applied to the face, a "minus" symbol will be displayed close to the cursor; if the face does not have a curtain system applied, a "plus" symbol will be displayed. Click to add or remove faces from the selection as required. Selected faces do not need to be adjacent, but do need to belong to the same mass family. While in the face selection mode two different tools are displayed on the Modify Tab:

    o **Clear Selection** – Remove all of the face from the selection set that will be used to create the curtain system.

    o **Recreate system** – Generate a curtain system for all currently selected mass faces.

- **Update to Face** – Recreate the curtain system when the mass element that was used to originally create the curtain system has changed. Update to face will be grayed out if the underlying mass element has not changed. Once the mass element is changed in some way, the update to face tool can be used. In some cases, the geometry of the mass element will have changed in a way to make it not possible to use the Update to Face command, or for it to not complete the operation. In these cases, you can delete the curtain system and recreate it.

Modify | Curtain Wall Grids

- **Add/Remove Segments** – Add or remove a curtain grid dividing segment. When a curtain grid line is placed, usually segments are added along the entire length of the grid line. A segment on a grid line will break the curtain wall panel at the segment and allow for a mullion to be placed at the segment. Without segments the curtain grid line is displayed as a dashed line. Add and remove segments from a curtain wall grid to create a non-regular pattern. Click the curtain wall grid line where a segment exists to remove it and click a curtain wall gridline where a segment does not exist to add it.

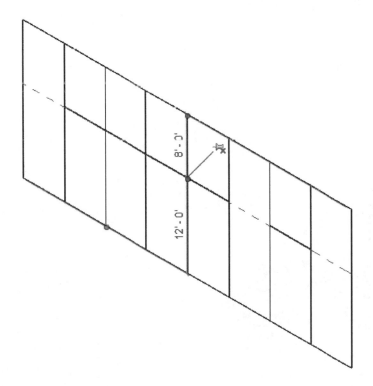

Modify | Curtain Wall Mullions

- **Make Continuous** – At adjacent ends the mullions are joined together so they appear graphically continuous. No lines are displayed between the adjacent mullions.

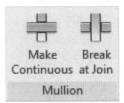

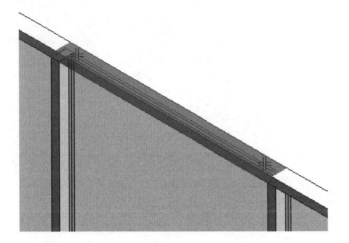

- **Break at Join** – At adjacent ends the mullions will have a line between them making the mullion appear graphically as an individual element.

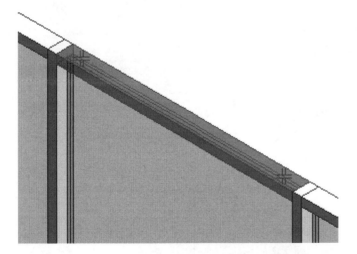

Modify | Detail Items

- **Detail Component**

  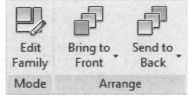

  - **Edit Family** - Opens the selected family in the family editor. The model where the family originated will remain open. Use Switch Windows to go back to a view in the open model.
  - **Bring to Front** – Adjust the draw order of the annotation element, placing it "on top" of all other annotation elements in the view.
  - **Bring Forward** – Adjust the draw order of the annotation element "one step" in front of other annotation elements in the view.
  - **Send to Back** - Adjust the draw order of the annotation element, placing it "beneath" all other annotation elements in the view. Annotation elements placed in a view are always drawn "on top of" model elements in a view. Annotation elements can never be positioned behind a model element.
  - **Send Backward** - Adjust the draw order of the annotation element "one step" in back of other annotation elements in the view.

- **Filled Region**

  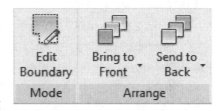

  - **Edit Boundary** – Start sketch editing mode where you can use tools from the draw panel to make edits to the boundary of the element. Boundaries must form closed loops in order to finish the boundary. Each segment in the boundary can be assigned a different line style using the line style dropdown. The "invisible lines" line style will create an edge for the boundary but will not be visible when the element is finished.
  - **Bring to Front** – Adjust the draw order of the annotation element placing it "on top" of all other annotation elements in the view.
  - **Bring Forward** – Adjust the draw order of the annotation element "one step" in front of other annotation elements in the view.
  - **Send to Back** - Adjust the draw order of the annotation element, placing it "beneath" all other annotation elements in the view. Annotation elements placed in a view are always drawn "on top of" model elements in a view. Annotation elements can never be positioned behind a model element.
  - **Send Backward** - Adjust the draw order of the annotation element "one step" in back of other annotation elements in the view.

- ## Masking Region

  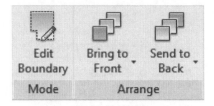

  - **Edit Boundary** – Start sketch editing mode where you can use tools from the draw panel to make edits to the boundary of the element. Boundaries must form closed loops in order to finish the boundary. Each segment in the boundary can be assigned a different line style using the line style dropdown. The "invisible lines" line style will create an edge for the boundary but will not be visible when the element is finished.

  - **Bring to Front** – Adjust the draw order of the annotation element, placing it "on top" of all other annotation elements in the view.

  - **Bring Forward** – Adjust the draw order of the annotation element "one step" in front of other annotation elements in the view.

  - **Send to Back** - Adjust the draw order of the annotation element, placing it "beneath" all other annotation elements in the view. Annotation elements placed in a view are always drawn "on top of" model elements in a view. Annotation elements can never be positioned behind a model element.

  - **Send Backward** - Adjust the draw order of the annotation element "one step" in back of other annotation elements in the view.

- ## Repeating Detail Component

  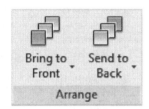

  - **Bring to Front** – Adjust the draw order of the annotation element, placing it "on top" of all other annotation elements in the view.

  - **Bring Forward** – Adjust the draw order of the annotation element "one step" in front of other annotation elements in the view.

  - **Send to Back** - Adjust the draw order of the annotation element, placing it "beneath" all other annotation elements in the view. Annotation elements placed in a view are always drawn "on top of" model elements in a view. Annotation elements can never be positioned behind a model element.

  - **Send Backward** - Adjust the draw order of the annotation element "one step" in back of other annotation elements in the view.

Modify | Dimensions

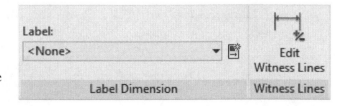

- **Label** – Assign or create a global parameter for the dimension value. If there are no global parameters in the model you can create a new one from the current dimension value. Global Parameters offer a way to globally control positions of elements.

- **Edit Witness Lines** – Add or remove references to a dimension string. Add references by selecting the new reference edges or points in the model. Click on an existing reference edge or point to remove it from the dimension string. Once reference points have been added and removed from the dimension string, click an empty space in the canvas to complete the dimension with the new references.

Modify | Doors

- **Edit Family** - Opens the selected family in the family editor. The model where the family originated will remain open. Use Switch Windows to go back to a view in the open model.

- **Pick New Host** – Change the host for the selected element. Depending on the element and the view you are using, valid hosts can vary. Look at the status bar to determine a valid host for the element.

- **Show Related Warnings** – Option appears when a warning exists involving the selected element. For example, the door has the same number (i.e. Mark) as another door. Click to see information about the warning and the elements involved—including their element ID numbers.

Modify | DWG Links (Imported or Linked)

- **Delete Layers** – Select and delete specific layers from the DWG import/link. Deleted layers are removed from the Revit model and cannot be restored unless the DWG is removed and reimported/linked. The DWG file is not changed.

- **Partial Explode** – (Imported DWG only) The imported DWG geometry is converted from one element to individually selectable elements (lines, arcs, regions, text, etc…) Line Styles, text types, etc. are created in the Revit model to maintain graphic quality as close to the original import as possible. Individual blocks in the original DWG file will remain together as an element in the Revit model.

- **Full Explode** - (Imported DWG only) The imported DWG geometry is converted from one element to individually selectable elements (lines, arcs, regions, text, etc…) Line Styles, text types, etc. are created in the Revit model to maintain graphic quality as close to the original import as possible. Individual blocks in the original DWG file will also be converted to individually selectable elements in the Revit model. If the Full Explode will result in more than 10,000 individual elements, the Full Explode is not allowed.

- **Query** – Select and then pick an element within the DWG link/import to see properties as shown in the example below. Use this information, as one example, to determine which Layer/Level an element is on and then turn it off in Visibility/Graphics Override for a given view (or just click the **Hide in View** button). **FYI:** the term "Layers" applies to AutoCAD files and "levels" applies to Microstation files, two popular CAD programs.

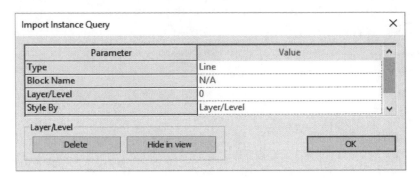

Modify | Elevations

- **Reference Other View** – Active only when the select elevation was set to reference another view during creation—meaning the selected view is not a live view of the Revit model.

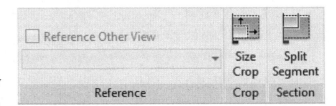

The checkbox status cannot be changed, but the other view being referenced may be changed via the drop-down.

- **Size Crop** – Set the Width, Height, and Annotation crop offsets for the crop region of the elevation (below next image). The crop region can be adjusted in the canvas using grips visually. The Size Crop tool allows you to be specific with the size of the crop. This can be useful when you have specific view size requirements for a presentation.

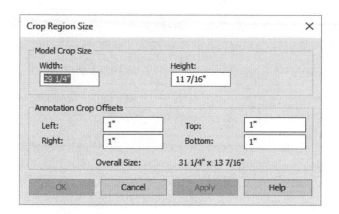

- **Split Segment** – Create jogs in the projection plane of the view. In the canvas use the knife cursor to create the split and then use the shape handles to adjust the position of the projection plane jog. Jogs in the projection plane are restricted to 90 degree jogs. To rejoin a split segment, use the shape handles to make the projection plane in a single line again.

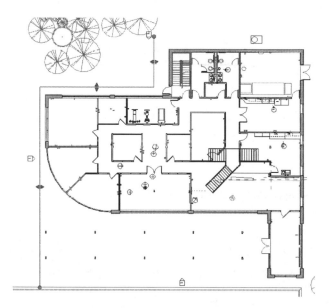

- **Edit Crop** – (When working in elevation view only.) Use when you want to create a non-rectangular crop. Enter a sketch mode and use tools from the Draw panel to create a closed boundary for the crop of the view. The edges of the crop boundary are restricted to straight segments. The annotation crop of the view will remain in a rectangular shape regardless of the shape of the edited crop.

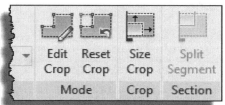

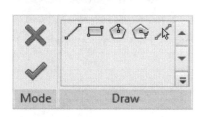

- **Reset Crop** – (Only if crop has been edited.) Resets the crop to a rectangular shape (bounding box) around the edited crop. In some cases, you may have to use the resulting rectangular crop's handles to resize the crop to fit the model as required.

> Editing the default, rectangular, boundary of an elevation often causes increased print times and large PDF file sizes.

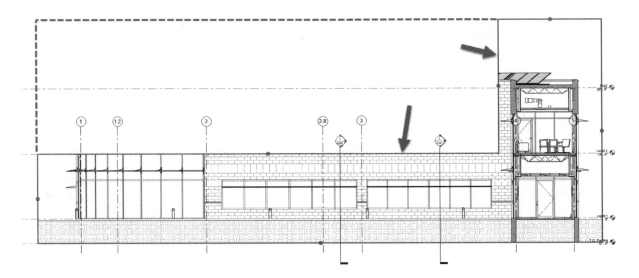

Example of Edited Crop in Elevation

Modify | Floors

The structural Analytical and Reinforcement features will not be covered given the architectural focus of this text.

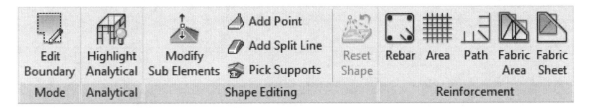

- **Edit Boundary** – Start sketch editing mode where you can use tools from the draw panel to make edits to the boundary of the element. Boundaries must form closed loops in order to finish the boundary.
- **Modify Sub Elements** – Enter a mode to edit the "Z" position of edges and points on a floor/roof element. Select a point or an edge and specify a Z value. If working in a 3d view you can use shape handles on the selection to visually edit the Z value.

- **Add Point** – Place a single point on the floor/roof to be used for shape editing. By default, corners and edges of a floor can be modified but additional points inside the boundary of the floor can be added with the tool.
- **Add Split Line** – Add a line to the inside of the floor/roof that can be used for shape editing. The Z value of the split line is modified in the same way points and edges of the floor are altered during shape editing.
- **Pick Supports**   Adds a split line to the floor/roof where a structural element is in the model. Select the structural element to generate the split line.
- **Reset Shape** – Returns all edges and points of the floor/roof to the original positions with no Z values assigned.

Shape editing tools will only be available when the floor/roof is flat, with no slope.

Modify | Generic Annotations (Symbols)

- **Edit Family** - Opens the selected family in the family editor. The model where the family originated will remain open. Use Switch Windows to go back to a view in the open model.
- **Leader Add/Remove** – Adds or removes a leader as defined in the symbol family type. Leaders are added to the default position on the symbol. Use grips in the canvas to change the leader position and where it is pointing to in the view. When you remove leaders, they are removed in the opposite order they were added in.

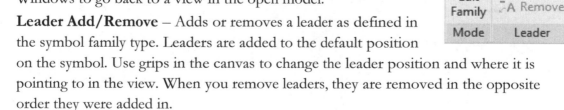

Modify | Generic Models

Modify | Decals

- Set the **Height** and **Width** of the decal numerically in the options bar. Sizes are in "real world" dimensions.
- **Lock Proportions** – Scale the decal when sizing.
- **Reset** – Set the height and width of the decal back to the original size as imported based on the image file being used for the decal.

Modify | Model Text

- **Edit Text** – Open the text editing dialog and change the contents of the model text.
- **Edit Work Plane** – Choose a new work plane for the model text. The work plane establishes where in space the model text is positioned. To edit the work plane, in the Work Plane dialog choose the mode to specify the work plane.
  - **Name** – Choose a work plane from a list of datums (levels, grids, or reference planes).
  - **Pick a Plane** – Graphically select a plane in the model. Only planes parallel with the current work plane can be selected.
  - **Pick a line and use the work plane it was sketched in** – Use a work plane as projected from a line selected in canvas.
  - **Pick New Host** – Any plane in the model can be selected and the model text will be moved to the plane of the clicked host and positioned at the cursor.
- **Pick New** – Pick a new work plane in the canvas. There are two modes to pick a new workplane.
  - **Face** – Pick any element face in the model to become the new host.
  - **Work Plane** – If your current work plane is different than the work plane of the model text, this option will move the model text to the current work plane and allow you to position it in the canvas by clicking.

Modify | Grids

- **Propagate Extents** – Transfer the visual (2d) position of a datum (level/grid) endpoint to another parallel view. Use this tool when the 2d positions of level/grid endpoints need to align across multiple views. For example, all grids on multiple plans need to line up; propagate extents can help do this. In the Propagate Extents dialog select the other parallel views where the extents from the current view should be applied.

Modify | Insulation Batting Lines

- **Bring to Front** – Adjust the draw order of the annotation element placing it "on top" of all other annotation elements in the view.

- **Bring Forward** – Adjust the draw order of the annotation element "one step" in front of other annotation elements in the view.

- **Send to Back** - Adjust the draw order of the annotation element placing it "beneath" all other annotation elements in the view. Annotation elements placed in a view are always drawn "on top of" model elements in a view. Annotation elements can never be positioned behind a model element.

- **Send Backward** - Adjust the draw order of the annotation element "one step" in back of other annotation elements in the view.

- **Width** – (Options Bar) Set the real world width of the insulation element.

Modify | Keynote Tags

- **Edit Family** - Opens the selected family in the family editor. The model where the family originated will remain open. Use Switch Windows to go back to a view in the open model.

- **Select Host** – Change the selection from the tag to the element it is tagging. This is a way you can see or modify the parameters of the tagged element.

- **Pick New Host** – Choose another element for the tag. Properties for the new element will be displayed in the tag.

- **Reconcile Hosting** – When tags used on elements form a linked file because of changes in the linked file, a tag can become out of position or orphaned. Reconcile Hosting opens a review palette so you can locate these tags and take the appropriate action. If the reconcile hosting palette is already open, this tool will be grayed out.

Contextual

- **Horizontal/Vertical** – (Options Bar) Choose the orientation of the tag as it relates to the view.
- **Leader** – Controls the visibility of the leader on the tag.
- **Attached/Free End** – Pulldown to set the end point of the leader. Attached will move when the tagged object moves but is limited to logical attachment points on the bounding box of the tagged element. Free allows the end point of the leader to be moved by dragging in the canvas and can be positioned anywhere. When the tagged element moves the free end does not move with it.

Modify | Legend Components

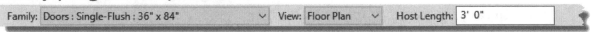

- **Family** – (Options Bar) Use the dropdown list to select a family to create the legend component. All loaded families in the model are listed in the dropdown.
- **View** – (Options Bar) Use dropdown to select from the available views of the element. Depending on the family selected, different views will be available for use as a legend component. For example, a wall can be placed as a floorplan view or as a section view while a window can be placed as floor plan view, front, or back elevation. Choose the view that makes the most sense for the kind of legend you are creating or the sheet you plan on using the legend on.
- **Host Length** – When placing a legend component that needs a host, i.e., door or window, set the length of the host segment shown in the legend. The element you are placing will be centered in the host. If the component being placed is a host, wall, floor, ceiling, etc.…this value will define how much of the host is created.

Modify | Levels

- **Propagate Extents** – Transfer the visual (2d) position of a datum (level/grid) endpoint to another parallel view. Use this tool when the 2d positions of level/grid endpoints need to align across multiple views. For example, all levels on multiple parallel elevations and sections need to line up; propagate extents can help do this. In the Propagate Extents dialog select the other parallel views where the extents from the current view should be applied.

Modify | Lines (Model)

- **Line Styles** – Use the Dropdown list to select a line style for the selected line. Line styles are defined as part of a template file. If you need to create a new line style you must first create the style in the model. (See Manage > Additional Setting > Line Styles.)

- **Convert Lines** – Changes the selected line(s) from a model line (visible in all

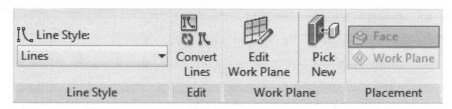

views) to detail line (visible in only the current view.)

- **Edit Work Plane** – Choose a new work plane for the line. The work plane establishes where in space the line is positioned. To edit the work plane, in the Work Plane dialog choose the mode to specify the work plane.
    - o **Name** – Choose a work plane from a list of datums (levels, grids, or reference planes).
    - o **Pick a Plane** – Graphically select a plane in the model. Only planes parallel with the current work plane can be selected.
    - o **Pick a line and use the work plane it was sketched in** – Use a work plane as projected from a line selected in canvas.
    - o **Pick New Host** – Any plane in the model can be selected and the model text will be moved to the plane of the clicked host and positioned at the cursor.
- **Pick New** – Pick a new work plane in the canvas. There are two modes to pick a new work plane.
    - o **Face** – Pick any element face in the model to become the new host.
    - o **Work Plane** – If your current work plane is different than the work plane of the line, this option will move the line to the current work plane and allow you to position it in the canvas by clicking.
- **Moves with Nearby Elements** – (Options Bar) Toggles the Moves with Nearby Elements parameter for the element. When this is set, the element will be constrained to other elements, typically walls, around it. When the element it is constrained to (wall) is moved, the element will also move. The element can be moved independently of the constraint to create a new relationship with a different nearby element. If an element is moving when walls in the model are moved, check to see if the Moves with Nearby Elements parameter is checked.

Modify | Lines (Detail)

- **Line Styles** – Use the dropdown list to select a line style for the selected line. Line styles are defined as part of a template file. If you need to

create a new line style you must first create the style in the model. (See Manage > Additional Setting > Line Styles.)

- **Convert Lines** – Changes the selected line(s) from a detail line (visible in only the current view) to model line (visible in all views).
- **Bring to Front** – Adjust the draw order of the annotation element, placing it "on top" of all other annotation elements in the view.
- **Bring Forward** – Adjust the draw order of the annotation element "one step" in front of other annotation elements in the view.
- **Send to Back** - Adjust the draw order of the annotation element, placing it "beneath" all other annotation elements in the view. Annotation elements placed in a view are always drawn "on top of" model elements in a view. Annotation elements can never be positioned behind a model element.
- **Send Backward** - Adjust the draw order of the annotation element "one step" in back of other annotation elements in the view.

Modify | Masses (In-Place Mass)

- **Edit In-Place** – Open the mass in the canvas and edit using in-place mass editing tools. Use this to add or change the geometry of the in-place mass in a significant way, change the position of an edge or point in the mass. In canvas shape handles

can be used to edit the position of the faces of the mass without using the edit in-place command.
- **Related Hosts** – If any of the building maker tools (wall by face, roof by face, etc...) were used on a face of the mass, use Related Hosts to select all elements related to faces of the selected mass.
- **Mass Floors** – Create a mass floor element for all level datums intersecting the selected mass. In the mass floor dialog select the datums you want to use to create mass floors. Mass floor elements can be scheduled. In order to use the Floor By Face command you must first create mass floors. When doing an energy analysis of building massing, mass elements with mass floors will be analyzed. Any mass elements without mass floors will not be analyzed and be counted as a "shading" device.
- **Show Related Warnings** – Option appears when a warning exists involving the selected element. For example, the door has the same number (i.e., Mark) as another door. Click to see information about the warning and the elements involved—including their element ID numbers.

Modify | Masses (Mass Family)

- **Edit In-Place** – Open the mass in the canvas and edit using in-place mass editing tools. Use this to add or change the geometry of the

in-place mass in a significant way, change the position of an edge or point in the mass. In canvas shape handles can be used to edit the position of the faces of the mass without using the edit in-place command.

- **Related Hosts** – If any of the building maker tools (wall by face, roof by face, etc…) were used on a face of the mass, use Related Hosts to select all elements related to faces of the selected mass.

- **Mass Floors** – Create a mass floor element for all level datums intersecting the selected mass. In the mass floor dialog select the datums you want to use to create mass floors. Mass floor elements can be scheduled. In order to use the Floor By Face command you must first create mass floors. When doing an energy analysis of building massing, Mass elements with mass floors will be analyzed. Any mass elements without mass floors will not be analyzed and be counted as a "shading" device.

- **Edit Work Plane** – Choose a new work plane for the line. The work plane establishes where in space the line is positioned. To edit the work plane, in the Work Plane dialog choose the mode to specify the work plane.
  - **Name** – Choose a work plane from a list of datums (levels, grids, or reference planes).
  - **Pick a Plane** – Graphically select a plane in the model. Only planes parallel with the current work plane can be selected.
  - **Pick a line and use the work plane it was sketched in** – Use a work plane as projected from a line selected in canvas.
  - **Pick New Host** – Any plane in the model can be selected and the model text will be moved to the plane of the clicked host and positioned at the cursor.

- **Pick New** – Pick a new work plane in the canvas. There are two modes to pick a new work plane.
  - **Face** – Pick any element face in the model to become the new host.
  - **Work Plane** – If your current work plane is different than the work plane of the line, this option will move the line to the current work plane and allow you to position it in the canvas by clicking.

Modify | Matchline

- **Edit Sketch** – Open the sketch editor mode and use the draw and modify tools to change the sketch. Cancel in the sketch mode to exit without changes. Click Finish in the sketch mode to apply the new sketch. Matchline sketches are limited to linear segments, but do not have to form a closed loop.

Modify | Material Tags

- **Edit Family** - Opens the selected family in the family editor. The model where the family originated will remain open. Use Switch Windows to go back to a view in the open model.

- **Select Host** – Change the selection from the tag to the element it is tagging. This is a way you can see or modify the parameters of the tagged element.
- **Pick New Host** – Choose another element for the tag. Properties for the new element will be displayed in the tag.
- **Reconcile Hosting** – When tags used on elements form a linked file because of changes in the linked file, a tag can become out of position or orphaned. Reconcile Hosting opens a review palette so you can locate these tags and take the appropriate action. If the reconcile hosting palette is already open this tool will be grayed out.
- **Horizontal/Vertical** – (Options Bar) Choose the orientation of the tag as it relates to the view.
- **Leader** – (Options Bar) Controls the visibility of the leader on the tag.
- **Free End** – (Options Bar) The pulldown is inactive for material tags. Material tags can only be placed with a free end. The end point of the leader defines what the material tag resolves with. The material under the end of the leader is used for the material tag value.

Modify | Model Group

- **Edit Group** – Open the group for editing. Edits made to a group will propagate to all other instances of the group in the model. In the edit mode you can make modifications to all elements in

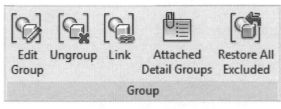

the group. The edit group panel opens in group edit mode with three additional commands:
  - o **Add** – Add new elements to the group definition. Click a model element to add it to the group.

- o **Remove** – Remove elements from the current group. Click an element in the group to remove it.
  - o **Attach** – Add a detail group to the model group. A detail group is made up of only annotation elements and attached to a model group.
- **Ungroup** – Elements making up the group are released from the group definition and will no longer update when the group definition is edited. Ungrouping elements does not remove the group definition from the model or ungroup other instances of the group in the model.
- **Link** – The instance of the group will be converted to a linked file. When you use this tool you have two options. You can save the group as a new RVT file and link that RVT file in place of the group, or you can replace the selected instance of the group with an existing RVT file. Use the first option if this is the first instance of the group you are linking. Use the second option when linking to a file that was already converted from a group.
- **Attached Detail Groups** – (Only visible if attached detail groups present for the model group.) Open the attached detail groups dialog where you can choose which attached detail groups you want to place with the model group. In the dialog check the detail groups you want to display and click ok. Removing the check mark in the dialog will turn off the display of the attached detail group.
- **Restore All Excluded** – (Only visible if elements have been excluded from the selected model group.) Click this option to restore any excluded model elements.
  - o **FYI:** Individual model elements, within a model group, can be selected (by first tapping the Tab key to highlight) and then Excluded via a right-click menu. Excluded elements are hidden in all views and schedule. This feature can be used to minimize the number of groups required in a project.

Modify | Detail Groups

- **Edit Group** – Open the group for editing. Edits made to a group will propagate to all other instances of the group in the model. In the edit mode you can make modifications to all elements in the group. Detail groups are made up of 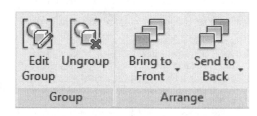 annotation elements. If tags are part of the group, the tags will report different instance based properties of the elements tagged. The edit group panel opens in group edit mode with two additional commands:
  - o **Add** – Add new elements to the group definition. Click a model element to add it to the group.
  - o **Remove** – Remove elements from the current group. Click an element in the group to remove it.

- **Ungroup** – Elements making up the group are released from the group definition and will no longer update when the group definition is edited. Ungrouping elements does not remove the group definition from the model or ungroup other instances of the group in the model.

Modify | "Opening by Face"

- **Edit Sketch** – Open the sketch editor mode and use the draw and modify tools to change the boundary sketch. Cancel in the sketch mode to exit without changes. Click Finish in the sketch mode to apply the new boundary.

Modify | Path of Travel Lines

- **Update** – If a design has changed since the path of travel line was placed, it may need an update to remain valid. The update button will

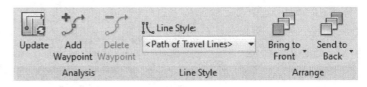

maintain the start and end points of the Path of Travel line but the analysis will be re-run calculating obstacles along the path again. The Path of Travel line will then be updated to accommodate the new conditions of the model.

- **Add Waypoint** – Add a point on the path of travel. The path of travel will intersect all waypoints on a path of travel line. Waypoints can be moved and positioned to control the path of travel route.

- **Delete Waypoint** – Remove waypoints added to a path of travel line. Delete Waypoint will not be available unless the selected path of travel has an added waypoint.

- **Line Style** – Change the line style of the path of travel line.

- **Bring to Front** – Adjust the draw order of the annotation element, placing it "on top" of all other annotation elements in the view.

- **Bring Forward** – Adjust the draw order of the annotation element "one step" in front of other annotation elements in the view.

Modify | Property Lines

- **Show Related Warnings** – Option appears when a warning exists involving the selected element. For example, the property line does not form a closed loop, so area will not be calculated. Click to see information about the warning and the elements involved—including their element ID numbers (see image below).

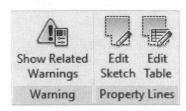

  - Warnings dialog

- **Show** will change the view to display the item selected in the dialog.
- **More Info** will show additional information if available.
- **Deleted Checked** will delete the element selected out of the model, not just delete the warning.
- **Export** will create an HTML file with the warning information.

- **Edit Sketch** – Open the sketch editor mode and use the draw and modify tools to change the sketch. Cancel in the sketch mode to exit without changes. Click Finish in the sketch mode to apply the new sketch. Property line sketches do not have to form a closed loop, but if they do not, you will receive a warning about the property line not being closed. If the property lines are table based, the Edit Sketch button will be grayed out.

- **Edit Table** – Open the property line table editor where distances and bearings can be directly entered to modify the property lines in the model. If the property lines are sketched before the button is clicked the sketch based property lines will be converted to table based property lines. Once a sketch based property line is changed to a table based property line, it cannot be changed back to a sketch based property line.

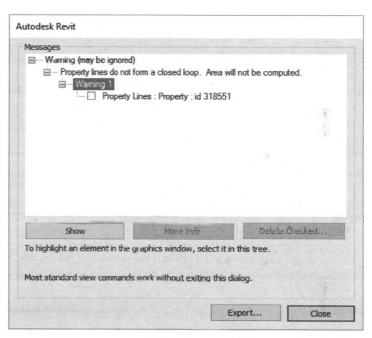

Modify | Railings

- **Edit Path** – Open the sketch editor mode and use the draw and modify tools to change the sketch. Cancel in the sketch mode to exit without changes. Click Finish in the sketch mode to apply the new sketch. Railing sketches do not have to form a closed loop.

- **Pick New Host** – Change the host for the selected element. Depending on the element and the view you are using, valid hosts can vary. Look at the status bar to determine a valid host for the element.

- **Reset Railing** – Any overrides applied to the railing (i.e., railing joins) are removed and the railing is reset to the instance and type parameters used to originally create the railing.

Modify | Ramps

- **Edit Sketch** – Open the sketch editor mode and use the draw and modify tools to change the sketch. Cancel in the sketch mode to exit without changes. Click Finish in the sketch mode to apply the new sketch. Ramp sketches must contain boundaries and riser lines.

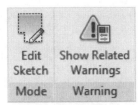

- **Show Related Warnings** – Option appears when a warning exists involving the selected element. For example, the property line does not form a closed loop, so area will not be calculated. Click to see information about the warning and the elements involved—including their element ID numbers.
  - o Warnings dialog (see next image)
    - **Show** will change the view to display the item selected in the dialog.
    - **More Info** will show additional information if available.
    - **Deleted Checked** will delete the element selected out of the model, not just delete the warning.
    - **Export** will create an HTML file with the warning information.

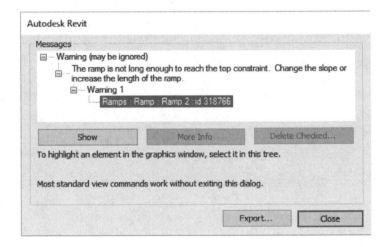

Modify | Reveals

- **Add/Remove Walls** – Add or remove adjacent walls to the sweep/reveal. Click an adjacent wall without the sweep/reveal to add. Click on a wall with a sweep/reveal to remove. Multiple segments of a single connected sweep/reveal placed on multiple walls are controlled by the same instance and type properties.
- **Modify Returns** – Return the sweep or reveal at the end of a wall. Click a sweep/reveal end point to create a return element and then use drag controls to define the length of the return. When creating

the return, use settings in the Options Bar to control the geometry of the return.

- o **Straight Cut** – Resets the sweep/reveal return to the original position before a return was applied. Use this to reset an applied return.
- o **Return** – Apply a return edge. Use the Angle control to define the angle the return will be applied to. Typically, the return angle would be 90 to return on the end of a wall. Once the return edge is applied, click modify and then use the grips in the canvas to set the length of the return.

If a Sweep is repositioned outside of its host, a warning will appear as shown in the next image.

- **Delete Element(s)** will remove the sweep(s) from the model.
- **OK** will leave the sweep floating in space, apart from its host. The sweep is still dependent on the host.
- **Cancel** will undo the change that prompted the warning.

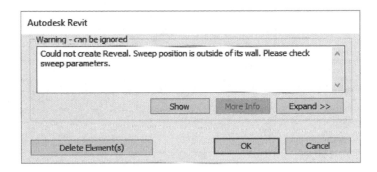

Modify | Revision Clouds

- **Edit Sketch** – Open the sketch editor mode and use the draw and modify tools to change the sketch. Cancel in the sketch mode to exit without changes. Click Finish in the sketch mode to apply the new sketch.

- **Revision** – (Options Bar) Use the dropdown list to select the revision the cloud belongs to. The dropdown list is populated from the revision list defined to the model.

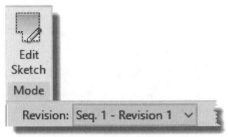

Modify | Roof Opening Cut (Vertical Opening)

- **Edit Sketch** – Open the sketch editor mode and use the draw and modify tools to change the sketch. Cancel in the sketch mode to exit without changes. Click Finish in the sketch mode to apply the new sketch.

Modify | Roofs (Footprint Roof)

- **Edit Footprint** – Open the sketch editor mode and use the draw and modify tools to change the sketch. Cancel in the sketch mode to exit without changes. Click Finish in the sketch mode to apply the new sketch.

Modify | Roofs (Extruded Roof)

- **Edit Profile** – Open the sketch editor mode and use the draw and modify tools to change the sketch. Cancel in the sketch mode to exit without changes. Click Finish in the

  sketch mode to apply the new sketch. Extruded roofs profiles are a single line. The thickness of the roof is generated above the sketched profile.

- **Vertical** – Sketch a vertical opening for the roof. The vertical opening is sketched on the reference level of the profile roof and projected to the roof as an opening.

- **Edit Work Plane** – Choose a new work plane for the roof profile sketch. The work plane establishes where in space the profile roof is positioned. To edit the work plane, in the Work Plane dialog choose the mode to specify the work plane.
    - **Name** – Choose a work plane from a list of datums (grids or reference planes).
    - **Pick a Plane** – Graphically select a plane in the model. Only planes parallel with the current work plane can be selected.
    - **Pick a line and use the work plane it was sketched in** – Use a work plane as projected from a line selected in canvas.

- Pick New Host – Any plane in the model can be selected and the model text will be moved to the plane of the clicked host and positioned at the cursor.
- **Pick New** – Pick a new work plane in the canvas. There are two modes to pick a new work plane.
  - **Face** – Pick any element face in the model to become the new host.
  - **Work Plane** – If your current work plane is different than the work plane of the line, this option will move the line to the current work plane and allow you to position it in the canvas by clicking.

Modify | <Room Separation>

- **Moves With Nearby Elements (Options Bar)** – Toggles the Moves with Nearby Elements parameter for the element. When this is set, the element will be constrained to other elements, typically walls, around it. When the element it is constrained to (wall) is moved, the element will also move. The element can be moved independently of the constraint to create a new relationship with a different nearby element. If an element is moving when walls in the model are moved, check to see if the Moves with Nearby Elements parameter is checked.

Modify | <Room Separation>   ☐ Moves With Nearby Elements

Modify | Room Tags

- **Edit Family** - Opens the selected family in the family editor. The model where the family originated will remain open. Use Switch Windows to go back to a view in the open model.

- **Show Related Warnings** – Option appears when a warning exists involving the selected element. For example, the Room tag is not within the host Room boundary. Click to see information about the warning and the elements involved—including their element ID numbers. This dialog also has a **Move to Room** button to help resolve this issue as shown in the image below.
- **Select Host** – Change the selection from the tag to the element it is tagging. This is a way you can see or modify the parameters of the tagged element.
- **Pick New Host** – Choose another element for the tag. Properties for the new element will be displayed in the tag.
- **Reconcile Hosting** – When tags used on elements form a linked file because of changes in the linked file, a tag can become out of position or orphaned. Reconcile Hosting opens a review palette so you can locate these tags and take the appropriate action. If the reconcile hosting palette is already open this tool will be grayed out.

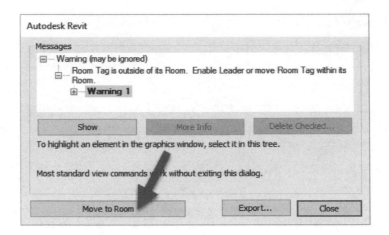

Modify | RVT Links

- **Bind Link** – Insert all elements from the linked model into the current model as a group. The instance of the link will be removed from the host model leaving behind the new group. To remove the link completely you must remove the link in the dialog as you bind the link or remove it later from the Manage Links dialog in the Bind Link Options dialog.

- **Manage Links** – Open the Manage Links dialog where you can reload the link or add new links to the model.
  - **Bind Link Options** dialog (see image below)
    - **Attached Details** – this option brings the view-specific elements such as notes and dimensions. Select the bound element, which is now a Group, and select **Attached Detail Groups** on the Ribbon.
    - **Levels** – be careful not to bring in levels that already exist. Multiple levels at the same elevation creates several problems.
    - **Grids** – check this option to include grids. Usually a good idea if the grids do not already exist in the host model.

    FYI: Binding a model with **Phasing** is not really possible. All the phases in the linked model are consolidated to a single phase (based on the current view's phase setting).

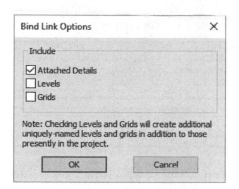

Modify | Shaft Opening

- **Edit Sketch** – Open the sketch editor mode and use the draw and modify tools to change the sketch. Cancel in the sketch mode to exit without changes. Click Finish in the sketch mode to apply the new sketch.

Modify | Slab Edges

- **Add/Remove Segments** - Add or remove segments to the slab edge element. Slab edges do not need to be adjacent or continuous but must be hosted to the same floor/slab. Click an edge that currently does not have a slab edge applied to add. Click an edge with a slab edge applied to remove. In order to

  remove a slab edge, you may need to first un-join the slab edge from the host floor/slab in order to make the edge of the floor/slab element visible.

- **Show Related Warnings** – Option appears when a warning exists involving the selected element. For example, one element is completely under another. Click to see information about the warning and the elements involved—including their element ID numbers.

- **Rebar** – Place planar rebar into the slab edge element. The rebar is placed according to cover instance parameters defined in the slab edge family. Rebar can't be placed in a 3d view.

Modify | Space Tags

- **Edit Family** - Opens the selected family in the family editor. The model where the family originated will remain open. Use Switch Windows to go back to a view in the open model.

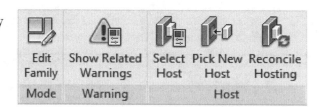

- **Show Related Warnings** – Option appears when a warning exists involving the selected element. For example, the Space tag is not within the host Space boundary. Click to see

Contextual

information about the warning and the elements involved—including their element ID numbers. This dialog also has a **Move to Space** button to help resolve this issue as shown in the image below.

- **Select Host** – Change the selection from the tag to the element it is tagging. This is a way you can see or modify the parameters of the tagged element.

- **Pick New Host** – Choose another element for the tag. Properties for the new element will be displayed in the tag.

- **Reconcile Hosting** – When tags used on elements form a linked file because of changes in the linked file, a tag can become out of position or orphaned. Reconcile Hosting opens a review palette so you can locate these tags and take the appropriate action. If the reconcile hosting palette is already open, this tool will be grayed out.

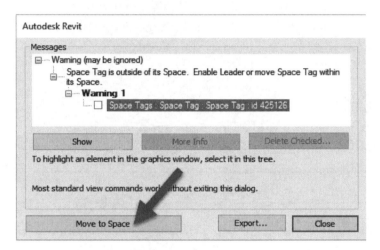

Modify | Spot Coordinate

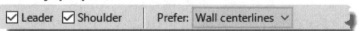

- **Leader** – (Options Bar) Turn the leader of the spot elevation on and off.
- **Shoulder** – (Options Bar) Turn the shoulder of the leader on and off. The leader shoulder allows a leader to be defined by 3 clicks. Without the shoulder the leader is created as a straight line defined by 2 clicks.
- **Prefer** – (Options Bar) Ignore this option as it is irrelevant.

Modify | Spot Elevations

- **Leader** – (Options Bar) Turn the leader of the spot elevation on and off.
- **Shoulder** – (Options Bar) Turn the shoulder of the leader on and off. The leader shoulder allows a leader to be defined by 3 clicks. Without the shoulder the leader is created as a straight line defined by 2 clicks.

- **Relative Base** – (Options Bar) This option is active when the selected Spot Elevation's type parameter Elevation Origin is set to Relative. When active, select from a list of Levels, which causes the spot elevation to relate to that level (positive or negative)
- **Display Elevations** – (Options Bar) Use the dropdown list to select the elevation of the selected element face. The spot elevation can display elevation of the top face, the bottom face, or both faces of the element.
- **Prefer** – (Options Bar) Ignore this option as it is irrelevant.

Modify | Spot Slopes

Slope Representation	Arrow ∨	Offset from Reference:	1/16"

- **Slope Representation** – Only active in elevation or section views. Use the dropdown to select the kind of symbol used to represent the slope. You can select from a triangle or an arrow. The triangle representation is best used when view is parallel to the sloping face.
- **Offset from Reference** – (Options Bar) Set the distance the slope marker will be positioned away from a parallel edge. The distance specified is the distance where the annotation will be positioned when printed (scaled with view scale.) When the spot slope annotation is placed on a perpendicular sloping face the Offset from Reference has no effect on positioning. See image below.
- **Prefer** – (Options Bar) Ignore this option as it is irrelevant.

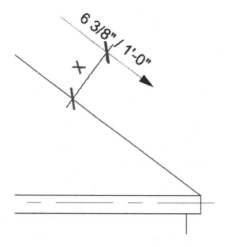

Modify | Stair Tread / Riser Numbers

- **Select Host** – Change the selection from the tag to the element it is tagging. This is a way you can see or modify the parameters of the tagged element.

Contextual

- **Start Number** – (Options Bar) Change the beginning number used for tread/riser numbers. If you want to start numbering at a different point other than 1 change this value.

Modify | Stairs

- **Edit Stairs** – Open the stair editing tools to make modifications to the runs and landings of the selected stair element.
- **Select Levels** – In a section or elevation view, use this command to create a **Multistory Stair**. Here are the basic steps:
  - Once the lowest level stair is created, select it in a section or elevation view and use the new **Select Levels** command on the Ribbon as shown in the image below.

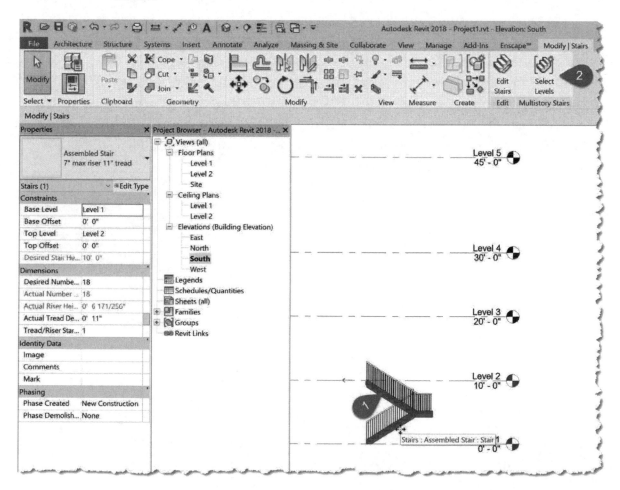

  - After clicking Select Levels, the Ribbon changes to allow the new MS Stair to be connected or disconnected from levels.

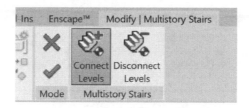

○ The result can be seen in the image below (on left). Notice the top run has a different floor to floor height.

○ All the runs with the same floor to floor height have been grouped together within the MS Stair. You can **Tab** into an individual run and change its type after **unpinning**. The image below (on right) has a different type applied to the first run (monolithic, where the rest are assembled).

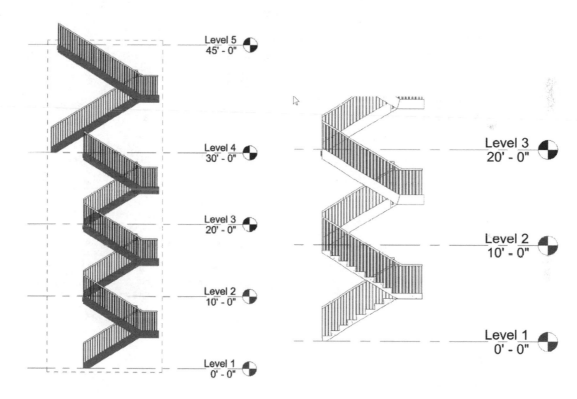

Modify | Structural Beam Systems

- **Edit Boundary** – Start sketch editing mode where you can use tools from the draw panel to make edits to the boundary of the element.

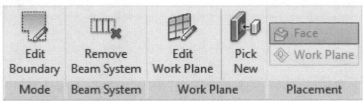

Boundaries must form closed loops in order to finish the boundary.

- **Remove Beam System** – Removes the rules system used to lay out beams in the beam system. When the beam system is removed the beams that were created remain, but further alterations to the beam layout at the "system" level will not be possible when the beam system is removed.

- **Edit Work Plane** – Choose a new work plane for the element. The work plane establishes where in space the element is positioned. To edit the work plane, in the Work Plane dialog choose the mode to specify the work plane.
    - o **Name** – Choose a work plane from a list of datums (grids or reference planes).
    - o **Pick a Plane** – Graphically select a plane in the model. Only planes parallel with the current work plane can be selected.
    - o **Pick a line and use the work plane it was sketched in** – Use a work plane as projected from a line selected in canvas.
    - o **Pick New Host** – Any plane in the model can be selected and the model text will be moved to the plane of the clicked host and positioned at the cursor.

- **Pick New** – Pick a new work plane in the canvas. There are two modes to pick a new work plane.
    - o **Face** – Pick any element face in the model to become the new host.
    - o **Work Plane** – If your current work plane is different than the work plane of the line, this option will move the line to the current work plane and allow you to position it in the canvas by clicking.

Modify | Structural Columns

- **Edit Family** - Opens the selected family in the family editor. The model where the family originated will remain open. Use Switch Windows to go back to a view in the open model.

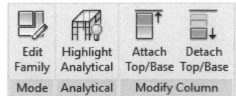

- **Highlight Analytical** – When the analytical model is visible in the view this will show the associated analytical line for the column selected. In some cases, the analytical line of the column is located in a different place than the physical location of the column.

- **Attach Top/Base** – Attach the top or the bottom of the column to a valid target. The column and the target will be constrained. When the target is moved, the column attachment will be maintained. Column must be able to physically intersect with the target in order to attach. These are valid targets for column attachment:

- o   Roof
- o   Floor
- o   Ceiling
- o   Beam
- o   Foundation
- o   Reference Plane
- o   Level

On the Options Bar select the options for the attachment.

- o   **Top/Base** – Choose the attachment point for the column.
- o   **Attachment Style** – Use the pulldown menu to choose the cutting behavior of the column and the attachment target. The column can be cut, the target can be cut, or neither will be cut.
- o   **Attachment Justification** – Use the pulldown to set how much of the column will intersect the attachment target. The Minimum intersection will stop the column attachment as soon as any part of the column geometry touches the target. Column midpoint will stop attachment at the mid-point of the column geometry. Maximum intersection will stop the attachment when all of the column geometry intersects the target.

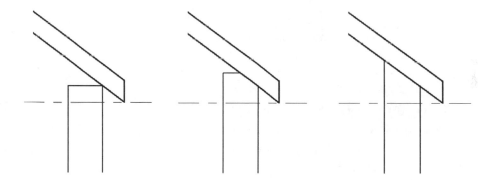

- o   **Offset from Attachment** – Set a distance away from the intersection point between the column and the target where the column stops attachment.
- **Detach Top/Base** – Remove an attachment target from a column. Click Detach top/Base and then select the target in the canvas. The column will no longer be constrained to the target, but will keep any height adjustments that were made when it was attached. Use the properties palette to reset any offsets.
- **Moves With Grid** – (Options Bar) If there is a nearby grid line, check this box to constrain the column to the grid line. When the grid line is moved the column will move as well.

Modify | Structural Foundations

Isolated

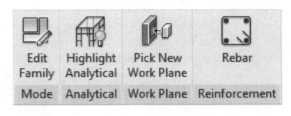

- o **Edit Family** - Opens the selected family in the family editor. The model where the family originated will remain open. Use Switch Windows to go back to a view in the open model.
- o **Highlight Analytical** – When the analytical model is visible in the view this will show the associated analytical line for the foundation selected. In some cases the analytical line of the foundation is located in a different place than the physical location of the foundation.
- o **Pick New Work Plane** – Pick a new work plane in the canvas. If your current work plane is different than the work plane of the foundation, this option will move the foundation to the current work plane and allow you to position it in the canvas by clicking.
- o **Rebar** – Place planar rebar into the element. The rebar is placed according to cover instance parameters defined in the family. Rebar can't be placed in a 3d view.
- o **Moves With Grid** – (Options Bar) If there is a nearby grid line, check this box to constrain the foundation to the grid line. When the grid line is moved the foundation will move as well.

Wall

- o **Rebar** – Place planar rebar into the element. The rebar is placed according to cover instance parameters defined in the family. Rebar can't be placed in a 3d view.
  - ▪ **TIP:** Use the in-canvas grips at each end of the footing to increase its overall length (beyond the wall above).

Slab

- o **Edit Sketch** – Start sketch editing mode where you can use tools from the draw panel to make edits to the boundary of the element. Boundaries must form closed loops in order to finish the boundary.
- o **Highlight Analytical** – When the analytical model is visible in the view this will show the associated analytical line for the foundation selected. In some cases, the

analytical line of the foundation is located in a different place than the physical location of the foundation.

- o **Rebar** – Place planar rebar into the element. The rebar is placed according to cover instance parameters defined in the family. Rebar can't be placed in a 3d view.
- o **Area** – Place an area of rebar. The area for rebar is defined by a sketched boundary. Linear rebar is placed in the boundary.
- o **Path** – Sketch a path for rebar element to be placed along. Shapes of rebar are defined to be repeated along the sketched path.
- o **Fabric Area** – Sketch an area where fabric reinforcement will be placed. The sketched boundary defines the extents of the reinforcement fabric.
- o **Fabric Sheet** - Place a sheet of fabric reinforcement using the extents of the slab as the boundary.

Modify | Structural Framing

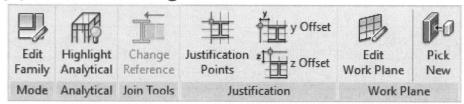

- **Edit Family** Opens the selected family in the family editor. The model where the family originated will remain open. Use Switch Windows to go back to a view in the open model.
- **Highlight Analytical** – When the analytical model is visible in the view this will show the associated analytical line for the foundation selected. In some cases, the analytical line of the foundation is located in a different place than the physical location of the foundation.
- **Change Reference** – Changes the position of the end point reference of the framing element as it relates and joins to an adjacent element. For example, the way a beam frames into a column. By default, the beam stops at the bounding box of the column. You can change the reference to adjust the beam end to the web of the column. This option will only be available when 2 framing elements join and is not available in coarse detail level views.
- **Justification Points** – Adjust the placement of the framing member as it relates to the points clicked when modeled. Justifies the geometry away from the selected justification line. Click the tool and then select a justification point on the framing member in the canvas.
- **Y Offset** – Offsets the framing geometry away from the placement line. Select the tool and then click 2 points to define the offset.

- **Z Offset** - Offsets the framing geometry away from the placement line. Select the tool and then click 2 points to define the offset.
- **Edit Work Plane** – Choose a new work plane for the element. The work plane establishes where in space the element is positioned. To edit the work plane, in the Work Plane dialog choose the mode to specify the work plane.
    - **Name** – Choose a work plane from a list of datums (grids or reference planes).
    - **Pick a Plane** – Graphically select a plane in the model. Only planes parallel with the current work plane can be selected.
    - **Pick a line and use the work plane it was sketched in** – Use a work plane as projected from a line selected in canvas.
    - **Pick New Host** – Any plane in the model can be selected and the model text will be moved to the plane of the clicked host and positioned at the cursor.
- **Pick New**– Pick a new work plane in the canvas. If your current work plane is different than the work plane of the element, this option will move the element to the current work plane and allow you to position it in the canvas by clicking.

Modify | Structural Trusses

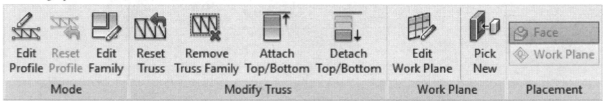

- **Edit Profile** – Open the sketch editor mode and use the draw and modify tools to change the shape of the truss. In the editor the **top chord** and the **bottom chords** can be adjusted from their default positions/shapes. When you complete the edits, the elements making up the truss will be adjusted to the new profile sketch. Cancel in the sketch mode to exit without changes. Click Finish in the sketch mode to apply the new profile.

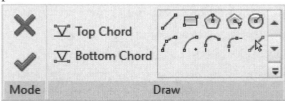

- **Reset Profile** – Resets any profile edits that were made to the truss family. The profile shape will be set back to the default shape based on the family definitions for the truss.
- **Edit Family** - Opens the selected family in the family editor. The model where the family originated will remain open. Use Switch Windows to go back to a view in the open model.
- **Reset Truss** – Individual framing members of a truss can be individually altered and swapped for different framing families besides those defined to the truss. Reset truss

removes those changes and resets the individual members of the truss back to the families defined to the truss.

- **Remove Truss Family** - Removes the rules system used to lay out framing members in the truss family. When the truss family is removed the framing elements that were created remain, but further alterations to the truss family will not be possible when the truss family is removed. All framing elements will act individually.

- **Attach Top/Bottom** – Attach the top or bottom chord of a truss to a floor or roof element. Select a roof or floor that intersects the truss and the profile of the truss will be altered to attach and constrain to the target element. Any profile modification to the truss will be discarded when attached. When an attached target moves, the truss will be altered to maintain the attachment.

- **Detach Top/Bottom** – Attachment constraint is removed from the truss and the profile is restored to the point before the attachment.

- **Edit Work Plane** – Choose a new work plane for the element. The work plane establishes where in space the element is positioned. To edit the work plane, in the Work Plane dialog choose the mode to specify the work plane.
  - **Name** – Choose a work plane from a list of datums (grids or reference planes).
  - **Pick a Plane** – Graphically select a plane in the model. Only planes parallel with the current work plane can be selected.
  - **Pick a line and use the work plane it was sketched in** – Use a work plane as projected from a line selected in canvas.
  - **Pick New Host** – Any plane in the model can be selected and the model text will be moved to the plane of the clicked host and positioned at the cursor.

- **Pick New** – Pick a new work plane in the canvas. There are two modes to pick a new work plane.
  - **Face** – Pick any element face in the model to become the new host.
  - **Work Plane** – If your current work plane is different than the work plane of the line, this option will move the line to the current work plane and allow you to position it in the canvas by clicking.

Modify | Tags (by category)

- **Edit Family** - Opens the selected family in the family editor. The model where the family originated will remain open. Use Switch Windows to go back to a view in the open model.

- **Select Host** – Change the selection from the tag to the element it is tagging. This is a way you can see or modify the parameters of the tagged element.

- **Pick New Host** – Choose another element for the tag. Properties for the new element will be displayed in the tag.

- **Reconcile Hosting** – When tags used on elements form a linked file because of changes in the linked file, a tag can become out of position or orphaned. Reconcile Hosting opens a review palette so you can locate these tags and take the appropriate action. If the reconcile hosting palette is already open this tool will be grayed out.

- **Horizontal/Vertical** – (Options Bar) Choose the orientation of the tag as it relates to the view.

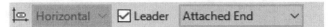

- **Leader** – (Options Bar) Controls the visibility of the leader on the tag.

- **Attached/Free End** – (Options Bar) Pulldown to set the end point of the leader. Attached will move when the tagged object moves but is limited to logical attachment points on the bounding box of the tagged element. Free allows the end point of the leader to be moved by dragging in the canvas and can be positioned anywhere. When the tagged element moves the free end does not move with it.

Modify | Text Notes

- **Leader** – On the leader panel choose to add or remove leaders. When adding a leader first decide which side of

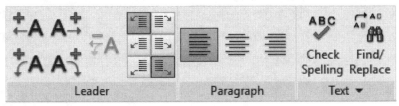

the text element the leader will be added, and then decide if you are going to place an arc leader or a leader with straight segments. Select the appropriate add button to place the desired leader side and shape (arc or straight). The justification controls set if the added leader will justify to the top middle or bottom of the text element as it is added. You can't mix arc and straight leaders on the same text element. Use the symbol with the "minus" symbol to remove leaders. Leaders will be removed from the text element in the reverse order they were created; i.e., the most recent leader added will be the first leader removed.

- **Paragraph** – Sets the Left/Center/Right justification for the text element.

- **Check Spelling** – checks spelling on the selected text element. Only one text element at a time can be spell checked.

- **Find/Replace** – Start the find and replace tool for text elements. Find and replace can perform a find/replace action on the current text element, text elements in the current view, or text elements in the entire project.

Modify | Title Blocks

- **Edit Family** - Opens the selected family in the family editor. The model where the family originated will remain open. Use **Switch Windows** to go back to a view in the open model.

Modify | Topography

- **Edit Surface** – Open the surface editing mode where points making up the toposurface can be edited, added, or removed (see image below). **Cancel** in the editing mode to exit without changes. Click **Finish** in the editing mode to apply the changes and adjust the resulting toposurface.

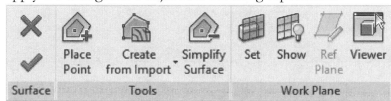

Modify | View Reference

- **Edit Family** - Opens the selected family in the family editor. The model where the family originated will remain open. Use Switch Windows to go back to a view in the open model.

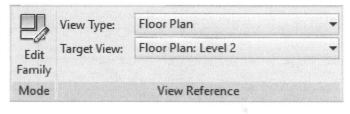

- **View Type** – Use the dropdown to select the view types filtered when assigning the view reference.

- **Target View** – Use the dropdown to select the target for the view reference. The list shown here is filtered to only show the types of views set in the View Type drop down. Once set, the view reference will become a hyperlink to the selected view. The values in the view reference will be resolved with the parameters from the selected view showing the view number and the sheet where the view is placed.

Modify | Views (Elevations/Sections/Callouts)

- **Reference Other View** – Only active if the view was originally created with this option checked. If active, use the drop-down list to change the referenced view.

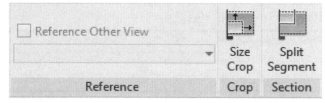

Contextual

- **Size Crop** – Set the Width, Height, and Annotation crop offsets for the crop region of the elevation. The crop region can be adjusted in the canvas using grips visually. The Size Crop tool allows you to be specific with the size of the crop. This can be useful when you have specific view size requirements for a presentation.

- **Split Segment** – Create jogs in the projection plane of the view. In the canvas use the knife cursor to create the split and then use the shape handles to adjust the position of the projection plane jog. Jogs in the projection plane are restricted to 90 degree jogs. To rejoin a split segment, use the shape handles to make the projection plane in a single line again.

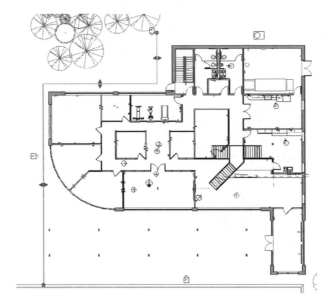

- **Edit Crop** – (When working in elevation view only.) Use when you want to create a non-rectangular crop. Enter a sketch mode and use tools from the Draw panel to create a closed boundary for the crop of the view. The edges of the crop boundary are restricted to straight segments. The annotation crop of the view will remain in a rectangular shape regardless of the shape of the edited crop.

- **Reset Crop** – (Only if crop has been edited.) Resets the crop to a rectangular shape (bounding box) around the edited crop. In some cases, you may have to use the resulting rectangular crop's handles to resize the crop to fit the model as required.

Modify | Wall Sweeps

- **Add/Remove Walls** – Add or remove adjacent walls to the sweep/reveal. Click an adjacent wall without the sweep/reveal to add. Click on a wall with a sweep/reveal to remove. Multiple segments of a single connected sweep/reveal placed on multiple walls are controlled by the same instance and type properties.

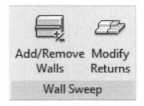

- **Modify Returns** – Return the sweep or reveal at the end of a wall. Click a sweep/reveal end point to create a return element and then use drag controls to define the length of the return. When creating the return use settings in the Options bar to control the geometry of the return.

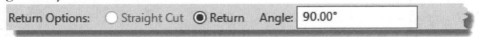

Return Options: ○ Straight Cut ● Return Angle: 90.00°

- ○ **Straight Cut** – Resets the sweep/reveal return to the original position before a return was applied. Use this to resct an applied return.
- ○ **Return** – Apply a return edge. Use the Angle control to define the angle the return will be applied to. Typically, the return angle would be 90 to return on the end of a wall. Once the return edge is applied, click modify and then use the grips in the canvas to set the length of the return.

Modify | Walls

- **Edit Profile** – Open the sketch editor where the elevation profile of the wall can be modified. An edited profile of a wall allows the wall to take on a shape other than rectangular as defined by the length and height of the wall. Profile edits must form closed loops. Walls that are not linear in plan cannot have the profile edited. Cancel in the sketch mode to exit without changes. Click Finish in the sketch mode to apply the new profile.

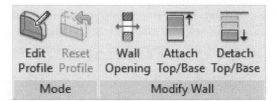

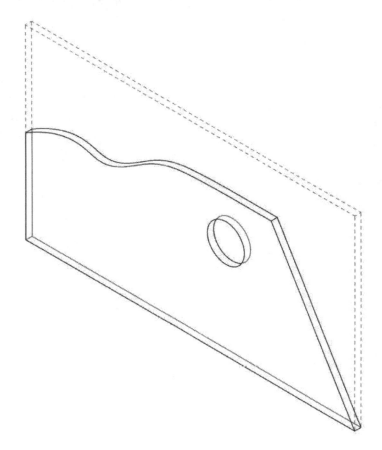

- **Reset Profile** – Once a wall has an edited profile use the reset profile tool to return the wall to the original rectangular profile.

- **Wall Opening** – Create a rectangular opening in a wall. When cursor is positioned over the wall click and drag to create the opening. Wall openings can be used to create an opening in a wall that is not linear in plan. Openings cannot be placed into a wall that has an edited profile.

- **Attach Top/Base** – Attach the top or the bottom of the wall to a valid target. The wall and the target will be constrained. When the target is moved, the wall attachment will be maintained. Wall must be able to physically intersect with the target in order to attach.

Set the attachment point (top or bottom) in the Options bar before you select the attachment target.

- These are valid targets for wall attachment:
    - o Roof
    - o Floor
    - o Ceiling
- **Detach Top/Base** – Attachment constraint is removed from the wall and the profile is reset if changed for attachment.

Modify | Windows

- **Edit Family** - Opens the selected family in the family editor. The model where the family originated will remain open. Use Switch Windows to go back to a view in the open model.
- **Pick New Host** – Choose another host wall for the window. Click the new host in the canvas to select.

Contextual

Notes:

Index

Notes: